FUNDAMENTALS OF AIR QUALITY SYSTEMS
Design of Air Pollution Control Devices

Kenneth E. Noll, Ph.D., P.E.
Illinois Institute of Technology

AN AMERICAN ACADEMY OF ENVIRONMENTAL ENGINEERS® PUBLICATION

American Academy of Environmental Engineers Staff

Editor: William C. Anderson, P.E., DEE

Production Manager: Yolanda Y. Moulden

Cover Design: William C. Anderson

The contents of this publication are not intended to be and should not be construed to be a standard of the American Academy of Environmental Engineers (The Academy) and are not intended for use as a reference in purchase specifications, contracts, regulations, statutes, or any other legal document.

No reference made in this publication to any specific method, product, process or service constitutes or implies an endorsement, recommendation, or warranty thereof by The Academy.

The Academy makes no representation or warranty of any kind, whether express or implied, concerning the accuracy, completeness, suitability or utility of any information, apparatus, product, or process discussed in this publication, and assumes no liability thereof. Anyone utilizing this information assumes all liability arising from such use, including but not limited to infringement of any patent or patents.

© 1999 by American Academy of Environmental Engineers.

All rights reserved.

Printed in the United States of America.

Library of Congress Cataloging-in-Publication Data

Fundamentals of air quality systems: design of air pollution control devices / by Kenneth E. Noll
    p. cm.
  Includes bibliographical references . (p.).
  ISBN 1-883767-25-3
  1. Air--Purification--Equipment and supplies--Design and construction. 2. Factory and trade waste--Purification--Equipment and supplies--Design and construction. 3. Gases, Asphyxiating and poisonous--Environmental aspects. 4. Particles--Environmental aspects. I. Title.
TD889.N645 1999
628'.5'3 -- dc21
                                            98-27184
                                                CIP

Contents

PREFACE vii

Chapter 1— Introduction 1
1.1 Air Pollution Laws for Regulation of Industrial Emissions 1
1.2 Types of Air Emission Standards Applicable to Stationary Sources of Pollution 1
1.3 Use of Emission Factors in Estimating Emission Quantities 6
1.4 Selection of Gas Cleaning Equipment 9
1.5 Estimating Costs for Air Pollution Control Devices 10
1.6 Pollution Prevention as a Regulatory Technique 13
1.7 Summary 19
1.8 Problem Set 20
1.9 References 22

Chapter 2 — Basic Concepts of Gases 25
2.1 Introduction 25
2.2 Physical Properties of Air 25
2.3 Gas Laws 30
2.4 Energy Concepts 39
2.5 Air/Water Vapor Mixtures 41
2.6 Cooling of Gaseous Effluents 45
2.7 Pressure 58
2.8 Problem Set 63
2.9 References 65

Chapter 3 — The Motion of Airborne Particles 67
- 3.1 Introduction 67
- 3.2 Parameters That Characterize Particulate Collection 80
- 3.3 Filter and Water Drop Collection Efficiency (Impaction, Interception, and Diffusion) 91
- 3.4 Problem Set 94
- 3.5 References 97

Chapter 4 — Fundamentals of Particulate Emission Control 99
- 4.1 General Concepts of Particulate Collection 99
- 4.2 Particle Size Distribution Functions 100
- 4.3 Collection Efficiency 107
- 4.4 Basic Modeling Concepts 110
- 4.5 Integrated Penetration Concept for the Prediction of Overall Efficiency 116
- 4.6 Design Summary 119
- 4.7 Problem Set 121
- 4.8 References 126

Chapter 5 — Cyclones 127
- 5.1 Introduction 127
- 5.2 Standard Cyclone Configuration 130
- 5.3 Pressure Drop 132
- 5.4 Prediction of Collection Efficiency 133
- 5.5 Effect of Particle Re-entrainment 143
- 5.6 Multiple Cyclones 144
- 5.7 Optimizing the Design 149
- 5.8 Cyclone Design 150
- 5.9 Problem Set 156
- 5.10 References 160

Chapter 6 — Fabric Filters 163
- 6.1 Introduction 163
- 6.2 Fabric Selection 165
- 6.3 Fabric Cleaning 167
- 6.4 Filtration Velocity and Air-to-Cloth Ratio 169
- 6.5 Pressure Drop 172
- 6.6 Experimental Measurements of K2 176
- 6.7 Theoretical Prediction of K2 178
- 6.8 Pressure Drop in Multi-Compartment Baghouses 181
- 6.9 Collection Efficiencies for Fibrous Fabrics 186
- 6.10 Fabric Filter Design Review 197
- 6.11 Problem Set 197
- 6.12 References 200

Chapter 7 — Wet Scrubbers 201
- 7.1 Introduction 201
- 7.2 Water Drop Formation 203
- 7.3 Design Procedure for Collection of Particles in Water Drop Scrubbers 205
- 7.4 Pressure Drop 232
- 7.5 Design Optimization 235
- 7.6 The Contact Power Theory 237
- 7.7 Mist Eliminators 241
- 7.8 Design Summary 256
- 7.9 Problem Set 257
- 7.10 References 262

Chapter 8 — Electrostatic Precipitators 265
- 8.1 Introduction 265
- 8.2 Resistivity 268
- 8.3 Electric Fields 271
- 8.4 Particle Charging Mechanisms 273
- 8.5 Design of Electrostatic Precipitators 277
- 8.6 Precipitator Design: Practical Aspects 295
- 8.7 Design Summary 299
- 8.8 Review Problem 300
- 8.9 Problem Set 304
- 8.10 References 308

Chapter 9 — Control of Volatile Organic Compounds 309
- 9.1 Introduction 309
- 9.2 Equilibrium Vapor Content 312
- 9.3 Evaporation Loss Sources 315
- 9.4 Control Device Selection 323
- 9.5 Design of Condensers to Remove VOCs 328
- 9.6 Chlorinated Solvents 337
- 9.7 Problem Set 340
- 9.8 References 345

Chapter 10 — Adsorption 347
- 10.1 Introduction 347
- 10.2 Adsorption Isotherms 350
- 10.3 Adsorption Equilibrium Relationships 352
- 10.4 Determination of Adsorption Capacity 355
- 10.5 Adsorption System Design 359
- 10.6 Adsorbent Regeneration 362
- 10.7 Maximum Adsorbent Bed Depth 364
- 10.8 Minimum Adsorbent Bed Depth 366
- 10.9 Service Time 367

10.10	Humidity Effects	369
10.11	Design Summary	370
10.12	Problem Set	378
10.13	References	382

Chapter 11 — Incineration 383

11.1	Introduction	383
11.2	Stoichiometric Combustion Air	385
11.3	Combustion Kinetics	386
11.4	Thermal Incinerator Design Principles	393
11.5	Catalytic Incineration Design Principles	401
11.6	Incineration of Chlorinated Hydrocarbons	407
11.7	Approximate Calculations	420
11.8	Summary-Incinerator Design	425
11.9	Problem Set	426
11.10	References	430

Chapter 12 — Absorption 433

12.1	Introduction	433
12.2	Solubility and Henry's Law	437
12.3	Absorption Design Theory	441
12.4	Design Procedures	444
12.5	Review of Design Procedure	460
12.6	Alkaline Absorption for SO_2	464
12.7	Problem Set	468
12.8	References	472

Chapter 13 — Control of Gaseous Emissions from Motor Vehicles 475

13.1	Introduction	475
13.2	Reactions in the Atmosphere	478
13.3	Engine Operation and Air Pollution Emissions	480
13.4	Combustion of Gasoline in Motor Vehicles	480
13.5	Nitrogen Oxide Emissions	484
13.6	Emission Standards	488
13.7	Control Devices	492
13.8	Design of Catalytic Converters Based on Mass Transfer Considerations	497
13.9	Problem Set	504
13.10	References	507

Chapter 14 — Air Quality Systems 509

14.1	Introduction	509
14.2	Selection of Air Pollution Control Devices for Application to Industrial Sources	512

14.3 Auxiliary Equipment 513
14.4 Problem Set 551
14.5 References 554

Appendix A — Practice Problems for the Air Quality Portion of the P.E. Examination in Environmental Engineering (with Solutions) 555

Appendix B — Selected Symbols and Acronyms 587

Appendix C — Conversion Factors 595

Appendix D — Physical Constants 603

Index 605

PREFACE

This book develops rational bases for the design of air pollution control devices for the removal of gases and particulate emissions from industrial sources. The practical aspects of design are emphasized by providing a detailed presentation of state of the art procedures for the design of each major air pollution control system in general use. The book describes the theory underling the design of each system as well as the philosophy for the design to aid in understanding of the subject. Because of this, the design concepts relate to air pollution control in a general sense.

The relevant body of information is presented in such a way that it may be useful as a textbook and also to the practicing engineer previously untrained in this area. The book has evolved from two courses taught for more than 25 years, from research activities, from consulting related to the design and construction of air pollution control devices, and from efforts to enforce air pollution control regulations while employed by the California Air Resources Board and as Chairman of Knox County Tennessee Air Pollution Control Board. The material is appropriate for upper-division undergraduate and graduate students in environment, chemical, civil, and mechanical engineering, and is written in a manner that would make it suitable as a text for a single-semester course in which all 14 chapters are covered broadly. The individual chapters are constructed with introductory material, design concepts, and general design equations presented first followed by alternate design methods and specific applications that would not be presented in a one semester course. Optimally, it can be used for a two-semester course in which the concepts are developed fully and all subjects in the book are considered in depth, for example, gas cleaning could be presented in one semester (chapters 1, 2, 9, 10, 11, 12, 13) and particle design in the other (chapters 3, 4, 5, 6, 7, 8, 14).

Another major purpose of this book is as a reference for practicing engineers preparing for the *Principles and Practice of Engineering Examination in Environmental Engineering*. There are numerous examples and end-of chapter problems that emphasize key points and design procedures. **An appendix provides problems that have multiple choice questions with unlinked answers, which is how the problems are presently handled on the P.E. Examination.**

The first chapter considers air pollution control regulations applicable to industrial sources and the procedures used to quantify the amount of material released from specific sources for comparison with the regulations. The next three chapters review basic concepts related to mass and heat transfer, small particle technology, and the general design approach used for air pollution control devices. The next eight chapters can stand alone. Four chapters cover the major particulate control devices and provide current design models for cyclones, fabric filters, wet scrubbers, and electrostatic precipitators. There is no redundancy in subject coverage and the chapters can be used in any sequence. However, there is some logic in teaching cyclone design before the other control devices because standard cyclone configurations are used for design; this aids the student in understanding the relationship between grade efficiency curves and the design variables that control collection efficiency.

Four additional chapters cover the principles used for the control of gases and vapors: condensation, adsorption, incineration, and absorption. The first three of these chapters deal with the control of organic matter. Chapter 9 covers the control procedures for volatile organic compounds (VOCs) and includes evaporation and condensation and should be covered before the chapters on adsorption (10) and incineration (11). This is because condensation is often used for initial control before adsorption or incineration. Chapter 12 presents gas transfer principles that relate to the control of inorganic emissions by absorption and concentrates on the design of packed columns for the removal of sulfur dioxide.

Chapter 13 discusses air pollution control applied to motor vehicles in which the concepts developed for the control of gases in chapters 9, 10, 11, and 12 are applied. This coverage is included because motor vehicles are a major source of air pollution in urban areas and control devices used on vehicles represent an important air pollution control system for gases (CO, NO_x, VOCs).

Chapter 14, the final chapter, discusses air pollution control systems for industrial sources in which air pollution control devices become part of the total emission control system. This chapter describes the procedure for selecting air pollution control devices for specific sources. It also includes a discussion of hood and duct design and procedures for selection of fans and motors. The material can be introduced briefly or used as the basis for a comprehensive discussion of process control.

A mixture of English and metric units is used throughout the text. This is due to the fact that many empirical formula and charts use one or the other of these units or use mixed units and are not easily changed. Many customary values are in either English or metric units and generally need to remain that way to be understood. Example problems are often given in mixed units or units are changed between the text and the example problem to facilitate an understanding of both sets of units where appropriate. For example, in Chapter 12, the example problems on sulfur dioxide removal are presented in metric units and repeated at the end of the chapter in English units for comparison.

I would be remiss if I did not recognize publications that have had a large impact on the material contained in this book. The *Air Pollution Engineering Manual*, first published by The Air Pollution Control District of Los Angeles County in 1963, and the USEPA Training Manuals for the control of particulate and gaseous air pollutants (1981), were used to provide basic information and example problems presented in many of the chapters. Werner Strauss authored the first comprehensive text, *Industrial Gas Cleaning,* in 1966 that provides fundamental information on the general principles of design with many practical applications. William Licht authored the text, *Air Pollution Control Engineering,* in 1980 that provides detailed information for the design of cyclones. Seymour Calvert provides details on the design procedure for scrubbers and the integrated penetration method in the *Handbook of Air Pollution Technology* (1984). These publications are typical of the type of references that will be found at the end of each chapter. Books and comprehensive review articles where the design procedures were first presented are identified. The references sited for the design of particulate control devices were generally published between 1965 and 1985 while the references for the gaseous control devices extend into the 1990s. A more recent text may contain similar information; however, credit was assigned to the original text. Individual articles can be obtained by referring to these references.

The author is indebted to a number of colleagues for ideas and assistance with the preparation of various chapters and for providing example problems, especially Mike Pilat of the University of Washington, Dale Lundgren of the University of Florida, Bill Franek of the Cook County Department of Environmental Control and Jerry Crowder of the Southwest Regional Air Pollution Control Training Center. Warawut Suadee, of Thammasat University (Thailand) provided valuable assistance and support throughout the preparation of this book. His efforts in typing and editing greatly improved the presentation of material. He is co-author of the solutions manual for the end-of-chapter problems. Additional typing and editing was provided by Uswama Shahin (Palestine), Ying-Kuang Hsu (Taiwan) and Puji Lestari of Bandung Institute of Technology (Indonesia) and Ali Oskouie of the Illinois Institute of Technology and is appreciated.

Reviewers for selected chapters of later stages of the manuscript included Larry Canter, University of Oklahoma; Wayne Davis, University of Tennessee; Sarina Ergas, University of Massachusetts; James Friend, Drexel University; Kumar Ganesan, University of Montana; Lynn Hildemann, Stanford University; Scott Lowe, Manhattan College; Thomas Overcamp, Clemson University; Kurtis Paterson, Michigan Technological University; Mark Rood, University of Illinois; and Michael Shelley, Air Force Institute of Technology. Their comments and guidance were most welcome and constructive.

This book is dedicated to Eliza for her support during its preparation.

Kenneth E. Noll
February, 1999

Chapter 1

Introduction

1.1 Air Pollution Laws for Regulation of Industrial Emissions

The Clean Air Act provides the legal basis of air pollution laws in the United States (Figure 1.1). The U.S. Environmental Protection Agency (US EPA) prepares and publishes detailed information showing how those laws shall be applied. Some of these regulations — New Source Performance Standards (NSPS) and National Emission Standard for Hazardous Air Pollutants (NESHAPS) — are applied nationwide, while others are developed and applied by State and Regional Air Pollution Control Programs. Table 1.1 provides some examples of emission limits applied to stationary sources of pollution.

1.2 Types of Air Emission Standards Applicable to Stationary Sources of Pollution

1.2.1 Particulate Matter Emission Standards

There are two types of regulations for the control of particulate matter which have traditionally been used. One type is based on the weight concentration of particles (mass emission) in the stack and the other relates the weight of emitted particles to the total weight of material processed (process weight). One of these types of regulations will be more restrictive as to the required degree of control for a

Figure 1.1
The Flow of Legal Authority Leading to Air Pollution Control in the United States (New Source Performance Standards; National Emission Standard for Hazardous Air Pollutants)

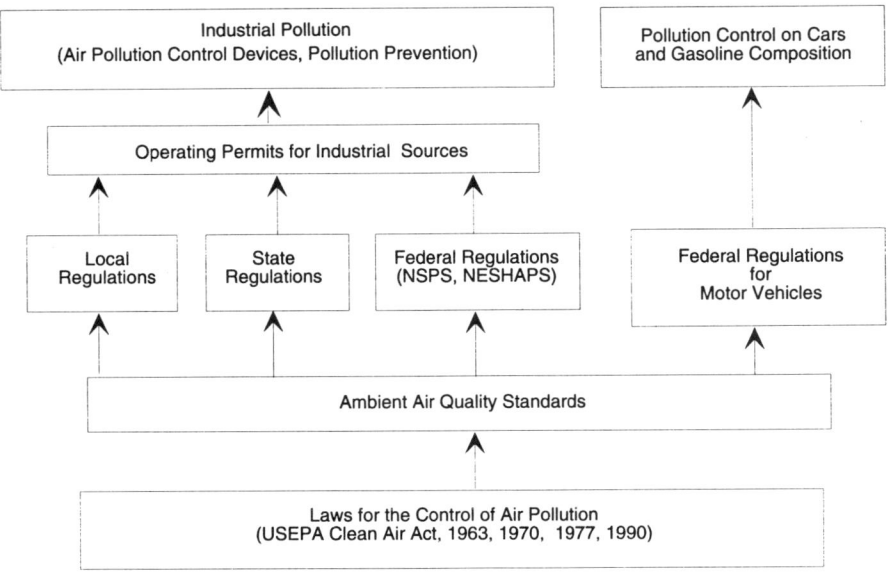

specific source and this regulation will govern the abatement control program. In practice, mass emission standards are more restrictive on combustion sources. Compliance with the process weight rule for material handling operations will generally provide compliance with the mass emission limit.

Mass Emission Standard

Mass emission standards usually limit particulate matter in a stack to some upper limit such as 0.2 grains/scf (7,000 grains/lb). This value may be different for different types of operations, such as incinerators, wood-waste burners, blast furnaces, open hearth furnaces, cupolas, and other metallurgical operations. Mass emission regulations are designed to apply to combustion sources, where it is possible to correct the concentration to some standard combustion condition, such as a certain percent excess air, CO_2, or O_2 in the exhaust gas.

Where combustion sources are involved, a standard may include not only the allowable concentration, but may specify the quantity of excess air the system may use while achieving this concentration. The standard for solid waste incinerators of 50 ton/day or greater capacity, particulate emissions not to exceed 0.08 grains/scf corrected to 12 percent carbon dioxide, is an example of this type of standard (see Table 1.1).

Chapter 1: Introduction

> **Table 1.1**
> **Federal Standards of Performance for New Stationary Sources (New Source Performance Standards)**
>
> This list is an excerpt from the 1991 version of 40CFR60. Standards are listed there for 68 industrial categories. New categories are added regularly, and existing ones modified. This excerpt shows the kind of regulations that are contained in that much larger compilation.
>
> 1. Coal-fired power plants whose construction started after September 18, 1978, may not emit the following to the atmosphere:
>   a. Particulate matter more than 0.03 lb/10^6 Btu, or 1% of the ash solids in the fuel, whichever is less.
>   b. Sulfur dioxide more than 1.2 lb/10^6 Btu, or more than 30% of the SO_2 that would be formed if all the sulfur in the coal were converted to SO_2, whichever is less.
>   c. Nitrogen oxides more than 0.6 lb/10^6 Btu for most coals, or 0.5 lb/10^6 Btu for sub-bituminous coal.
> 2. Large incinerators shall not emit to the atmosphere gas that contains particulates in concentrations greater than 0.08 grain/dry standard cubic foot, corrected to 12% CO_2.
> 3. Portland cement plants shall not emit to the atmosphere the following:
>   a. Gases from the kiln containing more than 0.30 lb/ton of kiln feed (dry basis).
>   b. Gases from the clinker cooler containing more than 0.10 lb/ton of feed to the kiln (dry basis).
> 4. Nitric acid plants shall not emit gases containing more than 3.0 lb of NO_2 per ton of nitric acid produced.
> 5. Sulfuric acid plants shall not emit gases containing more than 4 lb of SO_2 and/or 0.15 lb of sulfuric acid mist/ton of acid produced (100% basis).

This standard may also be written in terms of the total rate of heat input. A sliding scale is usually provided which requires more control for larger units (Figure 1.2). The sliding scale also allows a different standard for different sources. For example, waste-wood burners and small incinerators could have a heat input below 10 million Btu/hr and a maximum emission of 0.6 lb/10^6 Btu, while large furnaces would be allowed only 0.2 lb/10^6 Btu.

Process Weight Standard

Rules based on process weight limit the particulate emission on the basis of the pounds of material processed and thus preclude the use of dilution as a means of meeting a particulate emission limit. In this approach, permissible emission rates are related to material processed by a sliding scale as shown in Figure 1.3.

This standard usually governs the emissions from dust-producing operations, such as cement plants, rock-crushing operations, and lime kilns. For this regulation, a typical definition of process weight is "the total weight of all material introduced into any specific process which may cause any discharge into the atmosphere." Solid fuels are usually considered as part of the process, but liquid and gaseous fuels are not.

Figure 1.2
Allowable Emissions of Particulate Matter from Fuel-Burning Equipment

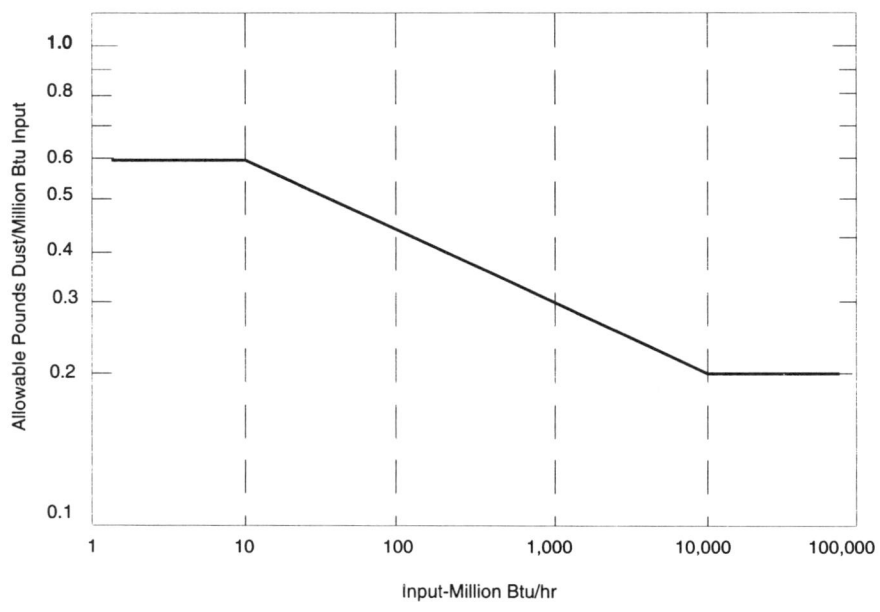

Figure 1.3
Typical Process Weight — Maximum Discharge Relationship

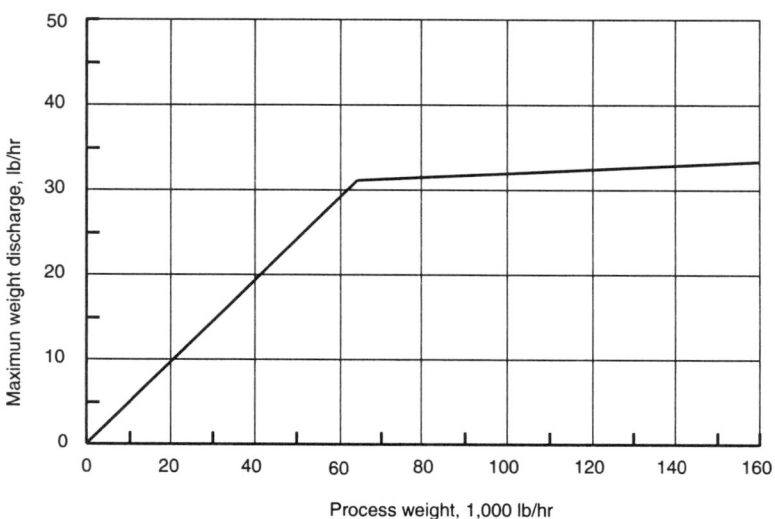

Chapter 1: Introduction

Example 1.1 Process Weight Factor Application

Figure 1.4 is a process rate standard for particulate emissions taken from a state's air quality control regulations.

Figure 1.4
Regulation for Allowable Particulate Emissions from Fuel-Burning Equipment

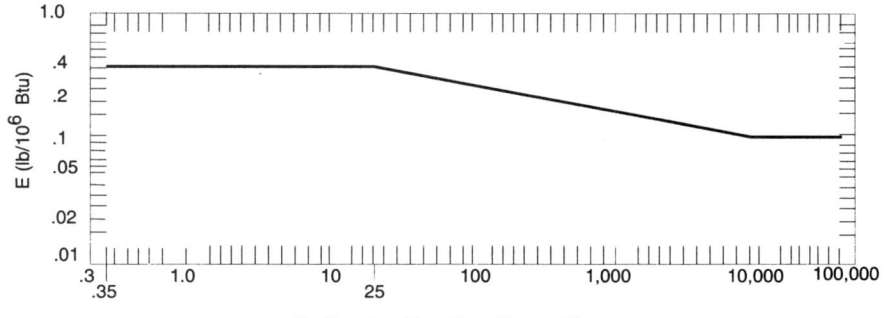

H = Total heat input in millions of Btu per hour

E = Maximum emissions in pounds of particulate matter per million Btu heat input.
$E = 0.8425 \, H^{-0.2314}$ (H = 25 to 10,000)

Given:

emission rate = 1800 g/sec

fuel — coal, heat value = 12,500 Btu/lb, is fed at 23 ton/hr

proposed abatement — an electrostatic precipitator with 99% rated collection efficiency.

Determine whether this plant meets the standard.

Solution:

Determine the process energy rate, H

$$H = \text{mass of coal} \bullet \text{energy value per unit mass}$$
$$= 23 \frac{\text{ton}}{\text{hr}} \bullet 12,500 \frac{\text{Btu}}{\text{lb}} \bullet \frac{2,000 \text{ lb}}{\text{ton}}$$
$$= 575 \bullet 10^6 \text{ Btu/hr}$$

Determine actual particulate weight rate

$$E_{actual} = \frac{1{,}800 \text{ g/sec} \cdot \dfrac{\text{lb}}{454 \text{ g}} \cdot 3{,}600 \dfrac{\text{sec}}{\text{hr}} \cdot (1-0.99)}{575 \cdot 10^6 \dfrac{\text{Btu}}{\text{hr}}}$$

$$E_{actual} = 0.25 \frac{\text{lb}}{10^6 \text{ Btu}}$$

Select the allowable emission rate from the graph on Figure 1.4 at $H = 575 \cdot 10^6$ Btu/hr

$E_{allow} = 0.19$ lb/10^6 Btu, or

Calculate from $E_{allow} = 0.8425 \cdot (575)^{-0.2314}$

$= 0.194$ lb / 10^6 Btu

Compare E_{actual} (0.25) E_{allow} (0.19)

Therefore, this unit does not conform.

1.2.2 Gas Emission Standards

Considerable attention has been given to the control of two gases, sulfur oxides and nitrogen oxides. For example, a typical SO_2 regulation for industrial sources requires that no person shall cause, suffer, allow, or permit the emission from any sources, gases containing more than 2,000 parts per million of SO_2. A typical standard for a combustion sources is 1.0 lb $SO_2/10^6$ Btu heat input. Hydrocarbon control can take the form of vapor pressure standards for the control of evaporation or vapor recovery system specifications or destruction efficiencies. Odor control can be obtained by limiting the concentration of emissions or by requiring special controls, such as the burning of gases from rendering plants.

1.3 Use of Emission Factors in Estimating Emission Quantities

The US EPA has compiled contaminant emission rate data (Emission Factors) for sources based on a summary of stack sampling data. Emission Factors are only estimates, however, since they represent, at best, the values for the particular processes studied. Emission Factor Rating, A, indicates that the values are quite reliable. Lower ratings indicate that the factors are based on fewer and/or poorer quality test data. Table 1.2, taken from "Compilation of Air Pollutant Emission Factors" (AP-42), gives emission factors for coal-combustion equipment without any controls. Using such a table, one can make estimates of uncontrolled emissions for a given plant once the type of unit is specified. The particulates from coal

Table 1.2
Emission Factors for Bituminous Coal Combustion Without Control Equipment
(Emission Factor Rating: A) (AP 42)

Furnace size 10⁶ Btu/hr heat input[a]	Particulates[b]		Sulfur oxides[c]		Carbon monoxide		Hydro carbons[d]		Nitrogen oxides		Aldehydes	
	lb/ton coal burned	kg/MT coal burned	lb/ton coal burned	kg/MT coal burned	lb/ton coal burned	kg/MT coal burned	lb/ton coal burned	kg/MT coal burned	lb/ton coal burned	kg/MT coal burned	lb/ton coal burned	kg/MT coal burned
Greater than 100 (Utility and large industrial boilers)												
Pulverized												
General	16A	8A	38S	19S	1	0.5	0.3	0.15	18	9	0.005	0.0025
Wet Bottom	13A[e]	6.5A	38S	19S	1	0.5	0.3	0.15	30	15	0.005	0.0025
Dry Bottom	17A	8.5A	38S	19S	1	0.5	0.3	0.15	18	9	0.005	0.0025
Cyclone	2A	1A	38S	19S	1	0.5	0.3	0.15	55	27.5	0.005	0.0025
10 to 100 (large commercial and general industrial boilers)												
Spreader stoker[f]	13A[g]	6.5A	38S	19S	2	1	1	0.5	15	7.5	0.005	0.0025
Less than 10 (commerical and domestic furnaces)												
Underfeed stoker	2A	1A	38S	19S	10	5	3	1.5	6	3	0.005	0.0025
Hand-fired units	20	10	38S	19S	90	45	20	10	3	1.5	0.005	0.0025

a. 1 Btu/hr = 0.252 kcal/hr.
b. The letter A on all units other than hand-fired equipment indicates that the weight percentage of ash in the caol should be multiplied by the value given. (Example: If the factor is 16 and the ash content is 10 percent, the particulate emissions before the control equipment would be 10 times 16, or 160)
c. S equals the sulfur contents (see footnote b above).
d. Expressed as methane.
e. Without fly ash reinjection.
f. For all other stokers, use 5A for particulate emission factor.
g. Without fly ash reinjection. With fly ash injection use 20A. This value is not an emission factor but represents loading reaching the loading equipment.

8 Fundamentals of Air Quality Systems

consist largely of carbon, silica (SiO_2), alumina (Al_2O_3) and iron oxide (Fe_2O_3 or Fe_3O_4), emitted as fly ash. Gaseous material is also emitted from coal-fired power plants. The type of unit refers to size with units greater than $100 \cdot 10^6$ Btu/hr taken to be power plants, units from $10 \cdot 10^6$ Btu/hr to $100 \cdot 10^6$ Btu/hr taken as industrial plants, and units less than $10 \cdot 10^6$ Btu/hr as domestic and commercial plants. Since high-quality coal produces about 13,000 Btu/lb, we can convert these to a Btu basis using $26 \cdot 10^6$ Btu/ton of coal burned. (AP-42)

Example 1.2 Use of Emission Factors

Consider the design of a new power plant consisting of three 750 megawatt units. The fuel is a low sulfur coal with the 8% ash, 0.5% sulfur, and 11,000 Btu/lb.

What will be the emissions of particulate matter, NO_x, SO_x? The coal-fired boiler units are to be of the "general-pulverized" type and the thermal efficiency of the plant is estimated to be 38%.

Solution:

Thermal Analysis: There will be 2,250 megawatts in all. The energy input required is:

$2,250 / 0.38 = 5,930 \cdot 10^6$ watts $= 20,200 \cdot 10^6$ Btu/hr

1 watt = 3.411 Btu/hr

The amount of coal required is:

$$\frac{20,200 \cdot 10^6 \text{ Btu/hr}}{11,000 \text{ Btu/lb}} = 1,834 \cdot 10^3 \text{ lb/hr} = 917 \text{ ton/hr} = 22,000 \text{ ton/day}$$

Determine the particulate and gaseous emissions factors using emission taken from Table 1.2 as:

Particulate (8% ash):

$$8\% \cdot 16 \frac{\text{lb}}{\text{ton coal}} \cdot 917 \frac{\text{ton coal}}{\text{hr}} = 117,376 \text{ lb/hr}$$

$$NO_2: 18 \frac{\text{lb}}{\text{ton coal}} \cdot 917 \frac{\text{ton coal}}{\text{hr}} = 16,506 \text{ lb/hr}$$

$$SO_2: 38 \frac{\text{lb}}{\text{ton coal}} (0.5\% \text{ sulfur}) \cdot 917 \frac{\text{ton coal}}{\text{hr}} = 17,400 \text{ lb/hr}$$

1.4 Selection of Gas Cleaning Equipment

In most cases, air pollution control equipment is installed on industrial sources to reduce emissions in order to meet regulations. However, it is possible to reduce emissions by other methods. Changing fuel sources, modifying or changing raw materials, or using alternative production procedures can also reduce emissions without adding control equipment. These methods are usually considered before installing expensive control equipment.

Air pollution control devices have reduced particulate and gaseous pollutants from various industrial sources for many years. Cyclones, baghouses, electrostatic precipitators, and wet scrubbers are used to reduce particulate emissions. Absorbers, adsorbers, combustors (incinerators), and condensers are used to control gaseous emissions. The use of a particular device depends on the physical and chemical properties of both the pollutant and the exhaust stream.

The proper choice of gas-cleaning equipment for any single problem depends on a number of variables. These can be grouped into four general areas:

- degree of reduction of emissions required to meet emission standards;
- process and effluent characteristics;
- equipment capacities and limitations; and
- capital investment and operation costs.

Important gas stream and particle characteristics of the process include: volume, temperature, moisture content, particle size, density, and explosiveness. High gas temperatures without cooling preclude the use of fabric filters; explosive gas streams prohibit the use of electrostatic precipitators.

The overriding factor in gas-cleaning equipment selection is the efficiency required or the penetration allowed. Simply setting the design efficiency normally narrows the selection of cleaning equipment to no more than a few types. Figure 1.5 provides expected efficiency characteristics for specific particulate control devices (Marchello and Kelly 1975). Multiclones exhibit a higher efficiency curve than cyclones, and venturi scrubbers show a higher efficiency than low-pressure drop scrubbers. It is also readily apparent that high-efficiency electrostatic precipitators and fabric filters are superior collection devices, especially in the smaller particle size ranges.

An important characteristic of control devices is the effect they have on the flow of exhaust gas in an industrial process. Pressure drop is a measure of the air resistance across a system. Collectors with large pressure drops require larger fans and, hence, have greater power requirements. Some control devices, such as venturi scrubbers, are designed to operate at high pressure drops (up to 100 in. H_2O). On

Figure 1.5
Comparison of Particulate Control Device Efficiencies

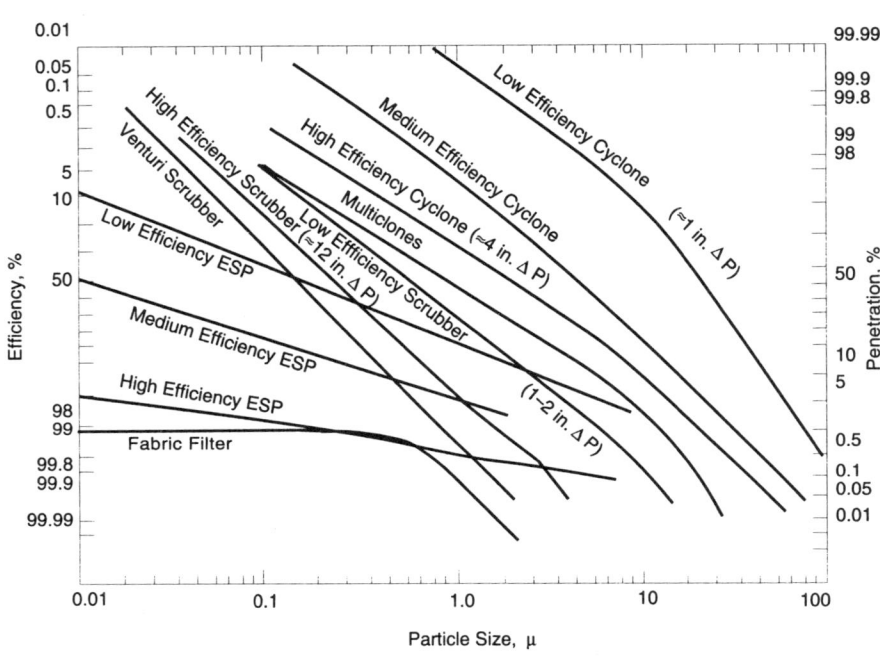

Source: Marchello and Kelly 1975

the other hand, electrostatic precipitators are designed to operate at much lower pressure drops (1.0 in. H_2O). The goal of control system design is to satisfy emission limitations at minimum cost with maximum reliability. The basic trade-offs involve decisions between collection efficiency, installation cost, and operating costs.

1.5 Estimating Costs for Air Pollution Control Devices

Selection of air pollution control devices requires estimates of cost. If two or more control devices can be used to control emissions, selection will be made based on an annual cost analysis.

The actual cost of installing and operating air pollution control equipment is a function of many direct and indirect factors. The principal costs of concern to the control equipment user are those directly associated with the capital investment, installation, and operation of control devices.

Capital investment costs can include the cost of:

- land acquisition;
- engineering design;
- control equipment;
- auxiliary equipment and replacement parts;
- structure modification;
- installation; and
- startup.

Installation costs depend upon factors, such as:

- plant age;
- transportation;
- space limitation;
- degree of preassembly;
- special equipment required for installation (e.g., cranes, helicopters); and
- labor rates or union contract requirements.

Maintenance cost is the expenditure required to sustain the operation of a control device at its designed efficiency with a scheduled maintenance program and necessary replacement of any defective parts.

Operating cost should include all direct costs incurred after installation except maintenance. The operating costs include power, labor, materials, utilities, disposal of collected materials, and absorbent, adsorbent, and catalyst renewals.

Power costs are the fan costs to move the air against the head loss in the equipment and ducts and the power to operate the control device and its auxiliary equipment. The power requirements for fans may be calculated using the equation

$$HP = \frac{Qh}{6,360 \, E} \quad [1.1]$$

where: HP = brake horsepower

Q = air flow rate, cfm

h = head loss, in. H_2O

E = mechanical efficiency of fan system.

Annual operating cost is the expense of operating a control device at its designed collection efficiency. This cost depends on the following factors:

- the gas volume cleaned;
- the pressure drop across the system;
- the operating time;
- the consumption and cost of electricity;
- the mechanical efficiency of the fan; and
- the scrubbing-liquor treatment/disposal and costs (where applicable).

Variations in the above cost components make estimates of control equipment costs site-specific and selection is made by comparison. One basis for comparing costs of air pollution control is the total annual cost. This cost includes the capital, operating, and maintenance costs.

Annual capital costs are calculated by

$$C_a = \left(\frac{P}{N}\right) \frac{(1+i)^n}{(1+i)^n - 1} \qquad [1.2]$$

where: C_a = annual cost

P = initial installed cost of equipment

N = lifetime or amortization base for equipment, yr

i = compound interest cost for money

n = number of interest periods

Example 1.3 Annual Cost Calculations

What is the annual capital cost for air pollution control equipment with an installed cost of $300,000 that is amortized over 10 years at 10%? Use Equation [1.2]

Solution:

$$C_a = \left(\frac{\$300,000}{10}\right) \frac{(1+0.10)^{10}}{(1+0.10)^{10} - 1}$$

$$= \$48,823.62$$

What is the annual cost of moving 1000 cfm of air against a head of 1 in. H_2O if electricity cost are $0.07 /kWh and the efficiency is 60% ?

Solution:

Determine horsepower required using Equation [1.1].

$$HP = \frac{1,000 \text{ cfm} \cdot 1 \text{ in. } H_2O}{6,360 \cdot 0.6}$$

$$= 0.262 \text{ HP}$$

Determine annual power cost.

$$0.262 \text{ HP} \cdot \frac{8,760 \text{ hr}}{\text{yr}} \cdot \frac{0.746 \text{ kWh}}{\text{HP hr}} \cdot \frac{\$0.015}{\text{kWh}} = \$120/\text{yr}$$

It should be kept in mind that large air flows and large head losses result in quite large power costs.

Some useful rules of thumb can give rough estimates of cost (± 25%) with a minimum of effort. Economy of scale is often estimated by the six-tenths rule (see Figure 1.6); the calculations are made using the formula:

$$\frac{C}{C_o} = \left(\frac{S}{S_o}\right)^{0.6} \quad [1.3]$$

where: C = cost of equipment with size S

C_o = cost of same equipment with size S_o

Sometimes a power of 0.7 is used with Equation [1.3], especially for total plant costs.

The installed cost of air pollution control equipment will usually run 2 to 3 times the base price of the equipment; therefore, the installed cost may be estimated as 200% of the purchase price for preliminary purposes. Table 1.3 presents a list of the relative capital cost factors for different control equipment devices. Table 1.4 presents typical direct and indirect operating costs for gas control equipment. More recent costs are available from US EPA (OAQPS Control Cost Manual 1990).

1.6 Pollution Prevention as a Regulatory Technique

The design of environmentally acceptable manufacturing plants generally means the use of end-of-pipe air pollution control devices through which effluent gases pass on their way to the environment. These devices are designed to meet government emission standards for particular compounds. This method of control is expensive and accompanying nonregulated substances often remain in these streams.

Figure 1.6
Cost Versus Size for Particulate Collectors

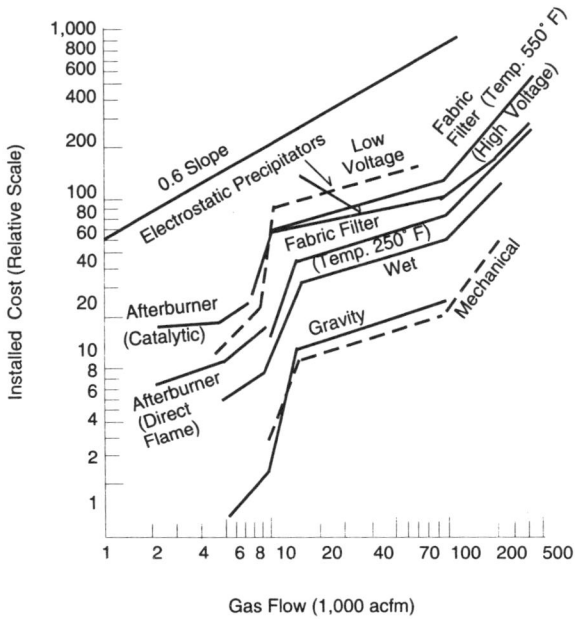

As an alternative, in-plant practices (as opposed to add-on devices) to reduce or eliminate waste is called waste reduction, pollution prevention, or production-integrated pollution control. *Waste reduction* can be defined as "in-plant processes that reduce, avoid, and eliminate" the generation of waste. Actions taken away from the manufacturing activity, including out-of-plant waste recycling, or treatment and disposal after the wastes are generated, are not considered waste reduction in this formulation, nor is concentrating wastes to reduce their volume. Avoiding formation of waste eliminates the need for treatment and disposal, both of which carry environment risk. Five methods of waste reduction listed in order of decreasing use are:

1. in-plant recycling;

2. changes in process technology;

3. change in plant operation (e.g., suppression of fugitive emission);

4. substitution of input (raw) materials; and

5. modification of end products to permit the use of less polluting upstream processes.

Table 1.3
Average Cost Factors for Estimating Capital Costs

	Precipitators	Scrubber	Fabric filters	Incinerators (combustion)	Adsorber
Direct Costs					
1. Purchase equipment costs					
a. Control Device	As req'd.	As req'd.	As req'd.	As req'd.	As req'd.
b. Auxiliary equipment	As req'd.	As req'd.	As req'd.	As req'd.	As req'd.
c. Instruments and controls	0.10	0.10	0.10	0.10	0.10
d. Taxes	0.03	0.03	0.03	0.03	0.03
e. Freight	0.05	0.05	0.05	0.05	0.05
Total	1.00	1.00	1.00	1.00	1.00
2. Installation direct cost					
a. Foundations and supports	0.04	0.06	0.04	0.08	0.08
b. Erection and handling	0.50	0.40	0.50	0.14	0.20
c. Electrical	0.08	0.01	0.08	0.04	0.08
d. Piping	0.01	0.05	0.01	0.02	0.05
e. Insulation	0.02	0.03	0.07	0.01	0.02
f. Painting	0.02	0.01	0.05	0.01	0.04
g. Site preparation	As req'd.	As req'd.	As req'd.	As req'd.	As req'd.
h. Facilities and buildings	As req'd.	As req'd.	As req'd.	As req'd.	As req'd.
Total	1.67	1.56	1.72	1.30	1.44
Indirect Costs					
3. Installation indirect costs					
a. Engineering and supervision	0.20	0.10	0.10	0.10	0.40
b. Construction and field expense	0.20	0.10	0.20	0.05	0.05
c. Construction fee	0.10	0.10	0.10	0.10	0.10
d. Start-up	0.01	0.01	0.01	0.02	0.02
e. Performance test	0.01	0.01	0.01	0.01	0.01
f. Model study	0.02	–	–	–	–
g. Contingencies	0.03	0.03	0.03	0.03	0.03
Total	2.24	1.91	2.17	1.61	1.75

Source: Neveril et al. 1978

Table 1.4
Basis for Estimating Annual Air Cleaner Operating Costs

	Cost factor
Direct Operating Cost	
1. Operating labor	$7.87/labor hr
a) Operator	15% of 1.a)
b) Supervisor	
2. Maintenance	
a) Labor	$8.66/labor hr
b) Material	100% of 2.a)
3. Replacement parts	As required
4. Utilities	
a) Electricity	$0.0432/kW hr
b) Fuel oil	$0.47/gal
c) Natural gas	$1.98/1,000 ft^3
d) Water	$0.25/1,000 gal
e) Stream	$5.04/1,000 lb
f) Compressed air	$0.02/1,000 ft^3
5. Waste disposal	$5 - 10/ton
Indirect Operating Costs	
6. Overhead	80% of 1.a) + 1.b) + 2.a)
7. Property tax	1% of capital costs
8. Insurance	1% of capital costs
9. Administration	2% of capital costs
10. Capital recovery costs	0.16275
a) Interest of 10%	
b) Equipment life of 10 years	
Credits	
11. Recovered Product	As required
Source: Neveril et al. 1978	

The 1990 Clean Air Act amendments (CAA amendments) require consideration of pollutant source reduction, process change, materials substitution, cross-media impacts, and other production modifications that could reduce or eliminate emissions in setting emissions standards for hazardous air pollutants. The amendments require the phaseout of chlorofluorocarbons (CFCs) and halons and ban the use of unsafe substitutes for these chemicals. They also provide the US EPA with authority to regulate gasoline additives. Maximum Achievable Control Technology (MACT) Standards required by the CAA amendments specify emission reduction strategies such as:

- enclosing process systems to eliminate emissions;
- collecting point or fugitive emissions and directing them to a suitable control device;
- substituting raw materials or changing processes to reduce emissions; and
- modifying work or operational practices.

For example, emission reduction strategies for the Hazardous Organic Industry (HON) include internal floating roofs with seals and gaskets for fixed roof tanks, a closed vent collection system with a thermal incinerator for loading transport

Chapter 1: Introduction

vehicles, and a leak detection and repair inspection program for equipment components (e.g., pumps, valves). Table 1.5 summarizes the general provisions of the HON MACT Standards. The HON production processes which are regulated include: styrene/butadiene, rubber, polybutadiene, chlorine, pesticides, pharmaceuticals, chlorinated hydrocarbons, and butadiene.

Table 1.5
General Provisions of MACT Standard for Hazardous Organic Chemical Manufacturing

Emission Point	Reference Control Technology	Reference Control Technology
Process vents	Combustion Flare Incinerator Boiler/Process heater	98% by weight
Storage vessels	Internal floating roof External floating roof Closed vent system with controls	95% by weight
Wastewater operations	Controlled transport and storage and steam stripper with controls for stripper vents	95-98% by weight
Transport operations	Combustion Flare Incinerator Boiler Recovery	98% by weight
Equipment leaks	Enhanced leak detection and repair equipment specifications	80 to 88

Regulatory pollution prevention need not be limited to mere expansion of traditional pollution control. Pollution prevention strategies can be used to attack particular pollutants in different media simultaneously. For example, some consumer products have major environmental consequences either during use (automobile or aerosol sprays) or when disposed (plastic containers, paints, and solvents). A case in point is plastic products, including containers and wrappings made from polymeric materials. For durability and low toxicity, these products are designed to be less reactive chemically. However, if incinerated, these plastics have two significant disadvantages. First, there is the destruction of a reclaimable resource. Second is the possible emission of hazardous air pollutants from incinerators burning chlorine-containing plastics, especially polyvinyl chloride. Thus, plastic consumer products are a part of a complex environmental problem that includes the consumer, the manufacturer, and the government. The selection and design of manufacturing processes and products should incorporate environmental constraints from the start, along with thermodynamic and economic factors. The goal is to reduce to a minimum the overall environmental impact of a product or process, and not simply address one aspect of that impact. This approach requires that environmental impacts be considered for the life of the product or process. As the shift is made from controlling pollution to preventing it, comprehensive approaches that focus on economics and the flow of material and energy are needed. Until these are fully developed and implemented, end-of-pipe treatment must be relied upon.

Example 1.5 Environmental Impact Analysis: Polythylene vs. Paper Grocery Sacks *(Modified from a problem by Allen, Bakshami and Rosselot, Pollution Prevention, AICHE 1992.)*

Compare the air pollution emissions and energy required to produce grocery sacks made from paper and polyethylene using the data from the following table.

	Air emissions, oz/ sack		Energy required, Btu/sack	
Life cycle stages	Paper	Polyethylene	Paper	Polyethylene
Materials manufacture plus Product manufacture plus Product use	0.0516	0.0146	905	464
Raw material acquisition Plus product disposal	0.0510	0.0045	724	185

Assume:

- two polyethylene sacks are required to hold as much as one paper sack
- use 1,000 lb of polyethylene
- polyethylene sacks have a mass of 0.2632 oz each
- paper sacks have a mass of 2.144 oz each
- 1.2 lb of petroleum (non-renewable resource) is required to manufacture 1 lb of polyethylene
- the energy for paper production comes from burning wood waste (renewable resource)

Which type of sack would you recommend being used in your local grocery store?

Solution:

The energy requirements and total mass of atmospheric pollutants for both paper and polyethylene (PE) grocery sacks calculated from the table are listed in the table below. All values pertaining to PE sacks are based on 1000 lb of product or 60,800 PE sacks. Values of the paper sacks are based on 60,800/2 = 30,400 sacks, the number required to hold an equivalent volume of groceries.

Sample calculation:

Air emissions for paper sacks

$$30,400 \text{ sacks} \bullet (0.516 + 0.0510) \bullet \frac{\text{oz}}{\text{sack}} \bullet \frac{1 \text{ lb}}{16 \text{ oz}} = 195 \text{ lb}$$

Energy required for paper sacks

$$30,400 \text{ sacks} \bullet (905 + 724) \bullet \frac{\text{Btu}}{\text{sack}} \bullet 4.95 \bullet 10^7 \text{ Btu}$$

Air emissions and energy requirements are similarly calculated for polyethylene sacks. The following table provides a comparative presentation for each sack type.

	Polyethylene 60,800 sacks	Paper 30,400 sacks
Energy required, Btu	$3.95 \bullet 10^7$	$4.95 \bullet 10^7$
Atmospheric pollutants, lbs	72.6	195

Conclusion: PE sacks require about 20% less energy than paper sacks. Atmosphere emissions are 60-70% lower for PE sacks when the use rate is 2 PE sacks/1 paper sack. PE sacks seem to be better for both energy and air pollution unless too many are substituted for 1 paper sack.

1.7 Summary

The basic types of emission control devices are mechanical collectors, wet scrubbers, baghouses, electrostatic precipitators, combustion systems, condensers, absorbers, and adsorbers. All of these have been used to some extent to control emissions from a variety of processes.

In view of the relatively high costs often associated with pollution control systems, engineers today are directing considerable effort towards process modification to eliminate as much of the pollution as possible at the source. This includes evaluation of alternative manufacturing and production techniques, substitution of raw materials, and improved process control methods. Unfortunately, if there is no alternative, the application of pollution control equipment must be considered. Considering the relatively high costs, proper design of this equipment is essential. Factors to consider to ensure proper design and performance are size and weight, fractional efficiency curves (in the case of particulates), mass transfer and/or contaminant destruction capability (in the case of gases or vapors), pressure drop, reliability and dependability, turndown capability, power requirements, utility requirements, temperature limitations, maintenance requirements, flexibility with regard to complying with more stringent air pollution regulations, etc.

This preceding list demonstrates that the proper selection of an air pollution control device for a specific application is complicated. The final choice in equipment selection is usually dictated by that equipment capable of achieving compliance with regulatory codes at the lowest uniform annual cost (amortized capital investment plus operation and maintenance costs).

1.8 Problem Set

1.1 What is the emission rate of SO_2 from a 1,000 megawatt power plant of 35% thermal efficiency burning 3% sulfur coal with Btu/lb heat value of 12,000? (1,000 watt = 3,413 Btu/hr)

1.2 A coal-fired power plant proposes to burn 3% sulfur coal with a heating value of 11,000 Btu/lb. If the plant is to meet the standard of performance given in Table 1.1, what percent cleanup is required for SO_2? What percent sulfur could the coal contain if the standards are to be met with no clean-up devices?

1.3 A 1,000 megawatt pulverized coal-fired unit (general type) of 40% thermal efficiency using 1.7% sulfur coal is to be built. The ash content of the coal is 10% and the heating value is 12,000 Btu/lb. Calculate the emissions from this plant for particulate, NO_x, and SO_x, if no controls are used. If uncontrolled, how much particulate of less than 5 µm in diameter would be emitted? Now assume that an electrostatic precipitator with an efficiency as noted in the following table is purchased for use with the plant. How much fly ash must be disposed each year? How much particulate of less than 5 µm diameter is released per day and per year?

Particle sizes, µm	0-5	5-10	10-20	20-44	>44
Emitted particles, % by weight	15	17	20	23	25
Efficiency of precipitator, %	75	94.5	97	99.5	100

Chapter 1: Introduction

1.4 Using information in the table below, estimate the SO_2 produced in a city of 500,000 which incinerates half of the solid waste generated and buries the rest. How does this figure compare to the emissions from a coal-fired power plant used to provide electricity for the city? (Assume 5 lb of solid waste per person per day and 1 kW of energy per person per day).

Sources	Emission factor, lb of SO_2/ton of refuse charged
Open-burning dumps and municipal incinerators	1.2 - 2.0
On-site commerical and industrial multiple chamber incinerators	1
On-site commercial and industrial single-chamber incinerators	2
On-site residential single-chamber incinerators	0.4
On-site residential flue-bed incinerators	0.2

1.5 An open-hearth steel plant operate 6 furnaces having a total capacity of 320 tons of steel in a nine-hour operating cycle. The charge is 7% more than the finished steel weight or 345 tons. The maximum emission rate occurs during a 30 min period and is 7.2 lb/hr per ton of steel. The gas volume is 62,000 scfm (70°F) and leaves at 1,350°F. The regulation for this plant is 0.01 grains/acf at 500°F. Calculate the maximum flow rate, emission flux in lb/hr and grains/acf, and efficiency required to meet the air pollution control regulation.

1.6 A 10,000 lb/hr continuous coffee bean roaster with a recirculation system exhausts about 4,000 scf/min from the roaster and 120 scf/lb coffee from the cooler. The emission factors for particulate are 4.2 lb/ton of beans from the roaster and 1.4 lb/ton of beans from the "stoner" cooler (a stoner removes any heavy objects from the coffee before grinding). Calculate the particulate loading in grains/scf in the two exhausts if no controls are applied (7,000 grains = 1 lb). If a simple cyclone collector is purchased with an overall efficiency of 70% for particulate from this plant, what will the emissions be in grains/scf?

1.7 Estimate the total annualized cost for a 60,000 cfm wet scrubber. The delivery cost is $17,000 and installation costs are expected to be 100% of the purchase cost. Annual depreciation, overhead, and interest costs will be 13% of the total capital cost. The operation costs are $16,000/yr due to a pressure drop of 15 in. of water for the wet scrubber. The lifetime of the scrubber is estimated to be 10 years.

1.8 Repeat the analysis in Example 1.5 using 3 polyethylene sacks to replace one paper sack.

1.9 The US EPA uses the average emission factors contained in the following table to estimate fugitive emissions from Synthetic Organic Chemical Manufacturing Industry (SOCMI) facilities.

Equipment	Service	Emission factor kg/hr/source
Valves	Gas	0.0056
	Light liquid	0.0071
	Heavy liquid	0.00023
Pumps	Light liquid	0.494
	Heavy liquid	0.0214
Compressors	Gas	0.228
Pressure relief valves	Gas	0.104
Flanges and other connectors	All	0.00083
Open-ended lines	All	0.0017
Sampling connections	All	0.015

Determine the emissions from an acrolein manufacturing facility (assume 87% of total volatile organic is acrolein). A count of plant equipment revealed 1,400 valves (168 gas service, 1,232 balance light liquid), 3,048 flanges, 27 pumps, 20 pressure relief valves, 21 open-ended lines and 20 sampling connections. Determine the acrolein emissions in lb/yr using the SOCMI emission factors. What can be done to control these emissions?

1.9 References

Calvert, S. and H. Englund (Eds.) *Handbook of Air Pollution Technology.* John Wiley & Sons. New York. 1984.

Marchello, J. M. and J. J. Kelly (Eds.). *Gas Cleaning for Air Quality Control.* Marcel Dekker. New York. 1975.

Neveril, R.B., J.U. Price, and K.L. Engdahl. *Capital and Operating Cost of Selected Air Pollution Control Systems.* V.J. Air Pollut. Control Assoc. 28, 1253-1256. 1978e.

U.S. Environmental Protection Agency (US EPA). "*Compilation of Air Pollutant Emission Factors, vol 1: Stationary Point and Area Sources,*" AP-42-SUB-B. U.S.

Environmental Protection Agency. Research Triangle Park, NC. 1985. (Available on the Internet at http://www.epa.gov/ttnchie1/ap42etc.html/)

U.S. Environmental Protection Agency (US EPA). *"OAQPS Control Cost Manual"* 4th ed. EPA 450/3-9-006. Research Triangle Park, NC. 1990. (Available on the Internet at http://www.epa.gov/ncepihom/)

Chapter 2

Basic Concepts of Gases

2.1 Introduction

In order to properly evaluate control devices, a thorough understanding of the process variables that affect a gas stream is essential. This chapter reviews a few basic concepts of gas behavior. Two important physical properties of gases — molecular weight and viscosity — are defined. Scientific laws used to predict the behavior of gas under varying process conditions including the ideal gas law and Henry's law are provided. Finally, enthalpy and pressure are discussed. Though these topics are widely divergent and covered with varying degrees of thoroughness, all of them will be used later in this text.

2.2 Physical Properties of Air

2.2.1 Molecular Weight

Standard air is defined as air with a density of 0.075 lb/ft^3 and an absolute viscosity of 1.225 • 10^5 lb$_m$/ft-sec (pounds-mass per foot second). This is equivalent to dry air at a temperature of 70°F and a pressure of 29.92 in. Hg. Atmospheric air is a mixture of dry air, water, and various impurities. Because of this, neither atmospheric air nor dry air have a true molecular weight. However, they do have an apparent molecular weight that can be calculated from their composition. Assuming dry air consists, by volume, of 78.09% nitrogen, 20.95% oxygen, 0.93% argon, and 0.03% CO_2, its apparent molecular weight may be calculated as:

Component	Volume fraction		Molecular weight		lb/lb mol
N_2	0.7909	•	28.016	=	21.873
O_2	0.2095	•	32.000	=	6.704
Ar	0.0093	•	39.944	=	0.371
CO_2	0.0003	•	44.010	=	0.013
Total	1.0000				28.966

Suppose the compositional information was available on a weight rather than a volume basis. If dry air consisted, by weight, of 75.52% nitrogen, 23.15% oxygen, 1.28% argon and 0.04% CO_2, its apparent molecular weight would be determined as follows:

Component	Weight fraction		Molecular weight		lb mol/lb
N_2	0.7552	÷	28.016	=	0.02696
O_2	0.2315	÷	32.000	=	0.00723
Ar	0.0128	÷	39.944	=	0.00032
CO_2	0.0005	÷	44.010	=	0.00001
Total	1.0000				0.03452

The apparent molecular weight of dry air with this composition is then 1/0.03452 = 28.969 lb/lb-mole.

For wet air, the apparent molecular weight (MW) may be calculated from the composition as shown above or by combining the molecular weights of the dry air and the water vapor on the basis of their respective volume fraction or mole fraction:

$$MW_{wet\ air} = (1 - x_{water}) \cdot (MW_{dry\ air}) + (x_{water}) \cdot (MW_{water}) \qquad [2.1]$$

2.2.2 Viscosity

Viscosity is associated with the fluid resistance to flow. Viscosity is the result of two physical phenomena: intermolecular cohesive forces and momentum transfer

Chapter 2: Basic Concepts of Gases

between flowing strata caused by molecular agitation perpendicular to the direction of motion. Between adjacent strata of a moving fluid, a shearing stress, (τ), occurs that is directly proportional to the velocity gradient. This is expressed in the equation:

$$\tau = \mu \frac{dv}{dy} \qquad [2.2]$$

where: τ = unit shearing stress between adjacent layers of fluid

dv/dy = velocity gradient, m/sec

μ = viscosity proportionality constant, centipoise

The proportionality constant, μ, is called the coefficient of viscosity, or merely, viscosity. It should be noted that pressure does not appear in Equation [2.2] which indicates that the shear, τ, and the viscosity, μ, are independent of pressure.

2.2.3 Kinematic Viscosity

The ratio of the absolute viscosity to the density of a fluid often appears in dimensionless numbers, such as the Reynolds numbers. The expression for kinematic viscosity is used to simplify calculations. Kinematic viscosity is defined as:

$$\upsilon = \frac{\mu}{\rho_g} \qquad [2.3]$$

where: υ = kinematic viscosity, cm^2/sec

μ = viscosity of the gas, g/cm • sec

ρ_g = density of the gas, g/cm^3

A typical inertial force per unit of fluid is $\rho v^2/L$; a typical viscous force per unit volume of fluid is $\mu v/L^2$. The first expression divided by the second provides the dimensionless ratio known as the Reynolds Number:

$$Re = \frac{\rho \bullet v \bullet L}{\mu} \qquad [2.4]$$

where: ρ = density of fluid, g/cm^3

v = velocity of the fluid, cm/sec

L = linear dimension, cm

μ = viscosity of the fluid, g/cm • sec

Re = Reynolds Number (dimensionless)

The viscosity of a gas at the given condition may be found from the following formula (Joseph and Beachler 1981):

$$\frac{\mu}{\mu_o} = \left(\frac{T}{273.2}\right)^n \qquad [2.5]$$

where: μ = viscosity of the gas for prevailing conditions

μ_o = viscosity at 0°C and prevailing conditions

T = absolute prevailing temperature, °K

n = an empirical exponent (n = 0.768 for air)

The viscosity of air and other gases at various temperatures and at a pressure of 1 atmosphere may be found in Table 2.1. One unit commonly used to describe viscosity has been the centipoise which equals 0.01 g/cm sec (6.72 • 10^{-4} lb/ft sec). In the SI system, this becomes a millipascal second (1 cp = 1 mPa sec).

For water at 1 atm pressure and 20°C = 68°F = 528°R

ρ = 62.3 lb_m/ft^3 = 998.2 kg/m^3

μ = 1.002 centipose = 1.002 • 10^{-5} Pa • sec = 1.002 • 10^{-3} kg/m sec

= 6.73 • 10^{-4} lb_m/ft sec = 2.09 • 10^{-5} $lb_f sec/ft^2$

$\upsilon = \mu/\rho$ = 1.004 • 10^{-6} m^2/sec

= 1.004 centistoke = 1.077 • 10^{-5} ft^2/sec

If the fluid is air at 1 atm pressure and 20°C = 68°F = 528°R

ρ = 0.075 lb_m/ft^3 = 1.20 kg/m^3

= 2.59 • 10^{-3} lb mol/ft^3 = 4.16 mol/m^3

μ = 0.018 centipoise

= 1.8 • 10^{-5} Pa • sec = 1.8 • 10^{-5} kg/m sec

$\upsilon = \mu/\rho$ = 1.613 • 10^{-4} ft^2/sec

= 1.488 • 10^{-5} m^2/sec

Table 2.1
Properties of Air

Temperature		Specific heat at constant pressure (C_p)	Absolute viscosity (μ)		Density (ρ)		Thermal conductivity (k)
°F	°C	Btu/lb °F or Cal/g. °C	$\frac{lb}{hr \cdot ft}$	$\frac{kg}{hr \cdot m}$	lb/ft³	kg/m³	Btu/hr · ft°F
0	-17.8	0.240	0.040	0.060	0.0863	1.382	0.0124
20	-6.7	0.240	0.041	0.061	0.0827	1.386	0.0128
40	4.4	0.240	0.042	0.062	0.0794	1.328	0.0132
60	15.6	0.240	0.043	0.064	0.0763	1.275	0.0136
80	26.7	0.240	0.045	0.067	0.0734	1.225	0.0140
100	37.8	0.240	0.047	0.070	0.0708	1.178	0.0145
120	48.9	0.240	0.047	0.070	0.0684	1.098	0.0149
140	60.0	0.240	0.048	0.071	0.0662	1.063	0.0153
160	71.1	0.240	0.050	0.074	0.0639	1.026	0.0158
180	82.2	0.240	0.051	0.076	0.0619	0.994	0.0162
200	93.3	0.240	0.052	0.077	0.0601	0.963	0.0166
250	121.1	0.241	0.055	0.082	0.0538	0.896	0.0174
300	148.9	0.241	0.058	0.086	0.0521	0.836	0.0182
350	176.7	0.241	0.060	0.089	0.0489	0.785	0.0191
400	204.4	0.241	0.063	0.094	0.0460	0.739	0.0200
450	232.2	0.242	0.065	0.097	0.0435	0.697	0.0207
500	260.0	0.242	0.067	0.100	0.0412	0.661	0.0214
600	315.6	0.242	0.072	0.107	0.0373	0.599	0.0229
700	371.1	0.243	0.076	0.113	0.0341	0.547	0.0243
800	426.7	0.244	0.080	0.119	0.0314	0.504	0.0257
900	482.2	0.245	0.085	0.126	0.0295	0.474	0.0270
1,000	537.8	0.246	0.089	0.132	0.0275	0.441	0.0283
1,200	648.9	0.248	0.097	0.144	0.0238	0.381	0.0308
1,400	760.0	0.251	0.105	0.156	0.0212	0.339	0.0328
1,600	871.1	0.254	0.112	0.167	0.0192	0.307	0.0346
1,800	982.2	0.257	0.120	0.179	0.0175	0.280	0.0360
2,000	1,092.3	0.260	0.127	0.189	0.0161	0.258	0.0370

Source: Danielson 1973

2.3 Gas Laws

2.3.1 Ideal Gas Law

The two precursors of the ideal gas law are Boyle's and Charles' Laws. Boyle found that the volume of a given mass of gas is inversely proportional to the absolute pressure if the temperature is kept constant:

$$P_1 V_1 = P_2 V_2 \quad [2.6]$$

where: V_1 = volume of gas at absolute pressure P_1 and temperature T

V_2 = volume of gas at absolute pressure P_2 and temperature T

Charles found that the volume of a given mass of gas varies directly with the absolute temperature at constant pressure.

$$\frac{V_1}{T_1} = \frac{V_2}{T_2} \quad [2.7]$$

where: V_1 = volume of gas at pressure P and absolute temperature T_1

V_2 = volume of gas at pressure P and absolute temperature T_2

Boyle's and Charles' Laws may be combined into a single equation in which neither temperature nor pressure need to be held constant:

$$\frac{P_1 V_1}{T_1} = \frac{P_2 V_2}{T_2} \quad [2.8]$$

Both Boyle's and Charles' Law are satisfied in the Ideal Gas Law:

$$P \bullet V = \frac{M \bullet R \bullet T}{MW} \quad [2.9]$$

where: P = absolute pressure

V = volume of a gas

M = mass of gas

MW = molecular weight of a gas

T = absolute temperature

R = universal gas constant

Chapter 2: Basic Concepts of Gases

The unit of R depends upon the units of measurements used in the equation. Some useful values are (Perry 1973):

$$1.987 \frac{\text{Btu}}{(\text{lb mole})(°R)} \qquad 1.987 \frac{\text{cal}}{(\text{g mole})(°K)}$$

$$1,544 \frac{(\text{lb})(\text{ft})}{(\text{lb mole})(°R)} \qquad 21.83 \frac{(\text{in. Hg})(\text{ft}^3)}{(\text{lb mole})(°R)}$$

$$554.6 \frac{(\text{mm Hg})(\text{ft}^3)}{(\text{lb mole})(°R)} \qquad 10.73 \frac{(\text{psi})(\text{ft}^3)}{(\text{lb mole})(°R)}$$

$$0.73 \frac{(\text{atm})(\text{ft}^3)}{(\text{lb mole})(°R)} \qquad 83 \frac{(\text{atm})(\text{cm}^3)}{(\text{g mole})(°K)}$$

$$8.3 \bullet 10^{-3} \frac{(\text{Pa})(\text{m}^3)}{(\text{g mole})(°K)} \qquad 8.3 \bullet 10^3 \frac{(\text{kPa})(\text{m}^3)}{(\text{kg mole})(°K)}$$

$$8.3 \frac{\text{Joule}}{(\text{g mole})(°K)}$$

Any value of R can be obtained by applying the fact, with appropriate conversion factors, that there are 22.414 liters per g mole (359 ft³/lb mole) at 0°C (32°F) and 101.3 kPa (29.92 in. Hg).

The Ideal Gas Law applies to mixtures of gases as well as a pure gas. Thus, a gram mole of ideal gas, or mixture of ideal gases, occupies 22.41 litres at 0°C and 101.3 kPa. This volume is called the gram-mole volume. Similarly, a pound mole of ideal gas occupies 359 ft³ at 32°F and 1 atm. The pound-mole volume at any other combination of temperature and pressure may be calculated by Equation [2.10].

$$V_2 = V_1 \frac{P_1}{P_2} \frac{T_2}{T_1} \qquad [2.10]$$

Other useful forms of the Ideal Gas Law are shown in Equations [2.11] and [2.12]. Equation [2.11] applies to gas flow rather than to gas confined in a container.

$$P \bullet Q = n \bullet R \bullet T \qquad [2.11]$$

where: Q = gas volumetric flow rate (ft³/hr)

P = absolute pressure (psia)

n = molar flow rate (lb mol/hr)

T = absolute temperature °R

R = 10.73 psia-ft³/lb mol °R

Equation [2.12] combines n and V from Equation [2.9] to express the law in terms of density.

$$P \cdot MW = \rho \cdot R \cdot T \qquad [2.12]$$

where: MW = molecular weight of gas (lb/lb mol)

ρ = density of gas (lb/ft³)

Volumetric flow rates are often given not at the actual conditions of pressure and temperature, but at arbitrarily chosen standard conditions, standard temperature and pressure (STP). To distinguish between flow rates based on the two conditions, the letters (a) and (s) are often used as part of the unit. The units (acfm) and (scfm) stand for actual cubic feet per minute and standard cubic feet per minute, respectively. The standard Ideal Gas Law can be used to convert from standard to actual conditions, but since there are many standard conditions in use, the STP being used must be known. Standard conditions most often used are shown in Table 2.2. When predicting the performance in designing equipment, the actual conditions must be employed.

Table 2.2
Standard Gas Conditions

System	Temperature	Pressure	Molar volume
SI	273°K	101.3 kPa	22.4 m³/kg mol
Universal scientific	0°C	760 mm Hg	22.4 L/g mol
Natural gas industry	60°F	14.7 psia	379 ft³/lb mol
American engineering	32°F	1 atm	359 ft³/lb mol
Hazardous Waste	60°F	1 atm	379 ft³/lb mol
Incinerator industry	70 F	1 atm	387 ft³/lb mol

Charles' Law can be used to correct flow rates from standard to actual conditions.

$$Q_a = Q_s (T_a/T_s) \qquad [2.13]$$

where: Q_a = volumetric flow rate at actual conditions (ft³/hr)

Q_s = volumetric flow rate at standard conditions (ft³/hr)

T_a = actual absolute temperature (°R)

T_s = standard absolute temperature (°R)

Chapter 2: Basic Concepts of Gases

Absolute temperatures and pressures must be employed in all ideal gas law calculations.

Example 2.1 Gas Laws Application — Molar Volume

Calculate the molar volume, V/n, for an ideal gas at 70°F and 1 atmosphere pressure.

Solution:

$$\frac{V}{n} = \frac{R \cdot T}{P} = \frac{\left(10.73 \frac{\text{psi ft}^3}{\text{lb mole }^\circ R}\right) \cdot 530\,^\circ R}{14.7 \text{ psia}}$$

$$\frac{V}{n} = 387 \frac{\text{ft}^3}{\text{lb mole}}$$

2.3.2 Density and Specific Volume

Density is the ratio of mass to the volume occupied, e.g., lb/ft^3 or g/cm^3. Specific volume is the volume occupied per mass and is equal to the inverse of density. Both of these quantities depend on the temperature and pressure of the system. Using the ideal gas law, and recognizing that the number of moles is given by mass divided by molecular weight, density (ρ) may be calculated from:

$$\rho = \frac{M}{V} = \frac{(P \cdot MW)}{R \cdot T} \qquad [2.14]$$

Density can also be determined from molecular weight and molar volume:

$$\rho = \left[\frac{MW}{387}\right] \cdot \left[\frac{530^\circ R}{T}\right] \cdot \left[\frac{P}{29.92}\right] \qquad [2.15]$$

where: MW = molecular weight

T = absolute temperature, °R

P = absolute pressure, in. Hg

Values for the density and other properties of air over a limited range of temperature are provided in Table 2.1.

Example 2.2 Gas Laws Application — Density

Determine the density of air at 75°C and 14.7 psia. The molecular weight of air is 29.

Solution:

$$\rho = \frac{P \bullet MW}{R \bullet T} = \frac{(14.7 \text{ psia}) \bullet (29 \text{ lb / lb mol})}{(10.73 \text{ ft}^3 \text{ psi / lb mol °R}) \bullet (75 + 460)°R}$$

$$= 0.0743 \text{ lb / ft}^3$$

Example 2.3 Gas Laws Application — Volume

Calculate the 1 lb mol volume (in ft³) of any ideal gas at 60°F and 14.7 psia.

Solve the ideal gas law for V and calculate the volume.

$$V = \frac{n \bullet R \bullet T}{P} = \frac{(1) \bullet (10.73) \bullet (60 + 460)}{(14.7)}$$

$$= 379 \text{ ft}^3$$

This result is an important number to remember in combustion calculation — 1 lb mol of any (ideal) gas at 60°F and 1 atm occupies 379 ft³

Example 2.4A Gas Laws Application — Flow Rate

Data from a hazardous waste incinerator indicates a volumetric flow rate of 10,000 scfm (60°F, 1 atm). Calculate the actual flow rate in actual cubic feet per minute for the incinerator if the operating temperature and pressure of the unit are 1,100 °F and 1 atm.

Solution:

Since the pressure remains constant, the actual cubic feet per minute can be calculated using Charles' Law.

$$Q_a = Q_s \left(\frac{T_a}{T_s}\right) = 10,000 \bullet \left(\frac{1,100 + 460}{60 + 460}\right)$$

$$= 30,000 \text{ acfm}$$

Example 2.4B Gas Laws Application — Flow Rate

Determine the final volumetric flow rate of a gas that is heated from 100° to 300 °F which has an initial volume of 3,500 scfm (60°F, 1 atm).

Solution:

$$Q_a = Q_s \left(\frac{T_a}{T_s}\right) = (3,500 \text{ scfm}) \cdot \left(\frac{100 + 460}{60 + 460}\right)$$

$$= 3,769 \text{ acfm}$$

Using Charles' Law again, calculate the final volumetric flow rate

$$Q_{a2} = Q_{a1}\left(\frac{T_2}{T_1}\right) = \frac{(3,769) \cdot (760)}{(560)}$$

$$= 5,115 \text{ acfm}$$

2.3.3 Concentration Units for Atmospheric Pollutants

Because concentrations of atmospheric pollutants are usually quite small, special concentration units simplify work in air pollution. The first is used for gaseous pollutants exclusively: parts per million by volume (ppm_v). It is defined by:

$$ppm_v = \frac{V_{pol}}{V_{air}} \cdot 10^6 \qquad [2.16]$$

where: V_{pol} = partial volume occupied by the pollutant in the mixture at total pressure P and temperature T

V_{air} = total volume occupied by the mixture at the same T and P

The concentration of either gaseous pollutants or particulate matter can be expressed in terms of micrograms of pollutant per cubic meter of air

$$C = \frac{\text{mass (micrograms)}}{\text{volume (cubic meters)}} = \frac{\mu g}{m^3}$$

To convert from parts per million by volume, ppm_v ($\mu L/L$), it is necessary to know the molar volume at the given temperature and pressure and the molecular weight of the pollutant.

Example 2.5 Application of Concentration Practice

The atmospheric concentration of SO_2 is reported as 2.5 ppm by volume.

(a) Determine the concentration in $\mu g/m^3$ at 25°C and 760 mm Hg.

(b) Determine the concentration in $\mu g/m^3$ at 37°C and 752 mm Hg.

Solution:

Let parts per million equal $\mu L/L$, then 2.5 ppm$_v$ = 2.5 $\mu L/L$. The molar volume at 25°C and 760 mm Hg is 24.46 L, and the molecular weight of SO_2 is 64.1 g/mole.

(a) $$C = \frac{2.5 \; \mu L}{L} \cdot \frac{1 \; \mu \text{ mole}}{24.46 \; \mu L} \cdot \frac{64.1 \; \mu g}{\mu \text{ mole}} \cdot \frac{1{,}000 \; L}{m^3}$$

$$= 6.5 \bullet 10^3 \; \frac{\mu g}{m^3}$$

at 25°C and 760 mm HG

(b) $$V = (24.46 \; \mu L) \bullet \left(\frac{310 \; K}{298 \; K} \right) \bullet \frac{760 \text{ mm Hg}}{752 \text{ mm Hg}} = 25.73 \; \mu L$$

$$C = \frac{2.5 \; \mu L}{L} \cdot \frac{1 \; \mu \text{ mole}}{25.73 \; \mu L} \cdot \frac{64.1 \; \mu g}{\mu \text{ mole}} \cdot \frac{1{,}000 \; L}{m^3}$$

$$= 6.2 \bullet 10^3 \; \frac{\mu g}{m^3}$$

at 37°C, 752 mm Hg

This example points out the need for reporting temperature and pressure when the results are presented on a weight-to-volume basis.

2.3.4 Dalton's Law of Partial Pressure

When gases or vapors are present as a mixture in a given space, the pressure exerted by a component of the gas mixture at a given temperature is the same as it would exert if it filled the whole space alone. The pressure exerted by one component of a gas mixture is called its partial pressure. For component A, Dalton's Law is expressed as (Perry 1973):

$$P_A = \frac{M_A \bullet R \bullet T}{MW_A \bullet V} \qquad [2.17]$$

Chapter 2: Basic Concepts of Gases

where: P_A = partial pressure of a component of air mixture

M_A = mass of A, g

MW_A = molecular weight of A, g mole

V = volume of mixture, m³

T = absolute prevailing temperature, °K

A useful concept from this and the ideal gas law is pressure percent = volume percent = mole percent. For example, if a mixture of air is 21 % oxygen and 79 % nitrogen by volume at 101.3 kPa (1 atm) and 0°C, then this mixture contains 0.21 mole fraction oxygen and 0.79 mole fraction nitrogen. Also, the partial pressure of oxygen is equal to 21.27 kPa (0.21 atm) and the partial pressure of nitrogen is 80.03 kPa (0.79 atm) at 0°C. In mathematical terms, the mole fraction in the gas phase is related to the partial pressure by:

$$y_A = \frac{P_A}{P_T} \qquad [2.18]$$

where: P_A = partial pressure of A, Pa

P_T = total pressure of the system, Pa

At a normal atmospheric pressure of 760 mm Hg, a mixture of air and solvent vapor will behave as if each exerted a pressure, the total of which would be 760 mm Hg. For instance, at temperature T, if a solvent has a vapor pressure of 200 mm Hg, there will be in that mixture of air and solvent vapor, enough solvent to represent 200/760 of the total volume. This also means that in a mole of the mixture there will be 200/760 moles of solvent and 560/760 moles of air. If the solvent has a molecular weight of 92 and the air a molecular weight of 29, there will be 24.2 lb $\left(\frac{200}{760} \bullet 92\right)$ of solvent and 21.4 lb $\left(\frac{560}{760} \bullet 29\right)$ of air, for a total weight of 45.6 lb or a weight concentration of 53 percent solvent and 47 percent air.

The foregoing concept is useful for calculating emissions measured in pounds of solvent per hour.

2.3.5 Gas-Liquid Relationships

Raoult's Law states that the partial pressure of a component over a solution is the product of the vapor pressure of that component and the mole fraction of that component. Raoult's Law is expressed as:

$$p_A = p_A^\circ x_A \qquad [2.19]$$

where: p_A = partial pressure of A, Pa

p_A = vapor pressure of A, Pa

x_A = mole fraction of A in solution

Raoult's Law applies over the entire concentration range from 0 to 1.0, but is applicable only if the solution behaves in an ideal manner. Unfortunately, few solutions behave ideally. A special case of Raoult's Law is Henry's Law. Henry's Law states that the concentration of a gas dissolved in solution is directly proportional to its partial pressure at a constant temperature.

$$p_A = H \cdot x_A \qquad [2.20]$$

where: H = Henry's Law constant, Pa/mole fraction

Henry's Law applies only if a plot of the partial pressure versus the mole fraction of component A is a straight line. This relationship holds true for most solutions in the dilute concentration range. The value and concentration range over which Henry's Law constant applies can be found only by experiment.

2.3.6 Vapor Pressure

Every liquid exerts a vapor pressure. *Vapor pressure* is defined as the pressure exerted by a pure component vapor in equilibrium with a flat surface of its pure component liquid at a certain temperature. Vapor pressure is a measure of the escaping tendency or volatility of the liquid; thus, we refer to it as the liquid's vapor pressure. For instance, at 150°F, n-decane (a component in kerosene) has a low vapor pressure (0.3 psi), whereas i-pentane (a component in gasoline) has a high vapor pressure (over 3 atm).

Vapor pressure increases rapidly with an increase in temperature. A common curve-fit equation for the vapor pressure/temperature relationship is the Antoine equation, which is:

$$\log P_{vi} = A_i - \frac{B_i}{C_i + T} \qquad [2.21]$$

where: P_{vi} = vapor pressure of pure liquid

T = temperature

A_i, B_i, C_i = curve-fit constants for component i

Vapor pressure data are often presented in tables or graphs (see Chapter 9).

The vapor pressure should not be confused with the partial pressure of a vapor in a mixture of gases. In general, the partial pressure can take on any value less than or equal to the vapor pressure, as long as no liquid is present. If the system is not at equilibrium, there are no theoretical ties between vapor pressure and partial pressure. When the system is at equilibrium with both vapor and liquid present, the vapor phase is said to be saturated, and the partial pressure is equal to the vapor pressure.

2.4 Energy Concepts

2.4.1 Heat Capacity

The *heat capacity* of a substance is defined as the quantity of heat required to raise the temperature of that substance by 1°C; the *specific heat capacity* is the heat capacity on a unit mass basis. Since the specific heat of water is approximately 1 cal/g °C or 1 Btu/lb °F, the term *specific heat* has come to imply heat capacity per unit mass. For gases, the addition of heat to cause the 1 degree temperature rise may be accomplished either at constant pressure or at constant volume. Since the amount of heat necessary is different for the two cases, subscripts are used to identify which heat capacity is being used, C_p for constant pressure, and C_v for constant volume. For liquids and solids, this distinction does not have to be made since there is little difference between the two. Heat capacities are often used on a molar basis instead of mass basis, in which case the unit become cal/gmol °C or Btu/lbmol °F. Average or mean heat capacity data over specific temperature ranges are available (see Table 2.3).

2.4.2 Enthalpy

Enthalpy is a measure of the thermal energy of a substance. Common units are Btu/lb or cal/g. The Btu, or British thermal unit, is defined as the amount of heat necessary to raise one pound of water from 59°F to 60°F at a pressure of one atmosphere. The calorie is the amount of heat required to raise one gram of water at one atmosphere from 14.5°C to 15.5°C. The enthalpy of a substance at a given temperature has no practical value except in relation to the enthalpy at another temperature or condition. Since enthalpy differences are arbitrary, datum temperatures may be chosen to define enthalpy at any other temperature:

$$h = C_p(T - T_{ref}) \qquad [2.22]$$

where: h = enthalpy, Btu/lb

C_p = heat capacity at constant pressure, Btu/lb °F

T = temperature of substance, °F

T_{ref} = reference temperature, °F

For gases, the reference temperature is typically 60°F, although this is highly variable; for water it is usually 32°F.

The enthalpy of water vapor is equal to the enthalpy of the water plus the latent heat of vaporization, λ_v:

$$h_{\text{water vapor}} = h_{\text{water}} + \lambda_v \qquad [2.23]$$

Like C_p, the latent heat of vaporization is also a function of temperature and can be found in a reference text. The enthalpy of an air-water vapor mixture is given by:

$$h_{\text{mixture}} = h_{\text{dry air}} + \phi\left(h_{\text{water vapor}}\right) \qquad [2.24]$$

where: $\quad \phi$ = absolute humidity, lb water/lb dry air

h_{mixture} = enthalpy of mixture, Btu/lb dry air

$h_{\text{dry air}}$ = enthalpy of dry air, Btu/lb dry air

Usually, the enthalpy difference is of interest.

$$\Delta H = h_2 - h_1 \qquad [2.25]$$

$$\Delta H = \left(C_p\right)_2 \left(T_2 - T_{\text{ref}}\right) - \left(C_p\right)_1 \left(T_1 - T_{\text{ref}}\right) \qquad [2.26]$$

For exact calculations, the heat capacity corresponding to each temperature condition must be used. Approximate results can be obtained, and the calculations considerably simplified, by using a heat capacity averaged over the range of temperature change. For air, this typically results in only a small error. Assuming T_{ref} is the same for both enthalpies, the relationship for enthalpy change then becomes:

$$\Delta H = \left(C_p\right)_{\text{avg}} \left(T_2 - T_1\right) \qquad [2.27]$$

An even simpler determination of enthalpy change can be made by using tabulated values of enthalpy as a function of temperature. One such tabulation has been included as Table 2.3. Thus, if one were interested in the amount of energy that must be removed in order to cool an air stream from 500°F to 100°F, one need only subtract the corresponding enthalpy values:

$$\Delta H = \left(h_{500} - h_{100}\right) \qquad [2.28]$$

$= 106.7 \text{ Btu / lb} - 9.6 \text{ Btu / lb}$

$= 97.1 \text{ Btu / lb}$

If enthalpy values from different information sources are used in this manner, it must be remembered that the choice of a reference temperature is arbitrary and may not be the same for all tabulations. Before subtracting two enthalpy values, it must be confirmed that they were determined for the same reference temperature.

Chapter 2: Basic Concepts of Gases

Table 2.3
Enthalpies of Various Gases in Btu/lb

Temp		CO_2	N_2	H_2O	O_2	Air
°F	°C					
100	38	5.8	6.4	17.8	8.8	9.6
150	66	17.6	20.6	40.3	19.8	21.6
200	93	29.3	34.8	62.7	30.9	33.6
250	121	40.3	47.7	85.5	42.1	45.7
300	149	51.3	59.8	108.2	53.4	57.8
350	177	63.1	73.3	131.3	64.8	70.0
400	204	74.9	84.9	154.3	76.9	82.1
450	232	87.0	97.5	177.7	87.2	94.4
500	260	99.1	110.1	201.0	99.5	106.7
600	316	124.5	135.6	248.7	123.2	131.6
800	427	176.8	187.4	346.4	171.7	182.2
1,000	538	231.9	240.5	447.7	221.6	234.1
1,200	649	289.0	294.9	552.9	272.7	287.2
1,400	760	347.6	350.5	661.3	324.6	341.5
1,600	871	407.8	407.3	774.2	377.3	396.8
1,800	982	469.1	464.8	889.8	430.4	452.9
2,000	1,093	531.4	523.0	1003.1	484.5	509.5
2,200	1,204	594.6	582.0	1130.3	538.6	567.1
2,400	1,316	658.2	642.3	1256.8	593.5	625.0
2,500	1,371	690.2	672.3	1318.1	621.0	654.3
3,000	1,649	852.3	823.8	1640.2	760.1	802.3

Note: The enthalpies tabulated for H_2O represent a gaseous system and the enthalpies do not include the latent heat of vaporization. It is recommended that the latent heat of vaporization at 60 °F (1,059.9 Btu/lb) be used where necessary.

Source: Danielson 1973

2.5 Air/Water Vapor Mixtures

The state of an air-water vapor mixture is completely defined by specifying the pressure, temperature, and humidity. The Gibbs-Dalton rule of partial pressures states that individual components in a mixture exert a pressure that would be the same as that exerted if the same mass of the component were present alone in the same total volume and at the same temperature. Thus, for an air-water vapor mixture:

$$p_{air} + p_{water} = p_{total} \qquad [2.29]$$

Relative saturation is then defined as the ratio of the partial pressure of water vapor present to that which would be present if the air were saturated:

$$\text{Relative saturation} = \frac{P_{water}}{P_{water\ at\ saturation}} \quad [2.30]$$

It should be noted that relative saturation is also equal to the ratio of the corresponding mole fractions. Relative humidity is simply relative saturation multiplied by 100 to express it in percent. Absolute or specific humidity is the weight of water vapor per weight of dry air, usually expressed as pounds of water per pounds of dry air.

Many of the values for air-water vapor mixtures are available in a graphical representation known as a psychrometric (humidity) chart (see Figure 2.1). A humidity chart is used to determine the properties of moist air and to calculate moisture content in air. The ordinate of the chart is the *absolute humidity* ϕ, which is defined as the mass of water vapor per mass of bone-dry air.

Equation [2.31] gives ϕ in terms of moles and also in terms of partial pressure.

$$\phi = \frac{18 n_{H_2O}}{29 \bullet (n_T - n_{H_2O})} = \frac{18 p_{H_2O}}{29 \bullet (P - p_{H_2O})} \quad [2.31]$$

where: n_{H_2O} = number of moles of water vapor

n_T = total number of moles in gases

p_{H_2O} = partial pressure of water vapor

P = total system pressure

Curves showing the relative humidity (ratio of the mass of the water vapor in the air to the maximum mass of water vapor the air could hold at that temperature, assuming the air was saturated) or humid air also appear on the charts. The curve for 100% relative humidity is also referred to as the saturation curve. The abscissa of the humidity chart is air temperature, also known as the dry bulb temperature (T_{DB}). The wet bulb temperature (T_{WB}) is another measure of humidity; it is the temperature at which a thermometer with a wet wick wrapped around the bulb stabilizes. As water evaporates from the wick to the ambient air, the bulb is cooled; the rate of cooling depends on how humid the air is. No evaporation occurs if the air is saturated with water, hence T_{DB} and T_{WB} are the same. The lower the humidity, the greater the difference between these two temperatures. On the psychrometric chart, constant wet-bulb temperature lines are straight with negative slopes. The values of T_{WB} correspond to the value of the abscissa at the point of intersection of this line with the saturation curve.

The humid volume is the volume of wet air per mass of dry air and is linearly related to the humidity. The product of the humid volume and the absolute humidity gives the volume of the moist air. The humid enthalpy is the enthalpy of the

Chapter 2: Basic Concepts of Gases   43

moist air on a bone-dry air basis. The enthalpy for saturated air can be read from the chart by extending the approximate wet-bulb temperature line upwards to the diagonal scale labelled enthalpy of saturation.

The following are some helpful points on the use of psychrometric charts.

- Heating or cooling at temperatures above the dew point (temperature at which the vapor begins to condense) correspond to a horizontal movement on the chart. As long as no condensation occurs, the absolute humidity remains constant.

- If the air is cooled, the system follows the appropriate horizontal line to the left until it reaches the saturation curve and follows it thereafter.

- In problems involving the use of a humidity chart, it is convenient to choose the mass of dry air as a basis, since the chart uses this basis.

In order to use the psychrometric chart, one must first locate the position on the chart that corresponds to the conditions of the air stream. This is done by knowing any two quantities and locating the point of intersection along their corresponding lines. Once this point is determined, values for the other quantities may be read from the appropriate scales.

One value that is of interest but cannot be read directly from the psychrometric chart is the density of the air-water vapor mixture. However, this can be determined from the values obtained from the chart, as follows:

$$\rho_{mixture} = \frac{(1+\phi)}{v} \qquad [2.32]$$

where: ϕ = absolute humidity, lb/lb dry air

v = specific volume, ft³/lb dry air

Example 2.6 Psychrometric Chart Application

Refer to the psychrometric chart (Figure 2.1) to answer the following:

(a) List key properties for humid air at a dry-bulb temperature of 160°F and a wet-bulb temperature of 100 °F.

(b) A stream of moist air is cooled and humidified adiabatically from T_{DB} of 100°F and T_{WB} of 70°F to T_{DB} of 80°F. Determine the amount of water added per pound of dry air.

Figure 2.1
Psychrometric Chart

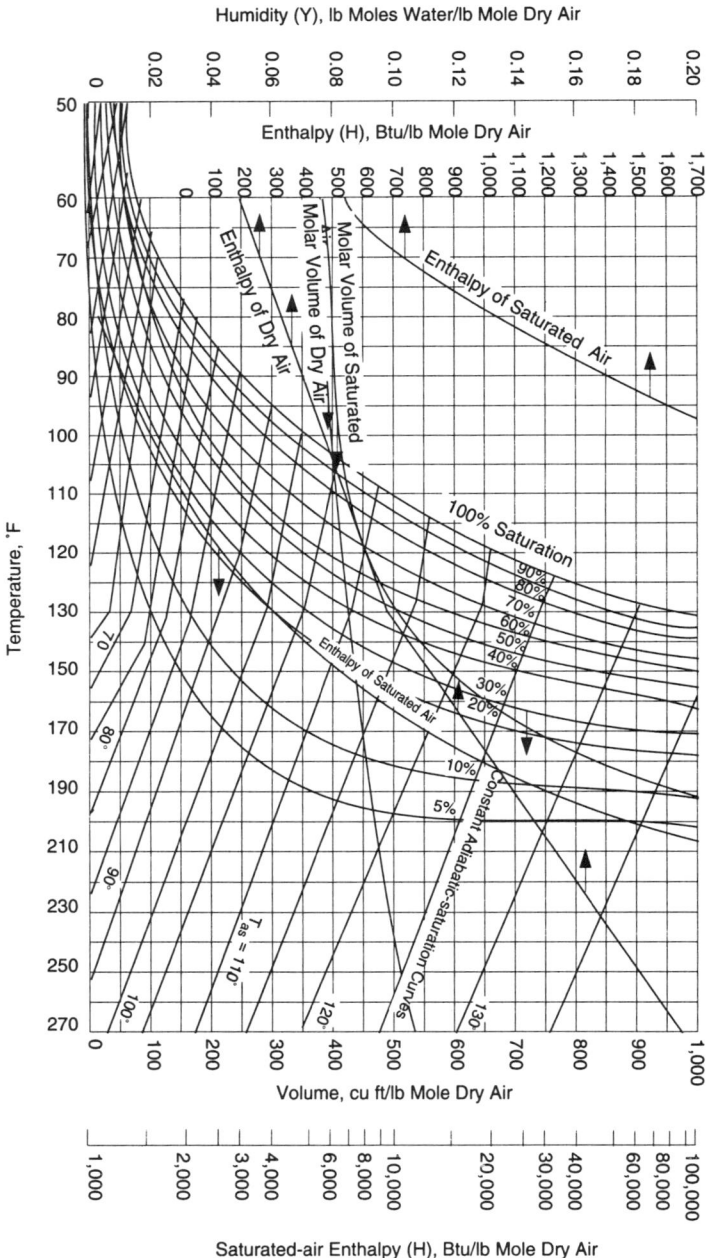

Solution:

(a) If the air were to be cooled until the moisture just begins to condense, the dew point would be reached. This is represented by a horizontal line at a constant humidity intersecting the saturation curve at a dew point of 87.5°F. The relative humidity is approximately 14 % (interpolating between the 15 % and 10% relative humidity lines). The absolute humidity is the horizontal line extended to the right intersecting the ordinate at the humidity of 0.0285 lb water/lb dry air. The humid volume is approximately 16.3 ft^3 moist air (interpolating between 16 and 17 ft^3 moist air volume). The enthalpy for saturated air at a T_{WB} of 100°F is 71.8 Btu/lb dry air. For the unsaturated air, the enthalpy deviation is -1.0 Btu/lb dry air; therefore, the actual enthalpy for the moist air at a T_{WB} of 100°F and a T_{DB} of 160°F is 70.8 Btu/lb dry air.

(b) Adiabatic cooling follows the wet-bulb temperature line upwards (toward the saturation curve). The difference in the final and initial humidity is the added moisture.

	ϕ
Initial	0.0090
Final	0.0133

$\Delta\phi = 0.0133 - 0.0090 = 0.0043$ lb water/lb dry air

2.6 Cooling of Gaseous Effluents

Many processes generate gas streams at temperatures that are too high for some air pollution control devices to accept. In most cases, the exhaust system must be cooled to temperatures below 500°F. Therefore, it is necessary to employ some type of cooling device to reduce gas temperature. Although there are several methods of cooling hot gases, those most commonly used in air pollution control systems are: (1) dilution with ambient air, (2) quenching with water, and (3) natural convection and radiation from ductwork. With the dilution method, the hot gaseous effluent from the process equipment is cooled by adding sufficient ambient air to result in a mixture of gases at the desired temperature. Natural convection and radiation occurs whenever there is a temperature difference between the gases inside a duct and the atmosphere surrounding it. Cooling hot gases by this method requires only the provision of enough heat transfer area to obtain the desired amount of cooling. The water quench method uses the heat of vaporization of water to cool the gases. Water is sprayed into the hot gases under conditions conducive to evaporation, and this cools the gases.

Cooling by dilution air is commonly used where conveying air volumes are low or where there is a large volume of dilution inherent in the hoods required to capture the air contaminants. If large gas volumes are necessary, and dilution air is not economical, then direct cooling with water quench chambers is generally favored over other cooling methods. This is due to the small space requirements, ease of operation, and low installation costs of the water quench chambers. When the characteristics of the gaseous effluent and the contaminants are such that water cannot be used, natural convection-radiation cooling is generally employed (Danielson 1973).

2.6.1 Dilution with Ambient Air

Cooling gases by dilution with ambient air is the simplest method that can be employed. With this technique, hot gas from a process is cooled by adding ambient air in sufficient quantity to produce a mixture with the desired temperature. The fundamental relationship governing performance may be developed through a heat balance on the dilution system.

$$m_1 h_1 + m_2 h_2 = m_3 h_3 \qquad [2.33]$$

where:
- m_1 = mass flow rate of hot gases
- h_1 = enthalpy of hot gases
- m_2 = mass flow rate of dilution air
- h_2 = enthalpy of dilution air
- m_3 = mass flow rate of gas mixture, $m_1 + m_2$
- h_3 = enthalpy of gas mixture

Solving the material and energy balance equation (assuming the heat capacity for air is virtually constant over the temperature range of interest) and applying the Ideal Gas Law leads to:

$$Q_d = Q_e \left(\frac{T_e - T_f}{T_f - T_d} \right) \left| \frac{T_d}{T_e} \right| \qquad [2.34]$$

where:
- Q_d = dilution air flow rate needed
- Q_e = exhaust air flow rate to be cooled
- T_e = temperature of exhaust air
- T_d = temperature of the dilution air
- T_f = final temperature of the mixed stream

The absolute value sign | | in Equation [2.34] serve as a reminder that, at this point, T_d and T_e must be expressed as absolute temperatures. Dilution air can be introduced through a branch into the hot duct. However, when the hot gas stream is large, the quantity of dilution air becomes excessively large, making the cost of the control device uneconomical.

Example 2.7 Dilutional Cooling

Calculate the flow rate of dilution air (in cfm at 90°F) needed to cool 50,000 cfm of air from 1,200°F to (a) 500°F, (b) 300°F, and (c) 150°F.

Flow diagram:

$Q_e = 50,000$ cfm, M_e, h_e
$T_e = 1,200\ °F$ → M_f, T_f, h_f

Q_d, M_d, $T_d = 90°F$, h_d

Using Equation [2.34]:

(a) $$Q_d = Q_e \left(\frac{T_e - T_f}{T_f - T_d} \right) \left| \frac{T_d}{T_e} \right| = 50{,}000 \bullet \left(\frac{1{,}200 - 500}{500 - 90} \right) \bullet \left(\frac{550}{1{,}660} \right)$$

$Q_d = 28{,}280$ cfm

(b) $$Q = 50{,}000 \bullet \left(\frac{1{,}200 - 300}{300 - 90} \right) \bullet \left(\frac{550}{1{,}660} \right)$$

$Q_d = 71{,}000$ cfm

(c) $$Q = 50{,}000 \bullet \left(\frac{1{,}200 - 150}{150 - 90} \right) \bullet \left(\frac{550}{1{,}660} \right)$$

$Q_d = 290{,}000$ cfm

2.6.2 Quenching with Water

When a large volume of hot gas is to be cooled and only a small volume of dilution is needed to capture the air contaminants, some method of cooling other than dilution with ambient air should be used. Since the evaporation of water requires a large amount of heat, the gas can be cooled simply by spraying water into the hot gas.

Cooling hot gases with a water quench is relatively simple and requires very little space. For controlling the gas temperature leaving the quench chamber, a temperature controller is generally used to regulate the amount of water sprayed into the quench chamber. Quench chambers are little more than enlarged portions of the ductwork equipped with water sprays. They are easy to operate and, with automatic temperature controls, only that amount of water is used that is needed to maintain the desired temperature of the gases at the discharge. Their installation and operating costs are generally considered to be less than for other cooling methods. A quench chamber should not be used when the gases to be cooled contain a large amount of gases or fumes that become highly corrosive when wet. This creates additional maintenance problems, not only in the quench chamber, but in the downstream control device. Also, the increased moisture content of the emission stream after water injection may be detrimental to the operation of some air pollution control devices, such as fabric filters.

The material and energy balance are:

$$M_a + M_w = M_f \qquad [2.35]$$

$$M_a(h_{ai} - h_{af}) = M_w(h_{wf} - h_{wi}) \qquad [2.36]$$

where: M = mass flow rates, lb_m/min

h = specific enthalpies, Btu/lb_m

a,w = subscripts indicating air and water, respectively

i, f = subscripts indicating initial and final states, respectively

The enthalpy change of the water in Equation [2.36] is due not only to temperature change, but also to the latent heat of vaporization of water. Equation [2.36] can be rewritten as:

$$M_a C_{pa}(T_a - T_f) = M_w[\Delta H_v + C_{pwv}(T_f - T_v) + C_{pwl}(T_v - T_w)] \qquad [2.37]$$

where: C_{pa} = average specific heat of air over the temperature range, Btu/lb_m°F

ΔH_v = heat of vaporization of water, Btu/lb_m

C_{pwv} = specific heat of water vapor, Btu/lb_m °F

C_{pwl} = specific heat of water liquid, Btu/lb_m °F

T_v = temperature at which the water vaporizes, °F

Chapter 2: Basic Concepts of Gases

For the required accuracy of these calculations, average values of C_p can be used. The values of C_{pa} range from 0.24 to 0.26 Btu/lb$_m$ °F over the range from 0°F to 2,000°F, and the values of C_{pwv} ranges from 0.44 to 0.51 from 60°F to 1,000°F. At 1 atm, ΔH_v ranges from 970 Btu/lb$_m$ (at 212°F) to 1,060 Btu/lb$_m$ (at 60°F). Note that T_v is often well below 200°F, depending on the partial pressure of the water in the system. Since enthalpy is a thermodynamic state function, there is no difference in the total enthalpy change calculated by (1) assuming that the water is completely vaporized at T_w and that the water vapor is then heated to T_f, or by (2) assuming that the water is heated as a liquid from T_w to T_f and then vaporized. For simplicity, we will assume that all the vaporization occurs at the inlet water temperature. Thus, the third term inside the brackets of Equation [2.37], C_{pwl} (T_v - T_w) becomes zero.

Equation [2.37] can be solved for the water injection rate required to achieve any desired T_f. Once the water injection rate is obtained, it can be converted to moles, then added to the number of moles of exhaust air. We can then calculate a final volumetric flow rate. Alternatively, a psychrometric chart can be used to estimate these values.

The following example will illustrate some of the factors that must be considered when designing a quench chamber to cool the gaseous effluent from a gray-iron-melting cupola.

Example 2.8 Water Quench

Given:

32-in. inside diameter cupola. Maximum temperature of gaseous effluent at cupola outlet is 2,000°F. Weight of gaseous effluent at cupola outlet is 216 lb/min. Volume of gaseous effluent at cupola outlet is 13,280 cfm at 2,000°F. This volume of effluent includes in-draft air at the charging door of the cupola. The temperature of 2,000°F is a maximum temperature.

Assume the effluent gases have the same properties as air. Consideration of the enthalpies of the gaseous constituents in the effluent gas stream will show that this is valid assumption.

Determine the water needed to cool the gaseous effluent to 225°F and the total volume of gases discharged from the quench chamber.

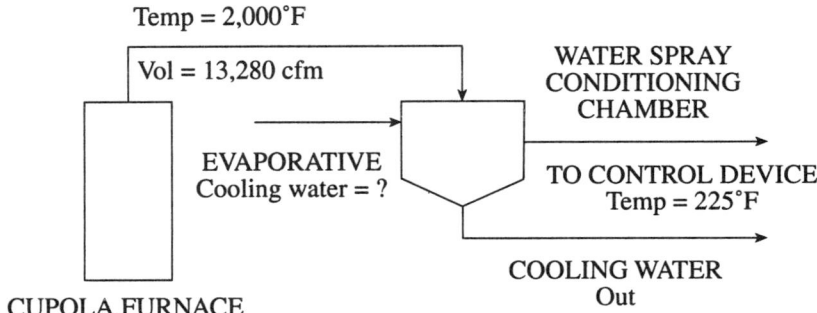

Solution:

1. Cooling required:

    Enthalpy of gas at 2,000°F = 509.5 Btu/lb

    Enthalpy of gas at 225°F = 39.6 Btu/lb

    ΔH = 469.9 Btu/lb

    (216)(469.9) = 101,300 Btu/min

2. Water to be evaporated:

    Heat absorbed per lb of water:

    $Q = h_g(225°F, 14.7\ psig) - h_f(60°F)$

    $= 1,156.8 - 28.06 = 1,128.7\ Btu/lb$

    Water required $= \dfrac{101,300}{1,128.7} = 90\ lb/min$

3. Volume of water evaporated at 225°F:

    $\dfrac{379}{18} \bullet \left(\dfrac{460+225}{460+60}\right) \bullet (90) = 2,510\ cfm$

4. Total volume vented from spray chamber at 225°F:

    $\text{Cupola} = (13,280) \bullet \left(\dfrac{225+460}{2,000+460}\right) = 3,700\ cfm$

    Water = 2,510 cfm

    Total = 6,210 cfm

2.6.3 Natural Convection and Radiation

When a hot gas stream flows through a duct, the duct becomes hot. As the air near the duct becomes heated, convection currents develop that carry the heat away. This phenomenon is referred to as *natural convection* and may be aided by forced convection due to wind motion. Heat may also be transferred from the duct surface by direct radiation. This behavior can be exploited to produce significant amounts of cooling by providing a section of duct that has a large surface area.

The rate of heat transfer is a function of the resistance to heat flow, the mean temperature difference between the hot gas and the air surrounding the duct, and the surface area of the duct. It may be expressed as:

$$Q = U \bullet A \bullet \Delta t_m \qquad [2.38]$$

where:
Q = rate of heat transfer, Btu/hr
U = overall heat transfer coefficient, Btu/hr ft² °F
A = heat transfer area, ft²
Δt_m = log-mean temperature, °F

$$\Delta t_m = \frac{(T_i - T_a) - (T_o - T_a)}{\ln \frac{(T_i - T_a)}{(T_o - T_a)}}$$

T_a = ambient temperature, °F
T_i = inlet temperature, °F
T_o = outlet temperature, °F

The rate of heat transfer is determined by the amount of heat to be removed from the hot gaseous effluent entering the exhaust system. For any particular basic process, the weight of gaseous effluent and its maximum temperature are fixed. The cooling system must, therefore, be designed to dissipate sufficient heat to lower the effluent temperature to the operating temperature of the air pollution control device to be used.

For radiation-convection cooling, the overall heat transfer coefficient is generally calculated based on the outside surface and is denoted by U_o. U_o is defined as:

$$U_o = \frac{h_{io} h_o}{h_{io} + h_o} \qquad [2.39]$$

where: h_{io} = inside film coefficient based on outside surface area, Btu/hr ft² °F

h_o = outside film coefficient, Btu/hr ft² °F

Figure 2.2 is a plot of j_H against Reynolds Number, where:

$$j_H = \frac{h_i D}{k}\left(\frac{C_p \mu}{k}\right)^{-\frac{1}{3}} \qquad [2.40]$$

where: D = duct diameter, ft

L = duct length, ft

Figure 2.2
Tube-Side Heat Transfer Curve

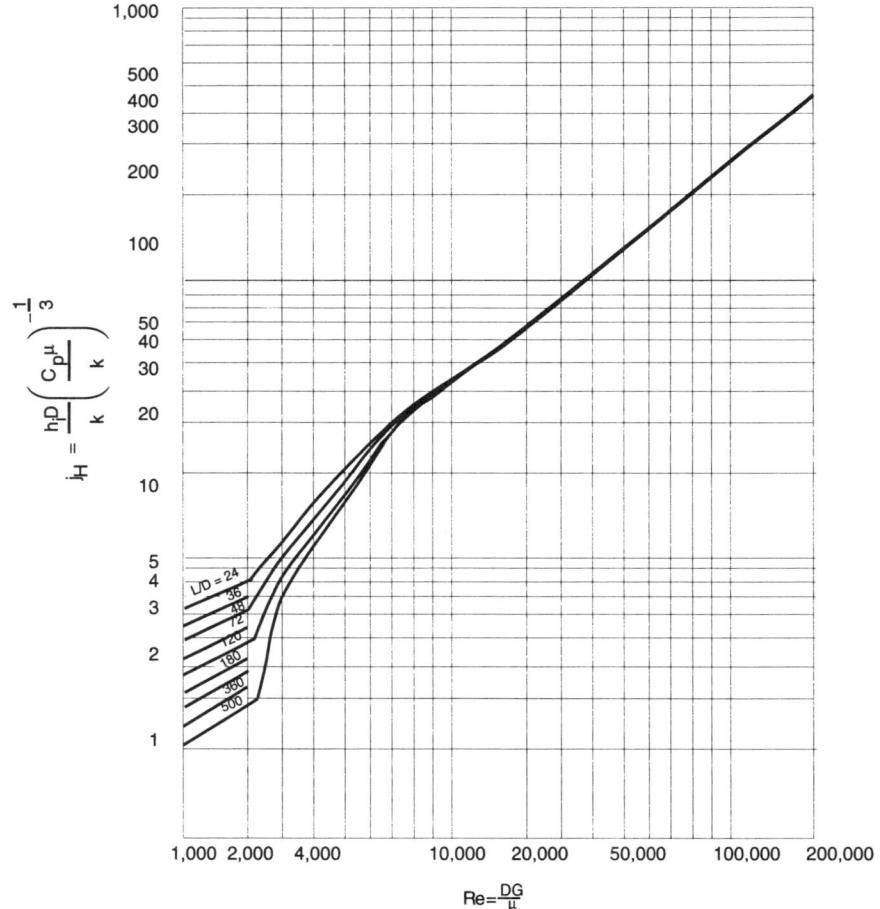

Source: Perry 1973

Chapter 2: Basic Concepts of Gases

k = thermal conductivity, Btu/hr ft °F

C_p = heat capacity, Btu/lb °F

μ = viscosity, lb/hr ft

G = gas mass velocity, lb/hr ft²

The values of k, C_p, and μ for air at specific temperatures are presented in Table 2.1.

The inside film coefficient, h_i, then can be calculated from Equation [2.40]. The h_i must be corrected to outside surface basis by:

$$h_{io} = h_i \frac{D_i}{D_o} \quad [2.41]$$

where: D_i = duct inside diameter

D_o = duct outside diameter

Outside film coefficient, h_o, includes the transfer due to both radiation and convection cooling to the atmosphere ($h_o = h_c + h_r$). Perry (1973) provides the following equation to calculate the overall heat transfer. For convection cooling, the heat transfer coefficient is

$$h_c = 0.27 \left(\frac{\Delta T}{D_o} \right)^{\frac{1}{4}} \quad [2.42]$$

D_o = outside duct diameter, ft

ΔT = log-mean temperature, °F

h_c = Btu/hr ft² °F

For radiant cooling, the overall radiant rate of transfer is given by:

$$h_r = 0.170\varepsilon \left[\left(\frac{T_1}{100} \right)^4 - \left(\frac{T_2}{100} \right)^4 \right] \quad [2.43]$$

where: ε = 1 for black duct

T_1 = duct temperature, °R

T_2 = surrounding temperature, °R

h_r = radiation coefficients of heat transfer, Btu/hr ft² °F

Figure 2.3 is a plot of Equation [2.43].

Figure 2.3
Radiation Coefficients of Heat Transfer, h_r

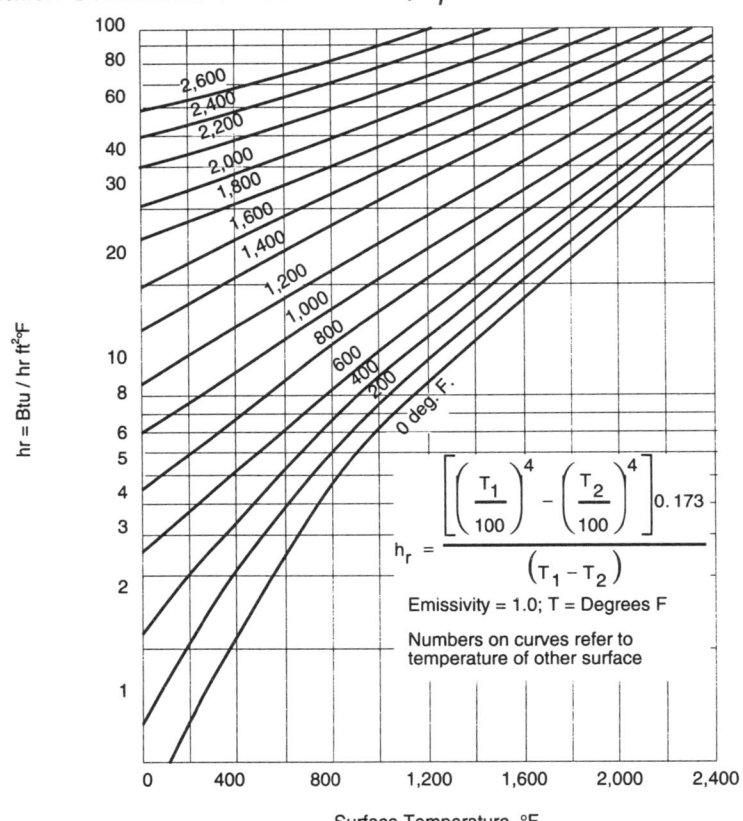

Source: Perry 1973

To calculate h_o, the temperature of the duct wall, T_w, must first be assumed and then checked. The following equation may be used for this purpose.

$$T_w = T_{ag} - \frac{h_o}{h_o + h_{io}}(T_{ag} - T_a) \qquad [2.44]$$

where: T_{ag} = average gas temperature, °F

T_a = ambient air temperature, °F

If calculated T_w is not the same as assumed T_w, estimate a new T_w and recalculate h_o. When the assumed T_w and calculated T_w are the same, use the corresponding h_o to calculate U_o. Typical values for U_o are in the 0.5 to 2.0 Btu/hr ft² °F range.

Chapter 2: Basic Concepts of Gases

With this kind of cooling, flexibility in controlling the gas temperature is limited. When either the gas stream or air temperatures, or both, are lower than design values, the gases discharged from the cooling device will be less than that calculated, and condensation of moisture from the effluent within the control device might result. Conversely, when design temperatures are exceeded, the temperature of the gases discharged from the cooling system could become too high. Radiant cooling is generally economical only when the gas temperature is above 500°F. Significant amount of space is required for the duct work.

Table 2.4 shows the relative volumetric changes required for various cooling methods for a combustion gas at an initial temperature of 1,800°F. The change in volume is not directly related to size, since velocity may be permitted to increase with the injection of water into the gas stream without increasing the pressure drop. Table 2.5 shows the approximate variation in the required size of the equipment if the pressure drop, expressed in in. H_2O, is kept constant. The tables show that water injection is preferred to air injection.

Table 2.4
Volume Effects of Cooling 1,800°F Combustion Products

Method	Relative Volumes		
	750°F	500°F	250°F
Indirect cooling	1.0	0.8	0.6
Water injection	2.4	2.0	1.6
Air injection	2.9	3.6	6.2

Table 2.5
Size Effects of Cooling 1,800°F Combustion Gases

Method	Relative Volumes		
	750°F	500°F	250°F
Indirect cooling	1.0	0.9	0.8
Water injection	1.7	1.6	1.5
Air injection	2.9	4.0	8.0

Example 2.9 Radiation Cooling

If natural convection-radiation is proposed to cool the flue gas from a combustion furnace before entering the control equipment, determine the length of duct needed to reduce the temperature of the flue gas to 400°F (modified from Danielson 1973).

Given:

gaseous flow rate	= 13,280 acfm
	= 216 lb/min
flue gas temperature	= 2,000°F
duct diameter	= 2.2 ft
duct wall thickness	= 0.141 in.
ambient temperature	= 100°F
average density of gas	= 0.024 lb/ft³
average heat capacity of gas	= 0.247 Btu/lb °F
gas viscosity	= 0.1 lb/hr ft
gas thermal conductivity	= 0.03 Btu/hr ft °F

Solution:

1. Determine total heat loss from gas stream.

    Using Equation [2.27] $\Delta H = (C_p)_{avg}(T_2 - T_1)$

    $Q = m \cdot \Delta H$

    $Q = (216 \text{ lb/min}) \cdot (0.247 \text{ Btu/lb°F}) \cdot (2{,}000 - 400)°F$

    $= 85{,}363 \text{ Btu/min}$

2. Determine heat loss by convection-radiation.

    Using Equation [2.38] $Q = U \cdot A \cdot \Delta t_m$

    $$\Delta t_m = \frac{(T_i - T_a) - (T_o - T_a)}{\ln \frac{(T_i - T_a)}{(T_o - T_a)}}$$

    $$\Delta t_m = \frac{(2{,}000 - 100) - (400 - 100)}{\ln \frac{(2{,}000 - 100)}{(400 - 100)}}$$

    $\Delta t_m = 867°F$

    $A = L \cdot 2.2 \cdot \pi = 6.91 \cdot L \text{ ft}^2$

    where L = duct length, ft

Chapter 2: Basic Concepts of Gases

3. Determine inside film coefficient (h_i).

   Use Figure 2.2 to determine j_H

$$Re = \frac{DG}{\mu}$$

$$G = (216 \text{ lb/min}) / \left(\pi (2.2 \text{ ft})^2 / 4\right)$$
$$= 56.8 \text{ lb/ft}^2 \text{ min}$$
$$= 3,409 \text{ lb/ft}^2 \text{ hr}$$

$$Re = \frac{2.2 \text{ ft} \bullet 3,409 \text{ lb/ft}^2\text{hr}}{0.1 \text{ lb/hr ft}} = 75,000$$

From Figure 2.2: $j_H = 200$

Using Equation [2.40] $j_H = \frac{h_i D}{k} \bullet \left(\frac{C_p \mu}{k}\right)^{-\frac{1}{3}}$

$$200 = \frac{h_i \; 2.2 \text{ ft}}{0.03 \text{ Btu/hr ft °F}} \bullet \left(\frac{0.247 \text{ Btu/lb°F} \bullet 0.1 \text{ lb/hr °F}}{0.03 \text{ Btu/hr ft°F}}\right)$$

$$h_i = 2.55 \text{ Btu/hr ft}^2\text{°F}$$

4. Convert h_i to inside film coefficient (h_{io}) based on outside surface area.

$$D_o = 2.2 + 0.141/12 = 2.224 \text{ ft}$$

   (10-gage wall thickness = 0.141 in.)

Using Equation [2.41] $h_{io} = h_i \frac{D_i}{D_o} = 2.55 \bullet \frac{2.2}{2.224} = 2.53 \text{ Btu/hr ft}^2 \text{ °F}$

5. Calculate the outside film coefficient (h_o) where $h_o = h_c + h_r$

Using Equation [2.42] $h_c = 0.27 \left(\frac{\Delta T}{D_o}\right)^{\frac{1}{4}}$

Assume a duct wall temperature of 500°F

$$h_c = 0.27 \left(\frac{500 - 100}{2.224}\right)^{\frac{1}{4}} = 0.99 \text{ Btu/hr ft}^2 \text{ °F}$$

From Figure 2.3: at surface temperature = 500°F and ambient temperature = 100°F

$h_r \cong 3.5$ Btu / hr ft² °F

$h_o = h_c + h_r = 0.99 + 3.5 = 4.49$ Btu / hr ft² °F

6. Check T_w.

Using Equation [2.44] $T_w = T_{ag} - \dfrac{h_o}{h_o - h_{io}}(T_{ag} - T_a)$

$T_{ag} = (2,000 + 400)/2 = 1,200$ °F

$T_w = 1,200 - \left(\dfrac{4.49}{4.49 + 2.53}\right)(1,200 - 100) = 496°F$

The assumption for T_w is valid.

7. Determine the overall heat transfer coefficient, U_o:

$U_o = \dfrac{h_{io} h_o}{h_{io} + h_o}$

$U_o = \dfrac{2.53 \bullet 4.99}{2.53 + 4.99} = 1.68$ Btu / hr ft² °F

8. Determine required duct length.

Heat balance required:

$85,363$ Btu / min • 60 min / hr $= (1.68$ Btu / hr ft² °F$) \bullet (6.91 \bullet L$ ft²$) \bullet (867°F)$

L = 509 ft

2.7 Pressure

As a fluid flows through a duct, its momentum and pressure may change. If the effects of frictional forces and compressibilities are ignored, Bernoulli's equation is applicable:

$$\dfrac{v^2}{2} + \dfrac{P}{\rho} + gz = \text{Constant} \qquad [2.45]$$

where: v = fluid velocity

P = fluid pressure

ρ = fluid density

g = gravitational acceleration

z = height above a reference datum

Although this equation strictly applies only to incompressible, inviscid fluids, it is of significant importance because it relates the pressure at a point in a fluid to its position and velocity and does so in a rather simple way.

Rearranging Equation [2.45] gives:

$$\frac{v^2}{2g} + \frac{P}{\rho g} + z = \text{Constant} \qquad [2.46]$$

In this form, each term represents energy per unit weight of fluid and has dimensions of length. Thus, each term may be regarded as representing a contribution to the total fluid head:

where: $v^2/2g$ = velocity head

$P/\rho g$ = pressure head

z = potential head

Consider the following situation (see Figure 2.4), in which an open tube has been inserted into a flowing fluid. The pressure of the flowing fluid causes the fluid to rise to a level, h, in the open tube.

Figure 2.4
Open Tube in Flowing Liquid

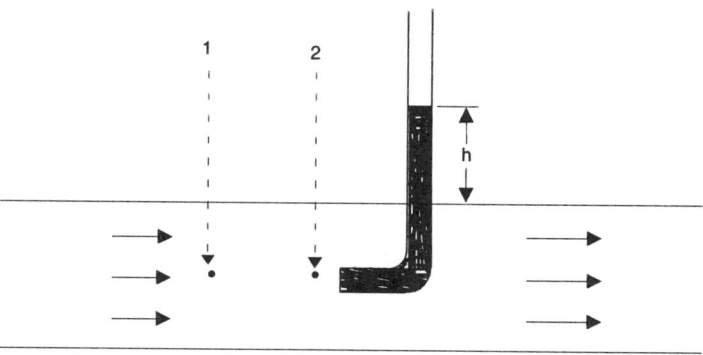

Since the terms in Bernoulli's equation sum to a constant, it may be written:

$$\frac{(v_1)^2}{2g} + \frac{P_1}{\rho g} + z_1 = \frac{(v_2)^2}{2g} + \frac{P_2}{\rho g} + z_2 \qquad [2.47]$$

The position of point 1 and 2 are on the same level, so $z_1 = z_2$. Also, the fluid at point 2, just at the entrance of the tube, is balanced by the fluid in the tube, so the velocity at this point is zero. Substituting gives:

$$\frac{(v_1)^2}{2g} + \frac{P_1}{\rho g} = \frac{P_2}{\rho g} \qquad [2.48]$$

or

$$\frac{(v_1)^2}{2g} = \text{velocity head}$$

$$\frac{P_1}{\rho g} = \text{pressure head}$$

$$\frac{P_2}{\rho g} = \text{total head}$$

Expressing these energy heads as pressures, we may write this relationship in its more common form (ACGIH 1988):

$$VP + SP = TP \qquad [2.49]$$

where: VP = velocity pressure

SP = static pressure

TP = total pressure

Thus, at any point in a flowing fluid the total pressure is the sum of the velocity pressure and the static pressure. This relationship is illustrated in Figure 2.5 for an air stream on either side of a fan. Here, the manometers with one leg connected to ports that are perpendicular to the flow streamlines, and the other leg open to the atmosphere, measure static pressure. The manometers with one leg connected to the tubes facing into the flow, and the other leg open to the atmosphere, measure total pressure, which is the sum of the static and velocity pressures. The manometers with one leg connected to the tubes measure the difference between total and static pressure, which is velocity pressure. It should be noted that while static and total pressures are usually negative upstream of a fan and positive downstream of a fan, velocity pressure is always positive.

In the above development, the velocity pressure, VP, was given by the term $v^2/2g$ in units of length of fluid. In measuring velocity pressure, typically the pressure of the flowing fluid is used to displace fluid in a manometer and is read in inches of water column. Converting the units of the velocity pressure term so that it has units of inches of water gives:

Chapter 2: Basic Concepts of Gases

Figure 2.5
Pressure Relationship

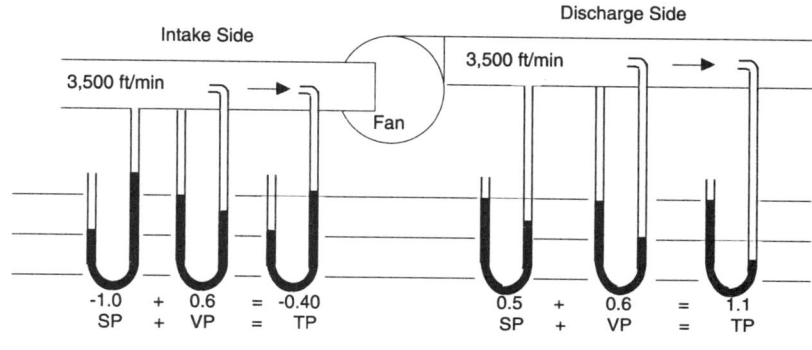

$$VP = \left[\frac{(v/60)^2}{2g}\right] \cdot \left(\frac{\rho_a}{\rho_w}\right)^{\frac{1}{2}} \quad [2.50]$$

where: v = air velocity, ft/min
  ρ_a = air density, lb/ft³
  ρ_w = water density, lb/ft³
  g = acceleration due to gravity, ft/sec²

Substituting a water density at 70°F of 62.302 lb/ft³ and a gravity acceleration at sea level of 32.174 ft/sec²:

$$VP = \rho_a \cdot \left(\frac{v}{1,096.7}\right)^2 \quad [2.51]$$

Since the velocity pressure is usually measured and used to calculate the air velocity, the more useful form of Equation [2.51] is:

$$v = 1,096.7 \cdot \left(\frac{VP}{\rho_a}\right)^{0.5} \quad [2.52]$$

For standard air, ρ_a = 0.075 lb/ft³ and Equation [2.52] becomes:

$$v = 4{,}005 \bullet (VP)^{0.5} \qquad [2.53]$$

Pressure drop in air pollution control devices can be related to energy usage and therefore operating costs by determining horsepower requirements for the system.

$$WK = Q \bullet (VP + SP) \qquad [2.54]$$

where: WK = work input rate

Q = volume flow rate of air or gas

TP = total pressure of system

If Q is expressed in units of cfm and VP in in. H_2O, WK in Equation [2.54] may be expressed in units of horsepower, HP, by:

$$HP = Q \text{ cfm} \bullet VP \text{ in.} H_2O \bullet \frac{0.036 \text{psi}}{1 \text{ in.} H_2O} \bullet \frac{1 \text{ Hp}}{550 \text{ ft lb/sec}} \bullet \frac{144 \text{ in}^2}{1 \text{ ft}^2} \bullet \frac{1 \text{ min}}{60 \text{ sec}}$$

$$HP \approx \frac{Q \bullet VP}{6{,}360} \qquad [2.55]$$

where: HP = horsepower, Hp

Q = air flow rate, cfm

VP = velocity pressure, in. H_2O

Example 2.10 Power Calculation

The exhaust system for a dryer needs to deliver 12,000 cfm at 600°F and 4 in. H_2O static pressure. Calculate the required horsepower.

Solution:

(Equation [2.55])
$$HP = \frac{Q \bullet VP}{6{,}360}$$

$$= \frac{12{,}000 \bullet 4}{6{,}360}$$

$$= 7.5 \text{Hp}$$

Chapter 2: Basic Concepts of Gases

2.8 Problem Set

2.1 Air is a mixture of several gases with an approximate volume relationship of 21% oxygen, 78.1 % nitrogen, and 0.9 % argon. Calculate the molecular weight.

2.2 Carbon dioxide gas (molecular weight = 44) flows through a duct that is three meters in diameter. Assume the following to be true:

$P_b = 1$ atm

$P_g = 0.1$ atm, vacuum

$T = 150°F$

$R = 0.82 \dfrac{(atm) \bullet (liters)}{(g\ mol) \bullet (°K)}$

kinematic viscosity = $1.1 \bullet 10^{-5}$ m²/sec

velocity = 0.5 m/sec

What are the values of the following?

    a) density, ρ

    b) absolute viscosity, μ

    c) Reynolds number, Re

2.3 If a 200 ml container of gas is heated from 40°C to 80°C at constant pressure, what is the volume of gas?

2.4 The specific volume of air is 12.4 ft³/lb at 32°F and 1 atm.

    a) Calculate the specific volume at 70°F.

    b) Using the ideal gas equation, determine the temperature in degrees Fahrenheit.

2.5 The concentration of carbon monoxide is 40 mg/m³. Convert this to ppm_v at standard temperature and pressure.

2.6 The exhaust from an incinerator shows a CO concentration with a partial pressure of 19.0 mm Hg. How many ppm_v is this ?

2.7 An air stream of 15,000 scfm contains 1 % by volume water vapor and 1,000 ppm H_2S.

    a) What is the partial pressure of the water vapor and H_2S?

b) If Henry's Law constant is 483 atm/mole fraction for H_2S dissolved in water, what is the maximum mole fraction of H_2S that can be dissolved in solution?

2.8 A gaseous stream at 25°C is known to contain only acetone and air. The total pressure is 760 mm Hg absolute. Determine the following:

a) The mole fraction of acetone present.

b) The ppm of acetone present.

2.9 The wet and dry bulb temperature for an air stream were measured as 60°F and 75°F. Determine the partial pressure of the water vapor.

2.10 1,000 cfm at 85°F dry bulb and 70 % saturated is cooled to a saturation temperature of 65°F. What quantity of heat is removed from the air per minute and what is the rate of condensation of moisture per minute? Use the Psychrometric Chart, Figure 2.1.

2.11 The moisture content of an air stream with a flow rate of 5,000 lb/min is 0.05 expressed as weight decimal.

a) What is the moisture content in terms of volume (percent)?

b) What is the flow rate expressed as cfm?

2.12 Given air at a dry bulb temperature of 80°F and a wet bulb temperature of 65°F, find its dew point temperature, percent saturation, and moisture content.

2.13 Air at 90°F is 66% saturated. When cooled to 65°F, what is its final moisture content?

2.14 Air passes through an air washer with the moisture being added adiabatically. The entering condition of the air is 85°F dry bulb and 70°F wet bulb. The final dry bulb temperature of the air is 75°F. How much moisture was added per pound of dry air?

2.15 What would be the dry bulb temperature, wet bulb temperature, dew point, and percent saturation if the following quantities of air were mixed: 5,000 ft³ at 90°F and 50% saturated; 3,500 ft³ at 75°F dry bulb and 65°F wet bulb temperature; and 2,500 ft³ at 50°F and saturated.

2.16 The emission from a 5,000 cfm process nitrogen stream (130°F and 1 atm) has a toluene mole fraction of 0.080. The stream will be cooled in a condenser to 80°F. Calculate the rate of condensation of toluene.

Chapter 2: Basic Concepts of Gases 65

2.17 A rotary dryer has an emission of 125,000 acfm gas at 500°F. The gas is cooled by water sprays to 200°F. Assume gas at 500°F is 10% H_2O Vapor. Calculate the new volume at 200°F.

2.18 The emissions from a brass melting furnace total 29 lb/min of exhaust gases at 2,500°F. Determine the amount of dilution air (100°F) required to reduce the temperature to 300°F.

2.19 A fan delivers 3,000 cfm of air at a static pressure of 1 in. H_2O. What is the horsepower required if the static fan efficiency is 60% at this load?

2.20 Determine the final flow volume for a system to cool 230,000 acfm of gas at 1350°F to 500°F. Compare dilution, water spray, and convection/radiation cooling systems. Calculate the amount of water and the duct size needed to provide the necessary cooling.

2.9 References

ACGIH. *Industrial Ventilation*. Twentieth Edition. Cincinnati, OH. 1988.

Danielson, J. (Ed). *Air Pollution Engineering Manual*, 2nd. ed. EPA AP-40. Research Triangle Park, NC. 1973.

Joseph, G. and D. Beachler. *Control of Gaseous Emissions*. Northrop Services, Inc. EPA 450/2-81-005. Research Triangle Park, NC. 1981.

Perry, J. (Ed). *Chemical Engineers Handbook*, 5th. ed. McGraw Hill. New York. 1973.

Theodore, L. and J. Reynolds. *Introduction to Hazardous Waste Incineration*. Wiley-Interscience. New York. 1978.

Chapter 3

The Motion of Airborne Particles

3.1 Introduction

This chapter reviews the important elements of mechanics of individual particle motion. It begins with the interaction of a moving particle with its surroundings, leading to the concept of a *drag force*. This force represents the resistance to particle motion when a force is applied to that particle. It is the competition between these two types of forces (i.e., "driving force" and "resistance force") which governs the outcome of the motion in terms of transport and deposition.

3.1.1 Drag Force on a Particle

When a body immersed in a fluid moves relative to the fluid, it experiences a force associated with the resistance (by the fluid) to relative motion (Figure 3.1). The drag force acting on the particle may be derived from the solution of the Navier-Stokes' equations for the air flow in the particle fluid system in question.

In engineering practice, the drag coefficient, C_D, is defined as

$$C_D = \frac{F_D}{A(\rho_a v_p^2 / 2)} \qquad [3.1]$$

Figure 3.1
The Force Acting on a Spherical Particle Moving in Air

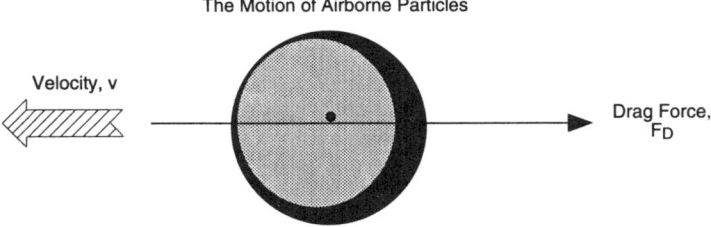

The Motion of Airborne Particles

where: A = area of the particle projected on a plane perpendicular to the direction of the particle motion, e.g., $A = \pi d_p^2 / 4$;

$\rho_a v_p^2 / 2$ = dynamic pressure on the particle; and

F_D = applied force on the particle

In the case of spheres, Equation [3.1] can be written as:

$$F_D = \frac{\pi d_p^2 \rho_a v_p^2 C_D}{8} \quad [3.2]$$

where: C_D = drag coefficient (unitless)

v_p = particle velocity (cm/sec)

d_p = particle diameter (cm)

ρ_a = the air density (g/cm^3)

The value of C_D is related to the velocity of the particle and the flow pattern of the fluid around the particle. The Reynolds number, Re_p, of the particle is used as an indication of this flow pattern. The Reynolds number of the particle is a function of the fluid density, particle diameter, particle velocity, and fluid viscosity. The Reynolds number is

$$Re_p = \frac{d_p \rho_a v_p}{\mu} \quad [3.3]$$

where μ is the fluid viscosity in g/cm sec.

In cgs units, at 20°C, $Re_p = 6.6\ v_p d_p$ (see Table 3.1) where v_p and d_p are measured in cm/sec and cm, respectively. If the air is moving, then v_p becomes the relative velocity between the particle and the air.

Table 3.1
Properties of Air at Standard Pressure

Properties	Temperature		units
	0°C	20°C	
Viscosity	$1.72 \cdot 10^{-4}$	$1.81 \cdot 10^{-4}$	$P(dyn\ sec/cm^2)$
Density	$1.29 \cdot 10^{-3}$	$1.2 \cdot 10^{-3}$	g/cm^3
Diffusion Coefficient	0.18	0.19	cm^2/sec

The particle Reynolds number depends on the relative velocity between an object such as an aerosol particle and the gas. Gas flow past a stationary particle is aerodynamically equivalent to a particle settling due to gravity through stationary air as long as the relative velocity is the same. Laminar flow around a particle occurs at low Reynolds numbers (Re<2) where viscous forces are much greater than inertial forces. It is characterized by a pattern of smooth streamlines that are symmetrical on the upstream and downstream sides of the particle, as shown in Figure 3.2 (Strauss 1975). As the Reynolds number increases above 20, eddies form downstream of a particle, gradually becoming more numerous and vigorous as the Reynolds number increases.

Figure 3.2
Flow Around a Sphere. (a) Laminar Flow, Re = 0.1 (b) Transition Flow (c) Turbulent Flow

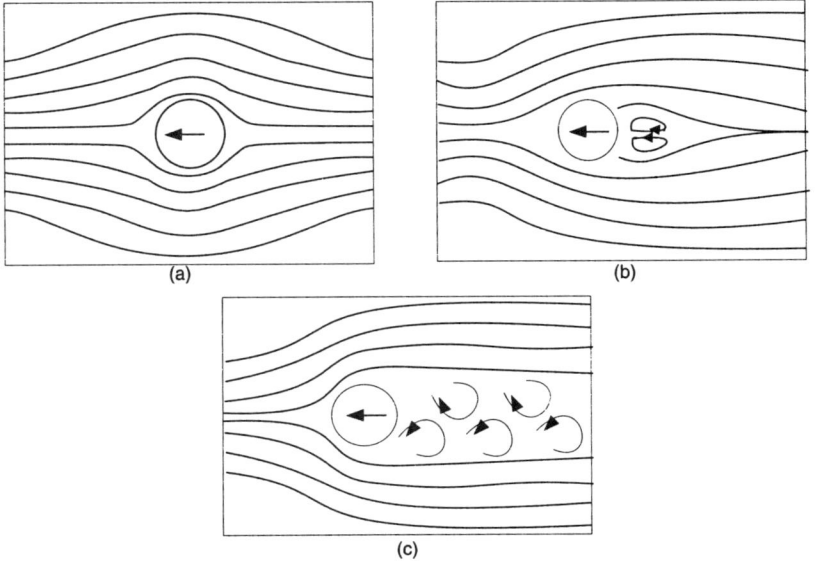

From experiment, it has been observed that three regimes exist (Figure 3.2). The three regimes are laminar (Stokes'), transition, and turbulent. These regimes are related to the Reynolds number of a particle, Re_p. The relationship of C_D versus Re_p is shown in Figure 3.3 (Perry 1973).

Figure 3.3
Drag Coefficient Versus Reynolds Number for Spheres

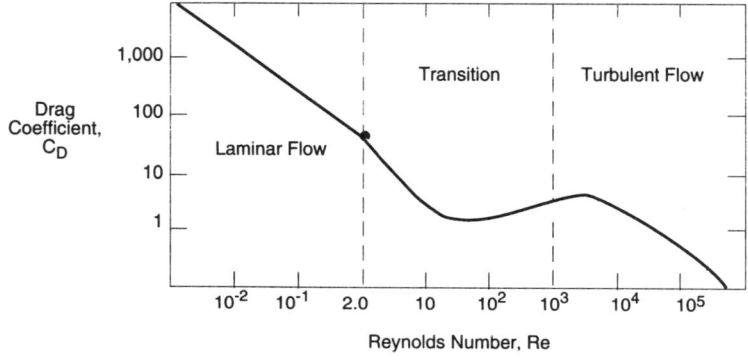

For low values of Reynolds number (Re <2), the flow is laminar. For much higher values of Reynolds number (Re>1,000), the flow is turbulent. For Reynolds numbers higher than 2 and lower than 1,000, the flow is transitional.

Mathematical expressions relating C_D and Re_p can be derived from Figure 3.3. Equations 3.4, 3.5, and 3.6 are used for determining C_D in the laminar, transition, and turbulent flow regimes, respectively (Beachles and Jahnke 1981):

$$C_D = \frac{24}{Re_p} \qquad Re_p < 2 \text{ (Laminar)} \qquad [3.4]$$

$$C_D = \frac{18.5}{(Re_p)^{0.6}} \qquad 2 < Re_p < 1,000 \text{ (Transition)} \qquad [3.5]$$

$$C_D = 0.44 \qquad 1,000 < Re_p < 200,000 \text{ (Turbulent)} \qquad [3.6]$$

In the laminar flow range (the straight line portion of Figure 3.3), when the drag coefficient is substituted into Equation 3.2, it becomes:

$$F_D = 3\pi\mu v_p d_p \qquad [3.7]$$

This is the particle drag force in the Stokes' Law range.

Chapter 3: The Motion of Airborne Particles

Example 3.1 Drag Force Calculation

A sample of oil mist is taken at a flow rate of 1.2 L/min through a horizontal tube 1 cm in diameter. The aerosol consists of 2 µm diameter oil droplets in air at standard conditions. The particles are moving through the tube with the air, but are settling at 0.01 cm/sec. What is (a) the flow Reynolds number and (b) the particle Reynolds number? (c) What is the drag force on the particle?

Solution:

(a) Flow Re $= \dfrac{\rho v d}{\mu} = 6.6 \, v \, d$

$v = Q/A$

$Q = 1.2 \cdot 1{,}000/60 = 20$ cm^3/sec, and

$A = \pi d^2/4 = 0.79$ cm^2

$v = 20/0.79 = 25$ cm/sec

Flow Re $= 6.6 \cdot 25 \cdot 1 = 165$

(Fluid flow should be laminar if the Reynolds number of the fluid is less than 2,100.)

(b) Particle Re$_p$ $= 6.6 \, v_p d_p$

$v_p =$ settling velocity

$d_p =$ diameter of the particle

Re$_p = 6.6 \cdot 0.01 \cdot 2 \cdot 10^{-4} = 1.3 \cdot 10^{-5}$

Since $1.3 \cdot 10^{-5}$ is less than 2, particle motion is laminar.

(c) Drag force $= 3 \pi \mu v_p d_p$

$= 3 \cdot 3.14 \cdot 1.81 \cdot 10^{-4} \cdot 0.01 \cdot 2 \cdot 10^{-4}$

$= 3.41 \cdot 10^{-9}$ dyn

3.1.2 Cunningham Correction Factor, C_c

An important assumption of Stokes' Law is that the relative velocity of the gas at the surface of the sphere is zero. This assumption does not apply to small particles whose size approaches the mean free path of the gas; such particles settle faster than predicted by Stokes' Law because there is slip at the surface of the particles.

At standard conditions, this error becomes significant for particles less than 1 μm in diameter. The factor, called the Cunningham Correction Factor, C_c, (also called the slip correction factor) is always greater than one and reduces the Stokes' drag force by:

$$F_D = \frac{3\pi\mu v_p d_p}{C_c} \qquad [3.8]$$

For particle size of 0.1 μm diameter and larger, C_c can be estimated by (Hinds 1982):

$$C_c = 1 + \frac{2.52\lambda}{d_p} \qquad [3.9]$$

where: λ = mean free path of the gas and is dependent on the gas temperature.

Note: λ_{air} at 1 atm and 20°C = 0.066 μm

C_c can also be estimated from the following equation for particles as small as 0.01 μm diameter (Beachles and Jahnke 1981):

$$C_c = 1 + \frac{6.21 \bullet 10^{-4} \bullet T}{d_p} \qquad [3.10]$$

where: T = absolute temperature of the gas, °K

d_p = particle diameter, μm

A compilation of values of C_c is provided in Figure 3.4 and Table 3.2 (Hinds 1982). For particles less than 1μm, the slip correction factor increases rapidly as size decreases (see Figure 3.4) and the Cunningham Correction Factor must be used. For accurate work it should be used for particles less than 10 μm.

3.1.3 External Forces on a Particle

One important application of Stokes' Law is the determination of the velocity of aerosol particles undergoing gravitational settling in still air. When a particle is released in air, it quickly reaches its terminal settling velocity, a condition where the drag force of the air on the particle will be exactly equal and opposite to the force of gravity, F_G. At this condition:

$$F_D = F_G = mg \qquad [3.11]$$

$$\frac{3\pi\mu v_p d_p}{C_c} = \frac{(\rho_p - \rho_g)\pi d_p^3 g}{6} \qquad [3.12]$$

Figure 3.4
Cunningham Correction Factor at 20°C and 1 atm

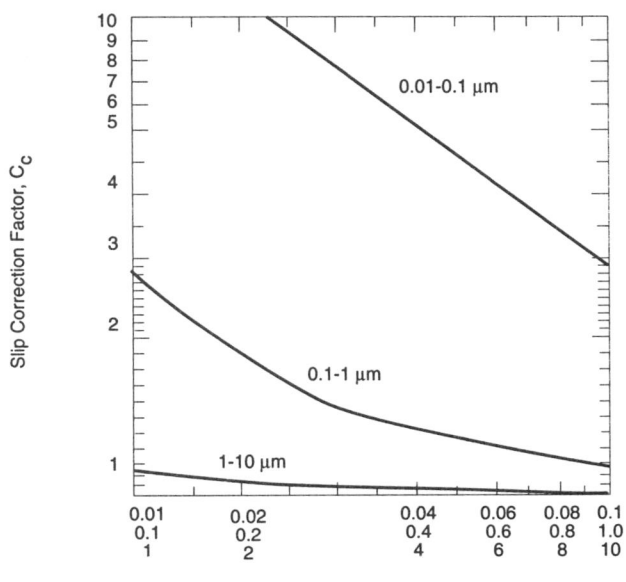

Source: Hinds 1982

Table 3.2
Cunningham Correction Factor

Particle diameter	Value of C_c at temperature of		
(μm)	70°F	212°F	500°F
0.1	2.88	3.61	5.14
0.25	1.682	1.952	2.528
0.5	1.325	1.446	1.711
1.0	1.16	1.217	1.338
2.5	1.064	1.087	1.133
5.0	1.032	1.043	1.067
10.0	1.016	1.022	1.033

where: g = acceleration of gravity

ρ_p = particle density

ρ_g = gas density

The gas density is included to account for the buoyancy effect, but this can usually be neglected because ρ_p is much greater than ρ_g in air (10^{-3}). Solving for the terminal settling velocity, v_{TS}, gives:

$$v_{TS} = \frac{\rho_p d_p^2 g}{18\mu} C_c \qquad [3.13]$$

The Stokes' settling equation, Equation [3.13], is of fundamental importance to the design of particulate control devices. Terminal settling velocity increases rapidly with particle size, being proportional to the square of particle diameter. As would be expected from the derivation, settling velocity in Stokes' region is inversely proportional to the viscosity and is not a function of gas density. Aerosol particles adjust to their terminal settling velocity almost instantly, and v_{TS} is appropriate for characterizing particle motion in most real situations.

For air at standard conditions and Stokes' flow, the velocity expression as can be simplified to:

$$v_{TS} = 0.003 \, \rho_p d_p^2 C_c = 0.003 \, d_p^2 \qquad [3.14]$$

where: v_{TS} = terminal settling velocity, cm/sec

               d_p = particle diameter for unit density spheres, μm

To convert the result to feet per minute, multiply by 1.96.

Example 3.2 Drag Force, Stokes' Region

What are the terminal settling velocity and drag force of a 2.5 μm diameter iron oxide sphere settling in still air? The density of iron oxide is 5.2 g/cm³.

Solution:

$$v_{TS} = \frac{\rho_p d_p^2 g}{18\mu} = \frac{5.2 \bullet (2.5 \bullet 10^{-4})^2 \bullet 980}{18 \bullet (1.81 \bullet 10^{-4})}$$

$$v_{TS} = 0.098 \text{ cm/sec}$$

Calculate the particle Reynolds number to ensure that motion is in the Stokes' region.

Chapter 3: The Motion of Airborne Particles

$$Re_p = 6.6vd_p = 6.6 \bullet 0.098 \bullet (2.5 \bullet 10^{-4})$$

$$Re_p = 1.61 \bullet 10^{-4}$$

Motion is well within the Stokes' region. The Stokes' drag force is

$$F_D = 3\pi\mu vd_p = 3\pi(1.81 \bullet 10^{-4}) \bullet 0.098 \bullet (2.5 \bullet 10^{-4})$$

$$F_D = 4.18 \bullet 10^{-8} \text{ dyn}$$

3.1.4 Particles Too Large for Stokes' Law

As the particle size becomes larger and larger, eventually the flow of the fluid around the sphere is no longer laminar. Thus, the Stokes' drag equation, which is based on that assumption, becomes inaccurate.

An equation that can be used to calculate the settling velocity for high Reynolds number regimes is (Beachles and Jahnke 1981):

$$v_{TS} = \frac{0.153 g^{0.71} d_p^{1.14} \rho_p^{0.71}}{\mu^{0.43} \rho_a^{0.29}} \quad (m/sec) \qquad [3.15]$$

This equation is applicable for the transition regime for which $C_D = 18.5/Re^{0.6}$. For the turbulent regime, $C_D = 0.44$ is substituted and V_{TS} can be calculated as:

$$v_{TS} = 1.74 \bullet \left(\frac{gd_p\rho_p}{\rho_a}\right)^{0.5} \quad (m/sec) \qquad [3.16]$$

To solve for the unknown particle settling velocity, the flow regime of particle motion must be determined. This is done to select the correct settling velocity equation. In other words, the Re_p and consequent flow regime cannot be determined because the velocity is unknown. The flow regime *can* be determined by the following equation (Beachles and Jahnke 1981):

$$K = d_p \left(\frac{g\rho_p\rho_a}{\mu^2}\right)^{0.33} \quad \text{(dimensionless)} \qquad [3.17]$$

Values of K that correspond to the different flow regimes are:

Laminar regime	K<3.3
Transition regime	3.3<K<43.6
Turbulent regime	K>43.6

Once the flow regime has been determined, the correct formula can be used to calculate the settling velocity of the particle.

In summary, Stokes' Law has been identified as the basic starting point for describing the aerodynamic drag forces acting on an aerosol particle moving in air. However, two corrections have been identified which could be important under certain conditions. These are the slip correction factor (Cunningham Correction Factor) for small particles (< 1 μm) and a flow correction factor for large particles (Re_p >2).

For quick reference, it is often more convenient to use a chart, such as Figure 3.5 (Wark and Warner 1981) portraying actual experimental settling velocities. Note that the top and left side axes must be used for larger particles and the bottom and right side axes are used for smaller particles.

Figure 3.5
Terminal Settling Velocity of Spherical Particles in Air at Standard Temperature and Pressure (particle density given in g/cm³)

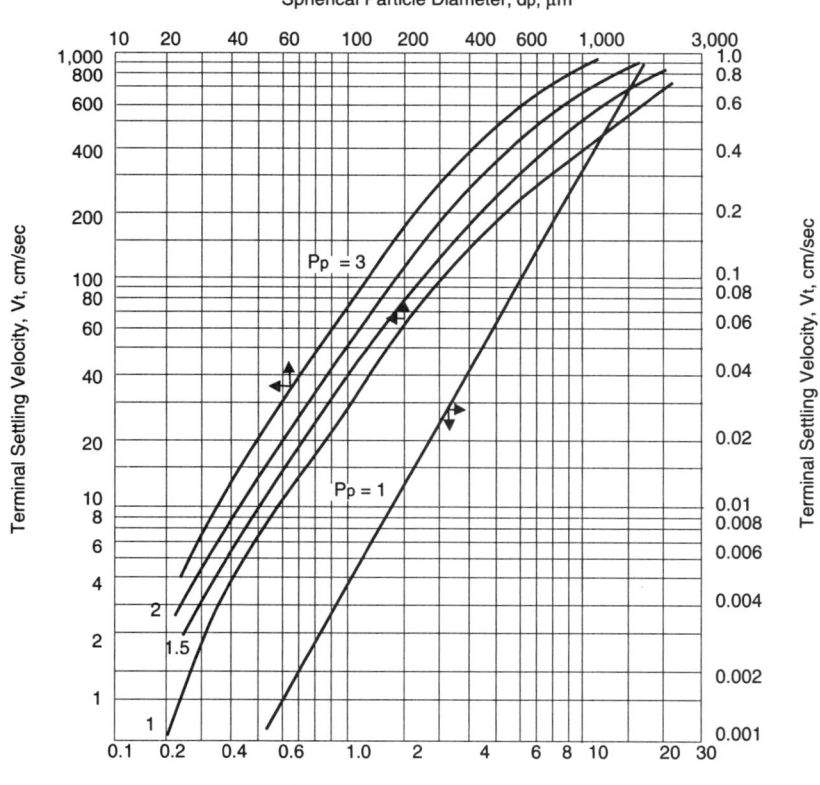

From *Air Pollution, Its Origin and Control*, 3rd edition, by Wark, Warner, and Davis; copyright 1998 by Addison-Wesley Longman, Inc. Reprinted by permission.

Example 3.3 Terminal Velocity, Flow Regime

Calculate the settling velocity of a particle moving in a gas stream. Assume the following information:

$$\rho_p = 0.899 \text{ g/cm}^3$$
$$\rho_a = 0.012 \text{ g/cm}^3$$
$$\mu(\text{air}) = 1.82 \cdot 10^{-4} \text{ g/cm sec}$$
$$g = 980 \text{ cm/sec}^2$$
$$d_p = 45 \text{ }\mu\text{m}$$
$$C_c = 1 \text{ (if applicable)}$$

Solution:

1. Calculate the K parameter to determine the proper flow regime.

$$K = d_p \left(\frac{g \rho_p \rho_a}{\mu^2} \right)^{\frac{1}{3}}$$

$$= 45 \cdot 10^{-4} \text{ cm} \left[\frac{980 \text{ cm/sec}^2 \cdot 0.899 \text{ g/cm}^3 \cdot 0.012 \text{ g/cm}^3}{(1.82 \cdot 10^{-4} \text{ g/cm sec})^2} \right]^{\frac{1}{3}}$$

$$= 3.07$$

Therefore, the flow regime is laminar.

2. The settling velocity is calculated using Equation [3.13].

$$v_{TS} = \frac{\rho_p d_p^2 g}{18 \mu} C_c$$

$$v_{TS} = \frac{0.899 \text{ g/cm}^3 \cdot (45 \cdot 10^{-4} \text{ cm})^2 \cdot 980 \text{ cm/sec}^2}{18 \cdot (1.82 \cdot 10^{-4} \text{ g/cm sec})} \cdot 1.0$$

$$= 5.38 \text{ cm/sec}$$

Example 3.4 Settling Velocity

Particles 20 microns in diameter at 70°F with a specific gravity of 1.8 flow in a duct. The density of H_2O is 62.4 lb/ft³, the density of air is 0.075 lb/ft³, and the viscosity of air is $1.23 \cdot 10^{-5}$ lb/ft sec.

(a) Calculate the settling velocity

(b) Calculate the drag force

Solution:

(a) Calculate the settling velocity

1. Convert to feet.

$$20 \ \mu m \cdot \frac{1 \ ft}{3.05 \cdot 10^5 \ \mu m} = 0.000065 \ ft$$

2. Calculate K to determine the regime.

$$K = d_p \left[\frac{g \rho_p \rho_g}{\mu^2} \right]^{\frac{1}{3}}$$

$$= 0.000065 \ ft \left[\frac{32.1 \ ft/sec^2 \cdot 1.8 \cdot 62.4 \ lb/ft^3 \cdot 0.075 \ lb/ft^3}{(1.23 \cdot 10^{-5} \ lb/ft \ sec)^2} \right]^{\frac{1}{3}}$$

$$= 0.788 < 3.3$$

Therefore, the flow regime is laminar.

3. Calculate the terminal velocity.

$$v_{TS} = \frac{\rho_p d_p^2 g}{18 \mu} C_c$$

$$v_{TS} = \frac{1.8 \cdot 62.4 \ lb/ft^3 \cdot (0.000065 \ ft)^2 \cdot 32.1 \ ft/sec^2}{18 \cdot (1.23 \cdot 10^{-5} \ lb/ft \ sec)} \cdot 1.0$$

$$= 0.0688 \ ft/sec$$

Chapter 3: The Motion of Airborne Particles

(b) Calculate the drag force.

$$F_D = 3\pi\mu v_p d_p$$

$$F_D = 3 \cdot \pi \cdot 1.23 \cdot 10^{-5} \text{ lb / ft sec} \cdot 0.0688 \text{ ft / sec} \cdot 0.000065 \text{ ft}$$

$$= 5.18 \cdot 10^{-10} \text{ lb}_m \text{ ft / sec}^2$$

convert to lb_f using g_c, $g_c = 32.1 \text{ ft lb}_m/\text{lb}_f \text{ sec}^2$

$$F_D = \frac{5.18 \cdot 10^{-10} \text{ ft lb}_m / \text{sec}^2}{g_c} = \frac{5.18 \cdot 10^{-10} \text{ ft lb}_m / \text{sec}^2}{32.1 \text{ ft lb}_m / \text{lb}_f \text{ sec}^2}$$

$$F_D = 1.61 \cdot 10^{-11} \text{ lb}_f$$

Example 3.5 Drag Coefficient and Settling Velocity

A spherical limestone particle is 400 μm in diameter with a specific gravity = 2.67. Calculate the drag coefficient, C_D, and the settling velocity in air at 70°F.

Solution:

1. Convert to feet.

$$400 \text{ μm} \cdot \frac{1 \text{ ft}}{3.05 \cdot 10^5 \text{ μm}} = 0.00131 \text{ ft}$$

2. Calculate K to determine the regime.

$$K = d_p \left[\frac{g\rho_p\rho_g}{\mu^2} \right]^{\frac{1}{3}}$$

$$= 0.00131 \left[\frac{32.1 \cdot 2.67 \cdot 62.4 \cdot 0.075}{(1.23 \cdot 10^{-5})^2} \right]^{\frac{1}{3}}$$

$$= 18.11$$

$3.3 < K < 43.6 \rightarrow$ transition region

For transition regime, $C_D = \dfrac{18.5}{Re^{0.6}}$

But it is not known until v_{TS} is determined. Thus, settling velocity must be calculated.

3. Calculate settling velocity.

$$v_{TS} = \frac{0.153 \bullet (g\rho_p)^{0.71} d_p^{1.14}}{\rho_a^{0.29} \mu^{0.43}}$$

$$v_{TS} = \frac{0.153 \bullet (32.1 \bullet 2.67 \bullet 62.4)^{0.71} (0.00131)^{1.14}}{(0.075)^{0.29} (1.23 \bullet 10^{-5})^{0.43}} = 76.8$$

$$= 9.62 \text{ ft / sec}$$

4. Calculate Re_p and C_D.

$$Re_p = \frac{\rho_a v_p d_p}{\mu} = \frac{9.62 \bullet 0.00131 \bullet 0.075}{1.23 \bullet 10^{-5}} = 76.8$$

as expected, $2 < Re\ (76.8) < 1{,}000$ for transitional regime

$$C_D = \frac{18.5}{Re_p^{0.6}} = 1.37$$

3.2 Parameters That Characterize Particulate Collection

3.2.1 Aerodynamic Diameter

For two spherical particles having different diameters, d_1 and d_2, and different densities, ρ_1 and ρ_2, their falling speeds in air will be the same provided:

$$d_1^2 \rho_1 = d_2^2 \rho_2 \qquad [3.18]$$

where, for simplicity, the slip and the Reynolds number corrections have been neglected.

Equation [3.18] leads directly to a definition of particle size based on falling speed: Stokes' diameter and aerodynamic diameter are two equivalent diameters that find wide application in aerosol technology. For any particle, they are defined as follows:

1. The Stokes' diameter, d_s, is the diameter of sphere that has the same density and settling velocity as the particle; and

2. The aerodynamic diameter, d_a, is the diameter of a unit density ($\rho_p = 1$ g/cm^3) sphere that has the same settling velocity as the particle.

The settling velocity can be written in terms of these two diameters as:

Chapter 3: The Motion of Airborne Particles

$$v_{TS} = \frac{\rho_b d_s^2 g}{18\mu} = \frac{\rho_o d_a^2 g}{18\mu} \qquad [3.19]$$

where: ρ_o = unit density

ρ_b = bulk material density or true particle density

Stokes' diameter standardizes particles of various shapes to spheres having the same settling velocity. Aerodynamic diameter standardizes not only the shape, but also density. An irregular particle and its equivalent spheres are compared in Figure 3.6 (Hinds 1982). The aerodynamic diameter, the most commonly used of the two, can be thought of as the diameter of a water droplet having the same aerodynamic properties as the particle. If a particle has an aerodynamic diameter of 1 μm, it behaves in an aerodynamic sense as a 1 μm water droplet, regardless of its shape, density, or physical size. Furthermore, it is aerodynamically indistinguishable from other particles of different size, shape, and density having an aerodynamic diameter of 1 μm. The Stokes' diameter is usually defined in terms of normal density of the bulk material of the particle, ρ_b. This eliminates the problem of defining the true density of the particle, which may be less then ρ_b due to porosity, occlusions, or agglomerated structure.

Figure 3.6
An Irregular Particle and its Equivalent Spheres

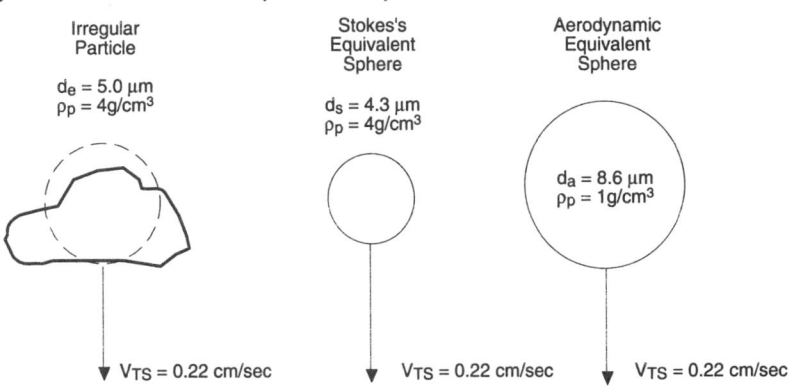

Both the Stokes' and the aerodynamic diameters are defined in terms of their aerodynamic behavior rather than their geometric properties. Aerodynamic diameter is the key particle property for characterizing the performance of many types of air pollution control devices. In many situations, it is not necessary to know the true size, shape factor, and density of the particle if its aerodynamic diameter is known. Instruments, such as the cascade impactor, use aerodynamic separation to measure aerodynamic particle size.

Rearranging Equation [3.19] gives:

$$d_a = d_s \left(\frac{\rho_b}{\rho_o}\right)^{0.5} \quad [3.20]$$

Figure 3.7
Relaxation Between Physical and Aerodynamic Diameter

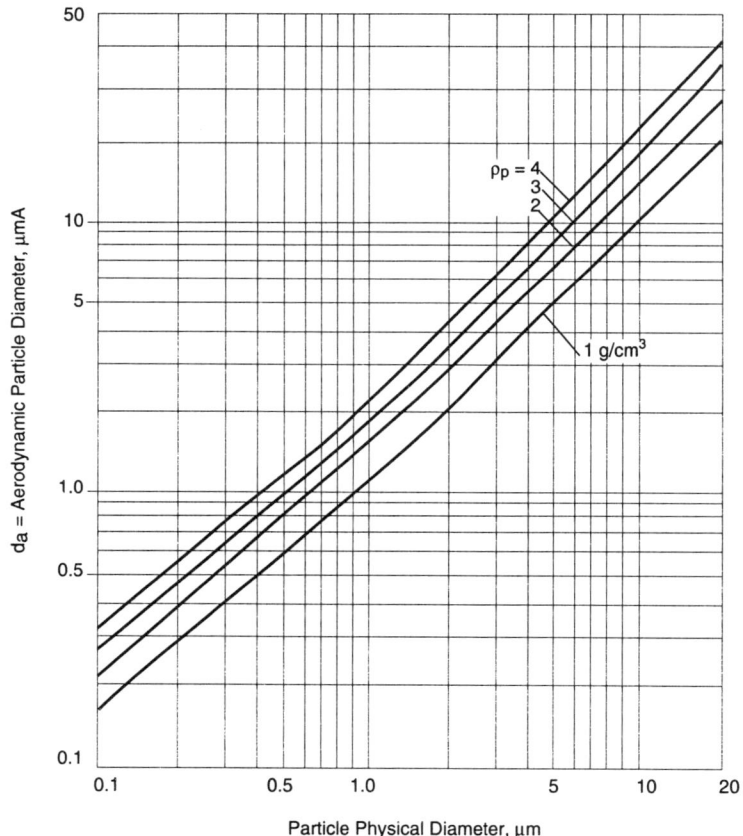

For the usual case of a sphere with a density greater than 1 g/cm^3, d_a is always greater than d_p (see Figure 3.7) (Calvert and Englund 1984).

3.2.2 Relaxation Time

The terminal settling velocity for a particle moving in the Stokes' regime is directly proportional to the external forces acting on the particle, and the constant of proportionality is the mechanical mobility, B.

$$v = BF \quad [3.21]$$

Chapter 3: The Motion of Airborne Particles

$$B = \frac{C_c}{2\pi\mu d} \quad [3.22]$$

The product of particle mass and mobility, m•B, occurs frequently in aerosol mechanics and is a useful quantity for the analysis of particle motion.

$$\tau = m \cdot B = \rho_P \frac{\pi}{6} d_P^3 \left(\frac{C_c}{3\pi\mu d_P} \right) = \frac{\rho_P d_P^2 C_c}{18\mu} \quad [3.23]$$

This quantity is called the *relaxation time* of the particle and is given the symbol, τ, and has units of time. Two seemingly different systems (with different particle sizes, densities, and different fluids) will behave in the same manner if τ is the same for both systems.

The relaxation time can be used for the calculation of terminal settling velocities. Substitution for τ in Equation [2.21] gives:

$$v_{TS} = \tau g \quad [3.24]$$

The term relaxation time is used because it characterizes the time required for a particle to adjust or relax its velocity to a new condition of forces. The relaxation time depends only on the mass and mobility of the particle and is not affected by the nature and/or the magnitude of the external forces acting on the particle. Although usually thought of as a quantity describing a particle property, the relaxation time includes viscosity and slip correction and is thus affected by temperature and pressure of the gas. The terminal settling velocity of a particle is the product of τ and the acceleration caused by the external forces, i.e., gravity, centrifugal, and electrostatic.

3.2.3 Stopping Distance (Stokes' Number)

Impaction is a key factor in particle removal from an air stream. In can be explained simply with the concept of stopping distance. If a sphere in Stokes' regime is projected with an initial velocity, v_o, into a motionless fluid, its velocity as a function of time (ignoring all but the drag force) is

$$v = v_o e^{-(t/\tau)} \quad [3.25]$$

The total distance traveled by the particle before it comes to rest is:

$$x_{stop} = \int_0^\infty v \, dt = v_o \tau \quad [3.26]$$

If the particle stops before striking the object, it can then be swept around the object by the altered fluid flow. Since τ is small, x_{stop} is also small. For instance, if a 1 μm particle with unit density is projected at 10 m/sec into air, it will stop after traveling 36 μm.

An impaction parameter, Ψ, called the Stokes' number can be defined as the ratio of the stopping distance of a particle to the diameter of the stationary object, D, or

$$\psi = \frac{x_{stop}}{D} = \frac{\rho_p d_p^2 v_o C_c}{18\mu D} \quad [3.27]$$

A particle is said to have a curvilinear motion when it follows a curved path rather than a straight line. Curvilinear motion is characterized by the Stokes' number. As the Stokes' number approaches zero, particles track the streamlines perfectly. As the Stokes' number increases, the particles resist changing their directions with the gas streamlines. The Stokes' number is used to characterize inertial impaction, the inertial transfer of particles onto the surfaces.

If Ψ is large, most of the particles will impact the object. If Ψ is small, most of the particles will follow the fluid flow around the object (Figure 3.8).

An empirical expression given by Hinds (1982) is accurate for calculating stopping distances for particles having an initial Reynolds number of between 1 and 400.

$$x_{stop} = \frac{\rho_p d_p}{\rho_g}\left[Re_o^{1/3} - \sqrt{6}\arctan\left(Re_o^{1/3}/\sqrt{6}\right)\right] \quad [3.28]$$

where: $Re_o = \rho_g v_o d_p/\mu$, and the arctan is in radians.

The collection efficiency equations of many air pollution control devices include a Stokes' number term (the ratio of the stopping distance to some dimension of the control device or to the size of the fiber and water drops that act as collector bodies).

Example 3.6 Stopping Distance (Stokes' Number)

A 1 micron diameter spherical particle with a specific gravity of 2.0 is ejected from a gun into a standard air at a velocity of 10 m/sec (32.8 ft/sec). How far does it travel before it is stopped by viscous friction?

Solution:

Here Newton's Law, F = ma applies. The drag force is the only force acting on the particle after it leaves the gun. It operates in the direction opposite the direction of motion and is given by the Stokes' drag resistance, Equation [3.7], as modified by the Cunningham Correction Factor, Equation [3.8]. Inserting these:

$$F = ma = \frac{3\pi\mu v_p d_p}{C_c} = \frac{\pi}{6} d_p^3 \rho_p \frac{dv_p}{dt}$$

$$\frac{dv_p}{dt} = \frac{18\mu v_p}{d^2 \rho_p C_c}$$

Substituting dx/v_p for dt, separating variables, canceling the two v_p terms, and integrating:

$$\int_{v=v_0}^{v=0} dv_p = -\frac{18\mu}{d_p^2 \rho_p C_c} \int_{x=0}^{x=x} dx$$

and

$$x_{\text{Stokes stopping}} = \frac{v_0 d_p^2 \rho_p C_c}{18\mu}$$

$$x_{\text{Stokes stopping}} = \frac{(10 \text{ m/sec}) \bullet (10^{-6} \text{m})^2 \bullet (2{,}000 \text{ kg/m}^3)1.12}{18 \bullet (1.8 \bullet 10^{-5} \text{ kg/m sec})}$$

$$= 6.9 \bullet 10^{-5} \text{m} = 69 \mu\text{m}$$

This value is surprisingly small; the particle stops in 0.07 mm = 0.0027 in. This makes clear that for particles of this size, and most particles of air pollution interest, the air is a very viscous fluid indeed.

3.2.4 Impaction Efficiency

Moving particles exhibit an inertial behavior, the tendency to continue to travel in the direction of their original motion. This tendency is greater the more massive the particle, the greater its approach velocity, and the more sharply the flow diverges.

Figure 3.8 shows a typical pattern of streamlines, together with corresponding trajectory patterns for particles of given aerodynamic diameter, d_a. Some trajectories intersect with the disc, which means that such particles will impact onto its surface. We may define a limiting trajectory inside which all particles will impact onto the disc, and outside which all will pass by. Impaction efficiency, η_I, is then defined as:

$$\eta_I = \frac{\text{number of particles arriving by impaction}}{\text{number incident on the obstacle}} \quad [3.29]$$

which, for particles uniformly distributed throughout the oncoming flow is equivalent to:

$$\eta_I = \frac{b''}{b'} \quad [3.30]$$

where b″ and b′ are the areas projected upstream by the limiting trajectory surface and the obstacle itself, respectively, as shown in Figure 3.8.

Figure 3.8
Ilustration of the Impaction Phenomenon Particles onto the Leading Surface of a Flat Disc Placed Normal to the Air Stream

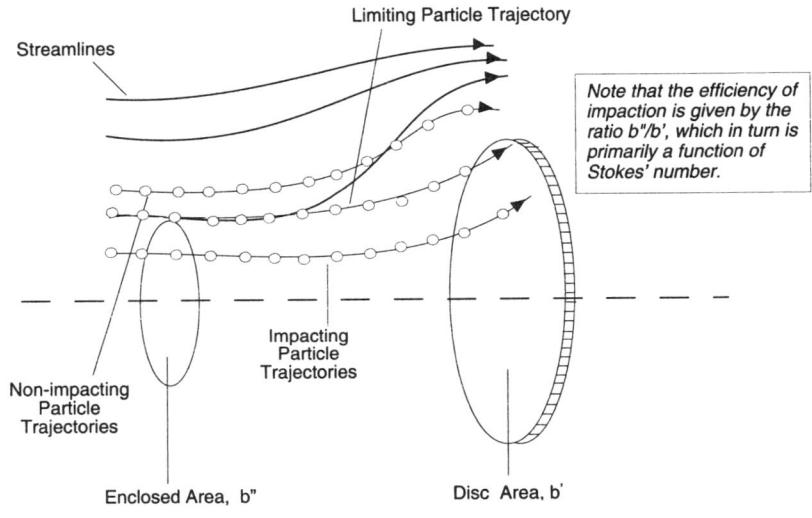

Particle trajectories are determined largely by the Stokes' number (St) and Reynolds number of the particle. Also, they will be dependent on the Reynolds number of the flow about the obstacle, since it governs the shape of the streamline pattern. η_I is a function of St, Re_p, and Re_c. That is (Strauss 1979):

$$\eta_I = f(St, Re_p, Re_c) \qquad [3.31]$$

The Reynolds number of the collecting body is:

$$Re_c = \frac{v_o \rho D}{\mu} \qquad [3.32]$$

where: v_o = undisturbed upstream fluid velocity

D = diameter of the collecting body

ρ = fluid density

μ = fluid viscosity

Inertial collection is a function of the impaction parameter and the flow regime, as defined by the Reynolds number. When $Re_c < 1$, the flow around the cylinder is considered to be viscous, while for $Re_c > 100$, the flow around the cylinder can be approximated by potential flow. Values between 1 and 100 represent transition conditions.

At collector Reynolds number of 0.2 (viscous flow), a 3 percent disturbance occurs at a distance of 100 D upstream, while when $Re_c = 2,000$ (potential flow), there is practically no fluid disturbance at a distance of 2 D upstream.

Values for Re_c in fiber filtration are low (about 0.2), therefore efficiency for fibrous filters does not consider impaction to be an important collection mechanism. However, for spray drops, impaction dominates the collection mechanisms because Re_c is large.

Although the physics of what has been described is simple, there are no analytical solutions of impaction efficiency. Instead, graphs, such as Figure 3.9 based on experimental results, are used to determine collection efficiency from Stokes' number.

The solution to the equations of motion will depend on the velocity field assumed. These equations have been solved for several collectors of different shapes under given boundary conditions, mostly by numerical methods. Figure 3.9 shows theoretical and experimental collection efficiency for a spherical collector for potential and viscous flow.

For potential flow and for values of ψ greater than 0.2, the experimental values of inertial collection efficiency for spheres can be approximated by the correlation (Calvert and Englund 1984):

$$\eta_I = \left(\frac{\psi}{\psi + 0.35}\right)^2 \qquad [3.33]$$

3.2.5 Brownian Motion

The terminal settling velocities of particles much smaller than the mean free path of the gas molecules are very low, the random bombardment of air molecules deflects these particles, and a random motion is superimposed on the settling motion. This random motion is called *Brownian motion*. Removal of these small particles is affected mainly by diffusion.

The diffusion coefficient, $\mathcal{D}_p$, is related to particle diameter by (Hinds 1982):

$$\mathcal{D}_p = \frac{RTC_c}{3\pi\mu_g d_p N} \qquad [3.34]$$

where: N = Avogadro's number

R = the universal gas constant

T = the absolute temperature

Figure 3.9
Estimated Collection Efficiencies as a Function of Inertial Parameter for Various Targets (potential flow)

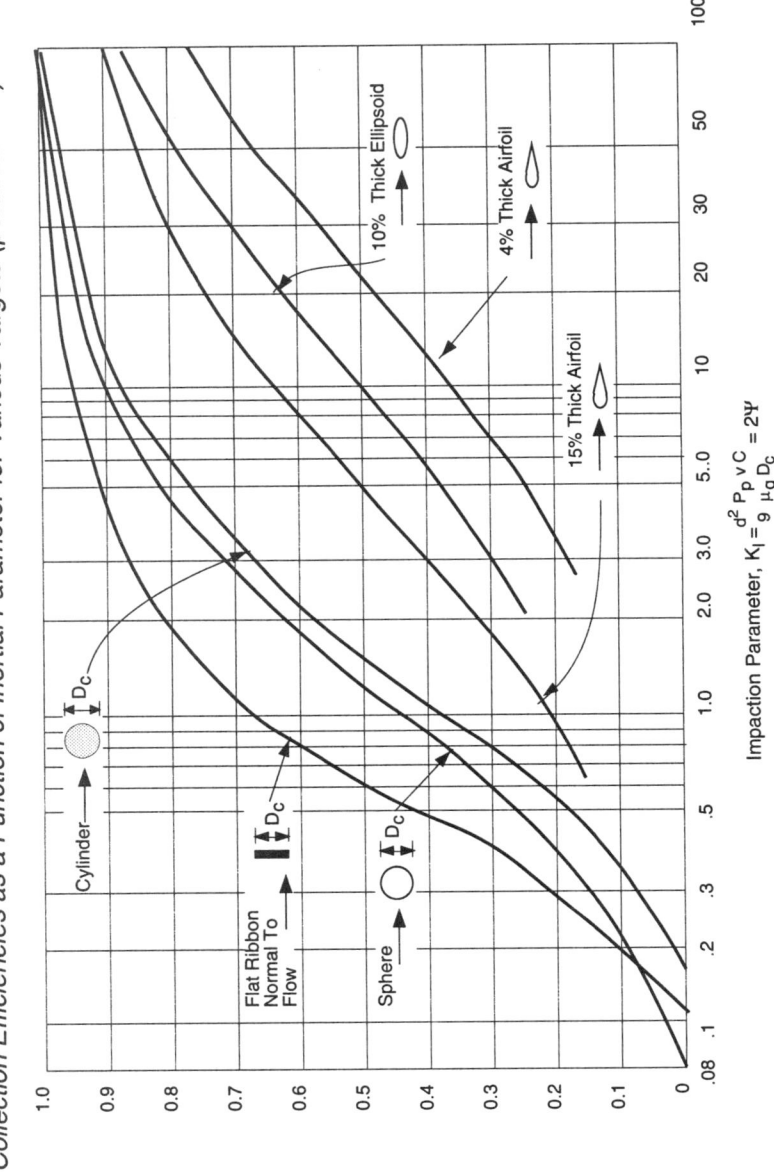

Source: Calvert and Englund 1984

Figure 3.10 shows the diffusivity as a function of particle size.

Figure 3.10
Particle Diffusivity in Air at 20°C and 760 mm Hg

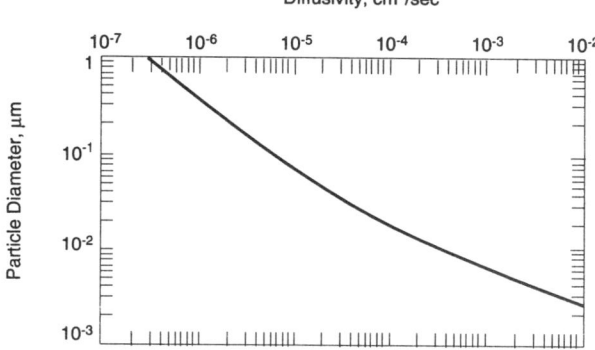

The mean square displacement of a particle, Δx^2, in a given interval of time, t, is:

$$\Delta x^2 = 2\mathcal{D}_p t \qquad [3.35]$$

This is an important result because it provides a basis for making quantitative estimates of the magnitude of the effect of molecular diffusion in given situations. The governing parameter is the inverse of the dimensionless diffusion coefficient:

$$\text{Pe} = \frac{vd_p}{\mathcal{D}_p} \qquad [3.36]$$

The scaling quantity (Pe) is known as the *Peclet number*. The smaller it is, the more pronounced the contribution due to diffusion. Therefore, diffusion is likely to contribute significantly only for very small particles in systems of small geometry where air flow velocities are low. Such conditions are found in filters.

Diffusivity has dimensions of (area)/(time). The dimensionless group which includes diffusivity is the *Schmidt number*, Sc:

$$\text{Sc} = \frac{\mu}{\rho_g \mathcal{D}_p} = \frac{\upsilon}{\mathcal{D}_p} \qquad [3.37]$$

where: υ = the kinematic viscosity, μ/ρ_g.

For a system involving a gas stream, velocity v, moving past a body, diameter d, then:

$$Pe = Re_c Sc = \frac{v \rho_g D}{\mu} \cdot \frac{\mu}{\rho_g \mathcal{D}_p} \qquad [3.38]$$

where: D = the collector diameter

$\mathcal{D}_p$ = the particle diffusion coefficient.

Collection efficiency by diffusion has the form (Strauss 1975):

$$\eta_D \propto Pe^{-n} \qquad [3.39]$$

where: $n = 2/3$

Figure 3.11 compares the root mean square displacement of a particle in 1 sec (Δx in Equation 3.35) in a given direction, with the distance a spherical particle of unit density falls in 1 sec in air at 760 mm Hg pressure and 20°C under gravity, calculated from Stokes' equation. It can be seen that particles of unit density smaller than approximately 0.5 µm in diameter have a larger Brownian displacement than displacement due to gravitational settling. The particle size that gives the minimum displacement, about 0.5 µm diameter, is an in-between size that is too large

Figure 3.11
Comparison Between Brownian Displacement and Settling Distance of Unit Density Spherical Particles in Air for a Time of One Second

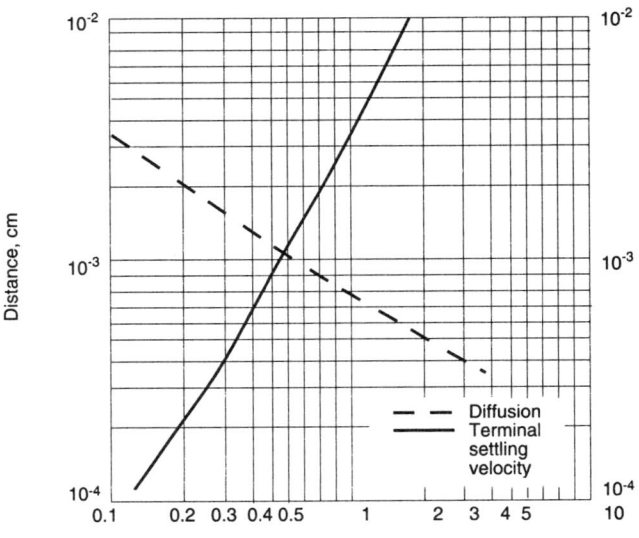

for diffusion collection to be effective and two small for inertial forces to be effective. A similar effect is obtained with respect to velocity, v_o. At low velocities, diffusion will predominate and at high velocities, inertia will be dominant. There will be a minimum collection efficiency determined primarily by direct interception at some intermediate velocity.

3.3 Filter and Water Drop Collection Efficiency (Impaction, Interception, and Diffusion)

The efficiency with which wet scrubber or filter fibers remove particles from an air stream is defined in terms of a single element efficiency, η_S. In a fabric filter, the target for particle capture is a fiber. In wet collectors, the target is a water droplet. Three separate mechanisms are responsible for particle removal: impaction, direct interception, and diffusion (Figure 3.12).

Impaction occurs when the particle is so large that it cannot follow the gas streamline around the object and impacts (Figure 3.12). Direct interception is a special case of the impaction mechanism. A collision will occur if the distance between the particle center and the collection surface is less than the particle radius. The mechanism of diffusion is responsible for the collection of particles which are so small that they become affected by collisions of molecules in the gas stream.

Because of their mass, particles do not always follow the streamlines which diverge around an obstacle, such as a droplet or fiber. An impacting particle is one which has broken through the streamlines and hits the object. The separation number characterizing inertial impaction (the impaction parameter) accounts for the effect that the mass of the particle has in penetrating streamlines. Interception accounts for the finite size of the particle as it nears the object. Interception occurs when the particle approaches the droplet by a distance less than $d_p/2$ (as measured from particle's center) — it will hit and be collected.

The combined efficiency for a single collector element can be expressed as (Strauss 1976):

$$\eta_{Combined} = 1 - (1 - \eta_{Impaction})(1 - \eta_{Interception})(1 - \eta_{Diffusion}) \qquad [3.40]$$

The combined single collector efficiency is then a function of three parameters:

$$\eta_{ICD} = 1 - [1 - f(\psi)] \bullet \left[1 - f\left(\frac{d_p}{d_o}\right)\right] \bullet \left[1 - f\left(\frac{1}{Pe}\right)\right] \qquad [3.41]$$

η_{ICD} = combined collection efficiency

where: I = impaction
C = interception
D = diffusion

Figure 3.12
Single-Fiber Collection by Impaction, Interception, and Diffusion

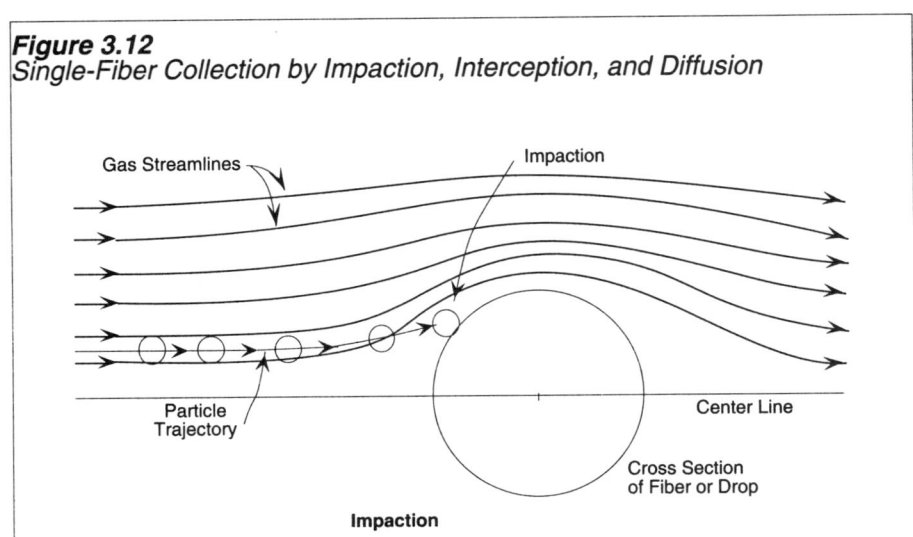

Impaction

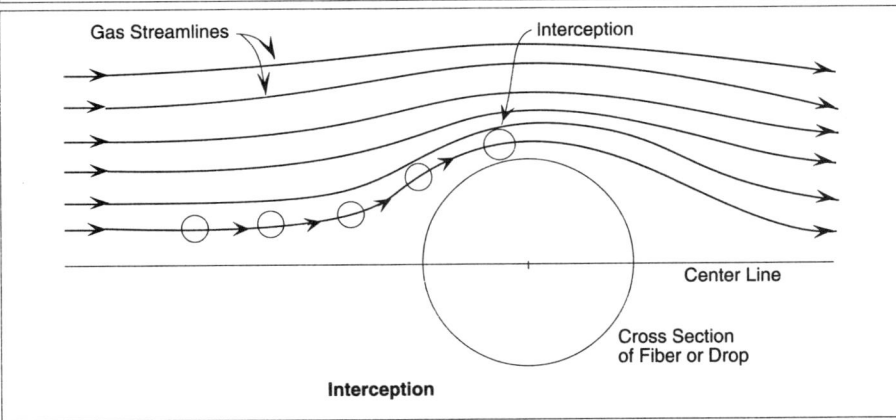

Interception

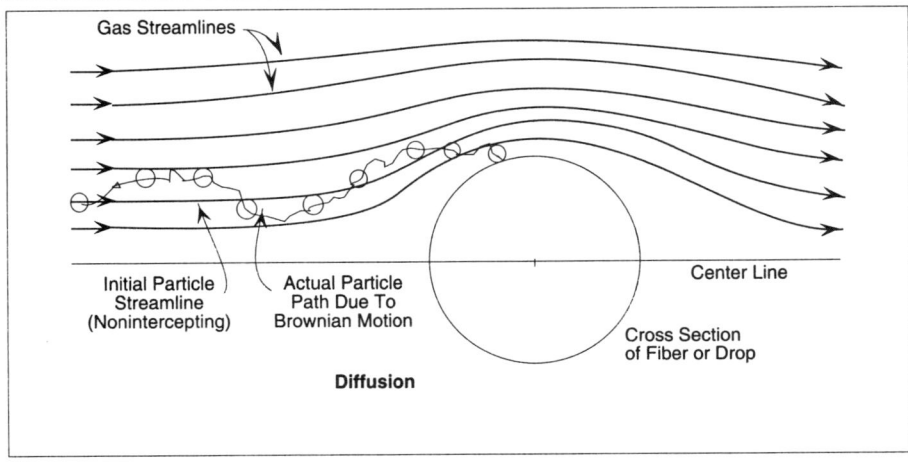

Diffusion

The capture efficiency due to any one mechanism is determined by the corresponding parameter through formulae or charts. In a given circumstance, particle size, target geometry and size, and fluid flow rate will determine which mechanism will predominate. Wet scrubbers have high velocities and inertial impaction and interception dominates the collection process (terms 1 and 2 in Equation [3.40]). Fabric filters have low velocities and interception and diffusion dominate the collection process (terms 2 and 3 in Equation [3.40]) (Figure 3.13).

Figure 3.13
Regions in Which Various Single Element Collection Mechanisms Contribute 20% or More to the Total Single-Fiber Efficiency

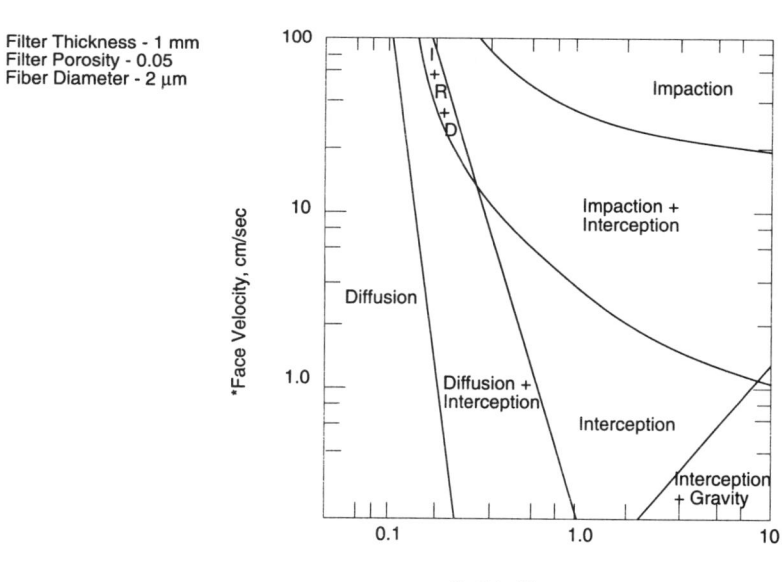

* Face velocity is the air flow through the fabric

For multiple collectors, the overall collection efficiency can be found from

$$\eta_o = 1 - (1 - \eta_{ICD})^n \quad [3.42]$$

where n is the number of collecting droplets or fibers encountered by the particles.

In most practical cases, n is large (greater than 25) and Equation 3.42 can be modified to

$$\eta_o = 1 - \exp(-n\eta_{ICD}) \quad [3.43]$$

3.4 Problem Set

3.1 The basic Stokes' Law drag force is given as $F_s = 3\pi\mu vd$. What type of corrections can be applied to extend the range of particle sizes to which the basic law can be applied? Why are these corrections required?

3.2 State at least one important way in which each of the following dimensionless numbers is used in connection with the behavior of small particles:

- Reynolds number
- Stokes' number
- Peclet number

3.3 Develop an expression for the free fall of a spherical particle.

3.4 Reynolds numbers for particles and for flow in ducts have different values for determination of the flow regime.

> a) 2.54 cm sphere moves through air with a velocity of 25 cm/min. Find the Reynolds number for the sphere and identify the flow regime (laminar, transitional, or turbulent).
>
> b) Air flows through a 16 in. duct at 3,500 ft/min. Find the Reynolds number for the gas and identify the flow regime (laminar, transitional, or turbulent).
>
> (Note: For flow in ducts: laminar, Re < 2,100; transitional, Re = 2,100 to 4,000; turbulent, Re > 4,000)

3.5 A 1 μm sphere with a density of 1 g/cm^3 moves through air with a velocity of 100 cm/sec. Compute the resisting force offered by the air. Assume normal temperature (20°C) and pressure (1 atm). Compare this to a 100 μm sphere with a density of 1 g/cm^3 moving through air with a velocity of 30 cm/sec.

3.6 Consider a particle with a density of 1 g/cm^3 and 50 μm diameter. Calculate the terminal velocity. Assume Stokes' flow and then check the Reynolds Number. Neglect the density of air relative to the particle density and use $\mu = 1.8 \cdot 10^{-4}$ g/cm sec.

3.7 Predict the terminal settling velocity of 1 mm diameter water droplets falling in the air at ambient conditions: temperature 20°C; pressure 1 atm. Use Equation [3.17]. Compare the predictions with the results presented in Figures 3.5 and 7.7.

3.8 Determine the free-fall velocity and Reynolds number for particles of specific gravity of 1, 2, and 3 and size 50, 100, 500, and 1,000 μm diameter, assuming Stokes' Law behavior. Compare to water droplet velocities given in Figure 7.7.

What is the percent difference between the calculated values and the values from Figure 7.7? Explain the difference. Where does Stokes' law begin to fail?

3.9 A droplet of water 200 μm in diameter is formed by condensation in a spray tower at 70°F. The gas is rising in the tower at a velocity of 1 m/sec. Which way will this drop move? How long will it take the drop to to travel vertically 1 m of the tower height?

3.10 Calculate the distances traveled by a spherical particle with a density of 1 g/cm³ and a diameter of 40 μm, falling freely in still air of 1 atm pressure and 20°C temperature, before it reaches 99 % of its terminal settling velocity.

3.11 For most practical situations, it is usually assumed that the acceleration of a particle is complete at t = 5τ. Solve Equation [3.25].

$$v/v_o = e^{-(t/\tau)}$$

For a period of time t = 5τ, determine the percent of its terminal velocity it will have attained if it started from rest. Determine the relaxation time for a 1 μm particle with a density of 1 g/cm³ at 20°C if 5τ is a significant time for acceleration to occur.

3.12 Spherical particles of density 2.0 g/cm³ are projected into dry air at 70°C and 1 atm with an initial horizontal velocity component of 8 ft/sec. Determine the size of particle which will have a stopping distance of 2 in.

3.13 The particles in problem 3.12 will be collected on a cylinder with a diameter 2 in. Determine the collection efficiency of the cylinder. Determine the collection efficiency if a cylinder of only 1 in. in diameter is used for collection.

3.14 Calculate the Stokes' Law stop distance for 0.5 μm diameter particles traveling at 200 mph. Compare this to the true stop distance for the particles (using Equation [3.28]). What size collector would be required to ensure collection of the particle for the Stokes' Law stop distance and the true stop distance? Which one would you use and why?

3.15 What is the true stop distance for a 50 μm diameter particle of unit density moving in air at 1,500 cm/sec? How does this compare to the stop distance obtained by assuming that the particle is in Stokes' Law drag force range? What causes the difference? What would be the required speed of the particle to just be inside the Stokes' drag range ($Re_p < 2.0$)?

3.16 What is the collection efficiency for collection by inertial impaction when the target is a 150 μm water drop moving at 5 cm/sec and the particle to be collected is 10 μm in diameter moving at 30 cm/sec in an air stream in the opposite direction to the water drops?

3.17 Which of the following targets has the greater capture efficiency? Target A is a sphere for which $v_o/D = 100$ /sec. Target B is a cylinder of 4 times the diameter of the sphere, and having 1/2 the approach velocity. The particles have a relaxation time of 0.005 sec for both targets. Stokes' Law may be used.

3.18 A vibrating orifice aerosol generator breaks up a fine stream of liquid flowing at 0.19 cm^3/min through a 23 µm diameter orifice to form monodisperse aerosol particles 21 µm in diameter. How far will one of these particles travel in still air?

3.19 A spherical steel particle, 20 µm in diameter, is thrown from the rim of a 7 in. diameter grinding wheel rotating at 3,450 rpm (density of steel is 7.8 g/cm^3).

   a) What is the stopping distance of this particle if Stokes' Law is assumed to hold?

   b) What is the actual stopping distance?

3.20 A particle has a Stokes' number of 5 and a relaxation time of 2 sec. What size is the particle and what is its velocity if the collector is 1 cm in diameter?

3.21 What is the relaxation time for a particle which is collected with 80% efficiency by a cylinder when the particle velocity is 600 cm/sec and the collector is 0.1 cm in diameter?

3.22 Estimate the aerodynamic capture efficiency of a cylindrical target 10 µm in diameter for particles 0.006 µm diameter (density 1.5 g/cm^3) in dry air at 20°C and 1 atm, flowing at 5 cm/sec.

3.23 The aerodynamic capture efficiency is 25 % for certain particles impinging from a jet onto a sphere (diameter = D) under a given set of conditions. What will the collection efficiency be for particles which are twice as big, on a plate of one-half the diameter, at a velocity one-quarter as much?

3.24 Show in which of the two cases A and B, described below, the aerodynamic capture efficiency due to inertial impaction will be greater. In both cases, the target is the same size and shape, and the relative velocity between target and particle is the same.

	A	B
Temperature of air	300°C	20°C
Particle diameter	3 µm	4 µm
Particle density	2.2 g/cm^3	1.2 g/cm^3

3.25 Calculate the Brownian diffusion coefficient $\mathcal{D}$, for particles of 0.1 μm and 0.001 μm diameter in air at 20°C and 1 atm.

3.26 Compare the displacement in 1 sec due to Brownian motion and gravity for a unit density sphere at standard conditions.

3.27 Calculate the mean thermal displacement for a 0.1 μm particle with a density of 3 g/cm^3 at a temperature of 300°K. What is the relaxation time for this particle? If this particle is projected into still air at 200 mph, what is the true stopping distance? What size collector would be required to ensure collection of the particle?

3.5 References

Beachles, D. and J. Jahnke. *Control of Particulate Emissions*. Northrop Services, Inc. EPA 450/2-80-066. Research Triangle Park, NC. 1981.

Calvert, S. and H. Englund (Eds.). *Handbook of Air Pollution Technology*. John Wiley & Sons. New York. 1984.

Hesketh, H. *Air Pollution Control for Traditional and Hazardous Pollutants*. Technomic Publishers. Lancaster, PA. 1991.

Hinds, W. *Aerosol Technology: Properties, Behavior, and Measurement of Airborne Particle*. John Wiley & Sons. New York. 1982.

Perry, J. (Ed.). *Chemical Engineering Handbook*. McGraw Hill. New York. 1973.

Strauss, W. *Industrial Gas Cleaning*, 2nd. ed. Pergamon Press. Oxford. 1975.

Wark, K., C. Warner, and Davis. *Air Pollution: Its Origin and Control*. 3rd. ed. Addison Wesley Longman. Menlo Park, CA. 1998.

Chapter 4

Fundamentals of Particulate Emission Control

4.1 General Concepts of Particulate Collection

The collection efficiency of particulate control devices depends on particle size. Because of this, the most important parameter for collector design calculations is the ability to predict the grade efficiency function for a given collector operating under specified conditions. The grade efficiency function is the efficiency with which particles are collected as a function of particle size. The central aspect of particulate collection theory is the development of mathematical models which show how grade efficiency depends on operating parameters such as flow rates, particle loading, dust properties, collector shape, and dimensions. With an appropriate model, an engineer may test alternative designs, and determine the effect of varying the operating parameters in order to arrive at an optimum design.

In order for particles to be removed from a gas (i.e., collected), the particles must come under the influence of a force that causes them to be diverted from the direction of flow. The types of collectors in common use may be classified according to the nature of the collecting forces and collecting surface as shown in Table 4.1 (Strauss 1975). There are two general categories of collecting surfaces. One type of collector surface may be a plane wall in a settling chamber or cyclone. The other is the outside of a target, such as a cylinder (fiber) or sphere (drops of liquid).

Table 4.1
Primary Collecting Force and Surface Geometry for Collectors in Common Use

Collector	Primary Collecting Forces	Collecting Surface
	Surface Collector	
Settling Chamber	Gravitational	Plane
Momentum	Gravitational & Inertial	Planes or Cylindrical
Cyclones	Centrifugal	Cylindrical
Electrostatic Precipitators	Electrostatic	Plane or Cylindrical
	Target Collectors	
Filters (media)	Diffussion & Interception	Cylindrical Fibers
Scrubbers	Inertia & Inception	Spherical Drops

A target collector consists of target elements, as in a filter made up of a mat of fibers, or in a scrubber consisting of suspended droplets.

In order to analyze or predict the performance of a collector, it is necessary to know how a given type of force acts upon a particle. The magnitude and direction of the force on a particle of given density and size must be predictable, so that a particle trajectory can be determined.

4.2 Particle Size Distribution Functions

Important particulate characteristics include size, size distribution, shape, density, stickiness, corrosivity, reactivity, and toxicity. From the viewpoint of design of air pollution control devices, the most important of these is the size distribution. The collection efficiency of a control device depends on particle size, with larger particles usually removed more efficiently. Thus, to calculate the overall collection efficiency of a device, it is imperative to have good information on the size distribution of particles. The most common industrial particles cover a wide range of sizes from 0.1 µm to 100 µm (Figure 4.1).

In addition to average particle concentration per unit atmospheric volume, it is important to note the size distribution by particle count and by mass. Such a distribution for a typical atmospheric particulate sample is shown in Table 4.2 (Wark and Warner 1981). Particles up to 1 µm in size constitute only 2% by mass. However, the number of particles in that range is overwhelmingly high compared with the rest of the sample.

In the design of air pollution control devices, particles having a variety of sizes exist. To discuss such particles and make useful calculations about their behavior in collection devices, some way of describing the particulate size distributions is required. The most often used distribution function is the Gaussian, or normal distribution function. It represents the observed distribution of many populations and is described by:

Chapter 4: Fundamentals of Particulate Emission Control   101

Figure 4.1
Range of Particle Sizes for Different Industrial Dusts

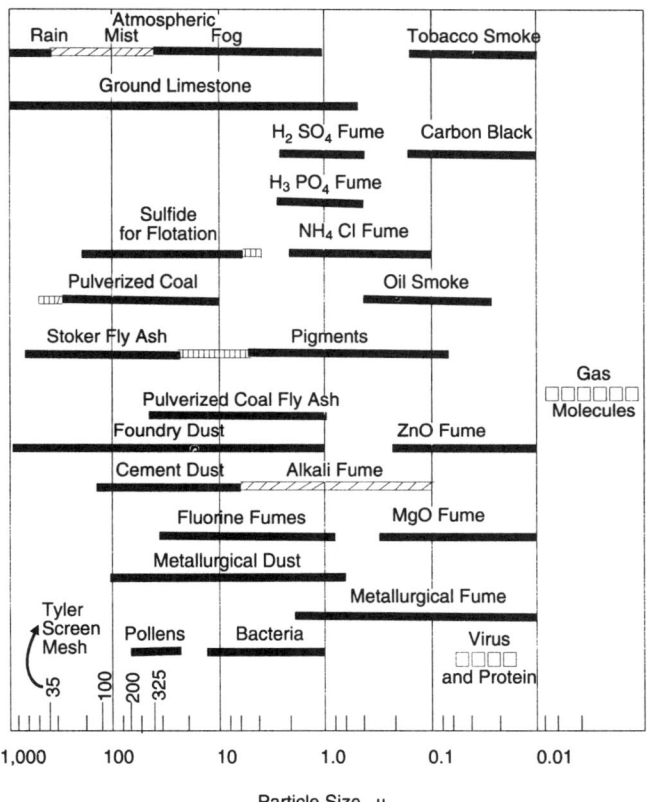

Table 4.2
Particle Distribution of a Typical Atmospheric Sample

Size range (μm)	Average size (μm)	Particle count	Mass percent
10-30	20	1	27
5-10	7.5	112	53
3-5	4	167	12
1-3	2	555	5
0.5-1	0.75	4,215	2
0-0.5	0.25	56,900	1

Source: Wark and Warner 1981

$$\frac{d\Delta}{dx} = \frac{1}{\sigma\sqrt{2\pi}} \exp-\left[\frac{(x-x_{mean})^2}{2\sigma^2}\right] \quad [4.1]$$

where: Δ = fraction of the cumulative total in the size range of interest

x = some suitable dimension or measure (diameter)

X_{mean} = average value of x

σ = variance

Particles from industrial sources generally are not well represented by the normal distribution function, but follow the log normal distribution function (Figure 4.2). In general, there is more skewness of the distribution and the probability curve follows a geometric or a logarithmic form. By plotting the log of the particle size against frequency, the skewed curve is converted into one of symmetrical type. The log probability distribution is expressed as (Hesketh 1977):

$$f(d_p)_n = \frac{\Sigma n}{\log \sigma_g \sqrt{2\pi}} \exp\left[\frac{-(\log d_p - \log d_g)^2}{2 \log^2 \sigma_g}\right] \quad [4.2]$$

where: $(d_p)_n$ = frequency of occurrence of diameter, d_p

n = total number of particles

σ_g = geometric standard deviation

d_g = geometric mean

The size distribution of a typical sample is shown in Figure 4.2a (Perkins 1974). In Figure 4.2b the log of the particle size is plotted as a function of frequency of occurence. The ordinate has units of number or mass of particles or units of mass/ unit log interval. If the log scale is used, then the curve is often approximated by a Gaussian distribution. This distribution is further plotted in Figure 4.2c on a log-probability scale. The resulting straight line indicates that the size distribution is log normal and it is a characteristic of particle size distributions that are log normal.

A log normal distribution can be characterized by two parameters: the geometric mean and the geometric standard deviation. In the case of a log normal distribution, the relationships between the mean and the standard deviation are:

$$\log(d_{84.1}) = \log(d_{50}) + \log \sigma_g \quad [4.3]$$

Chapter 4: Fundamentals of Particulate Emission Control

Figure 4.2
a) Particle-Size Distribution. b) Particle-Size Distribution, Logarithmic Size Scale, c) Logarithmic Probability Plot of Particle Size

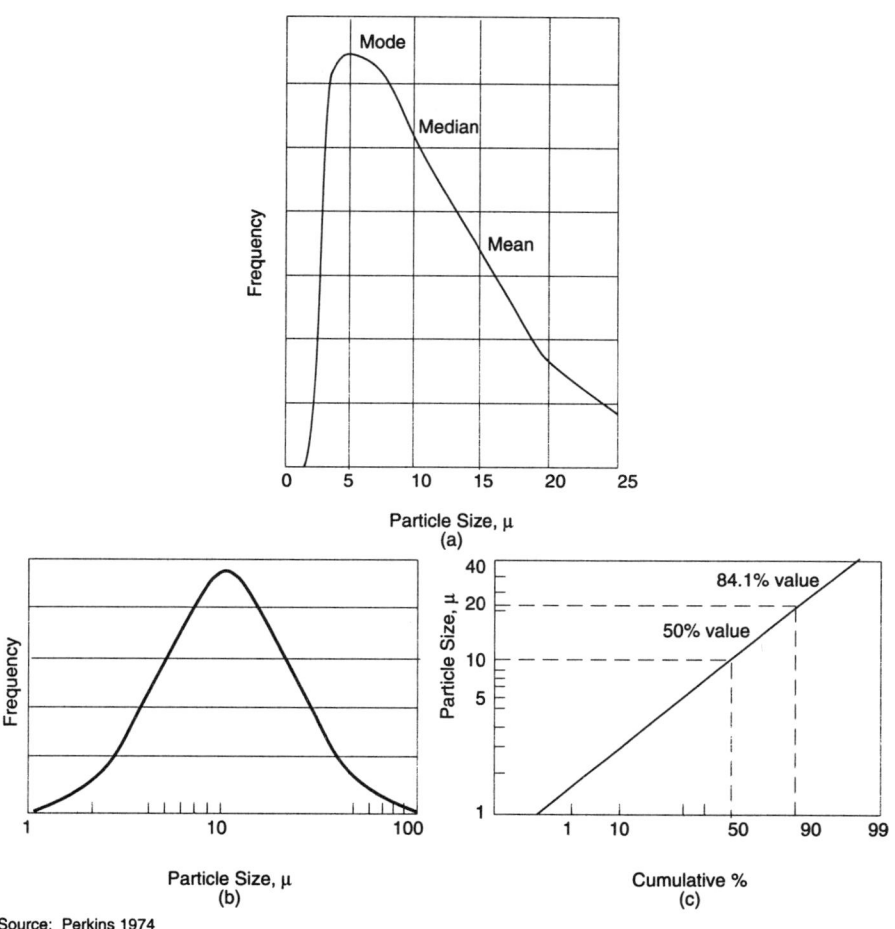

Source: Perkins 1974

where: $d_{84.1}$ = diameter such that particles constituting 84.1 % of the total mass are smaller than this size

d_{50} = geometric mean diameter with 50% of the mass smaller than this size

Equation [4.3] can be rewritten as:

$$\sigma_g = \frac{d_{84.1}}{d_{50}} = \frac{d_{50}}{d_{15.9}} \qquad [4.4]$$

The geometric mean diameter, d_g, and standard deviation, σ_g, are determined from the particle size distribution as measured with a particle-sizing instrument, such as a cascade impactor.

The particle size data shown in Table 4.3 (Calvert and Englund 1984) were determined from a cascade impactor measurement of an emission source. The particle diameters given in this table are the aerodynamic particle size, μm A, since this is the measured particle diameter by a cascade impactor. The particle size distribution is then plotted on a log-probability paper, as shown in Figure 4.3.

Table 4.3
Particle Size Distribution Data

Aerodynamic particle diameter (μm)	Cumulative percentage	Cumulative mass concentration (g/Nm³)
	100	2.29
4.5	92	2.1
2.5	70	1.6
1.4	40	0.92
0.8	15	0.34
0.5	4	0.09

The aerodynamic diameter is plotted versus the cumulative mass percent of the particles smaller than that size. Fifty percent of the particles' mass is smaller than the mass median diameter (MMD), d_{gm}. From Figure 4.3, MMD is $d_{gm} = 1.7$ μm and the geometric standard deviation, σ_g, is:

$$\sigma_g = \frac{d_{84.1}}{d_{50}} = \frac{3.4}{1.7} = 2$$

The aerodynamic diameter is the proper diameter to apply to particles when discussing collection efficiencies (see section 3.2.1). The aerodynamic diameter automatically results from aerodynamic classifiers, such as cascade impactors (Hinds 1982). In a cascade impactor, air and particles are drawn through a series of stages that consist of slots or holes and impaction plates. Each successive stage has narrower slots and closer plates so that each successive stage captures increasingly smaller particles. The masses of particles collected on all stages are then used to determine the size distribution of the particulate stream.

Figure 4.3
Particle Size Distribution

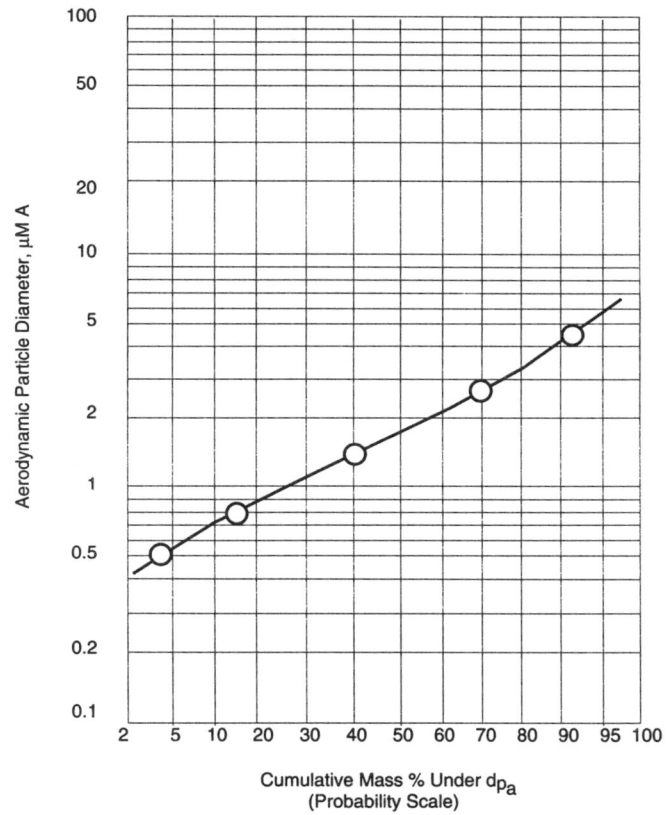

Example 4.1 Particle Size Distribution Analysis *(Reprinted by permission of Waveland Press, Inc. Cooper and Alley,* Air Pollution Control: A Design Approach. *Prospect Heights, IL: Waveland Press, Inc. 1994. All rights reserved.)*

The following data were obtained from a cascade impactor (Cooper and Alley 1990). Show that the distribution is log normal, and find d_{gm} and σ_g.

Size range, μm	0-2	2-5	5-9	9-15	15-25	>25
mass, mg	4.5	179.5	368	276	73.5	18.5

Solution:

From the data collected, a table, as shown below, is prepared of size range versus cumulative percent less than the stated size.

Size range, μm	mass fraction in the size range, m_i	cumulative percent less than top size
0-2	0.0049	0.50
2-5	0.195	20.0
5-9	0.400	60.0
9-15	0.300	90.0
15-25	0.080	98.0
>25	0.020	100

A plot of these data is presented in Figure 4.4 from which values for $d_{84.1}$, d_{gm}, $d_{15.9}$ can be read directly as 13.5, 7.8, and 4.6 microns.

σ_g = 13.5/7.8 = 1.73, or 7.8/4.6 = 1.7.

Therefore, d_{gm} = 7.8 mm, σ_g = 1.72, the average of the calculated two values. The d_{gm} value is the *Mass Median Diameter* (MMD).

Figure 4.4
Plot of Data for Showing a Log-Normal Distribution

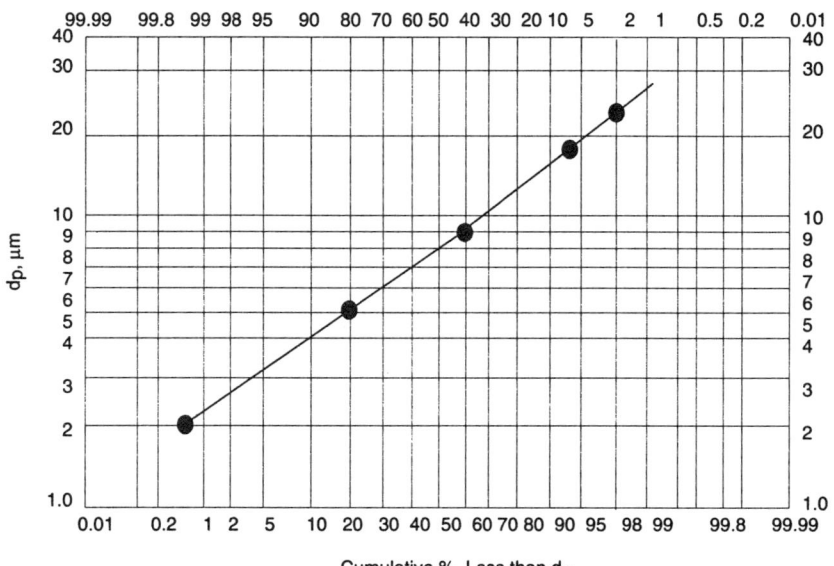

4.3 Collection Efficiency

Many factors influence the choice of a control device used to reduce industrial particulate emissions. The control equipment should be designed to meet emission limits at minimum cost with maximum reliability. The basic trade-off involves decisions between collection efficiency, pressure drop, installation cost, and operating cost. Of these, the principal one is the trade-off between collection efficiency and pressure drop (which can be translated into power requirements) across the control device.

Collection efficiency (by weight) can be defined by the following formula:

$$\text{collection agency} = \frac{\text{inlet loading - outlet loading}}{\text{inlet loading}} \cdot 100\% \quad [4.5]$$

Pressure drop describes the pressure loss between the inlet and outlet sections of the control device. Collectors with large pressure drops require larger fans (and greater power requirements) to either push or pull the exhaust gas through the system. Some control devices, such as venturi scrubbers, are designed to operate at high pressure drops (as great as 100 in. H_2O). On the other hand, electrostatic precipitators are designed to operate at much lower pressure drops (usually less than 1.0 inches of water) for similar collection efficiencies as venturi scrubbers.

4.3.1 Overall Efficiency

When the particulate size distribution is known and the efficiency of the device is known as a function of particle size, the overall efficiency can be predicted as follows:

$$\eta = \sum \eta_i m_i \quad [4.6]$$

where: η_i = efficiency of collection for the i th size range

m_i = mass percent of particles in the i th size range

The penetration of a device is the mass fraction that is not collected (that is, the fraction that penetrates through the device). Thus,

$$P_t = 1 - \eta \quad [4.7]$$

where: P_t = penetration

4.3.2 Grade Efficiency

The relationship between collection efficiency and particle size is called the grade efficiency or fractional efficiency. For particle control devices, there is a unique grade efficiency curve for a particular set of operating conditions. This function will depend upon such parameters as the nature and design dimensions of the collector, rate of flow, and loading of gas stream, temperature, nature of collecting forces, etc. When the grade efficiency relationship is known, it may be used to predict the overall collection efficiency of an aerosol with a given size distribution.

Experimentally determined grade efficiency curves for various kinds of collecting devices reveals that each kind of device tends to have a typical performance curve. The curves in Figure 4.5 are selections of graphical presentations of typical performance associated for certain types of equipment (Licht 1980). Qualitatively, they show, for example, that cyclones are generally less efficient than other kinds of control devices. Venturi scrubbers are more effective on finer particles than other scrubbers, and that there may be a certain particle size for which collection is at a minimum.

Figure 4.5
Grade Efficiency Curves for Various Types of Dust-Collecting Equipment

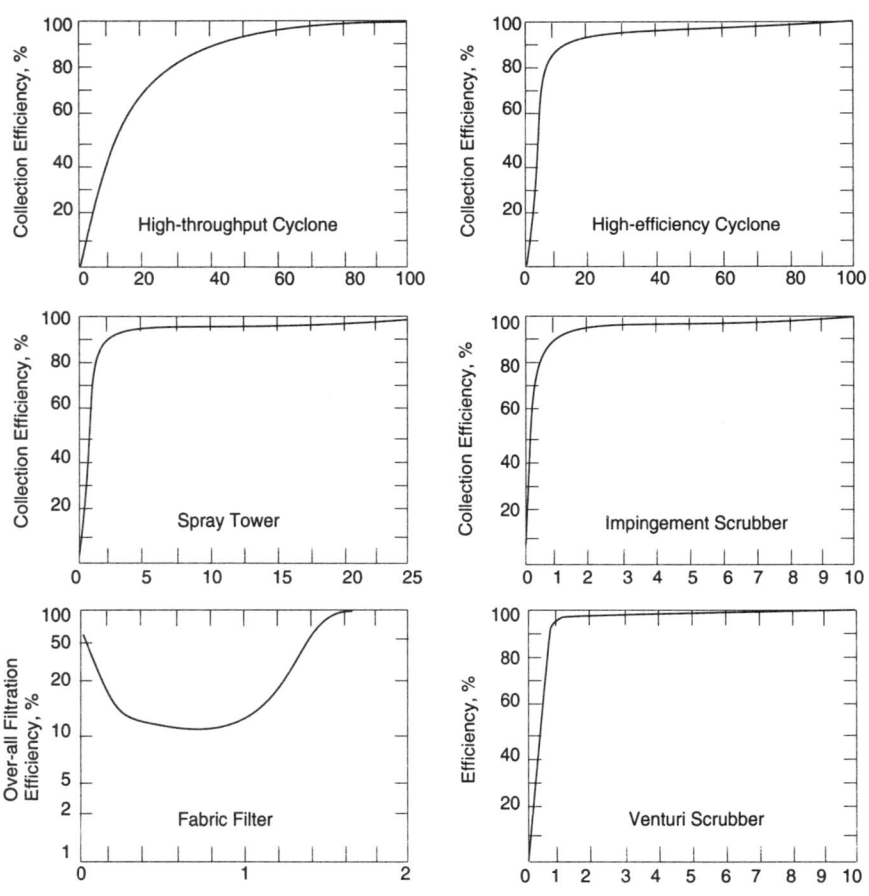

Particle Size, μ

Source: Licht 1980

Chapter 4: Fundamentals of Particulate Emission Control

Example 4.2 Grade Efficiency *(Reprinted by permission of Waveland Press, Inc. Cooper and Alley,* Air Pollution Control: A Design Approach. *Prospect Heights, IL: Waveland Press, Inc. 1994. All rights reserved.)*

The efficiency of an air pollution control device has the grade efficiency relationship shown in the table below. If the particulate size distribution of the source to be controlled is given by the data in Example 4.1, calculate the overall efficiency of the device.

Size range, μm	$d_{i(avg)}$, μm	η_I, %
0-2	1	10
2-4	3	30
4-6	5	60
6-10	8	80
10-14	12	90
14-20	17	96
20-30	25	98
30-50	40	99
>50	>50	100

Solution:

From Example 4.1, determine the fraction of particles in each size range. Multiply the fraction by the corresponding efficiency for that size and sum the results to give an overall efficiency for the device on the particulate matter released from the source.

The table below is a summary of these calculations.

Size range, μm	$d_{i(avg)}$, μm	η_I, %	Cumulative % less than the top size	m_i, %	$\eta_I m_i$, %
0-2	1	0.10	0.5	0.5	0.05
2-4	3	0.30	10	9.5	2.85
4-6	5	0.60	30	20	12.0
6-10	8	0.80	67	37	29.6
10-14	12	0.90	86	19	17.1
14-20	17	0.96	96	10	9.5
20-30	25	0.98	99.4	3.4	3.33
30-50	40	0.99	99.8	0.4	0.4
>50	>50	1.00	100	0.2	0.2
					$\eta = 75\%$

4.3.3 Cut Diameter

An important point on the grade efficiency function is the *cut size diameter*. This value will be frequently used in discussing collector performance. Cut diameter, d_{pc}, is defined as the diameter for which collection efficiency (and penetration) equals 50%. In the collection of particles, this becomes a convenient method of indicating the particle removal efficiency of a collector for a dust with a specific particle size distribution. Cut diameter is most applicable to systems such as scrubbers or cyclones that collect particles mainly by a single mechanism. For these two devices, the particle collection efficiency is a continuous function of the particle size. A filter which operates with one mechanism for larger particles and another for smaller particles may not be categorized as accurately by the cut diameter efficiency relationship.

4.4 Basic Modeling Concepts

4.4.1 Residence Time of Particles

A fundamental factor which influences whether a given aerosol particle will be collected is the length of time during which that particle comes under the influence of the collecting forces. This time must be sufficient for the particle to travel at the collecting velocity from its point of entrance into the zone of influence of the collecting force to a point of collision before the particle is swept out of the zone of influence by the air stream velocity.

The length of time a particle is within the zone of influence is called the residence time, t_r. For collision to occur, the residence time must be equal to or greater than the collecting time. The collecting time, t_c, may be defined as:

$$t_c = \frac{\text{distance travelled from point of entry to collecting surface}}{\text{collecting velocity}} = \frac{H}{v_i} \quad [4.8]$$

An overall average residence time for all particles in a stream may be calculated from the free volume of the collector and the volumetric flow rate.

$$t_{r_{avg}} = \frac{\text{volume of the zone of influence in collector}}{\text{volumetric flow rate of air stream}} = \frac{V}{Q} \quad [4.9]$$

or

$$t_{r_{avg}} = \frac{(\text{inlet area}) \bullet (\text{length, L})}{(\text{inlet area}) \bullet (\text{inlet velocity, v})} = \frac{L}{v} \quad [4.10]$$

The actual residence time of an individual particle may be quite different from this value, depending upon the nature of the fluid velocity profile. In the case of laminar or turbulent flow, there will be a fluid velocity pattern over the cross section of the collection zone perpendicular to the direction of the flow. The precise description of this pattern must be known in order to calculate the residence time for particles entering along different stream lines. When the t_c is equal to t_r, then the collection efficiency, $\eta_i = 1$. When the t_c is greater than t_r, then the collection efficiency will be less than 1.

4.4.2 Elementary Models

There are two flow patterns which are generally used as inputs to basic models that describe the particulate collection processes: (1) plug flow with no radial or axial mixing of uncollected particles and (2) turbulent flow with complete radial (lateral) mixing of uncollected particles, but no axial mixing (Licht 1980). Each of these models may be used as the basis for predicting a corresponding form of the grade efficiency relationship.

Plug Flow, No Mixing

This model may be illustrated in terms of a rectangular two-dimensional view of the zone of influence (Figure 4.6). It is assumed that the particles are distributed uniformly across the section, that all particles are of the same size and kind, d_{pi}, that the collecting force is operating only in one direction, and that the trajectories of all particles comprise a family of parallel straight lines. Whether a given particle is collected, then, is determined solely by the distance, h_c, from its point of entry to the collecting surface. All those particles for which this distance h_c is less than h_c will collide with the surface.

Figure 4.6
Plug Flow Model Trajectory Pattern

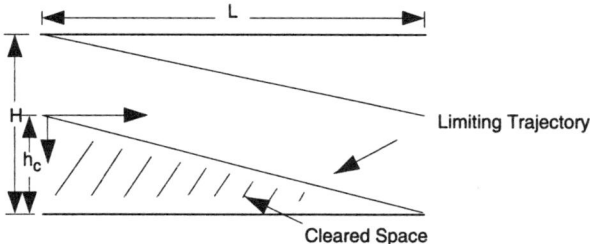

The efficiency of collection, by number or by mass, will then be given by:

$$\eta_i = \frac{t_r}{t_c} = \frac{h_c}{H} = \frac{u_i L}{vH} \qquad [4.11]$$

The grade efficiency relationship will be determined by the way in which the particulate collection velocities, u_i, depends on d_{pi}. If $\eta_I \propto dp_i^2$, ($\eta_i = \left(\frac{\rho_p d_{pi}^2 g}{18\mu}\right) \cdot \frac{L}{vH}$), which is a common case, then η_i versus d_{pi} will be a parabolic graph with $\eta_i = 1$ for all particles large enough so that $u_i \geq \dfrac{vH}{L}$ (as in Figure 4.8)

Turbulent Flow with Lateral Mixing

This model assumes that the uncollected particles are well mixed by lateral turbulence in the gas stream, hence their concentration is always uniform across any lateral or radial section perpendicular to the collecting surface (Figure 4.7).

Figure 4.7
Lateral Mix (Turbulent) Model in Vertical Direction with Uniform Plug Flow in Axial Direction

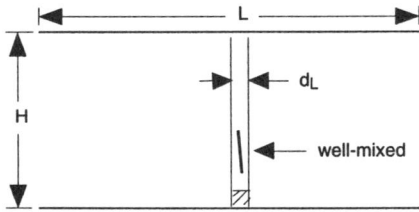

As the stream flows over a differential distance, dL, during a period of time, dt, all particles of the ith grade within a distance, $u_i dt$, of the collecting surface will move into the surface. There will occur a fractional reduction in particle concentration which is the same everywhere along the direction of flow. It is given by:

$$-\frac{dn_i}{n_i} = \frac{u_i dt}{H} = \frac{u_i}{vH} dL \qquad [4.12]$$

This may be integrated over the length of the collecting zone to give:

$$-\int_{n_o}^{n_i} \frac{dn_i}{n_i} = \frac{u_i}{vH} \int_0^L dL = \frac{u_i}{v} \frac{L}{H} \qquad [4.13]$$

$$\ln \frac{n_i}{n_o} = -\frac{u_i L}{vH} = \ln P_t \qquad [4.14]$$

The collection efficiency is then:

$$\eta_i = 1 - \exp\left(-\frac{u_i L}{vH}\right) = 1 - \exp\left(-\frac{u_i t_r}{H}\right) \qquad [4.15]$$

where t_r is the residence time.

Chapter 4: Fundamentals of Particulate Emission Control

Equation [4.15] shows that the collection efficiency is dependent on two parameters. One is the velocity ratio, u_i/v, which is determined by the properties of the particle, the fluid, and by operating conditions. The other is the geometric ratio, L/H, which represents the physical shape of the collector space. It will be seen that these two parameters will occur in the efficiency model for all of the particulate control devices.

The grade efficiency relationship will depend upon how u_i is related to d_{pi}, but in any event, the exponential function will require that η_i approach 1 asymptotically as d_{pi} increases. If $u_i \propto d_{pi}^\beta$ the equation becomes:

$$\eta_i = 1 - \exp(-A d_{pi}^\beta) \qquad [4.16]$$

where A represents all the factors which do not depend upon d_{pi} and particle interaction with the carrier gas determines the value of the exponent, β.

This is the most commonly used model for the design of air pollution control devices. If $\beta = 2$ (inertial collection in the Stokes' range of Reynolds number, see Chapter 2), this expression possesses a point of inflection at

$$d_{p\,\inf}\sqrt{\frac{1}{2\beta}} \text{ and } \eta_{\inf} = 0.394 \qquad [4.17]$$

The graph, therefore, will be an "S" shape as shown in Figure 4.8.

Figure 4.8
Generalized Grade Efficiency for a Gravity Collector

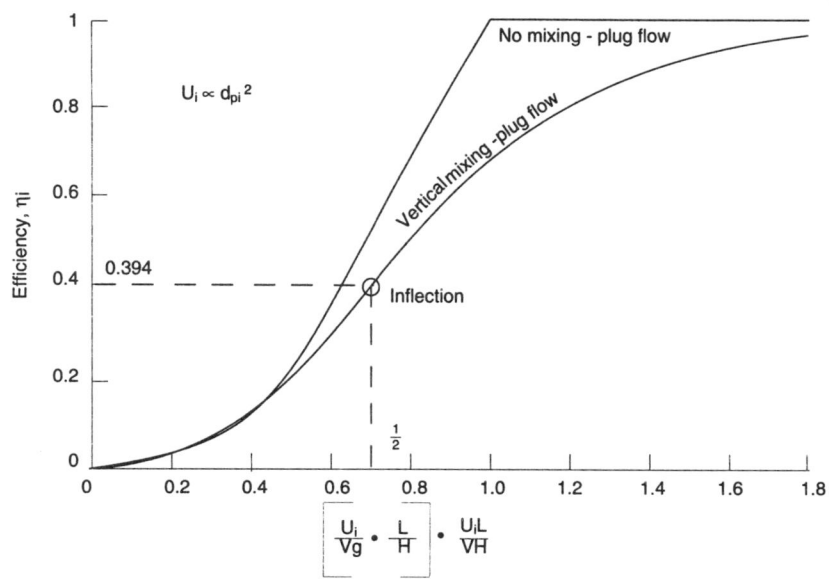

For small particles for which the calculated collection efficiencies are small, the lateral mixed and plug flow models give practically the same answer. For larger particles, the difference in the calculated collection efficiencies become larger, and the two models give different answers. It is important to realize that the plug flow model provides higher overall collection efficiencies estimates than the lateral mixed model and that experimental data usually fit the lateral mixed model better (deNevers 1995).

Example 4.3 Collection Efficiency Determination *(from deNevers, Air Pollution Control Engineering, 1995. Reproduced with permission of the McGraw-Hill Companies.)*

Compute the collection efficiency relation for a gravity settler that has H = 2 m, L = 10 m, v_{avg} = 1 m/sec for both the plug and lateral mixed flow models assuming Stokes' Law applied and the particle density is 2,000 kg/m³.

Solution:

First, compute the plug flow efficiency for a 1 μm particle,

$$\eta = \frac{Lgd^2 \rho_{part}}{18\mu H v_{avg}} = \frac{(10 \text{ m})(9.81 \text{ m/sec}^2)(10^{-6} \text{m})(2,000 \text{ kg/m}^3)}{(18)(1.8 \bullet 10^{-5} \text{ kg/m sec})(2 \text{ m})(1 \text{ m/sec})} = 3.03 \bullet 10^{-4}$$

For 1 μm particles the plug flow assumption leads to an efficiency of $3.03 \bullet 10^{-4}$.

The lateral mixed assumption leads to practically the same result.

$$\eta_{mixed} = 1 - \exp(-3.03 \bullet 10^{-4}) = 3.029 \bullet 10^{-4}$$

The efficiencies for other particle sizes are shown in the table below and were determined by ratios.

Particle Diameter, μm	η_{plug}	η_{mixed}
1	0.000303	0.00303
10	0.0303	0.0298
30	0.273	0.239
50	0.76	0.53
57.45	1.0	0.63
80		0.86
100		0.95
120		0.99

Figure 4.9 graphically shows the difference between these two models. For example, with a particle size of 57.45 μm, the plug flow model has an efficiency of 1 while the corresponding efficiency calculated from the lateral mixed flow model is only 63%.

Figure 4.9
Comparison of the Efficiencies for a Gravity Settler, Calculated by the Plug and Laterial Mixed Models (see Example 4.3)

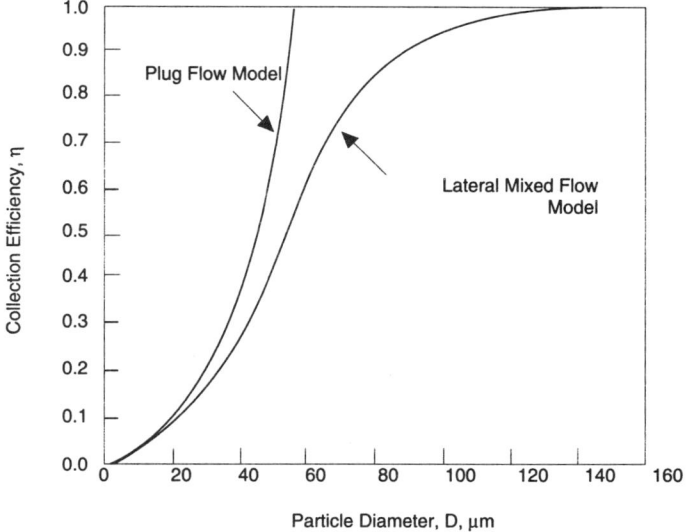

Application of the Models

It is evident that the exact nature of the grade efficiency relationship corresponding to each model will depend upon the manner in which the collecting velocity is related to particle size. It involves an identification of all the forces acting upon a particle and the use of Newton's law of motion to predict the path taken by the particle under the combined action of these forces (see Chapter 2). This leads to the functional relation between u_i and d_{pi}.

In many cases, $u_i \propto d_{pi}^2$. This corresponds to conditions in which the motion of the particle is governed by Stokes' Law. However, other relationships, such as $u_i \propto d_{pi}^\beta$, are possible. There is a range of conditions just outside of these corresponding to Stokes' Law in which the value of β may lie between 1.6 and 2 due to Navier-Stokes' Law drag forces acting on the particle. A general treatment of the models can be made by taking $u_i \propto d_{pi}^\beta$ where β may range from -1 to +2 in value.

With different values of β, the appearance of the grade efficiency graphs will change. For example, the model graphs may not exhibit a point of inflection. The efficiency of the point of inflection can be calculated using the lateral mix model as follows:

$$\eta_{inf} = 1 - \exp\left[\frac{1-\beta}{\beta}\right] \qquad [4.18]$$

Using Equation [4.18], the efficiency at the inflection point can be calculated (Table 4.4) (Licht 1980). For different values of β, the appearance of the grade-efficiency curves will change markedly. For example, for cyclones, β is about 0.67, and therefore there is no inflection point in the grade efficiency curve (see Figure 4.5).

Table 4.4
Calculation of Efficiency at Inflection Point in Lateral Mixed Grade Efficiency Model

β	Efficiency at the Inflection Point
-1	Negative
0.5	Negative
1	0
1.5	0.283
1.6	0.313
1.7	0.338
1.8	0.359
1.9	0.377
2.0	0.393

4.5 Integrated Penetration Concept for the Prediction of Overall Efficiency

The design of air pollution control devices requires the prediction of overall collection efficiency from the particle size distribution of the dust and the fractional efficiency of the air pollution control device. The traditional means of estimating overall collection efficiency from collector fractional efficiency and particle size distribution is to perform the integration graphically, i.e., to break the size distribution into segments (or histograms) multiplied by the collection efficiency at the mean of the segments and sum this over the range of particle sizes (see Example 4.2). This procedure is somewhat laborious and frequently inaccurate since it involves a large number of values read from a graph. The calculation often needs to be repeated due to the need for trial and error solutions to arrive at the required collection efficiency. If the particle size distribution of the dust is lognormal, the collector performance (overall collection efficiency) can be determined from the integration of Equation [4.6] (Calvert and Englund 1984):

$$\eta_T = \Sigma \eta_i m_i \qquad [4.19]$$

where η_i is given by Equation [4.16]

$$\eta_i = 1 - \exp(-Ad_p^\beta) \qquad [4.20]$$

and m_i is given by Equation [4.2]

$$m_i = \frac{\sum n_i}{\sqrt{2\pi} \log \sigma_g} = \exp\left[\frac{-\left(\log d_p - \log d_{gm}\right)^2}{2 \log^2 \sigma_g}\right] \quad [4.21]$$

Note: where $d_{gm} = d_g$ = MMD (mass median diameter)

Figure 4.10
Overall Penetration Generalization

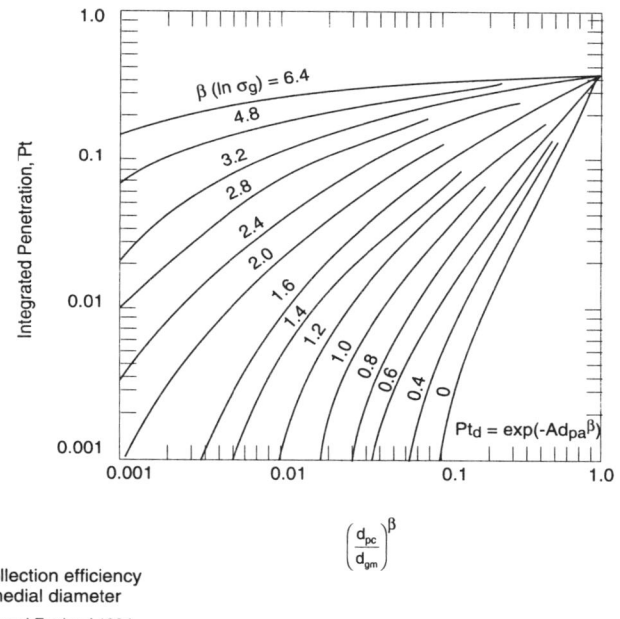

Note:
d_{pc} = 50% collection efficiency
d_{gm} = mass medial diameter
Source: Calvert and Englund 1984

Charts such as Figures 4.10 and 4.11 are available which give the value of η as a function of MMD (d_{gm}), σ_g, β, and d_{pc}, where d_{pc} or d_{p50} is the particle if collected with 50% efficiency will allow the collector to meet the integrated total efficiency, η_T. It is the cut diameter required to achieve the required overall collection efficiency.

$$\eta_{50} = \exp\left(-A d_{p50}^\beta\right) \quad [4.22]$$

where A is the appropriate set of collector dimensions to meet the overall collection efficiency required by the air pollution regulations.

Figure 4.11
Overall Penetration, Special Case

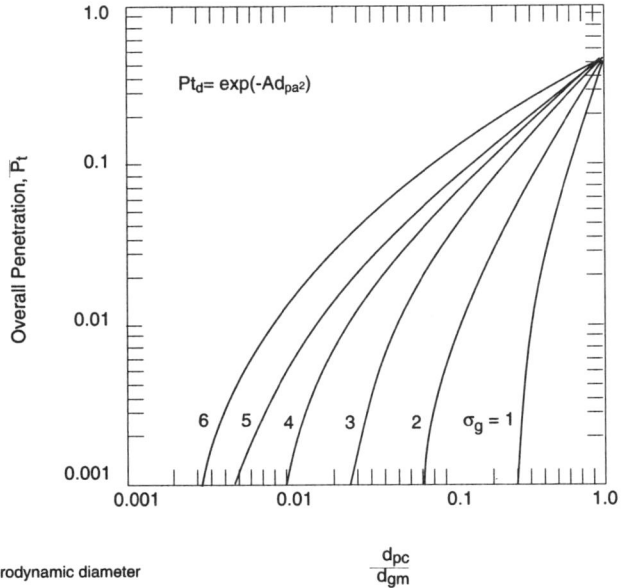

Note: d_{pa} = aerodynamic diameter

Source: Calvert and Englund 1984

Example 4.4 Cut Diameter Calculation

A particle size analysis indicated that d_{gm} =12 µm and σ_g = 3. If a collection efficiency of 99% is required to meet emission standards, what is the required cut diameter of the device?

$$P_t = 1 - \eta$$
$$= 1 - 0.99$$
$$= 0.01$$

Using Figure 4.10 for Pt and σ_g = 3

$$\frac{d_{pc}}{d_{gm}} = 0.63$$

Since the MMD or d_{gm} = 12 µm, then the scrubber must be able to collect particles of size 0.63 • 12 = 0.76 µm with at least 50% efficiency to achieve an overall scrubber efficiency of 99%. This is the aerodynamic particle size (see section 3.2.1).

Figure 4.11 was developed for inertial impaction dominated collectors using Equation [4.20], where the exponent $\beta = 2$. For other values of β, Figure 4.10 may be used.

4.6 Design Summary

The concepts discussed in this chapter may be summarized by a general system flow chart shown in Figure 4.12 (Licht 1980). This shows that the principal inputs to the collector system performance are:

(1) the type of collector and its specific dimensions;

(2) the operating conditions (gas flow rate, temperature, particle loading, etc.); and

(3) the size distribution of the particles to be collected.

Figure 4.12
General System Analysis of Particulate Collector Performance

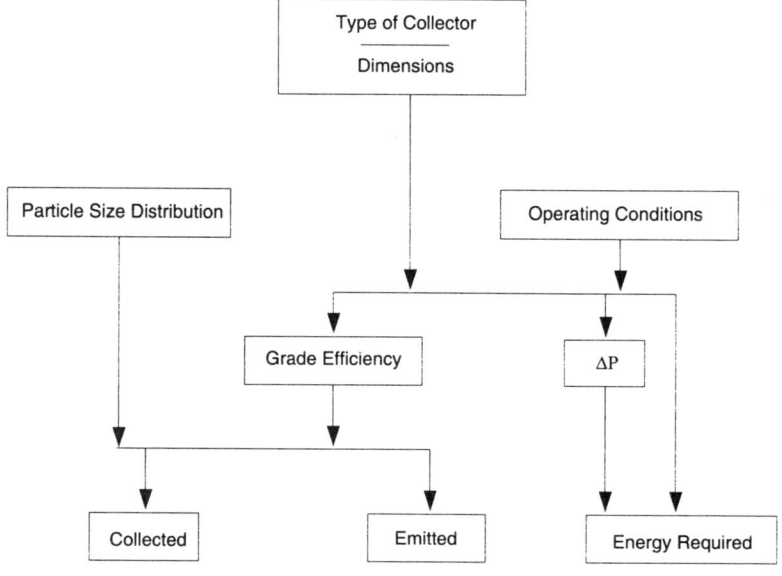

The interaction between (1) and (2) produces a grade efficiency that interacts with (3) to determine the output of the system: particulate matter collected and particulate matter emitted. Finally, (1) and (2) also determine the energy requirements, including the pressure drop.

It is helpful, particularly in making a preliminary survey or choice of collecting equipment, to have some general method of overall comparison of performance.

A convenient method of comparing the grade efficiency performance of various collectors is to replot data, such as given by Figure 4.5, as a semi-log plot of log (1-η_i) vs. d_{pi}. As an illustration of this procedure, a few selected cases taken from Figure 4.5 are shown in Figure 4.13.

Figure 4.13
Comparison of Collector Performance

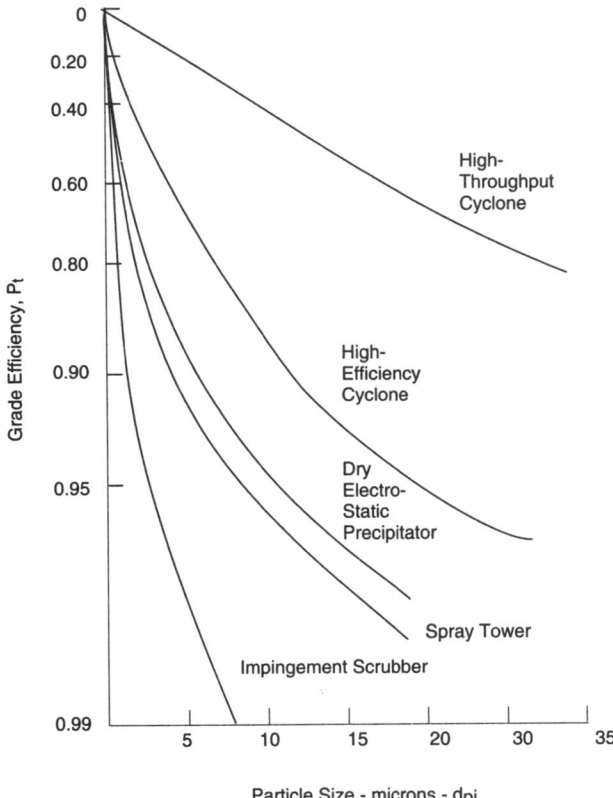

The method of plotting for performance comparison gives all curves the common point in the upper left corner where $d_{pi} = 0$ and $\eta = 0$ or $P_t = 1.00$. It is really a plot of log P_t vs. d_p. The less efficient performance is represented by a curve with a smaller slope, i.e., extending farther to the right.

There is no reason to expect that such plots should be straight lines. It may easily be shown that a straight line would correspond to a special case of the lateral mixed model with $\beta = 1$, i.e., $\eta_i \propto d_{pi}$.

From Equation [4.16]

$$\ln(1-\eta_i) = \ln P_i = -Ad_{pi}$$

or 2.303 log(1-η_i) = -Ad$_{pi}$, so that -A/2.303 is the slope of this line which is always negative. Such a case as this rarely occurs. If $\beta \neq 1$, the line will be curved.

It must be remembered that the performance curve for a specific type of collector depends upon operating parameters. This means that each curve in Figure 4.13 could shift in either direction due to variations in gas flow rate, nature of dust, temperature, and other conditions. Therefore, it cannot be concluded that all scrubbing collectors are more efficient than all electrostatic precipitators.

When the lateral mixing model is used in design, the resulting grade efficiency function is of the form:

$$\eta = 1 - \exp\left[-Ad_p^\beta\right] \qquad [4.23]$$

Here, β (usually $0 < \beta < 2$) is given by the model theory, and A is a function of the collector design and operating conditions. The most common problem in the design of particulate air pollution control devices is to find the value of A to meet a required efficiency determined from air pollution control regulations.

4.7 Problem Set

4.1 You are given the following data obtained by sequential sieving of a sample of granite dust.

Sieve Number	Size Opening Size, μm	Mass Trapped on Sieve, g
1	200	4.0
2	100	21.6
3	50	38.4
4	40	8.0
Final filter	0	8.0
	Total	80.0

Determine the mass median diameter (MMD) and geometric standard deviation, σ_g, of this distribution.

4.2 Given the following distributions obtained from size differentiating equipment:

Particle sizes dp, μm	Distribution A μg/m³	Distribution B μg/m³
<0.62	25.5	8.5
0.62- 1.0	33.15	11.05
1.0 - 1.2	17.85	7.65
1.1 - 3.0	102.0	40.8
3.0 - 8.0	63.75	15.3
8.0 - 10.0	5.1	1.692
> 10.0	7.65	0.008

a. Is either distribution A or distribution B log-normal?

b. If so, what is the geometric mean and standard deviation?

4.3 An aerosol having an MMD of 5 μm and σ_g = 2.0 and particle density = 2.0 g/cm³ was placed in a current of air at 100°C and 1 atm flowing upward with a velocity of 0.3 cm/sec. What fraction by weight of the particles will be carried upward?

4.4 The lateral mix model requires particles to be uniformly mixed over a vertical cross section (see Figure 4.7). For turbulent gas flow velocities, v_o, near 1 m/sec, the eddy velocities are of the order of 3-10 cm/sec. Compare these mixing velocities with the terminal settling velocities of unit density spheres. What size particles will be uniformly mixed due to eddy motion?

4.5 Plot the grade efficiency curve for a settling chamber that represents the lateral mix model. Use β values from 0.5 to 2.0. Locate the inflection points (see Figure 4.8).

4.6 Calculate the minimum size particle with 100% theoretical collection efficiency for a settling chamber 30 ft long and 6 ft high with a gas velocity of 5 ft/sec if the air temperature is 100°F and the particulate density is 2 g/cm³.

4.7 A hydrochloric acid mist in air at 25°C is to be collected in a gravity settler. The unit is 30 ft wide, 20 ft high, and 50 ft long. The actual volumetric flow rate of the acidic gas is 50 ft³/sec. Calculate the smallest mist droplet (spherical in shape) that will be entirely collected by the settler. The specific gravity of the acid is 1.6. Assume the acid concentration to be uniform through the inlet cross section of the unit.

4.8 A gravity settler 5 m wide, 10 m long, and 2 m high is used to trap particles with diameter of 10 μm. The gas flow rate is 0.4 m³/sec . Calculate the operating efficiency of a settling chamber for the data given below. Assume Stokes' Law regime and a Cunningham Correction Factor of 1.0.

$\rho_p = 1.10$ g/cm^3, $\rho = 1.2 \cdot 10^{-3}$ g/cm^3, $\mu = 1.8 \cdot 10^{-4}$ g/cm sec

4.9 For a gravitational settling chamber, determine the length required to collect 50 μm diameter particles if the flow rate is 30,000 cfm, the chamber depth is 15 ft, the chamber height is 5 ft, and the particle specific gravity is 2.5. Assume the air temperature is 80°F. Good design practice calls for a bulk velocity of less than 10 ft/sec. Does this design meet the requirement?

4.10 A settling chamber is 30 ft long, 6 ft high, and 12 ft wide. The air flow rate at 500°F and 29.92 in. Hg is 4,320 acfm and the entrained particles have a specific gravity of 2.5. What is the diameter in μm of the smallest particle that can be removed with 10% efficiency if the effective settling velocity in the chamber is 1/2 the Stokes' settling velocity?

4.11 Air at 298°K and 1 atm laden with acid fog from a manufacturing process flows through a square horizontal settling chamber 8 m long and 50 cm high. The fog can be considered to consist of spherical droplets of diameter 0.8 mm and density 1 g/cm^3. It is desired to remove 90% of the fog from the air stream. Find the volumetric flow rate, in m^3/hr, which will allow 90% removal.

4.12 Improvement in settling chamber efficiency (Figure a) can be obtained by decreasing the height that a particle must fall before being collected. One technique for decreasing this distance is by the use of trays as shown in Figure b. The multitray settling chamber has 20 trays spaced 1 ft apart and handles 200,000 cfm of air at 20°C. Each tray is 15 by 15 ft.

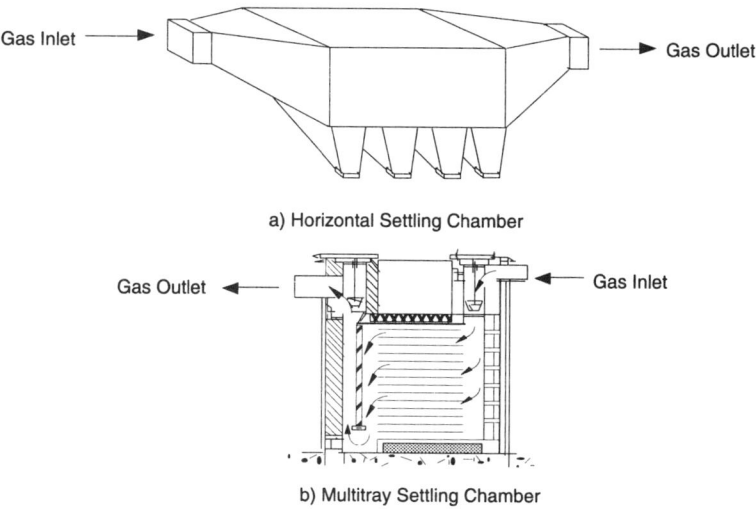

a) Horizontal Settling Chamber

b) Multitray Settling Chamber

Estimate the particle collection efficiency of particles with a density of 2.0 g/cm^3 having diameters of 100, 50, and 20 μm.

4.13 Given

$$\eta_i = 1 - \exp(-Ad_p^\beta) \qquad \text{Equation [4.20]}$$

and $\overline{P_t} = \int_0^\infty P_t f(d_p) d(\ln d_p)$

$$f(d_p) = \frac{1}{\sqrt{2\pi} \log \sigma_g} \exp\left[\frac{-(\log d_p - \log d_{gm})^2}{2 \log \sigma_g}\right] \qquad \text{Equation [4.2 or 4.21]}$$

where:
- d_p = particle diameter (aerodynamic)
- d_{pc} = d_{p50} (cut diameter)
- d_{pg} = d_{gm} = d_g = geometric mean diameter
- σ_g = geometric standard deviation

Explain these equations in words and combine them to allow a plot of $\overline{P_t}$ vs $\left(\dfrac{d_{p50}}{d_{pg}}\right)$

4.14 A 16 in. diameter Stairmand high efficiency cyclone is proposed to clean a polluted air stream under the following conditions: the particle size distribution is shown in the table below, the air flow rate is 240,000 m³/hr (STP), and the dust concentration and density are 10 g/m³ (STP) and 2.7 g/cm³, respectively. The efficiencies of the particles are given in column 5 in the table. Find the overall efficiency as the sum of the fractional efficiency calculation and also from the integrated penetration method presented in Figure 4.10 to 4.11 and compare the results.

Dust size μm	d_{pi} (average) μm	Amount %	Cumulative Distribution %<upper size	Fractional efficiency %
150-104	127	3	100	100
104-75	89.5	7	97	100
75-60	67.5	10	90	100
60-40	50	15	80	100
40-30	35	10	65	100
30-20	25	10	55	100
20-15	17.5	7	45	100
15-10	12.5	8	38	100
10-7.5	8.7	4	30	98.1
7.5-5.0	6.25	6	26	92.5
5.0-2.5	3.75	8	20	74
>2.5	1.25	12	12	11

Chapter 4: Fundamentals of Particulate Emission Control

4.15 Given the following cascade impactor data from the dust to be collected and cyclone grade efficiency curve:

Size range, μm	0-4	4-8	8-16	16-30	30-50	>50
mass, mg	25	125	100	75	30	5

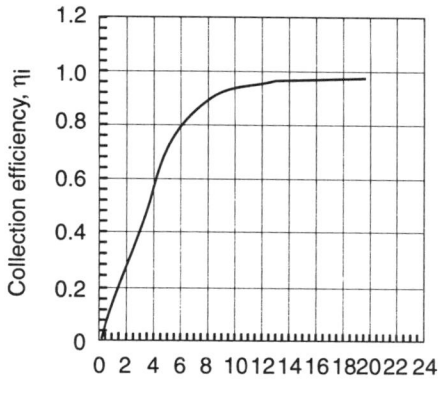

a. Determine the MMD and σ_g for this dust.
b. What is the value of A for 95% collection efficiency for this cyclone?
c. Use the integrated penetration theory (Figure 4.11) to determine the d_{p50} design particle for the grade efficiency curve to obtain 95% collection efficiency (5% penetration) for this dust.
d. Explain how you can provide a design with this particle size (d_{p50}) to achieve overall efficiency and the advantage of this design procedure.

4.16 It is desired to install settling chambers on two small heating plants, one using a traveling grate stoker and the other a spreader stoker. The conditions of operation are listed in the following table.

	Traveling grate	Spreader
Chamber width, ft	10.2	10.8
Chamber height, ft	2.46	2.46
Chamber length, ft	15.0	15.0
Volumetric flow rate, scf/sec	70.6	70.6
Flue gas temperature, °F	0.0	0.0
Flue gas pressure, in. Hg gauge	0.23	1.22
Dust concentration g/scf	72.0	57.0
Mass mean diameter, μm	1.95	4.06
Geometric standard deviation	2.65	2.65
Note: Standard condition = 32°F and 29.92 in. Hg		

Estimate the overall collection efficiency assuming actual settling velocity = 1/2 the theoretical Stokes' settling velocity.

4.17 The installation of a gravity settler has been proposed to help remove limestone particles (density = 2.7 g/cm^3) from air at ambient conditions. The inlet dust loading is 35 grains/ft^3, the volumetric flow rate is 980 ft^3/min. The inlet size distribution of the limestone is given in the table below.

Size range, μm	Weight, %
0.5	2
5-10	4
10-20	12
20-60	40
40-80	28
80-120	10
>120	4

The chamber is 30 ft wide, 25 ft high, and 80 ft long. Calculate and plot the average collection efficiency in each size range and also calculate the overall collection efficiency. Compare this to calculations using the integrated penetration method (Figure 4.11).

4.18 Design a compact multiple-tray settling chamber to remove almost all (99%) 10 μm diameter droplets from a 2,000 cfm gas flow stream. Assume both a laminar flow and a turbulent flow model and compare the plate lengths.

4.8 References

Beachles, D. and J. Jahnke. *Control of Particulate Emissions*. Northrop Services, Inc. EPA 450/2-80-066. Research Triangle Park, NC. 1981.

Calvert, S. and H. Englund (Eds.). *Handbook Of Air Pollution Technology*. John Wiley & Sons. New York. 1984.

Cooper, C. and F. Alley. *Air Pollution Control: A Design Approach*. Waveland Press. Prospect Heights, IL. 1994.

deNevers, N. *Air Pollution Control Engineering*. McGraw-Hill. New York. 1995.

Hinds, W. *Aerosol Technology: Properties, Behavior, and Measurement of Airborne Particle*. John Wiley & Sons. New York. 1982.

Licht, W. *Air Pollution Control Engineering: Basic Calculations for Particulate Collection*. Marcel Dekker. New York. 1980.

Perkins, H. *Air Pollution*. McGraw Hill. New York. 1974.

Strauss, W. *Industrial Gas Cleaning*, 2nd. ed. Pergamon Press. Oxford. 1975.

Wark, K., C. Warner, and Davis. *Air Pollution: Its Origin and Control*. 3rd. Ed. Addison Wesley Longman. Menlo Park, CA. 1998.

Chapter 5

Cyclones

5.1 Introduction

The cyclone is a simple mechanical device commonly used to remove relatively large particles from gas streams. Cyclones have a distinctive form (Figure 5.1) and are used extensively in the wood product industry, feed mills, cement plants, smelters, and many other industrial sites. Single cyclones or banks of small cyclones (multiclones) can be used to effectively remove particles having diameters of approximately 5 to 10 μm. Cyclones provide one of the least expensive methods of removing relatively large particles from gas streams. They can be made of almost any type of material. They have been operated at temperatures higher than 1,000°C using refractory linings, and have been constructed to operate at high pressures. Capital costs are low and operating costs are also low, compared to those associated with more complicated systems.

Cyclones use the centrifugal force created by a spinning gas stream to separate particles from a gas. The particulate-laden gas enters tangentially near the top of the device (Figure 5.1). The cyclone's shape forces the gas to flow into a downward spiral. Centrifugal force causes the particles to move outward, collide with the outer wall, and then slide downward to the bottom of the device. Near the bottom of the cyclone, the gas reverses its downward spiral and moves upward in a smaller inner spiral. The gas exits from the top through a tube and the particles exit from the bottom of the cyclone.

In the cyclone, the forces applied to the particles result from the centrifugal acceleration of particles derived from the rotation of the body of the particle-laden air as it passed through the device. The magnitude of the centrifugal force is determined by parameters (such as geometry, dimensions, and air flow rate) over which the designer has control. As a result, forces much greater than gravitational are possible and more efficient removal of particles can be achieved.

The outward centrifugal force on a particle of mass, m, located at radius, R, is given by:

$$F_{cent} = \frac{mv_T^2}{R} = m\omega^2 R \qquad [5.1]$$

where:
- ω = angular velocity = v_T/R
- v_T = tangential velocity taken as the inlet average air velocity
- R = cyclonic radius

The important dimensions needed in the development of an empirical model of a cyclone performance are shown in Figure 5.1.

Figure 5.1
Reverse-Flow Cyclone Schematic

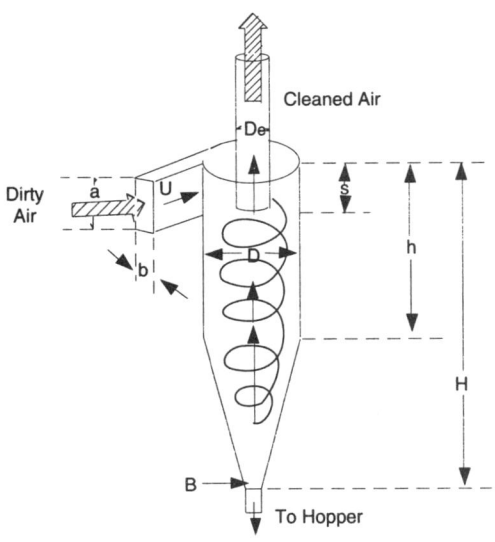

Example 5.1 Cyclone Forces

A particle is traveling in a cyclone gas stream with a velocity of 60 ft/sec and a radius of 1 ft. What is the ratio of centrifugal force to gravity force (separation factor) acting on it?

Solution:

$$F_{cent} = mv^2/R = m \cdot 60 \cdot 60/1 = 3{,}600 \cdot m$$

$$F_{gravity} = mg = 32.2 \cdot m$$

$$\text{Ratio} = (3{,}600 \cdot m)/(32.2 \cdot m) = 111.8$$

At the modest velocity and common cyclone size used in Example 5.1, the centrifugal force is two orders of magnitude larger than the gravity force. This is why cyclones are much more efficient than gravity settlers.

Equation [5.1] can be rewritten in the following form:

$$F_{cent} = \frac{\rho_p d_p^3 v_T^2}{R} \qquad [5.2]$$

where ρ_p is the particle density and d_p is the particle diameter.

The centrifugal force is proportional to the square of the tangential velocity and inversely proportional to the radius of curvature of the gas trajectory. It is also proportional to the cubic diameter of the particle so that removal efficiency increases rapidly for large particles. The cyclone efficiency increases as the diameter of the device decreases. To achieve higher efficiencies dictates the use of smaller cyclones. However, the pressure drop through the cyclone increases rapidly as the tangential velocity increases. A way to maintain high efficiencies with a moderate pressure drop is to use large numbers of small cyclones placed in parallel.

Unfortunately because the flow field in cyclones is complicated and usually accompanied with strong turbulence, it is very difficult to proceed directly from Equation [5.1] to the determination of realistic quantitative estimates of performance. However, several empirical or semi-empirical models have been proposed. An understanding of the tangential gas velocity patterns in the cyclone is necessary to assess factors that contribute to particle motion and to pressure drop. Gas entering the cyclone forms a confined vortex in which the tangential velocity, v_T, is related to the distance from the cyclone axis, R, by (Strauss 1975):

$$v_T R^n = \text{constant} \qquad [5.3]$$

where: n = a constant called the vortex exponent (dimensionless)

In the outer vortex, tangential velocity increases from a minimum near the wall to higher values at the edge of the vortex core. The vortex exponent n ranges from 0.5 to 0.9. Alexander (1949) found that the exponent is given by:

$$n = 1 - \left(1 - 0.67 D_c^{0.14}\right)\left(\frac{T}{283}\right)^{0.3} \qquad [5.4]$$

where: D_c = the cyclone body diameter, m

               T = the carrier gas temperature, °K

This empirical equation is used to select an n value when D_c is given. It is not appropriate to use this empirical equation to determine a value for D_c.

5.2 Standard Cyclone Configuration

The removal efficiency of a cyclone for a given size particle is largely dependent upon the cyclone dimensions. The pressure drop at a given volumetric flow rate is most affected by the diameter. The overall length determines the number of turns of the vortex. The greater the number of turns, the greater the efficiency. The length and the width of the inlet are also important, since the smaller the inlet, the greater the inlet velocity becomes. A greater inlet velocity gives greater efficiency but also increases the pressure drop.

Extensive work has been done to determine how the relative dimensions of cyclones affect their performance. A number of configurations has been proposed and studied sufficiently to be regarded as standards (Shepherd and Lapple 1939; Stairmand 1951; Swift 1969). Cyclones are designed with geometric similarity such that the ratio of the dimensions remains constant at different diameters and those dimensions can be expressed in terms of the body diameter. Table 5.1 presents standard dimension ratios along with the values of geometric configuration parameter, K, and a constant, N_H, relating the pressure drop through the cyclone to the inlet velocity head. Figure 5.1 illustrates the various dimensions in Table 5.1 (Licht 1980).

Selection of one of these standard cyclones may be regarded as the starting point in a design procedure, and they may also be used as standards for comparison with other proposed configurations. High efficiency cyclones generally have smaller inlet and exit areas with a smaller body diameter and longer overall length. The exit tube is an important consideration in the design of any cyclone. Its length must extend beyond the inlet so that the eddies created in the annulus between the tube and the walls do not entrap the particles in the cyclone exhaust.

As shown in Figure 5.1, eight dimensions are required to specify a tangential entry cyclone. Seven dimensions ratios, that is K_a = a/D, K_b = B/D and so on, will

Table 5.1
Cyclone Design Configuration

			High efficiency			General purpose	
Term	Description		Stairmand	Swift	Shepherd & Lapple	Swift	Peterson & Whitby
D	Body diameter		1	1	1	1	1
a	Inlet height	$K_a = a/D$	0.5	0.44	0.5	0.5	0.583
b	Inlet width	$K_b = b/D$	0.2	0.21	0.25	0.25	0.208
S	Outlet length	$K_S = S/D$	0.5	0.5	0.625	0.6	0.583
D_e	Gas outlet diameter	$K_{de} = D_e/D$	0.5	0.4	0.5	0.5	0.5
h	Cylinder height	$K_h = h/D$	1.5	1.4	2	1.75	1.33
H	Overall height	$K_H = H/D$	4	3.9	4	3.75	3.17
B	Dust outlet diameter	$K_B = B/D$	0.375	0.4	0.25	0.4	0.5
K	Configuration		551.3	699.2	402.9	381.8	342.3
N_H	Inlet velocity head		6.4	9.24	8	8	7.76
surf	Surface parameter		3.67	3.57	3.78	3.65	3.2
K/N_H surf			23.5	21.2	13.3	13.1	13.8

specify the configuration and one dimension, chosen as D, will specify the size. The ratios for the five standard configurations in Table 5.1 have been found to be practical and effective. A designer might select other configurations but should keep in mind certain useful guidelines :

- a < S to prevent short circuiting of incoming dust to the outlet tube;
- b < (D-D_e)/2 to avoid excessive pressure drop;
- H > 3D to keep the tip of the vortex inside the cone;
- Angle of cone between 7° and 8° for ready slippage of dust; and
- D_e/D = 0.4 to 0.5, H/D_e = 8 to 10, S/D_e = 1 for maximum efficiency.

K is the efficiency parameter and N_H is the pressure drop parameter for the cyclone configuration. The ratio K/N_H is therefore an optimization parameter. Table 5.1 shows that while the Swift design has the highest efficiency, the Stairmand design is optimum by this criteria.

Cyclone dimensions are determined by a selection of the configuration ratios coupled with the selection of the design gas velocity at the inlet. For high collection efficiency, the inlet velocity should be as large as possible without causing excessive rebounding or reentrainment of particles, or excessive pressure drop. Cyclone design consists of selecting a configuration, then determining the size, grade efficiency, pressure drop, and power requirement of the cyclone to be used. These determinations are based on given gas flow rate, composition, temperature, pressure, and particle loading, together with data on the particle size distribution in the feed. The design will also predict overall collection efficiency and emission rate.

5.3 Pressure Drop

Cyclone design usually consists of choosing an acceptable standard design that will meet gas clean-up requirements at a reasonable pressure drop. These procedures usually trade collection efficiency against pressure drop. While forcing the gas through the cyclone at higher velocities results in improved removal efficiencies, to do so increases the pressure drop and operating costs. There is ultimately an economic tradeoff between efficiency and operating cost. The pressure drop in conventional tangential cyclones can range from 2 to 16 inches of water. The pressure drop is usually the limiting factor in cyclone design. To maintain the pressure drop within acceptable levels, the removal efficiency for small particles must remain relatively low.

Pressure drop, expressed as the number of gas inlet velocity heads, N_H, is based on the inlet and outlet dimensions of the cyclone.

$$N_H = K_c \left(ab / D_e^2 \right) \qquad [5.5]$$

where a, b, and D_e are the dimensions shown in Figure 5.1 and N_H values are given in Table 5.1 for standard cyclones. Shepherd and Lapple (1939) found that $K_c = 16$ for a normal tangential inlet and $K_c = 7.5$ for a cyclone with an inlet vane. The pressure drop is related to the velocity head as follows:

$$\Delta p = 0.5 \, \rho_g v_g^2 N_H \qquad [5.6]$$

where: Δp = pressure drop

ρ_g = gas density

v_g = gas velocity

The units of Δp in Equation [5.6] depend on units used for the variable ρ_g and v_g. When ρ_g is in lb/ft^3 and v_g in ft/sec, then the pressure drop is

$$\Delta p \text{ (in. } H_2O) = 0.003 \, \rho_g v_g^2 N_H \qquad [5.7]$$

The term $0.5 \, \rho_g v_g^2$ is referred to as the velocity head. Therefore, N_H is the number of inlet velocity heads. Inlet velocities for cyclones range from 20 to 80 ft/sec although common velocities range from 50 to 60 ft/sec. At velocity greater than 80 ft/sec, scouring of the cyclone by the particles will increase.

A practical limit of Δp is about 12 inches of water, but the usual working range is considerably less. For v_g values between 50 and 75 ft/sec and N_H = 6 to 9, the value of Δp for air (ρ_g = 0.075 lb/ft^3) will range from 3.4 in. to 11.4 in. of water.

Common ranges for pressure drops are:

low efficiency cyclone	2 to 4 in. of water.
medium efficiency cyclone	4 to 8 in. of water.
high efficiency cyclone	8 to 10 in. of water.

Some of the energy due to the radial motion of the ascending gases can be recovered by the application of scroll devices or by placing outlet drums on top of the exit tube. These are essentially flow straighteners and can effectively reduce the pressure drop across the cyclone without reducing the collection efficiency.

Notice in Table 5.1 that, for a given set of operating conditions and body diameter, the Swift high efficiency standard configuration is more efficient (higher value of K), but results in higher pressure drop (higher value of N_H). The Lapple configuration, with relatively high pressure drop, is not nearly as efficient as the other two.

5.4 Prediction of Collection Efficiency

Collection efficiency is a function of particle size and cyclone diameter. The collection efficiency of a single particle size can be determined by either a semi-empirical approach (Lapple 1951) or by one of several more theoretical approaches that have since been developed (Licht 1980).

5.4.1 Semi-Empirical Approach (Laminar flow)

The basis of the simplest theoretical derivation of efficiency is as follows: particles enter with the gas stream, but tend to move outward under the influence of the centrifugal force. This is resisted by the drag force acting in the radial direction. The radial velocity is found by equating the centrifugal and drag forces. To be collected, the particle should reach the outer wall before the gas leaves the outer vortex. The gas residence time depends on the gas inlet velocity, radius of the cyclone, number of turns in the vortex, and the maximum distance to be traveled by the particles, i.e., the distance between the inlet and outlet of the cyclone. Assuming a laminar flow, a grade efficiency equation is derived that relates the collection efficiency to the cyclone parameters and operation conditions (Lapple 1951; Lapple 1963):

$$\eta = \frac{\pi N_e \rho_p d_p^2 v_g}{9\mu b} \qquad [5.8]$$

where: η = efficiency

N_e = effective number of turns

ρ_p = particle density

d_p = particle diameter

v_g = gas velocity

μ = gas viscosity

b = inlet width

This design method is usually employed when a cyclone size and configuration has already been selected and also works to check the efficiency of operation. This equation indicates the strong dependence of efficiency on particle diameter (squared), the dependence on the number of vortex turns (related to the length of the cyclone) and the inlet velocity, and inverse dependence on the inlet width, which is proportional to the body diameter.

The number of vortex turns, N_e, can be estimated from the cyclone dimensions, since it depends on the height of one turn of the vortex and the length of the cyclone

$$N_e = \frac{1}{a}\left[h + \frac{H-h}{2}\right] \qquad [5.9]$$

Cyclones can be optimized for high collection efficiency by using small diameters, long cylinders, and high inlet velocity. The semi-empirical approach uses the concept of cut size, d_{p50}, defined as the size of the particle that is collected with 50% efficiency. The value is a characteristic of the control device and operating conditions, not of the size range of the dust particles. Using experimental particle collection efficiency versus the particle size data, corrected for the cyclone standard proportions, a generalized curve can be plotted as shown in Figure 5.2 (Danielson 1973).

A particle whose efficiency is to be found is characterized by the ratio of its diameter, d_p, to the cut size, d_{p50}, which gives the value of the abscissa of the graph, and the collection efficiency is read from the ordinate. The value of the cut size is calculated from Equation [5.8] by setting the efficiency equal to 0.5 and solving for d_p, which is d_{p50} by definition (Lapple and Kamack 1955):

$$d_{p50} = \left[\frac{9\mu b}{2\pi N_e \rho_p v_g}\right]^{1/2} \qquad [5.10]$$

Figure 5.2
Cyclone Efficiency Versus Particle-Size Ratio

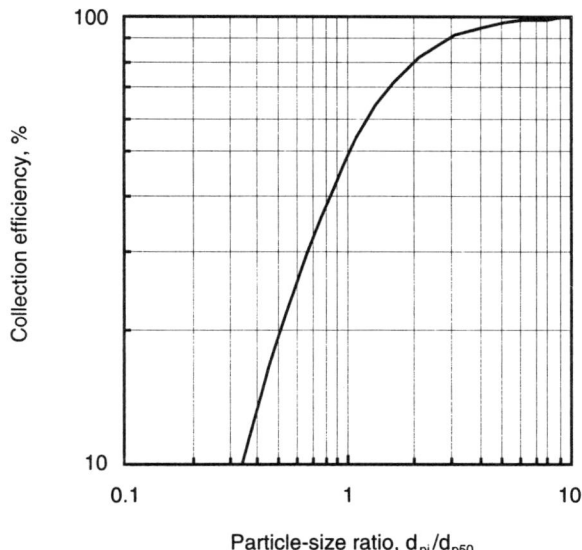

Source: Danielson 1973

Lapple's graph was found to fit an algebraic equation which makes it more convenient for computer applications (Theodore and DePaola 1980):

$$\eta_i = \frac{1}{1 + \left[\dfrac{d_{p50}}{d_{pi}}\right]^2} \qquad [5.11]$$

where: d_{pi} = particle diameter in the range i; and

η_i = efficiency for particle size in the i range.

Example 5.2 Lapple Cyclone Efficiency

For the following particle size distribution, calculate the efficiency of a Lapple standard conventional cyclone with a body diameter of 1.6 ft. The particles' density = 76 lb/ft^3, the gas density = 0.057 lb/ft^3, the gas viscosity ≈ 1.14 •10^{-5} lb/ft sec, temperature = 32°F, and the inlet gas velocity = 80 fps.

particle size range, μm	mass percent in size range
0-4	3
4-10	10
10-20	30
20-40	40
40-80	15
>80	2

Solution:

Using Equation [5.10]: $$d_{p50} = \sqrt{\frac{9\mu_g b}{2\pi N_e v_g \rho_p}}$$

Table 5.1 for standard Lapple cyclone: a = 0.5D, b = 0.25D, h = 2D, H = 4D

$$N_e = \frac{1}{a}\left[h + \frac{H-h}{2}\right] = \frac{1}{0.5D}\left[2D + \frac{4D-2D}{2}\right]$$

= 6 (for all standard Lapple Cyclones)

$\mu_g = 1.14 \cdot 10^{-5}$ lb/ft sec, b = 0.25 • 1.6 = 0.4 ft

$v_g = 80$ ft/sec

$\rho_p = 76$ lb/ft^3

$$d_{p50} = \sqrt{\frac{9 \cdot 1.14 \cdot 10^{-5} \text{ lb/ft sec} \cdot 0.4 \text{ ft}}{2 \cdot \pi \cdot 6 \cdot 80 \text{ ft/sec} \cdot 76 \text{ lb/ft}^3}}$$

$d_{p50} = 1.336 \cdot 10^{-5}$ ft = 4.07 μm

From Figure 5.2 or Equation [5.11]

Size range, μm	d_{pi}, μm	d_{pi}/d_{50}	η_i (Eq. 5.11)	m_i, %	$\eta_i m_i$, %
0-4	2	0.49	0.1937	3	0.58
4-10	7	1.72	0.7464	10	7.46
10-20	15	3.68	0.9311	30	27.93
20-40	30	7.35	0.9818	40	39.27
40-80	60	14.63	0.9954	15	14.93
80+	80	19.51	0.9974	2	1.99
					92.18%

$$\eta_i = \frac{1}{1+(d_{50}/d_{p_i})^2}$$

$$\eta = \Sigma \eta_i m_i = 92.18\%$$

5.4.2 Theoretical Approach

The laminar flow model has limitations, as gas flow in a cyclone is not laminar, and the model usually overpredicts collection efficiency. Also, it is difficult to design a cyclone to achieve desired collection efficiencies without application of a laborious trial and error procedure. Licht (1980) developed an equation for collection efficiency from theoretical considerations that take into account the back mixing of the uncollected particles and determines an appropriate average residence time for the gas in the cyclone. The grade efficiency equation has the lateral mix model form (see Chapter 4):

$$\eta = 1 - \exp\left[-2(K\Psi)^{\frac{1}{2n+2}}\right] \quad [5.12]$$

It uses a parameter K, which is a function of the cyclone's dimensional ratios, and ψ which is a modified inertial impaction parameter. The parameters in Equation [5.12] can be estimated by

$$\Psi = \frac{\rho_p d_p^2 Q}{18\mu_g D_c^3}(n+1) \quad [5.13]$$

$$N = \frac{1}{(n+1)} \quad [5.14]$$

(N = β in Equation [4.20])

Using Equation [5.4] $n = 1 - (1 - 0.67 D_c^{0.14})\left(\frac{T}{283}\right)^{0.3}$ with the cyclone diameter, D, in inches, and temperature, T, in °Kelvin.

Equation [5.12] is plotted in Figure 5.3.

This figure shows the following:

- Efficiency is not sensitive to small changes in the value of n for efficiencies greater than 75%.

- KΨ has values between 0.1 and 1.0 for typical cyclone efficiencies between 80 and 95%.

Figure 5.3
Generalized Cyclone Efficiency Curve

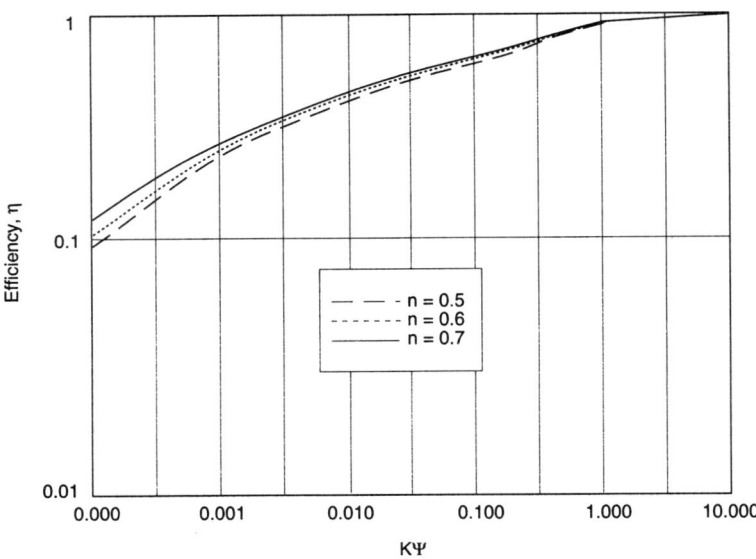

Based on the results in Figure 5.3, n ~ 0.7 is usually assumed for cyclone efficiencies greater than 60% when the cyclone diameter has not been determined. K is the cyclone design number reflecting the physical shape of the cyclone and is given in Table 5.1. The geometric configuration parameter, K, for the cyclone is calculated from the dimension ratios only, and is independent of the size of the cyclone. The model gives a conservative estimate of the performance, such as would be obtained on fairly dilute particle/gas streams, e.g., at low particle loadings (below 10 g/m^3) (Licht 1980). The limited data available indicate that efficiency improves as particle loading increases.

For selection of the size of the cyclone where $v_g = \dfrac{Q}{a \bullet b}$

$$A = 2\left[\frac{KQ\rho_p(n+1)}{18\mu D_c^3}\right]^{\frac{N}{2}} \quad [5.15]$$

Combine Equations [5.12], [5.13], [5.15] and N = 1/(n+1) :

$$\eta = 1 - \exp\left[-Ad_p^{\frac{1}{(n+1)}}\right] = 1 - \exp\left[-Ad_p^N\right] \quad [5.16]$$

Chapter 5: Cyclones

When the efficiency is set at 50% ($\eta = 0.5$), the 50% cut diameter, d_{p50}, can be solved from Equation [5.16].

$$d_{p50} = \left(\frac{0.693}{A}\right)^{n+1} \quad [5.17]$$

The parameter A can be solved from Equation [5.17] as,

$$A = 0.693 \bullet \left[\frac{1}{d_{p50}}\right]^{\frac{1}{(n+1)}} \quad [5.18]$$

When A is substituted into Equation [5.16], the following equation is obtained:

$$\eta_i = 1 - \exp\left[(0.693)\left(\frac{d_p}{d_{p50}}\right)^{\frac{1}{n+1}}\right] \quad [5.19]$$

The development and testing of this model is presented by Leith and Licht (1972). The model reproduces experimental grade efficiency curves more satisfactorily than other models in the literature. Through the parameter, ψ, it correctly indicates that grade efficiency tends to increase for higher flow rate, denser particles, lower viscosity, and smaller cyclone size for a given cyclone geometry, K.

Overall particle collection efficiency has been calculated using Equation [5.12] for cyclones with diameters from 6 in. to 36 in. with an inlet velocity of 60 fps at 20°C for a Lapple conventional cyclone. The particle distribution was assumed to be log normal with a geometric standard deviation of 2.0 and the dust was assumed to have a density of 2.5 g/cm³. Figure 5.4 shows the theoretical relationship between overall collection efficiency and mass median diameter.

Efficiency is seen to increase with increasing particle mass mean diameter and with decreasing cyclone diameter. The observation that the typical efficiency ranges from 70% to 90% further illustrates that cyclones are generally not used as final primary control devices, but as precleaners.

Example 5.3 Design Using Leith and Licht Method

Redo Example 5.2 using:
    a) Leith and Licht design method; and
    b) The Integrated Penetration Theory from Chapter 4

Figure 5.4
Overall Collection Efficiency Versus Particle Mass Mean Diameter for a Range of Cyclone Sizes at 70°F

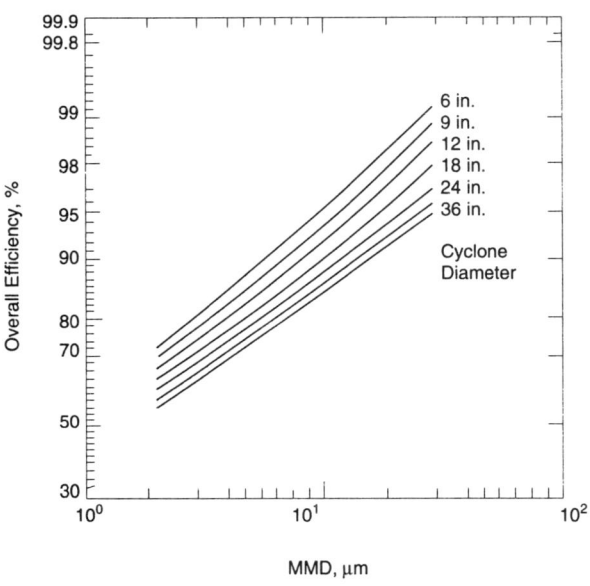

Solution:

a) Leith and Licht design method

$\mu_g = 1.14 \cdot 10^{-5}$ lb/ft sec (at 32°F = 273°K)

$D = 1.6$ ft $= 0.488$ m

$a = 0.5D = 0.8$ ft

$b = 0.25D = 0.4$ ft

$v_g = 80$ ft/sec

$Q = v_g \cdot a \cdot b = 25 \cdot 0.8 \cdot 0.4 = 25.6$ ft³/sec

$\rho_p = 76$ lb/ft³

From Table 5.1, K = 402.9 for Shepherd & Lapple high efficiency cyclone.

Using Equation [5.4] $\quad n = 1 - \left[1 - 0.67 D_c^{0.14}\right] \left(\dfrac{T}{283}\right)^{0.3}$

$$n = 1 - \left[1 - 0.67(0.488)^{0.14}\right]\left(\frac{273}{283}\right)^{0.3}$$

$$n = 0.61$$

Using Equation [5.14] $N = \dfrac{1}{n+1}$

$$N = \frac{1}{1+0.61} = 0.621$$

Using Equation [5.15] $A = 2\left[\dfrac{KQ\rho_p(n+1)}{18\mu D_c^3}\right]^{\frac{N}{2}}$

$$A = 2\left[\frac{402.9 \bullet 25.6 \text{ ft}^3/\text{sec} \bullet 76 \text{ lb}/\text{ft}^3(1+0.61)}{18 \bullet 1.14 \bullet 10^{-5} \text{ lb}/\text{ft sec} \bullet 1.6^3 \text{ ft}^3}\right]^{\frac{0.621}{2}}$$

$$A = 1,398.80$$

Using Equation [5.17] $d_{p50} = \left(\dfrac{0.693}{A}\right)^{n+1}$

$$d_{p50} = \left(\frac{0.693}{1,398.80}\right)^{0.61+1}$$

$$d_{p50} = 4.77 \bullet 10^{-6} \text{ ft} = 1.45 \text{ μm}$$

Using Equation [5.19] $\eta_i = 1 - \exp(0.693)\left(\dfrac{d_{pi}}{d_{50}}\right)^{\frac{1}{n+1}}$

The values in the following table are calculated.

Size range, μm	d_{pi}, μm	d_{pi}/d_{50}	η_i (Eq. 5.11)	m_i, %	$\eta_i m_i$, %
0-4	2	1.38	0.5709	3	1.71
4-10	7	4.83	0.8415	10	8.42
10-20	15	10.34	0.9480	30	28.44
20-40	30	20.69	0.9894	40	39.58
40-80	60	41.38	0.9991	15	14.99
80+	80	55.17	0.9998	2	2.00
The efficiency is 95.1%.					95.13%

b) Integrated Penetration Theory Method

The overall efficiency is determined by graphical integration (Figure 4.10, Chapter 4).

Using Equation [5.13] $\quad \Psi = \dfrac{\rho_p d_p^2 Q}{18 \mu_g D_c^3}(n+1)$

$$\Psi = \dfrac{76 \text{ lb/ft}^3 \bullet (d_p\, \mu m \bullet 10^{-6}\, m/\mu m \bullet 3.28\, ft/m)^2 \bullet 25.6\, ft^3/sec}{18 \bullet 1.14 \bullet 10^{-5}\, lb/ft\, sec \bullet (1.6\, ft)^3} \bullet (0.61+1)$$

$\Psi = 4.01 \bullet 10^{-5} d_p^2 \quad (d_p \text{ in } \mu m)$

Using Equation [5.12] $\quad \eta = 1 - \exp\left[-2(K\Psi)^{\frac{1}{2n+2}}\right]$

$\qquad\qquad\qquad\qquad = 1 - \exp\left[-2(402.9 \bullet 4.01 \bullet 10^{-5} d_p^2)^{0.310}\right]$

$\qquad\qquad\qquad\qquad = 1 - \exp\left[-0.556\, d_p^{0.62}\right]$

Using Equation [5.16] $\quad \eta_i = 1 - \exp(-A d_p^N)$

$\qquad\qquad\qquad\therefore A = 0.556,\ N = 0.62$

Using Equation [5.17] $\quad d_{50} = \left(\dfrac{0.693}{A}\right)^{n+1}$

$\qquad\qquad\qquad\qquad d_{50} = \left(\dfrac{0.693}{0.556}\right)^{1.61}$

$\qquad\qquad\qquad\qquad d_{50} = 1.44$

From the graph on the following page, MMD = d_{pg} = $d_{50\%}$ = 18 μm, $d_{84\%}$ = 41 μm

$\sigma_g = d_{84\%}/d_{50\%} = 41\,\mu m / 18\,\mu m = 2.28$

Using Figure 4.10

The value of the ordinate in the figure $\left(\dfrac{d_{pc}}{d_{pg}}\right)^\beta = \left(\dfrac{1.44}{18}\right)^{0.62} = 0.209$.

The curve β (ln σg) = 0.62 • ln 2.28 = 0.51.

Read P_t = approximately 0.04 from Figure 4.10.

Using Equation [4.7] $P_t = 1 - \eta$

$\eta = 1 - P_t$

$= 1 - 0.04$

$= 0.96$ or 96% efficiency

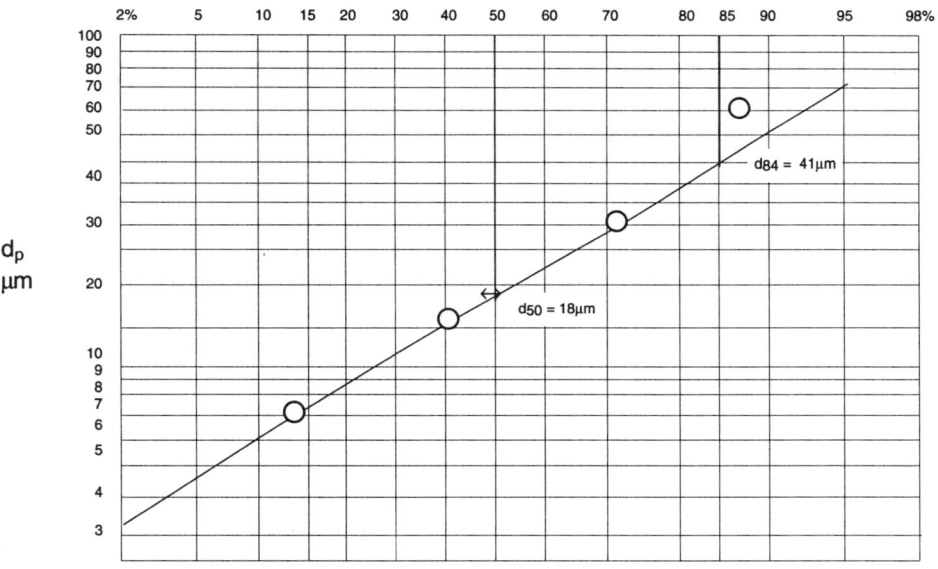

5.5 Effect of Particle Re-entrainment

Equation [5.12] indicates that the grade efficiency for a cyclone will increase with the inlet velocity. However, there is an upper limit above which an increase in inlet velocity will produce a decrease in efficiency due to particle re-entrainment (saltation). Experimental studies (Licht 1980) have found that inlet velocities in the range of 50 to 90 ft/sec in typical cyclones with b/D values near 0.2 and particle densities between 1 and 2.5 g/cm³ will not produce significant re-entrainment. Using these criteria, an empirical relationship has been developed for selection of the cyclone size based on the gas flow rate and cyclone configuration:

$$D = 0.0502 \left[\frac{Q\rho_g^2}{\mu \rho_p} \cdot \frac{(1-K_b)}{(K_a)(K_b)^{2.2}} \right]^{0.454} \quad [5.20]$$

This is an important equation because it limits the inlet velocity that is allowed in a cyclone. The equation permits selection of a cyclone size based on flow rate and often is the starting point for the design of a cyclone. For industrial sources with large flow volumes and a restriction on the inlet velocity to the cyclone, the size of the cyclone will be large. This will decrease collection efficiency because larger cyclones have lower efficiencies for the same inlet velocity and pressure drop. This leads to the use of multiple cyclones instead of one large cyclone.

5.6 Multiple Cyclones

The centrifugal force is proportional to the square of tangential velocity and inversely proportional to the radius of curvature of the gas trajectory. Therefore, the efficiency of a cyclone increases as the diameter of the device is reduced. To achieve higher efficiencies dictates the use of smaller cyclones. However, the pressure drop through the cyclone increases rapidly as the tangential velocity increases. A way to maintain high efficiencies with a moderate pressure drop is to use a large number of small cyclones placed in parallel (see Figure 5.5). If N_c is the number of cyclones in parallel in a multiple cyclone system, the gas volumetric flow rate is Q/N_c. The grade efficiency Equation [5.15] can be rewritten as:

$$A_{N_c} = 2\left[\frac{KQ\rho_p(n+1)}{18\mu N_c D_c^3}\right]^{N/2} \qquad [5.21]$$

The pressure drop through a multiple cyclone system becomes:

$$\Delta P = \frac{1}{2}N_H \rho_g \left(\frac{Q}{N_c a \bullet b}\right)^2 \qquad [5.22]$$

Two or more cyclones of the same configuration and size may be used in parallel, the flow being divided equally between them. In this case, each cyclone can be of smaller diameter than if only one used. The collection grade efficiency for each cyclone is the same, and is greater than if only one larger one was used. A large number of very small cyclones in parallel may be used to increase efficiency for very large flows of gas without increasing the pressure drop. The following example shows that a multiple cyclone system can achieve a higher removal efficiency for small particles at the same pressure drop as a single cyclone. If the required overall efficiency is not reached by a certain design, a number of cyclones in parallel can be used to increase the efficiency without increasing the pressure drop.

Figure 5.5
Multiple Cyclones Arranged in Parallel in Housing

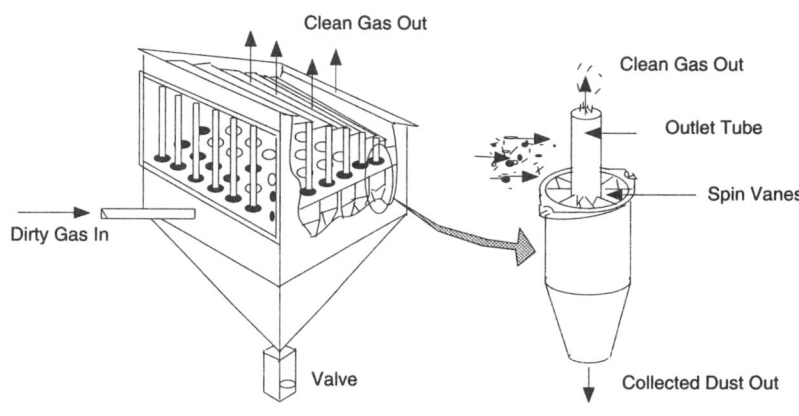

Example 5.4 Comparison of Cyclone Size Effects

A single cyclone handling 6,000 cfm has a diameter of four feet and other proportions as a Stairmand high efficiency cyclone. In an attempt to increase the efficiency, a group of new cyclones is to be designed with the same geometrical proportions and the same pressure drop as the single cyclone. The diameter of the small cyclone is to be six inches. How many cyclones are required and what are the old and new inlet velocities? How will the new collection efficiency compare to the old collection efficiency?

Given:

1. Single Cyclone:

    $Q = 6,000 \text{ ft}^3/\text{min} = 100 \text{ ft}^3/\text{sec}$

    $D = 4 \text{ ft}$

    Stairmand high efficiency (Table 5.1)

    $D = 4 \text{ ft} = 1.22 \text{ m}, \quad a = 2 \text{ ft}, \quad b = 0.8 \text{ ft}, \quad S = 2 \text{ ft},$

    $h = 6 \text{ ft}, \quad N_H = 6.4, \quad H = 16 \text{ ft}, \quad K = 551.3$

2. Small Cyclone:

    pressure drop: same as single cyclone

    $D = 6 \text{ in.} = 0.5 \text{ ft} = 0.15 \text{ m}$

$a = 0.25$ ft, $b = 0.1$ ft, $Q = 100$ ft³/sec

Solution — Number of Cyclones:

Assume: $\rho_p = 1$ g/cm³ $= 62.43$ lb/ft³

temp $= 20°C$

$\mu = 1.81 \cdot 10^{-4}$ g/cm sec $= 1.22 \cdot 10^{-5}$ lb/ft sec

$\rho_g = 0.0749$ lb/ft³

Calculate the pressure drop of single cyclone using Equation [4.7]:

$$\Delta p = 0.003 \rho_g v_g^2 N_H$$

$$v_g = \frac{Q}{a \bullet b} = \frac{100 \text{ ft}^3/\text{sec}}{(2 \bullet 0.8)\text{ft}^2} = 62.5 \text{ ft/sec}$$

$N_H = 6.4$ (Table 5.1)

$\Delta p = 0.003 \bullet 0.0749 \bullet (62.5)^2 \bullet 6.4 = 5.62$ in. H_2O

Use Equation [5.7] to calculate number of cyclones, N:

$$\Delta p = 5.62 = 0.003 \bullet \rho_f \bullet v_g^2 N_H$$

At small cyclone

$Q' = 100/N$ ft³/sec

$$v_g = \frac{Q'}{a \bullet b} = \frac{100 \text{ ft}^3/\text{sec}}{N(0.25 \text{ ft} \bullet 0.1 \text{ ft})} = \frac{4,000}{N} \text{ ft/sec}$$

$\Delta P = 5.62 = 0.003 \bullet 0.0749 \bullet (4,000/N)^2 \bullet 6.4$

$(4,000/N) = 62.51$

$N = 64$ cyclones at $D = 0.5$ ft, $\Delta p = 5.62$ in. H_2O

Solution — Inlet velocities:

The inlet velocities do not depend on the number of cyclones.

old cyclone $v_g = 62.5$ ft/sec

new cyclone $v_g = 62.5$ ft/sec

Chapter 5: Cyclones

Solution — Collection Efficiency:

1. Single Cyclone:

   Use Equation [5.12]:

   $$\eta_i = 1 - \exp\left[-2(K\psi)^{\frac{1}{2n+2}}\right] \text{ with}$$

   $$K = 551.3$$

   $$\Psi = \frac{\rho_p d_p^2 Q}{18\mu_g D^3}(n+1) = \frac{\rho_p d_p^2 v_g \bullet 0.2 \bullet 0.5}{18\mu_g D}(n+1)$$

   From Equation [5.4]

   $$n = 1 - \left[1 - 0.67 D^{0.14}\right]\left(\frac{T}{283}\right)^{0.3}$$

   $$n = 1 - \left[1 - 0.67(1.22)^{0.14}\right]\left(\frac{293}{283}\right)^{0.3}$$

   $$n = 0.686$$

   $$\Psi = \frac{1 \text{ g/cm}^3 \bullet 62.5 \text{ ft/sec} \bullet d_p^2 \text{ cm}^2 \bullet 0.2 \bullet 0.5}{18 \bullet 1.81 \bullet 10^{-4} \text{ g/cm sec} \bullet (4 \text{ ft})^3}(1.686)$$

   $$= 8.086 \bullet 10^{-6} d_p^2 \quad (d_p \text{ in } \mu m)$$

   $$\eta_{old} = 1 - \exp\left[-2(K\Psi)^{\frac{1}{2n+2}}\right] \quad \textit{(Note: } \eta_{old} \textit{ for Single Cyclone)}$$

   $$= 1 - \exp\left[-2\left(551.3 \bullet 8.086 \bullet 10^{-6} d_p^2\right)^{\frac{1}{2\bullet 0.686+2}}\right]$$

   Using Equation [5.16] with:

   $$\eta_i = 1 - \exp\left(-A d_p^N\right)$$

   $A = 0.402$, $N = 0.59$

   $$\eta_{old} = 1 - \exp\left[-0.402 d_p^{0.59}\right]$$

2. Small Cyclone:

$$\eta_i = 1 - \exp\left[-2(K\psi)^{\frac{1}{2n+2}}\right]$$

$$K = 551.3$$

$$n = 1 - \left[1 - 0.67D^{0.14}\right]\left(\frac{T}{283}\right)^{0.3}$$

$$n = 1 - \left[1 - 0.67(0.15)^{0.14}\right]\left(\frac{293}{283}\right)^{0.3}$$

$$n = 0.51$$

$$\Psi = \frac{1\ g/cm^3 \bullet 62.5\ ft/sec \bullet d_p^2\ cm^2 \bullet 0.2 \bullet 0.5}{18 \bullet 1.81 \bullet 10^{-4}\ g/cm\ sec \bullet (0.5\ ft)^3} \quad (1.51)$$

$$= 5.79 \bullet 10^{-5} d_p^2 \quad (d_p\ in\ \mu m)$$

$$\eta_{new} = 1 - \exp\left[-2(551.3 \bullet 5.79 \bullet 10^{-5} d_p^2)^{\frac{1}{2\bullet 0.51+2}}\right]$$

Using Equation [5.16] with:

$$A = 0.64,\ N = 0.66$$

$$\eta_i = 1 - \exp(-Ad_p^N) \quad \text{(Note: } \eta_{new}\text{ for Small Cyclone)}$$

$$\eta_{new} = 1 - \exp\left[-0.64 d_p^{0.66}\right]$$

If any value of d_p is used in both the η_{old} and η_{new} equations, η_{new} will be greater than η_{old}. Therefore,

$$\eta_{new} \geq \eta_{old}$$

The following shows the comparison of collection efficiency between single (old) and multiple small (new) cyclones:

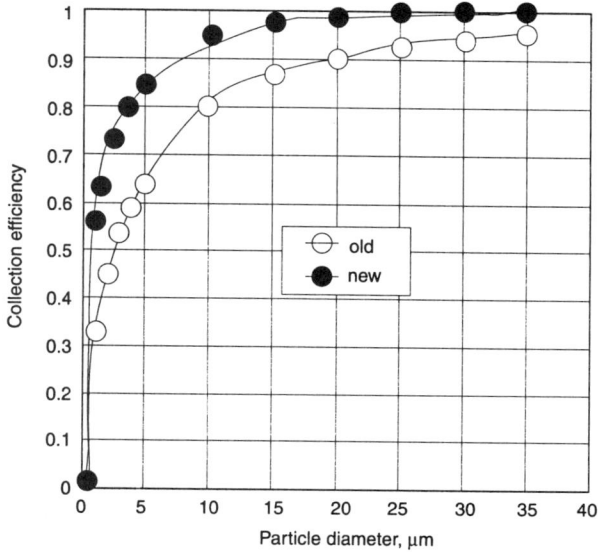

5.7 Optimizing the Design

A configuration parameter may be constructed which is proportional to the benefit/cost ratio for a given cyclone configuration. The benefit or accomplishment of collection may be taken to be proportional to the configuration parameter K as given in Table 5.1. Note that this is inversely related to the cut diameter. Cost is made up of two parts: (1) initial cost of construction, which is essentially proportional to the total surface area of the material of construction used, and (2) cost of operation, which is proportional to power consumption. Power consumption also reflects, roughly, the initial cost of the gas moving system.

A useful optimization parameter (OP) may be conceived as (Calvert and Englund 1984, Ramachandran and Leith 1991):

$$OP \propto \frac{K}{\text{surface area} \bullet \text{Power}} \qquad [5.23]$$

Total surface area may be calculated by using ($\pi D^2 \bullet$ Surf), where Surf is a function of the configuration only. Values for Surf are given in Table 5.1 for standard configurations and the method of calculation for other shapes is given by Licht (1980). For a given set of operating conditions, and dropping the various constants:

$$OP = \frac{K}{\left(\pi D^2 \bullet \text{Surf}\right)\left(Qv_g^2 N_H\right)} \qquad [5.24]$$

The factor $(D^2 Q v_g^2)$ may be replaced by $Q^3 / K_a K_b$, resulting in:

$$OP = \frac{K K_a K_b}{N_H \bullet Surf \bullet Q^3} \quad [5.25]$$

For a given gas flow, OP depends only on the configuration parameters, K, K_a, K_b, N_H, and Surf. Using these as given in Table 5.1, values for OP can be calculated as follows:

**Table 5.2
Calculated OP Parameter for Standard Cyclone Configuration**

Configuration	OP
Stairmand	2.35
Swift (high eff.)	1.96
Lapple	1.66
Swift (general)	1.64
Peterson and Whitby	1.67

This calculation suggests that the Stairmand standard cyclone configuration will generally provide a higher collection efficiency for a given pressure drop among the configurations given in Table 5.2.

5.8 Cyclone Design

This section provides a procedure that can be used to determine the diameter and operating characteristics that can meet a required collection efficiency with a reasonable pressure drop without a long trial and error procedure. The design process is helpful in the design of cyclones for industrial application. It involves the following steps:

1. Select a design configuration, for example, the Stairmand configuration as given in Table 5.1.
2. Determine the cyclone factors from Table 5.1.
3. Determine an initial cyclone diameter using Equation [5.20].
4. Calculate n using Equation [5.14].
5. Determine the collection efficiency using a d_{50} determined from Figure 4.10, if such efficiency will satisfy regulatory requirements.
6. Determine all body dimensions according to configuration ratios in Table 5.1.

Chapter 5: Cyclones

7. Determine v_g from Q/ab (this value should be within the acceptable range of 50 to 90 fps).

8. Determine the pressure drop using Equation [5.6] or [5.7].

9. Evaluate the velocity, pressure drop, and diameter. If any of these parameters are excessively large, consider using multiclones.

10. For multiclones, repeat steps 2 through 6 above with the flow rate divided by the number of cyclones, N, used.

It may also be desired to design a cyclone to have a specified cut diameter. In this case, Equation [5.16] may be used to determine A, using an estimated value of n, say 0.7. After selecting a design configuration K, the value of Q/D^3 may be calculated using Equation [5.15]. This may then be combined with Equation [5.20] to find both Q and D individually. Examples 5.4 and 5.5 apply these design procedures.

Example 5.5 Cyclone Design

Design a cyclone to collect 2.5 μm particles with 75% collection efficiency. The pressure drop is limit to 10 in. of water and the total flow volume is 3,500 cfm at 170°F. The particles have a density of 1.5 g/cm³.

Solution:

At 170°F = 349.67°K, μ_g = 0.0505 lb/ft hr = 2.0876 • 10^{-4} g/cm sec

ρ_g = 0.0629 lb/ft³

d_p = 2.5 μm = 2.5 • 10^{-4} cm

1. Select cyclone configuration.

    Stairmand high efficiency is selected:

    $K_a = a/D_c = 0.5$, $K_b = b/D = 0.2$, K = 551.3, N_H = 6.4 (Table 5.1)

2. Calculate A using Equation [5.16].

    $$\eta = 1 - \exp\left[-A d_p^{\frac{1}{(n+1)}}\right]$$

    η = 0.75, dp = 2.5μm = 2.5 • 10-4 cm, assume n = 0.7

$$0.75 = 1 - \exp\left[-A \bullet (2.5 \bullet 10^{-4})^{\frac{1}{(0.7+1)}}\right]$$

$$0.25 = \exp\left[-A \bullet (7.606 \bullet 10^{-3})\right]$$

$$7.606 \bullet 10^{-3} \bullet A = 1.386$$

$$A = 182.22$$

3. Calculate Q/D_c^3 using Equation [5.15] or $Q/N_c D_c^3$ using Equation [5.21].

$$A = 2\left[\frac{KQ\rho_p(n+1)}{18\mu D_c^3}\right]^{\frac{N}{2}}$$

$$182.22 = 2\left[\frac{551.3 \bullet Q \bullet 1.5 \text{ g/cm}^3 \bullet (0.7+1)}{18 \bullet 2.0876 \bullet 10^{-4} D_c^3}\right]^{\frac{1}{2(1+.7)}}$$

$$91.11 = \left[374,117.8 \frac{Q}{D_c^3}\right]^{0.294}$$

$$\frac{Q}{D_c^3} = 12.365 \text{ or } \frac{Q}{N_c D_c^3} = 12.365$$

where Q is in cm³/sec and D_c is in cm

$Q = 3,500$ cfm $= 3,500 \bullet (30.48)^3/60 = 1,651,816$ cm³/sec

$$\frac{1,651,816}{ND_c^3} = 12.365$$

4. Calculate gas velocity for a 10 in. H₂O pressure drop.

$$10 = 0.003\ \rho_g v_g^2 N_H$$

$$10 = 0.003 \bullet 0.0629 \bullet (v_g)^2 \bullet 6.4$$

$$v_g = 91 \text{ fps}$$

5. Find cyclone size and number of cyclones by solving two equations as follows.

   (a) 3. above provides an equation based on the required efficiency of 75%.

$$\frac{Q}{N_c D_c^3} = 12.365$$

   (Note: Q = cm³/sec D = cm)

(b) Equation [5.22] provides an equation based on the pressure drop limit of 10 in. H$_2$O:

$$V_g = \frac{Q}{N_c(a \bullet b)} = \frac{Q}{N(0.10^2)} = 91 \text{ fps}$$

(Note: Q = cfm, D$_c$ = ft)

when Q = 3,500 cfm, there are two equations and two unknowns

Based on Efficiency:

$$N_c D_c^3 = 4.7 \quad \text{(Note: } D = ft\text{)}$$

Based on Pressure Drop:

$$N_c D_c^2 = 6.41 \quad \text{(Note: } D = ft\text{)}$$

Solving these equations provide

D$_c$ = 0.75 ft

N$_c$ = 12

The required cyclone system will be 12 cyclones each with a diameter of 0.75 ft and a gas velocity of 91 fps (Δp = 10 in. H$_2$O, efficiency = 75%)

Example 5.6 Cyclone Diameter and Pressure Drop Determination

A dust has a log-normal size distribution with σ_g = 2.2, MMD = 8.0 μ, and density of 4 g/cm^3. The collection efficiency is to be 80%. Temperature is 70°F. (a) What is the diameter and pressure drop for a Stairmand high efficiency cyclone operating with an inlet velocity of 70 ft/sec on this source? (b) What is the collection efficiency and pressure drop for a cyclone that is one-half the diameter and has an inlet velocity of 90 ft/sec?

Solution — (a):

σ_g = 2.2, MMD = 8.0 μm, η = 80%, Temp = 70°F, ρ_p = 4.0 g/cm^3

at this condition: μ_g = 1.81•10^{-4} g/cm sec,

v_g = 70 ft/sec = 2,133.6 cm/sec

assume n = 0.7 (for cyclone efficiency greater than 60%)

$$\beta = \frac{1}{(n+1)} = \frac{1}{(0.7+1)} = 0.588 \quad \text{(Equation [5.14])}$$

$\beta (\ln \sigma_g) = 0.588 \bullet \ln 2.2 = 0.463$

From Figure 4.10, for $\eta = 80\%$ then $P_T = 0.2$ and $\beta (\ln \sigma_g) = 0.463$

$$\left(\frac{d_{p50}}{d_{pg}}\right)^\beta = 0.4$$

$$\left(\frac{d_{p50}}{8}\right)^{0.588} = 0.4$$

$$d_{p50} = 1.7 \;\mu m$$

$$\Psi = \frac{\rho_p d_p^2 Q}{18 \mu_g D_c^3}(n+1) \quad \text{(Equation [5.13])}$$

For Stairmand cyclone, $a = 0.2D$, $b = 0.5D$ (Table 5.1)

$$Q = v \bullet a \bullet b = v \bullet 0.2D \bullet 0.5D = 0.1 \; v \; D^2$$

$$\Psi = \frac{4.0 \;g/cm^3 \bullet (1.7 \bullet 10^{-4} \;cm)^2 \bullet 0.1 \bullet 2{,}133.6 \;cm/\sec \bullet D^2}{18 \bullet 1.81 \bullet 10^{-4} \;g/cm\sec \bullet D_c^3}(0.7+1)$$

$$\Psi = 1.287 \bullet 10^{-2} / D_c \quad (D_c \text{ in cm})$$

$$\eta = 1 - \exp\left[-2(K\Psi)^{\frac{1}{2n+2}}\right] \quad \text{(Equation [5.12])}$$

For Stairmand cyclone, $K = 551.3$ (Table 5.1)

$$\eta = 1 - \exp\left[-2\left[\frac{551.3 \bullet 1.287 \bullet 10^{-2}}{D_c}\right]^{\left(\frac{1}{2 \bullet 0.7+2}\right)}\right]$$

$$0.8 = 1 - \exp\left[-2\left(\frac{7.095}{D_c}\right)^{(0.294)}\right]$$

$$\left(\frac{7.095}{D_c}\right)^{0.294} = 0.805$$

$D_c = 14.85 \;cm = 0.5 \;ft$

$\Delta P = 0.003 \rho_g v_g^2 N_H \quad \text{(Equation [5.7])}$

Chapter 5: Cyclones

$N_H = 6.4$ (Table 5.1), $\rho_g = 0.075$ lb/ft³, $v_g = 70$ ft/sec

$\Delta P = 0.003 \cdot 0.075 \cdot 70^2 \cdot 6.4$

$ = 7.06$ in. H_2O

Solution — (b):

$$d_{p50} = \left(\frac{0.693}{A}\right)^{n+1} \quad \text{(Equation 5.17)}$$

$$\frac{(d_{p50})_1}{(d_{p50})_2} = \left(\frac{A_2}{A_1}\right)^{(n+1)}$$

$$A = \left[\frac{KQ\rho_p(n+1)}{18\mu_g D_c^3}\right]^{\frac{N}{2}} \quad \text{(Equation 5.16)}$$

$Q = 0.1 v_g D^2$, $N = 1/(n+1)$

$$A = \left[\frac{K \cdot 0.1 \, v_g \cdot \rho_p (n+1)}{18\mu_g \cdot D_c}\right]^{\frac{1}{2(n+1)}}$$

$$\frac{(d_{p50})_1}{(d_{p50})_2} = \left(\frac{(v_g)_2 \cdot D_1}{(v_g)_1 \cdot D_2}\right)^{\frac{1}{2}}$$

$(v_g)_1 = 70$ ft/sec, $(v_g)_2 = 90$ ft/sec, $D_2 = 0.5 \, D_1$

$$\frac{1.7 \, \mu m}{(d_{p50})_2} = \left(\frac{90 \cdot D_1}{70 \cdot 0.5 D_1}\right)^{\frac{1}{2}} = 1.6$$

$$(d_{p50})_2 = 1.06 \, \mu m$$

$$\left(\frac{d_{pc}}{d_{pg}}\right)^{\beta} = \left(\frac{1.06}{8}\right)^{0.558} = 0.32$$

$\beta (\ln \sigma_g) = 0.588 \cdot \ln 2.2 = 0.463$

From Figure 4.5, the efficiency is near 85%

$\Delta P = 0.003 \cdot 0.075 \cdot 90^2 \cdot 6.4$

$= 11.66$ in. H_2O

5.9 Problem Set

5.1 Develop the expression for the "Separation Factor", $v_T^2/R\rho_g$, used in Example 5.1

5.2 A particle size analysis was run on a cyclone with the following results:

The overall efficiency was found to be 77.5%.

a. Determine the fractional efficiencies.

b. Construct a fractional efficiency curve for the cyclone.

Size range, μm	Hopper, %	Outlet gas, %
0-5	12.2	53.3
5-10	36.6	41.9
10-20	13.2	3.8
20-30	15.1	0.7
>30	22.9	0.3

5.3 The size, mass, and cyclone collection efficiency data for a gas containing limestone dust are given below. Calculate the overall collection efficiency of the unit.

Particle diameter, μm	Wt, %	Collection efficiency, %
0-5	2	4
5-10	8	6
10-20	13	20
20-30	26	32
30-50	12	78
50-75	11	89
75-100	9	95
100-200	8	98
>200	11	99+

5.4 The following table provides the particle size distribution of dust from a cement kiln and the grade efficiency data for high and low efficiency cyclones. Compare the overall collection efficiencies for the two cyclones collecting the cement dust.

Diameter μm	Portland cement kiln emissions Wt, fraction	Collection efficiency, %	
		High-efficiency cyclone	Low-efficiency cyclone
<1 (0.3)	0.03	3	0.01
1-5 (2.2)	0.20	60	0.5
5-10 (7.1)	0.15	85	34
10-20 (14.1)	0.20	95	69
20-30 (24.5)	0.16	100	88
30-40 (34.6)	0.10	100	96
40-50 (44.7)	0.06	100	100
50-60 (54.7)	0.03	100	100
>60 (77.5)*	0.07	100	100
	1.00		
* log-mean diameter for the interval			

5.5 A cyclone 8 in. in diameter carrying a gas (essentially air) at 340°F with an inlet velocity of 50 ft/sec removes a 5 μm particle (ρ_p = 2.5 g/cm³) with 50% efficiency. If this cyclone were operated at 170°F with an inlet velocity of 25 ft/sec, estimate the size of particle removed with 50% efficiency.

5.6 An efficiency test on a Stairmand high efficiency unit test cyclone produced the following data:

    Efficiency = 80%

    Specific gravity of dust = 2.0

    Inlet concentration = 3 grains/ft³

    Temperature = 70°F

    Cut diameter = 4.0 μm

A new cyclone is to operate at 60% of the velocity and be one-half the diameter of the test cyclone (but with similar geometrical dimensions). Estimate the cut diameter for the new cyclone.

5.7 A cyclone has been specified for a small industrial operation to operate on 5,000 cfm of air at approximately 70°F and 1 atm pressure. The diameter is 16 in. and the capital cost is 20 cents/cfm. Capital and operating costs need to be determined if the cyclone requires 0.50 Hp/in. H_2O pressure drop per 1,000 cfm; industrial electricity costs 1.5 cents per kW hr, and the cyclone will operate 8 hours per day.

5.8 The cut diameter for a Swift high-efficiency design cyclone operating under a certain set of conditions is 2.0 μm and the pressure drop is 3.0 in. H₂O. What would be the cut diameter and the pressure drop for a Stairmand design of the same diameter, D, operating at the same flow rate, temperature, particle-loading, etc.?

5.9 What is the collection efficiency and pressure drop for a high-efficiency Stairmand cyclone given the information:

$Q = 0.06$ m3/sec

$T = 20°C$

$\rho_p = 2$ g/cm^3

$d_{50} = 1.4$ μm

$V_T = 16$ m/sec

Particle size (average size in range), μm	Wt, %
1	3
5	20
10	15
20	20
30	16
40	10
50	6
60	3
>60	7

5.10 The particle size distribution of a dust from a cement kiln is provided below:

The following information is also known:

Gas viscosity	0.02 centipoise (cp)
Particle specific gravity	2.9
Inlet gas velocity	50 ft/sec
Effective number of turns within cyclone	5
Cyclone diameter	10 ft
Cyclone inlet width	2.5 ft

a) Determine the cut size particle diameter, i.e., diameter of particle collected at 50% efficiency, and estimate the overall collection efficiency using Lapple's method.

b) Determine the collection efficiency using the integrated penetration procedure and compare the results with the cumulative efficiency determined in a).

c) Determine the collection efficiency using the Leith and Licht method and compare the results with the answers to a) and b) above.

5.11 Design a cyclone system to collect 2.5 μm particles with a 50% collection efficiency. The pressure drop limit is 8 in. H_2O and the total flow volume is 3,500 cfm at 170°F. The particles have a density of 1.5 g/cm³.

5.12 The Dirty Dust Company has a 15,000 cfm emission stream at 100°C which contains 10 μm particles. Design a cyclone system for 98.0% collection efficiency. The Air Pollution Control Agency desires a cyclone system with 99.5% collection efficiency; design this system also. The owner of the facility does not want the cyclone to have a pressure drop of more than 8 in. H_2O.

5.13 A stream of air flowing at the rate of 3,000 acfm at 25°C contains 25 grains/ft³ of particulate solids. The particle specific gravity is 2.0 and a particle size of 5 μm may be assumed. Determine whether one large cyclone or a multiple-cyclone system should be provided to remove the particulate matter from the air stream. Assume an inlet gas velocity to the cyclone of 50 ft/sec and use the Leith-Licht Model. Assume that at least 90% collection efficiency is required.

5.14 Prepare a preliminary design for a system to collect dust from a plant under the following conditions:

> sample of dust: 50% by wt. below 30 μm
> 80% by wt. below 70 μm
> density 2.5 g/cm³
> gas flow rate: 30,000 acfm ± 5,000
> particle loading: 0.5 to 4 grains/ft³, avg = 1.5 grains/ft³
> emission limit: 1.5 lb/hr

Design a cyclone collection system which might meet the emissions limits. Provide a reasonable pressure drop through the cyclone (i.e., 8 to 10 in. H_2O).

5.15 a) Calculate the collection efficiency and pressure drop for a 4 μm diameter particle of 2 g/cm³ density flowing with a 5,000 cfm air volume through a Stairmand medium efficiency cyclone which is 4.5 ft in diameter. The temperature is 20°C.

b) Replace the above cyclone with a group of cyclones which provide an 80% collection efficiency for this particle at a pressure drop of 3 in. H_2O.

5.16 A system of four Stairmand high-efficiency cyclones in parallel, each 1,250 mm in diameter, was proposed to collect entrained dust from a fluidized bed combustor under the following design conditions:

Gas flow	41,035 acm/hr
Gas temperature	250°C
Particle density	1.6 g/cm^3
Solids feed rate	220 kg/hr
Pressure drop limit	140 mm H_2O
Required efficiency	94 %
Particle size distribution	

Particle size (average size in range), μm	Wt, %
0-5	18.0
5-10	16.0
10-15	8.0
15-20	8.0
20-30	6.0
>30	<u>4.0</u>
	Total 100.0

Determine whether the proposed design seems reasonable and likely to meet the requirements. If it is not satisfactory, recommend changes which should be made in the design.

5.10 References

Alexander, R. "Fundaments of Cyclone Design and Operation" Proc. Austral. Inst. Min. and Met. 152:202. 1949.

Danielson, J.(Ed). Air Pollution Engineering Manual. AP-40 US EPA, Research Triangle Park, NC. 1973.

Calvert, S. and H. Englund(Eds.). *Handbook of Air Pollution Technology.* John Wiley & Sons. New York. 1984.

Lapple, C. "Processes Use Many Collection Types" *Chem. Eng.* 58:145-51. 1951.

Lapple, C. "Dust & Mist Collection" *Chemical Engineering Handbook.* Perry, J.(Ed.). McGraw-Hill. New York. 1963.

Lapple, C. and H. Kamack. "Performance of Wet Dust Collectors" *Chem. Eng. Prog.* 51:110-121. 1955.

Leith, D. and W. Licht. "The Collection Efficiency of Cyclone Type Particle Collectors- A New Theoretical Approach" *AIChE Symp. Series*, (68) 126. 1972.

Licht, W. *Air Pollution Engineering: Basic Calculations for Particulate Collection.* Marcel Dekker. New York. 1980.

Ramachandran, G. and D. Leith. "Cyclone Optimization Based on a New Empirical Model for Pressure Drop" *Aerosol Sci. and Tech.* (15) 135. 1991.

Shepherd, C. and C. Lapple. "Flow Patterns and Pressure Drop In Cyclone Dust Collectors," *Ind. and Eng. Chem.* 31(8). 1939.

Stairmand, C. "The Design and Performance of Cyclone Separators" *Trans. Instn. Chem. Engrs*, (29)356. 1951.

Strauss, W. *Industrial Gas Cleaning*, 2nd. ed. Pergamon Press. Oxford. 1975.

Swift, P. "Dust Control in Industry," *Steam Heating Eng.* 38, 453. 1969.

Theodore, L. and V. De Paola, "Predicting Cyclone Efficiency", *J. of Air Pollution Control Assn.* 30 (10), Oct. 1980.

Chapter 6

Fabric Filters

6.1 Introduction

Fabric filters remove the dust from a gas stream by passing the gas through a fabric. Dust particles form a porous cake on the surface of the fabric and aid in the filtration process. Fabric filter systems are developed for industrial application as baghouse systems. A *baghouse* consists of the following components (Figure 6.1): filter medium and support, filter cleaning device, collection hopper, and shell. Baghouses are usually constructed using many cylindrical bags that hang vertically in the baghouse. Bags vary in both length and diameter depending on baghouse design and manufacturer. The number of bags can vary from a few hundred to a thousand or more depending on the size of the baghouse. When a dust layer has built up to a sufficient thickness, the bags are cleaned, causing the dust particles to fall into the collection hopper.

Woven and felted fabrics are used to make bag filters. Woven filters are made of yarns with a definite repeated pattern. Felted filters are composed of randomly placed fibers compressed into a mat and attached to some loosely woven backing material. Woven filters are used with low energy cleaning methods, such as shaking and reverse air. Felted fabrics are usually used with higher energy cleaning systems, such as pulse-jet cleaning.

The filtering surface for the woven filter is not the fabric itself, but the dust layer or filter cake. The fabric simply provides the surface for capture of larger particles.

Figure 6.1
Schematic Diagram of a Shaker Type Dust Collector (baghouse)

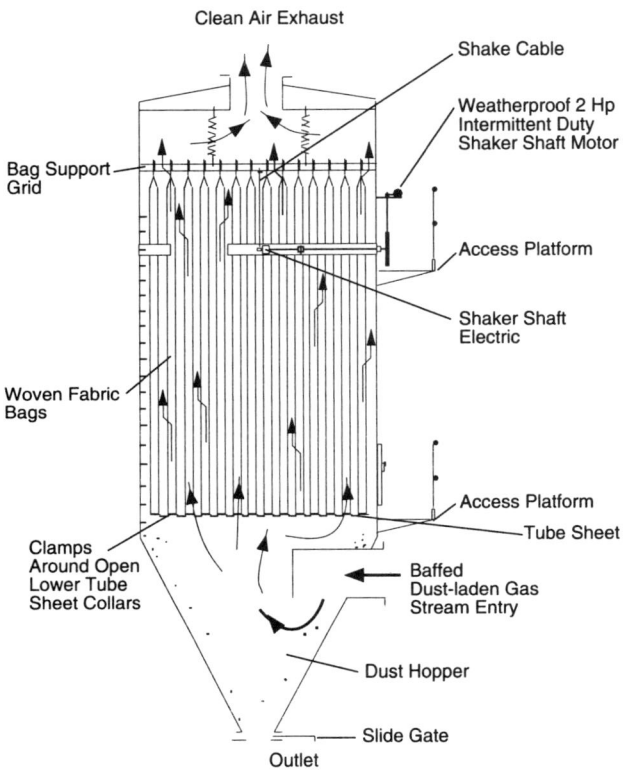

Particles are collected by impaction or interception, and the open areas in the weave are closed. This process is referred to as sieving (Figure 6.2). Some particles escape through the filter until the cake is formed. Once the cake builds up, effective filtering will occur until the bag becomes plugged and cleaning is required.

Felted filters are made by needle-punching fibers onto a woven backing called a scrim (Figure 6.3). The fibers are randomly placed as opposed to the definite repeated pattern of the woven filter. The felts are attached to the scrim by chemical, heat resin, and stitch-bonding methods. The felted filters are generally 2 to 3 times thicker than woven filters. Each individual randomly oriented fiber acts as a target for particle capture by impaction and interception.

Figure 6.2
Sieving

Figure 6.3
Felted Fabric Filter

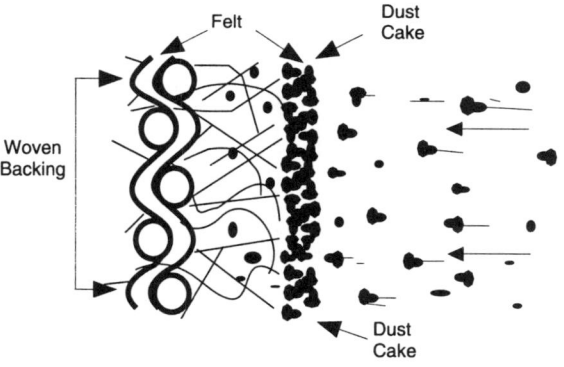

6.2 Fabric Selection

Baghouse operating costs are reduced if the baghouse has a high air-to-cloth ratio, a low pressure drop, and a long life. In each case, the filter media selected is crucial. The primary media selection criteria are the compatibility of the selected fiber with the gaseous environment and the physical configuration of the fiber and resulting fabric as it affects filtration performance. The fiber material from which the fabric is made must have adequate strength characteristics at the maximum gas temperature expected and adequate chemical compatibility with both the gas and the collected dust.

Several types of natural and synthetic fabrics are used in baghouse systems. Characteristics such as temperature, acidity, alkalinity, and particulate matter properties (e.g., abrasiveness and hygroscopsity) determine the fabric type to be used. In many instances, several fabrics will be appropriate, and a final selection will be based on the cleaning method and the desired air-to-cloth ratio.

Most of the principal synthetic fibers have been adapted for use as filtering fabrics while the only natural fibers in common use are cotton and wool. Some of the more common synthetic fibers in commercial use are nylon (aromatic and polyamide), acrylic, polyester, polypropylene, and fiberglass. Natural fibers can be used for gas temperatures up to 200°F and have only moderate resistance to acids and alkalis contained in the gas stream. Synthetics can operate at temperatures up to 550° F and generally have greater chemical resistance. Table 6.1 presents information on the maximum continuous operating temperature and resistance characteristics of commonly used filter fabrics (Sink 1991).

Table 6.1
Characteristics of Several Fibers Used in Fabric Filtration

Fiber type	Maximum operation temperature (°F)	Resistance					Specific gravity
		abrasion	mineral acids	organics acids	alkalis	solvent	
Cotton	180	G	P	G	G	E	1.5
Wool	200	F/G	F	F	P/F	G	1.32
Modacrylic (Dynel)	175	F	VG	VG	G	G	1.30
Polypropylene	200	E	E	E	E	G	0.90
Nylon Polamide	220	E	P	F	VG	E	1.14
Acrylic (Orlon)	260	G	G	G	F	E	1.18
Polyster (Dacron, Creslan)	275	E	G	G	G	E	1.38
Nylon Aromatic (Nomex)	450	E	F	F	VG	E	1.38
Fluorocarbon (Teflon, TFE)	500	F/G	E	E	E	E	2.10
Fiber glass	550	P, G	VG	E	P	E	2.54
P = Poor, F = Fair, G = Good, VG = Very Good, E = Excellent							

A critical criteria for media selection is the ability of the fabric to release dust during the cleaning cycle. This ability depends largely on the mode and intensity of cleaning, but also on the adhesive character of the fabric. The way in which fabric construction relates to cake release has not been completely determined, but it is known that a smooth fabric surface releases dust more readily than does a fuzzy surface. Some fabric surface treatments are specifically intended to enhance dust cake release.

For baghouses currently operating on coal-fired boilers, the bags are fabricated from glass fibers and can tolerate temperatures of 500 to 550°F. Special finishes have been developed for the glass fabrics used in high temperature filtration. The basic purpose of the finish is to protect the glass fibers from abrading themselves, but it can also enhance dust release characteristics.

A special surface treatment involves the application of a Gore-Tex® membrane to the fabric surface. The Gore-Tex® membrane is expanded polytetrafluoroethylene (PTFE) deposited as a thin film. The fabrics finished with Gore-Tex® membrane have the tensile strength characteristic of the woven fabric, but permeability is lower. The membrane improves cleanability and reduces residual dust buildup in the fabric.

6.3 Fabric Cleaning

As the pressure drop increases across the filters due to dust buildup, the filter bags must be cleaned to reduce the dust cake resistance. This cleaning must be timed and performed so as to accomplish the following: (1) keep the pressure drop within a reasonable limit; (2) clean bags as gently as possible to minimize bag wear and to maximize efficiency; and (3) leave a sufficient dust layer on the bags to maintain filter efficiency.

The three methods used to accomplish fabric cleaning are mechanical shaking, reverse air flow, and pulse-jet cleaning (Danielson 1973). The first method uses the fabric flexing mechanism; the latter two methods use a combination of reverse air flow and fabric flexing. With mechanical shaking, bags are hung on an oscillating framework that periodically shakes the bags at predefined time intervals (Figure 6.1). The bags are off-line during cleaning. Because of the stress produced, only woven fibers are used in mechanically cleaned collectors. The dust cake is allowed to build up on the fabric for periods of 30 minutes to two hours or longer before cleaning.

Reverse air flow cleaning is used to flex or collapse the filter bags by allowing a large volume of low pressure air to pass countercurrent to the direction of flow during filtration. Reverse air cleaning usually occurs with the unit off-line. Reverse air cleaning allows the use of fragile bags, such as fiberglass. As with mechanical shaking, woven fabrics are used, cleaning cycles are 30 minutes or longer, and the dust cake effectively filters the air stream.

In pulse-jet cleaning, a high pressure air pulse is introduced into the bag from the top through a compressed air jet (Figure 6.4). This rapidly expands the bag, and

Figure 6.4
Pulse-Jet Fabric Filter

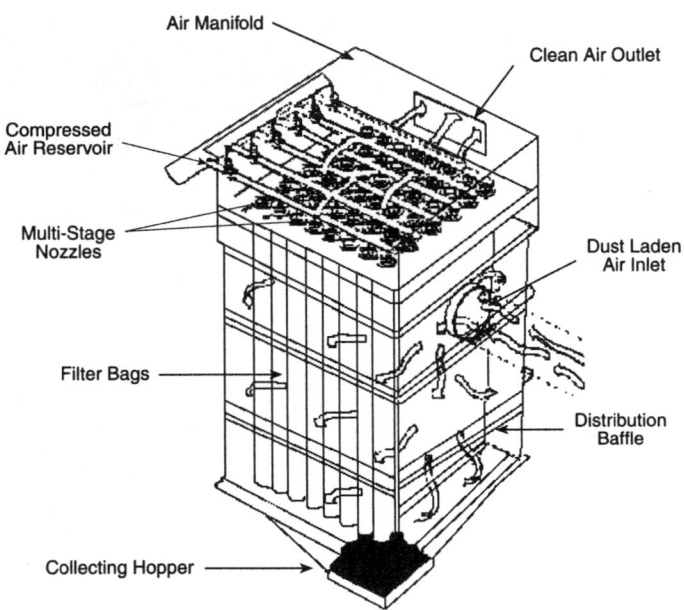

dislodges the particles. The pulse of air cleans so effectively that no dust cake remains on the fabric to contribute to particulate collection. Because such a cake is essential for effective collection on woven fabrics, felted fabrics are used in pulse-jet cleaning fabric filters. Because the cleaning air pulse is of such a relatively high pressure (up to 100 psi) and short duration (0.1 sec), cleaning is usually accomplished with the unit on line. Extra bags are not necessary, therefore, to compensate for bags off line during cleaning. Cleaning occurs more frequently than with mechanical shaking or reverse air flow cleaning (2 to 15 minute intervals), which permits higher air velocities for the same pressure drop (higher air-to-cloth ratios) than other cleaning methods. These features allow pulse-jet cleaned fabric filters to be installed in a smaller space than fabric filters cleaned by the other methods.

Most pulse-jet baghouses use bags that are 4 to 6 inches in diameter. The length of each bag is typically 8 to 10 feet. The shaker and reverse air baghouses use larger bags. The bags in these units are 6 to 18 inches in diameter and up to 40 feet in length.

Blinding of a filter fabric occurs when the fabric pores are blocked and effective cleaning cannot occur. Blinding can result when moisture blocks the pores or increases the adhesion of the dust, or because a high velocity gas stream embeds the particles too deeply in the fabric. Particles can be removed from the surface of woven fabrics (surface collection) more easily than from felted fabrics where particles are embedded in the fabric. Therefore, mechanical collectors are used when particle/fabric interaction may make cleaning difficult. Although pulse-jet collectors can be designed to have a high collection efficiency, mechanically shaken and reverse air cleaning collectors will usually have a higher efficiency if they have a good cake layer.

6.4 Filtration Velocity and Air-to-Cloth Ratio

The terms *filtration velocity* (face velocity) and *air-to-cloth ratio* can be used interchangeably. The formula used to express filtration velocity is

$$v_f = \frac{Q}{A} \qquad [6.1]$$

where: v_f = filtration velocity, ft/min
 Q = volumetric flow rate, ft^3/min
 A = area of cloth filter, ft^2

Air-to-cloth (A/C) ratio is defined as the ratio of the gas filtered in cubic feet per minute to the area of filtering media in square feet. Typical A/C ratios for shakers,

Table 6.2
Typical Air-to-Cloth Ranges

Baghouse Cleaning Method	Air-to-Cloth Ratio
Shaking	2-6 (ft^3/min)/ft^2
Reverse air	1-3 (ft^3/min)/ft^2
Pulse jet	5-15 (ft^3/min)/ft^2

reverse air, and pulse-jet baghouses are listed in Table 6.2 (Joseph and Beachler 1981).

Tables 6.3 and 6.4 summarize the ranges of typical A/C ratios categorized by bag cleaning methods for many dusts and for various industrial processes (Danielson 1973; Joseph and Beachler 1981; Sink 1991). These values should be used as a guide for design. The lower design values need to be used if the dust loading is high or the particle size is small. When compartmental baghouses are used, the design A/C ratio must be based upon having enough filter cloth available for filtering while one or two compartments are off line for cleaning.

Table 6.3
Recommended Air-to-Cloth (A/C) Ratios for Various Dusts by Cleaning Method (ft/min)

	A/C Ratios recommended for cleaning method, ft/min		
Dust	**Shaker**	**Reverse Air**	**Pulse-jet**
Abrasive	2.0-3.0	*	9
Alumina	2.25-3.0	*	*
Aluminum	3.0	*	16
Aluminum Oxide	2.0	*	*
Asbestos	2.5-4.0	*	9-16
Bauxite	2.25-3.2	*	8-10
Blast Cleaning	3.0-3.5	*	*
Carbon	1.2-2.5	*	5-7
Carbon Black	1.5-2.5	1.1-1.5	8-12
Chrome	1.5-2.5	*	9-12
Coal	2.0-3.0	*	12-16
Coke	2.5	*	9-12
Dyes	2.0	*	10
Fertilizer	2.0-3.5	1.8-2.0	8-10
Flint	2.5	*	*
Fly Ash	2.0	2.1-2.3	9-10
Foundry	*	*	8-12
Grass	2.5	*	*
Graphite	1.5-3.0	1.5-2.0	7-9
Gypsum	2.0-3.5	1.8-2.0	10-16
Iron Ore	2.0-3.5	*	11-12
Iron Oxide	2.0-3.0	1.5-2.0	8-16
Iron Sulfate	2.0-2.5	1.5-2.0	6-8
Lead Oxide	2.0-2.5	1.5-1.8	6-9
Leather	3.5-4.0	*	15-20
Lime	2.0-3.0	1.5-2.0	10-16
Limestone	2.0-3.3	*	8-12
Machining	3.0	*	16
Manganese	2.25	*	*
Metal Fumes	1.5	1.5-1.8	6-9
Metal Powders	2.0	*	9-10
Mica	2.25-3.3	1.8-2.0	9-11
Paint Pigments	2.0	*	*
Paper	3.5-4.0	*	10-12
Perchlorates	*	*	10
Plastics	2.0-3.0	*	7-10

Table 6.3 (cont.)
Recommended Air-to-Cloth (A/C) Ratios for Various Dusts by Cleaning Method (ft/min)

Dust	A/C Ratios recommended for cleaning method, ft/min		
	Shaker	Reverse Air	Pulse-jet
Polyethylene	*	*	10
PVC	*	*	7
Resin	2.0	*	8-10
Silica	2.25-2.8	1.2-1.5	7-12
Silica Flour	2.0-2.5	*	*
Silicates	*	*	9-10
Silicon Carbine	*	*	10
Slate	2.5-4.0	*	12-14
Starch	2.25	*	*
Talc	2.25	*	*

*No information available

Table 6.4
Typical A/C Ratios [(ft³/min)/ft²] for Selected Industries

Industry	Fabric Filter, air-to-cloth ratio		
	Reverse air	Pulse jet	Mechanical shaker
Basic oxygen furnaces	1.5-2.0	6-8	2.5-3
Brick manufacturing	1.5-2.0	9-10	2.5-3.2
Castable refractories	1.5-2.0	8-10	2.5-3.0
Clay refractories	1.5-2.0	8-10	2.5-3.2
Detergent manufacturing	1.2-1.5	5-6	2.0-2.5
Electric arc furnaces	1.5-2.0	6-8	2.5-3.0
Feed mills	-	10-15	3.5-5.0
Ferroalloy plants	2.0	9	2.0
Glass manufacturing	1.5	-	-
Grey iron foundries	1.5-2.0	7-8	2.5-3.0
Iron and steel (sintering)	1.5-2.0	7-8	2.5-3.0
Lime kilns	1.5-2.0	8-9	2.5-3.0
Phosphate fertilizer	1.8-2.0	8-9	3.0-3.5
Phosphate rock crushing	-	5-10	3.0-3.5
Polyvinyl chloride production	-	7	-
Portland cement	1.2-1.5	7-10	2.0-3.0
Secondary aluminum smelters	-	6-8	2.0
Secondary copper smelters	-	6-8	-

Example 6.1 Baghouse Bag Sizing

Estimate the amount of baghouse cloth needed for a process flow rate of 10,000 ft³/min and a filtration velocity of 2 ft/min.

Use area $= \dfrac{Q}{v_f}$

$= \dfrac{10{,}000 \text{ ft}^3 / \text{min}}{2 \text{ ft} / \text{min}}$

$= 5{,}000 \text{ ft}^2$ of cloth

To determine the number of bags required in the baghouse, use the formula:

$A_b = \pi \, d \, h$

where: A_b = bag area
  d = bag diameter
  h = bag height

If the bag diameter is 8 in. and the bag height is 12 ft, the area of each bag is

$A_b = 3.14 \bullet 2/3 \bullet 12$

$= 25.13 \text{ ft}^2$

The calculated number of bags in the baghouse:

$N = \dfrac{\text{Total cloth area}}{\text{Bag area}} = \dfrac{5{,}000 \text{ ft}^2}{25.13 \text{ ft}^2} = 200 \text{ bags}$

6.5 Pressure Drop

Pressure drop describes the resistance to air flow across the baghouse. Pressure drop is a function of the pressure drop across both the filter and the deposited dust cake. The equation used to predict pressure drop across the filter is derived from Darcy's law governing the flow of fluids through porous media and is given as

$\Delta P_f = K_1 v_f$ [6.2]

where: ΔP_f = pressure drop across a clean fabric, in. H_2O
  K_1 = fabric resistance, in. H_2O /ft min
  v_f = filtration velocity, ft/min

Chapter 6: Fabric Filters

The term K_1 is the fabric resistance and is a function of the exhaust gas viscosity and filter characteristics, such as thickness and porosity, where porosity describes the volume of voids in the filter.

The pressure drop across the deposited dust cake can be estimated using Equation [6.3]. This formula is also derived from Darcy's law:

$$\Delta P_c = K_2 C_i v_f^2 t \qquad [6.3]$$

where: ΔP_c = the pressure drop across the cake, in H_2O

K_2 = resistance of the cake, in. $H_2O/(lb/ft^2 \bullet ft/min)$

C_i = dust concentration loading, lb/ft^3

t = filtration time, min

v_f = filtration velocity, ft/min

The term $C_i \, v_f \, t$ is the mass per unit fabric area of the dust cake (lb/ft^2) and is usually represented by W = the dust mass areal density. The term K_2 is the dust filter resistance coefficient. It is usually called the specific cake resistance. It has dimensions as follows:

$$K_2 = \frac{\Delta P}{v_f W} = \left[\frac{\text{in. } H_2O}{\frac{ft}{min} \bullet \frac{lb}{ft^2}} \right] = \text{in. } H_2O/\left(lb/ft^2 \bullet ft/min\right) \qquad [6.4]$$

The K_2 units are unusual, having been developed from experimental determination of K_2 using dust loading in lb/ft^2, filtration velocity in ft/min, and pressure drop in in. H_2O. The coefficient is dependent on the gas viscosity, particle density, and dust porosity. The total pressure drop equals the pressure drop across the filter plus the pressure drop across the cake and is given as

$$\Delta P_T = \Delta P_f + \Delta P_c \qquad [6.5]$$

$$\Delta P_T = K_1 v_f + K_2 C_i v_f^2 t \qquad [6.6]$$

Filter drag is the resistance across the fabric-dust layer. It is a function of the quantity of dust accumulated on the filter and is given as:

$$S = \frac{\Delta P}{v_f} \qquad [6.7]$$

where: S = filter drag, in. $H_2O/(ft/sec)$

ΔP = pressure drop across the filter and dust cake, in. H_2O

v_f = filtration velocity, ft/sec

Filter drag essentially provides the pressure drop occurring per unit velocity.

Drag, rather than pressure drop, is the measure of filter aerodynamic resistance preferred by design engineers, because its use preserves the concept of media layer property independent of flow velocity. Pressure drop is analogous to voltage in an electrical circuit. The filter drag is a measure of aerodynamic resistance analogous to electrical resistance.

A filter performance curve for a single bag or compartment of a fabric filter is shown in Figure 6.5. The drag is plotted versus the dust mass deposited on the filter. S_R is the residual drag in a single compartment when it is first brought back on line after cleaning. S_T is the drag in a single compartment at the end of the filtration cycle.

Figure 6.5
Schematic Presentation of Basic Performance Parameters for a Single Compartment Fabric Filter

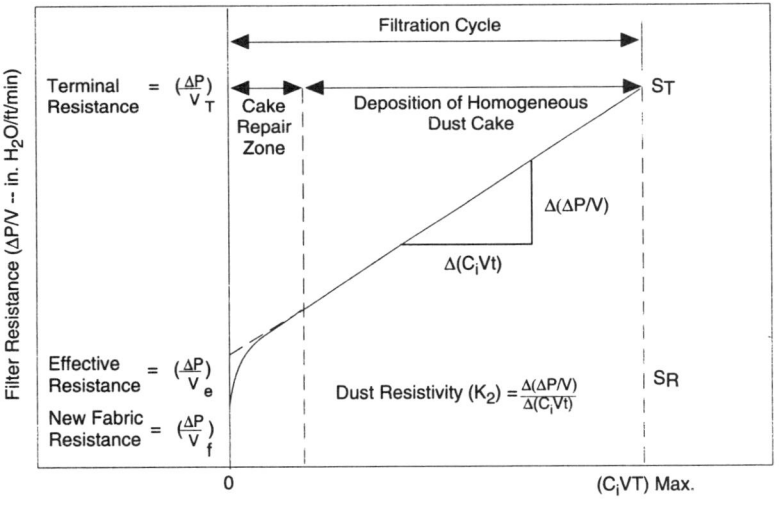

The filter drag first increases exponentially and then at a constant rate. The exponential portion of the curve is the period of cake repair and initial cake buildup. Effective filtration takes place when the filter drag begins to increase at a constant rate. The slope of the straight line portion of the curve is equal to K_2 and represents the resistance to filtration through the dust layer. When the total pressure drop reaches a value set by the system design, bag cleaning is initiated. At this point, the pressure drop decreases to the initial point. Cake repairs begin when the filter cycle repeats.

Example 6.2 Baghouse Design

A plywood mill plans to install a fabric filter as an air cleaning device. How many bags, each 8 inches in diameter and 12 ft long, must be used to treat the exhaust gas which has a particulate loading of 2 grains/ft^3 and the exhaust fan is rated at 7,000 ft^3/min? Estimate the pressure drop after four hours of operation if the resistance coefficients of the filter and dust cake are, respectively, K_1 = 0.8 in. H$_2$O/(ft/min) and K_2 = 3 in. H$_2$O/(lb dust/ft^2 cloth area)(ft/min, filtering velocity).

Solution — Number of Bags:

Assume a filtration velocity of 2 ft/min

1. Total area required = $\dfrac{7{,}000 \text{ ft}^3 / \text{min}}{2 \text{ ft} / \text{min}}$

    = 3,500 ft^2

2. Area of each Bag = $\pi \bullet d \bullet H$

    = $3.14 \bullet 8 \text{ in.} \dfrac{1 \text{ ft}}{12 \text{ in.}} \bullet 12 \text{ ft}$

    = 25.13 ft^2/bag

3. # bags required = $\dfrac{\text{total area}}{\text{area ea. bag}}$

    = $\dfrac{3{,}500 \text{ ft}^2}{25.13}$

    = 139 bags

Solution — Pressure Drop:

$\Delta P_T = \Delta P_f + \Delta P_c$ (Equation [6.5])

= $K_1 v_f + K_2 C_1 v_f^2 t$ (Equation [6.6])

= $\dfrac{0.8 \text{ in. H}_2\text{O}}{(\text{ft/min})} \bullet \dfrac{2 \text{ ft}}{\text{min}} + \dfrac{3 \text{ in. H}_2\text{O}}{(\text{lb dust/ft}^2)(\text{ft/min})} \bullet \dfrac{2 \text{ grains/ft}^3}{7{,}000 \text{ grains/lb}} \bullet \left(\dfrac{2 \text{ ft}}{\text{min}}\right)^2 \bullet 4 \text{ hr} \bullet \dfrac{60 \text{ min}}{\text{hr}}$

= 2.42 in. H$_2$O

Example 6.3 Baghouse Cleaning Frequency

A plant emits 50,000 acfm of gas with a dust loading of 5 grains/ft^3. The dust is collected by a fabric filter at 98% efficiency when the average filtration velocity is 10 ft/min.

where:
$$K_1 = 0.2 \frac{\text{in. H}_2\text{O}}{\text{ft / min}}$$

$$K_2 = 5 \frac{\text{in. H}_2\text{O}}{\frac{\text{lb dust}}{\text{ft}^2} \bullet \frac{\text{ft}}{\text{min}}}$$

The system is designed to begin cleaning when the pressure drop reaches 8 inches of water. How frequently should the bags be cleaned?

Solution:

$\Delta p = 0.2 v_f + 5 C_i v_f^2 t$, solving for t:

$$t = \frac{\Delta p - 0.2 v_f}{5 C_i v_f^2}$$

$$= \frac{8 \text{ in. H}_2\text{O} - \frac{0.2 \text{ in. H}_2\text{O}}{\text{ft / min}} \bullet \frac{10 \text{ ft}}{\text{min}}}{\frac{5 \text{ in. H}_2\text{O}}{\text{lb dust / ft}^2 \bullet \text{ft / min}} \bullet \frac{5 \text{ gr / ft}^3}{7,000 \text{ gr / lb}} \bullet (10 \text{ ft / min})^2}$$

$= 16.8$ min. between cleaning

This high face velocity and short cleaning cycle would indicate that this is a pulse-jet baghouse (see Table 6.3 and 6.4).

6.6 Experimental Measurements of K_2

Many researchers have conducted laboratory and pilot scale fabric tests to measure K_2. Based on this work, K_2 was determined for different A/C ratios and particulate mass median diameters (MMD) (Figures 6.6 and 6.7, respectively) (Buonicore and Davis 1992). The data reported are from eight different sources for fly ash, mica, and talc at 2 to 6 ft/min. The data clearly indicates a strong dependence of K_2 on particle size.

It is evident from these data that velocity also has an effect on K_2. While this observed effect may be partially attributed to the effect of velocity on dust cake packing and Reynolds number, most researchers have reported that K_2 is a function of velocity such that:

Figure 6.6
K_2 vs. MMD and Face Velocity

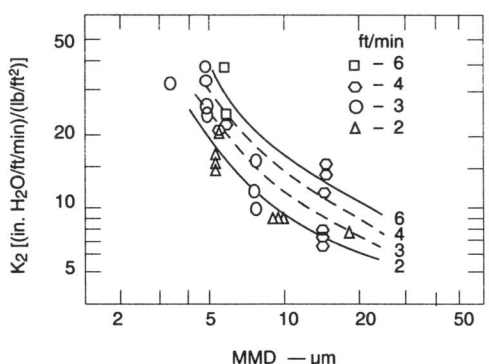

Source: Buonicore and Davis, *Air Pollution Engineering Manual*, 1992. Reprinted with permission of John Wiley & Sons. Inc.

$$K_2 = kv^x \qquad [6.8]$$

Davis and Frazier (1982), in a series of tests on fly ash using 11 different filter materials, reported an average exponent value of 0.7 for fly ash. The data in Figure 6.7 were normalized to a velocity of 3 ft/min assuming an average value for x of 0.6.

Figure 6.7
K_2 Normalized to 3 ft/min vs. MMD

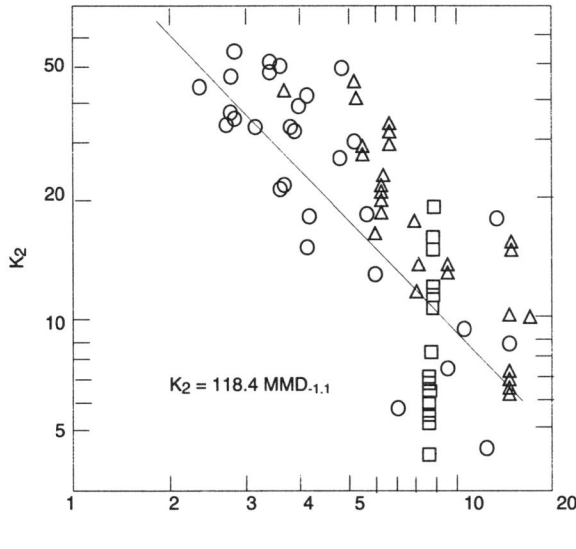

Source: Buonicore and Davis, *Air Pollution Engineering Manual*, 1992. Reprinted with permission of John Wiley & Sons. Inc.

A best fit equation was determined for the data (Davis and Frazer 1982):

$$K_2 = 118.4 \, MMD^{-1.1} \qquad [6.9]$$

where K_2 is measured in the English system (in. $H_2O/(lb/ft^2 \bullet ft/min)$ and MMD is in microns. The best fit equation predicts the K_2 value within a factor of two. The data is excellent considering that measurements obtained under carefully controlled laboratory conditions for a constant particle size distribution have shown a factor of two variation within a single laboratory.

Example 6.4 Determination of K_2

What is the K_2 value for a fly ash dust that has an MMD of 8 microns, and is collected on a fabric filter with an A/C ratio of 2.0 ft/min?

Solution:

$$K_{2(v=3 \, fpm)} = 118.4 \, MMD^{-1.1}$$

$$= 118.4 \bullet 8^{-1.1}$$

$$= 12.02 \text{ in. } H_2O / \left(lb / ft^2 \bullet ft / min\right)$$

$$K_2 = k \, v^x$$

For fly ash, $x = 0.7$

$$\frac{K_{2(v=3 \, fpm)}}{K_{2(v=2 \, fpm)}} = \left(\frac{3}{2}\right)^{0.7}$$

$$K_{2(v=2 \, fpm)} = 9.05 \text{ in. } H_2O / \left(lb / ft^2 \bullet ft / min\right)$$

6.7 Theoretical Prediction of K_2

Stokes' Law gives the drag force on a spherical particle (see Chapter 3) moving with velocity v through a fluid medium, expressed as:

$$F = 3 \pi \mu v d \qquad [6.10]$$

where d is the particle diameter. By equating the pressure drop across a particle layer as the sum of the drag force on each separate particle, $K_2 \, v_f$ (from Equation [6.4]) can be written as:

$$K_2 v_f = \frac{\Delta P}{W} = \frac{\Sigma \text{drag force/area}}{\Sigma \text{mass/area}} = \frac{3 \pi \mu v_f d N}{\pi d^3 \rho_p N / 6} \qquad [6.11]$$

Chapter 6: Fabric Filters 179

where: N = the number of particles/area in the dust cake

ρ_p = the particle density

From Equation [6.11], K_2 becomes:

$$K_2(\text{Stokes}) = \frac{18\mu}{\rho_p d^2} = \sec^{-1} \qquad [6.12]$$

This value of K_2 is the Stokes' K_2 calculated on the assumption of a non-interacting layer (Donovan 1985); that is, the particles are assumed to be far apart. It has units of reciprocal time. As particle spacing decreases, the layer aerodynamic resistance increases. This contribution can be accounted for by the addition of a resistance factor R,

$$K_2 = \frac{18\mu}{\rho_p d^2} R = K_2(\text{Stokes}) R \qquad [6.13]$$

In Equation [6.13], R is a layer-specific number greater than one, which when multiplied by the Stokes' K_2 defined in Equation [6.12], gives the specific cake resistance of that layer. Typically, R depends on the porosity. Perhaps the best known expression for R is that derived from the Carman-Kozeny equation:

$$R = \frac{2k(1-\varepsilon)}{\varepsilon^3} \qquad [6.14]$$

where: k = the Kozeny constant, an empirical constant = 4.8

ε = the dust cake porosity

$\equiv \dfrac{\text{total void volume of the layer}}{\text{the total volume of the layer}} \equiv \dfrac{\rho_p - \rho_R}{\rho_p}$

ρ_p = true density of a single particle

ρ_R = density of the dust layer (bulk density)

Equation [6.14] is a semi-empirical relationship based on modeling a porous media as a collection of parallel capillaries (Figure 6.8).

Example 6.5 Theoretical Prediction of K_2

What is the value of K_2 in units of in. $H_2O/(lb/ft^2)(ft/min)$ with the following data:

k = 5, μ_g = 1.18 • 10^{-4} g/cm sec, ε = 0.5, ρ_p = 2 g/cm^3, d_p = 8 μm

Solution:

Using Equation [6.14]:

$$R = \frac{2k(1-\varepsilon)}{\varepsilon^3}$$

$$R = \frac{2 \cdot 5 \cdot (1-0.5)}{0.5^3} = 40$$

Using Equation [6.13]:

$$K_2 = K_{2(Stokes)} R = \frac{18\mu}{\rho_p d^2} \cdot R$$

$$K_2 = \frac{18 \cdot (1.81 \cdot 10^{-4} \, g/cm \, sec)}{(2.0 \, g/cm^3)(8 \cdot 10^{-4} \, cm)^2} \cdot 40$$

$$= 101,812.5 \, sec^{-1}$$

$1 \, sec^{-1} \approx 1 \cdot 10^{-4}$ in. $H_2O/(lb/ft^2)(ft/min)$

$K_2 = 10.18$ in. $H_2O/(lb/ft^2)(ft/min)$

This value of K_2 is comparable to the value determined in Example 6.4 for an 8 μm diameter particle when the face velocity was 2 ft/min.

Figure 6.8
Resistance Factor Model

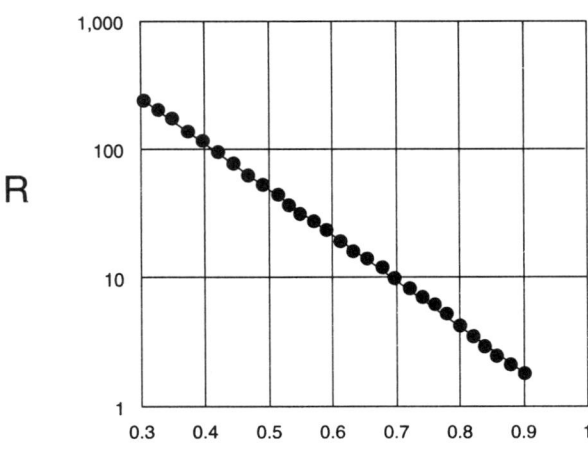

6.8 Pressure Drop in Multi-Compartment Baghouses

In multi-compartment baghouses where the various compartments are cleaned one at a time, the performance curve takes a different shape than for single compartment baghouses (see the insert in Figure 6.9). In this case, the change in the curve is less pronounced than for a single compartment. The performance curve has a sawtooth shape for the net pressure drop across the entire baghouse. Each of the minima points on the curve represent the cleaning of an individual compartment. For optimum filtration rate and collection efficiency, the baghouse should be designed to operate at a pressure drop that approaches a constant value. This can best be accomplished by dividing the baghouse into compartments and cleaning each compartment at a different time. The more compartments that are employed, the smaller the pressure change will be when a compartment is taken off line and cleaned.

Equations [6.5] through [6.7] describe the instantaneous pressure drop after some definite period of filtration through an area of fabric. When the area is distributed across several compartments operated in parallel, the compartments are generally cleaned sequentially. The total pressure drop across the baghouse, in terms of drag, is analogous to a set of electrical resistances in parallel (Robinson, Harrington, and Spaite 1967).

$$\frac{n}{S_e} = \frac{1}{S_1} + \frac{1}{S_2} + \ldots + \frac{1}{S_n} \qquad [6.15]$$

where: S_e = total (multicompartment) baghouse drag at any time
$= \Delta p / v_{avg}$

v_{avg} = average gas velocity

$= \frac{Q}{A} = \frac{\text{total volume flow rate of gas}}{\text{total cloth area}}$

n = total number of filter compartments

$S_1, S_2, \text{--}$ = individual component drag

e = multi-compartment

If one knows the drag in each compartment at a given time, the drag can be calculated for the entire system.

The terminal and residual drag in the baghouse are usually described by S_{eT} and S_{eR} and for each compartment by S_T and S_R.

where: $S_{eT} = \dfrac{\Delta P_{MAX}}{V_{avg}}$

S = drag

T = terminal

R = residual

The pressure drop for a baghouse with multiple compartments can be calculated if the relationship between S_T/S_R and S_{eT}/S_R for each value of n (the number of compartments) is approximated by a straight line (Figure 6.9). The slope of each line can then be calculated (designated as m below) and an approximate relation between S_{eT}/S_R and S_T/S_R developed so that:

$$S_{eT} = S_R + m\Delta S \qquad [6.16]$$

where: S_{eT} = total (maximum) drag (multicompartment) immediately before a compartment is taken off line for cleaning

S_R = (single compartment) drag immediately after a clean compartment is brought on line

$$\Delta S = \frac{K_2 C_i v_f^2 t}{v_f}$$

$$S_{eT} = S_R + mK_2 C_i v_f t = S_R + mK_2 W \qquad [6.17]$$

The m values for different values of n are listed in Table 6.5 below.

Table 6.5
Values for m as a Function of the Number of Compartments (n) in a Baghouse

n	m	n	m
1	1.0	11	0.57
2	0.78	12	0.56
3	0.72	13	–
4	0.68	14	0.55
5	0.65	15	–
6	0.63	16	–
7	0.61	17	0.54
8	0.6	18	–
9	0.59	19	–
10	0.58	20	0.53

Chapter 6: Fabric Filters 183

Some confusion is possible regarding the (weight of dust)/(square foot of fabric) that is to be used in Equation [6.17]. This can be calculated as follows:

$$W = C_i v_{f(avg)} t \qquad [6.18]$$

where: t = total filtration time, i.e., the interval between time a clean compartment is put on stream and time taken off stream

$v_{f(avg)}$ = Q/A for the total baghouse

To complete this discussion, it is necessary to develop equations that will allow for a field determination of residual (maximum) velocity (v_{fR}) in the baghouse. The maximum velocity, v_{fR}, can be defined as that velocity which exists in a cleaned compartment immediately after it has been put into service. As such, it may be several times greater than $v_{f(avg)}$ and, therefore, may be important in determining filter efficiency, dust permeability, and filter blinding.

$$v_{fR} = \frac{S_{eR}}{S_R} v_{f(avg)} \qquad [6.19]$$

S_{eR}/S_R can be determined from Figure 6.9, where v_f average is equal to Q/A for the baghouse.

Example 6.6 Baghouse Design

Design an 8-compartment baghouse that does not exceed an 8-inch pressure drop for a dust with:

$$MMD = 5 \ \mu m$$
$$C_i = 4 \ \text{grains/scf}$$
$$Q = 100{,}000 \ \text{cfm}$$
$$T = 100°C$$
$$\rho_p = 1 \ \text{g/cm}^3$$
$$v_{f(avg)} = 2 \ \text{fpm}$$

What is the maximum face velocity and cleaning cycle time? Assume residual drag $S_R = 2$ in. H_2O/fpm.

Figure 6.9
Basic Design Curves for Multi-Compartment Filter Units

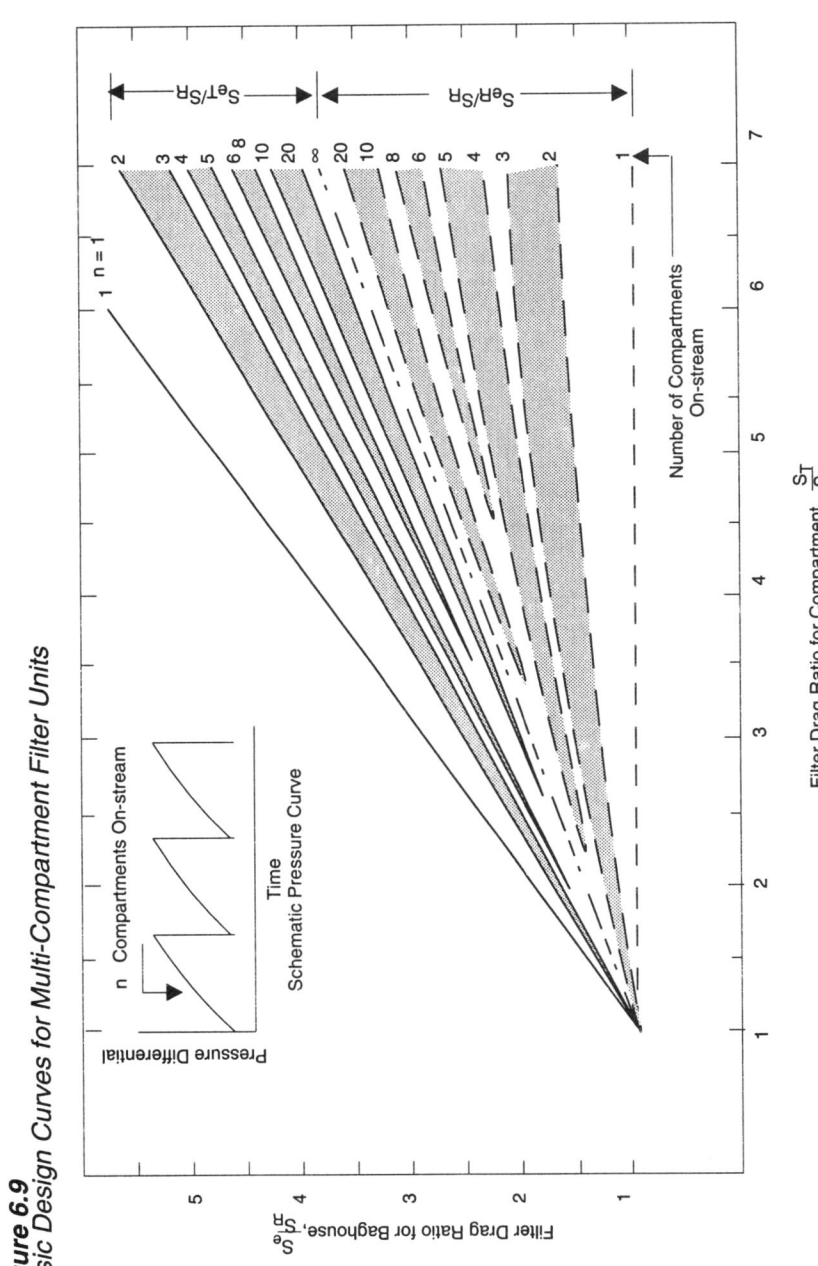

Source: Robinson, Harrington, and Spaite 1967

Chapter 6: Fabric Filters

Solution:

1. Calculate K_2:

    Using Equation [6.9] $K_{2\ (v=3\ fpm)} = 118.4$ MMD $^{-1.1}$

    $= 118.4 \cdot 5^{-1.1}$

    $= 20.16$ in. $H_2O/(lb/ft^2 \cdot ft/min)$

    Correct to 2 fpm using Equation [6.8]:

    $K_2 = k\ v^x$ with $x = 0.6$

    $$\frac{K_{2(v=2\ fpm)}}{K_{2(v=3\ fpm)}} = \left(\frac{2}{3}\right)^{0.6}$$

    $K_{2(v=2\ fpm)} = 0.748 \cdot 20.16 = 15.8$ in. $H_2O/(lb/ft^2 \cdot ft/min)$

2. Determine maximum face velocity in compartment immediately after cleaning:

    $v_{f(avg)} = 2$ fpm, $\Delta P_{max} = 8$ in. H_2O, $S_R = 2$ in. H_2O/fpm

    $S_{eT} \cdot v_{f(avg)} = \Delta P_{max}$

    $S_{eT} \cdot 2$ fpm $= 8$ in. H_2O

    $S_{eT} = 44$ in. H_2O/fpm

    $S_{eT}/S_R = 4/2 = 2$

    From Figure 6.9 at $S_{eT}/S_R = 2 \Rightarrow S_T/S_R = 2.8$, $S_{eR}/S_R = 1.8$

    (Note: The top portion of Figure 6.9 provides a group of S_{eT}/S_R values on the y axis as a function of S_T/S_R and n. The bottom portion of the figure provides a graph of S_{eR}/S_R values on the y axis as a function of S_T/S_R and n.)

    $S_{eR} = 1.8 \cdot S_R = 1.8 \cdot 2 = 3.6$ in. H_2O/fpm

    $v_{f(max)} = (S_{eR}/S_R) \cdot v_{f(avg)} = 1.8 \cdot 2 = 3.6$ fpm

    $v_{f(max)} = 3.6$ fpm

3. Determine cleaning cycle time:

    $S_{eT} = S_R + mK_2C_pv_ft = S_R + m\Delta S$

    From Table 6.5, for 8-compartment baghouse (n = 8), m = 0.6 so that:

$$\Delta S = \frac{S_{eT} - S_R}{m} = \frac{4 \text{ in. } H_2O/fpm - 2 \text{ in. } H_2O/fpm}{0.6}$$

$\Delta S = 3.33$ in. H_2O/fpm

$\Delta S = K_2 C_i v^f t$

$$3.33 \frac{\text{in. } H_2O}{fpm} = 15.8 \frac{\text{in. } H_2O}{(lb/ft^2)(fpm)} \cdot 4 \frac{\text{grains}}{ft^3} \cdot \frac{1 \text{ lb}}{7,000 \text{ grains}} \cdot 2 \frac{ft}{min} \cdot t$$

$t = 184$ min ≈ 3 hr

4. Calculate number of bags per compartment:

   If 8-in. diameter bags that are 12 ft long are used, the area per bag = 25.13 ft²

$$\text{The } \frac{\text{\# bags}}{\text{compartment}} = \frac{Q}{v_{avg} \cdot \text{area of bag} \cdot \text{\# of compartments}}$$

$$= \frac{100,000 \text{ cfm}}{2 \text{ fpm} \cdot 25.13 \text{ ft}^2 \cdot 8}$$

$$\approx 250 \text{ bags/compartment}$$

Design note:

1. *Because the cycle time is 184 minutes, 23 minutes can be allowed to clean each compartment (184/8 ≈ 23). This would provide for 2 mechanical shaking periods of 4 minutes each with approximately 8 minutes to allow the dust to settle into the hopper after each shaking interval and before returning the clean compartment to service.*

2. *With one compartment continuously off line for cleaning, an extra compartment will be required to maintain the average face velocity at 2 fpm.*

6.9 Collection Efficiencies for Fibrous Fabrics

6.9.1 Single Fiber Collection Efficiency

The collection of particles by fibers in a filter is essentially the capture of particles by cylindrical collecting bodies. The gas stream passing through the filter takes the particles close to these collecting bodies and then a number of mechanisms accomplish the actual collection. In any particular case, the relative importance of these mechanisms varies with the relative size and velocity of the particles and

Chapter 6: Fabric Filters

with the collecting body. The basic mechanisms are: inertial impaction, interception, and diffusion (see Chapter 3).

Mathematical models have been developed for each individual mechanism. In many cases, one or two mechanisms predominate the collection process. Thus, for particles in the micron size range and larger, inertial impaction and interception predominates, while diffusion is of much greater importance for sub-micron sized particles.

The collection efficiency due only to interception and diffusion is usually applied to fibrous filters and they are combined together as follows (Figure 3.13):

$$\eta_{DC} = 1 - (1 - \eta_D)(1 - \eta_C) \quad [6.20]$$

where: η_D = Collection efficiency by diffusion

η_C = Collection efficiency by interception

Friedlander (1977) developed the following relationship based on Equation [6.20] which is generally applied to the design of filters.

$$\eta_{DC} = 6 Sc^{-\frac{2}{3}} Re^{-\frac{1}{2}} + 3R^2 Re^{\frac{1}{2}} \quad [6.21]$$

where:

$$Sc = \frac{\mu}{\rho \mathcal{D}} \text{ with } \mathcal{D} = \frac{kT}{3\pi\mu d_p} \quad [6.22a]$$

$$Re = \frac{\rho v D_f}{\mu} \quad [6.22b]$$

$$R = \frac{d_p}{D_f} \quad [6.22c]$$

The following define the terms used in the above equations:

μ = viscosity of fluid (air)

ρ = density of fluid (air)

$\mathcal{D}$ = diffusion coefficient

k = Boltzman's constant = $1.4 \cdot 10^{-16}$ g cm^2/sec^2

T = temperature

d_p = particle diameter

v = gas velocity in filter

D_f = collecting body diameter

Data for Equation [6.21] are shown in Figure 6.10.

Equation [6.21] can be rewritten in a more convenient form as:

$$\eta_{DC} = \frac{3D_f^2 v^{1/2}}{\upsilon^{1/2} d_p^{3/2}} + \frac{6\mathcal{D}^{2/3}}{\upsilon^{1/6} d_p^{1/2} v^{1/2}} \quad [6.23]$$

where: υ = kinematic viscosity

The first term on the right is for interception, whereas the second term is for collection by diffusional collection efficiency.

Figure 6.10
Combined Collection Efficiency Based on Experimental Results

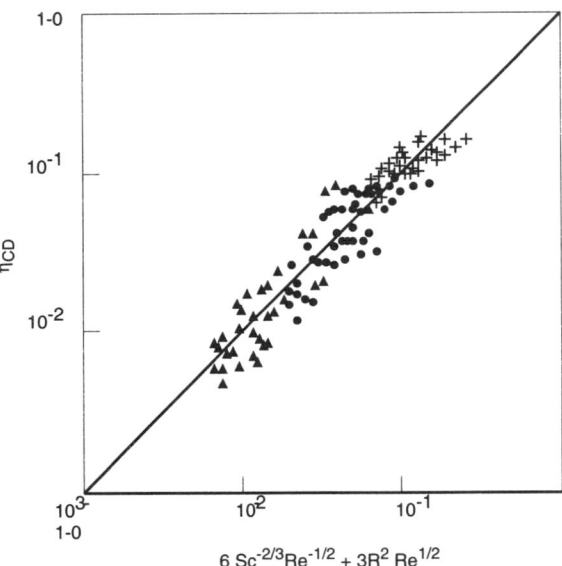

Source: Friedland 1975

6.9.2 Total Collecting Efficiency for Fabric Filters (Multiple Collecting Fibers)

The extension of single body collection efficiencies to realistic multi-body systems is discussed here. In fabric filters, the particles encounter a number of collectors during their passage through the fabric. When the collection efficiency of a single fiber has been found, the overall efficiency, η_T, can be found (Strauss 1975).

$$\eta_T = 1 - (1 - \eta_{DC})^n \quad [6.24]$$

where: n = number of fibers encountered by the particles.

Chapter 6: Fabric Filters

In most practical cases, n is large (greater than 25), and Equation [6.24] can be modified to:

$$\eta_T = 1 - e^{-n\eta_{DC}} \qquad [6.25]$$

This equation assumes that the filter consists of a number of identical cylindrical collectors, evenly spaced, with no inter-fiber interference, and at right angles to the air flow. Actually, fibrous filters consists of randomly oriented fibers with a range of diameters. It may be assumed, for fabric filters of the type used in bag filters and other industrial filters with low resistance to the gas flow, the fibers are relatively far apart and staggered in relation to one another.

The number of fiber elements in fabric filter, n, is usually calculated from the fabric weight in ounces/square yard and the fiber size used to construct the fabric. If the average diameter of the fiber is D_f, the total length of fibers is L, and the single fiber efficiency within the bed is η_{DC}, then the number of particles removed by the filter in unit time is

$$\eta_T = 1 - P_T = \frac{N}{N_o} = 1 - \exp\left\{-\frac{H}{1-\alpha} D_f L \eta_{DC}\right\} = 1 - e^{-n\eta_{DC}} \qquad [6.26]$$

where:
η_T = total fabric efficiency
P_T = total penetration
N_o = initial concentration
N = final concentration
H = bed depth
L = fiber length
D_f = fiber diameter
α = packing density of the fibers
η_{DC} = single fiber collection efficiency
n = number of fibers

The packing density is the actual volume of fiber in a unit volume of filter:

$$\alpha = \pi D_f^2 L / 4 \qquad [6.27]$$

Substituting for L from Equation [6.27] into Equation [6.26] gives

$$\eta_T = \frac{N}{N_o} = 1 - \exp\left\{-\frac{4H}{\pi D_f}\left(\frac{\alpha}{1-\alpha}\right)\eta_{DC}\right\} \qquad [6.28]$$

Combining Equations [6.28] and [6.24], it can be seen that n is defined by:

$$n = \frac{4H}{\pi D_f}\left(\frac{\alpha}{1-\alpha}\right) = S_f \qquad [6.29]$$

where: S_f = alternate symbol for n referred to as the solidarity factor

An alternative expression for the packing density is:

$$\alpha = \frac{\text{volume of fibers}}{\text{volume of filter}} = \frac{W_f}{\rho_f H} \qquad [6.30]$$

where: W_f = weight of fabric per area of fabric (areal density)

ρ_f = density of the fibers

H = thickness of the fabric

Most fabric filters utilize woven fabrics that range from 5 to 18 ounces/square yard. With substitution of Equation [6.30] into [6.29], the solidarity factor can be written as:

$$S_f = n = \frac{4W_f}{\pi \rho_f D_f (1-\alpha)} \qquad [6.31]$$

For fibrous filters, α is on the order of 0.1 or less, representing a relatively porous open packing of fibers that offer small aerodynamic resistance to air flow. Woven filters, on the other hand, usually have an α value of 0.2 to 0.3 and must generally operate at much lower face velocity. Because of the significance difference in α, the particle collection process differs between fabric and fibrous filters. The high α of fabric filters means that most of the incoming particles are collected on, or in the vicinity of, the fabric surface where they form a particle layer (cake). This cake itself acts as a filtration media for subsequently arriving particles.

Low α fibrous filters, on the other hand, rely more on fiber bed depth to achieve high efficiency. Low α filters do not depend upon cake formation for high efficiency operation and once the early stages of cake formation are reached, the aerodynamic resistance of the media is so high that filter cleaning is required.

When fibers are close together in a filter bed, there will be a change in flow pattern around a fiber because of the neighboring fibers. This effect increases the collection efficiency. The effect of the fiber on collection efficiency for values of α less than 0.1 has been experimentally determined (Strauss 1975) as:

$$\eta_\alpha = \eta_T(1 + 4.5\alpha) \qquad [6.32]$$

Table 6.6
Fabric Specifications for Typical Fabrics Used in Industrial Baghouses

Fabric Code	Fabric Characteristics	W_f, Weight oz/yd^2	D_f, Fiber Diameter, μ	α, Packing Density	H Thickness, μ
A	Fiberglass Filament Bulk Warp	14.5	8	0.296	635
B	Polyester All-Spun	12.7	12.5	0.344	2,092
C	Nomex-Needled Felt	8.67	15	0.083	2,553
D	Polyester Filament (knitted)	8.1	15	0.355	559
E	Polyester Filament Warp-Spun	6.7	25	0.393	381
F	Polyester Filament	5.4	25	0.58	228

Example 6.7 Fabric Collection Efficiency

Calculate the collection efficiency for the fabrics in Table 6.6 that are proposed to be used in a new baghouse collecting 0.1 and 1.0 diameter particles with a face velocity of 2 fpm.

Use Equation [6.28]

$$\eta_T = \frac{N}{N_o} = 1 - \exp\left\{-\frac{4H}{\pi D_f}\left(\frac{\alpha}{1-\alpha}\right)\eta_{DC}\right\}$$

$$\eta_{DC} = 6Sc^{-2/3}Re^{-1/2} + 3R^2Re^{1/2}$$

μ = viscosity of air = $1.8 \cdot 10^{-4}$ g/cm sec

ρ = density of air = $1 \cdot 10^{-3}$ g/cm^3

v = gas velocity = 2 fpm = 1.02 cm/sec

d_p = particle diameter 0.1 μm and 1 μm ($1 \cdot 10^{-5}$ and $1 \cdot 10^{-4}$ cm)

D_f = fiber diameter, cm

T = temperature = 293°K

k = Boltzman's constant = $1.4 \bullet 10^{-16}$ g cm²/sec²°K

$$Re = \frac{\rho v D_f}{m} = 1 \bullet 10^{-3} \bullet 1.02 \bullet D_f / (1.8 \bullet 10^{-4}) = 5.76 D_f$$

$$\mathcal{D} = \frac{kT}{3\pi\mu d_p} = \frac{1.4 \bullet 10^{-16} \bullet 293}{3\pi 1.8 \bullet 10^{-4} d_p} = 2.417 \bullet 10^{-11} \bullet d_p^{-1}$$

$$R = d_p/D_f = d_p \bullet D_f^{-1}$$

$$Sc = \frac{\mu}{\rho\mathcal{D}} = \frac{1.8 \bullet 10^{-4}}{1.0 \bullet 10^{-3} \bullet 2.417 \bullet 10^{-11} \bullet d_p^{-1}} = 7.45 \bullet 10^9 \bullet d_p$$

$$\eta_{DC} = 6(7.45 \bullet 10^9 d_p)^{-\frac{2}{3}} (5.67 D_f)^{-\frac{1}{2}} + 3(d_p \bullet D_f^{-1})^2 (5.67 D_f)^{\frac{1}{2}}$$

For Fabric A and particle diameter, $d_p = 0.1$ μm ($1 \bullet 10^{-5}$ cm), and $D_f = 8$ μm ($8 \bullet 10^{-4}$ cm)

$$\eta_{DC} = 6(7.45 \bullet 10^9 \bullet 1 \bullet 10^{-5})^{-\frac{2}{3}} (5.67 \bullet 8 \bullet 10^{-4})^{-\frac{1}{2}} + 3\left(\frac{1 \bullet 10^{-5}}{8 \bullet 10^{-4}}\right)^2 (5.67 \bullet 8 \bullet 10^{-4})^{\frac{1}{2}}$$

$$\eta_{DC} = 0.0502$$

$$\eta_T = 1 - \exp\left\{-\frac{4H}{\pi D_f}\left(\frac{\alpha}{1-\alpha}\right)\eta_{DC}\right\}$$

For Fabric A: H = 635 μm or $6.35 \bullet 10^{-2}$ cm and $\alpha = 0.296$

$$\eta_T = 1 - \exp\left\{-\frac{4 \bullet 6.35 \bullet 10^{-2}}{\pi \bullet 8 \bullet 10^{-4}}\left(\frac{0.296}{1-0.296}\right) 0.0502\right\}$$

$$\eta_T = 1 - e^{-2.133} = 0.8815$$

Table 6.7 provides the total collection efficiency for the fabrics described in Table 6.6.

Fabric A has a high collection efficiency because it has the smallest fiber diameter (8 μm) and the largest weight (14.5 oz/yd²). Fabric B has a high collection efficiency because it is thick (2,092 μm) and also has a large weight (12.7 oz/yd²). Fabric F has a low collection efficiency because it has a small weight, large fiber diameter, and is thin.

Table 6.7 demonstrates that when selecting a fabric for high particle collection, thicker fabrics with smaller diameter fibers usually have a higher collection efficiency. These types of fabrics generally have a larger weight (oz/yd²).

Chapter 6: Fabric Filters

Table 6.7

Particle Diameter	Fabric Code	Packing density, α	Thickness μm	Fiber dia. μm	Sc	Re	R	η_{CD}	S_f	η_T
0.1 μm	A	0.296	635	8	74886.22	0.0045	0.013	0.0502	42.514	0.882
	B	0.344	2092	12.5	74886.22	0.0071	0.008	0.0401	111.799	0.989
	C	0.083	2553	15	74886.22	0.0085	0.007	0.0366	19.625	0.513
	D	0.355	559	15	74886.22	0.0085	0.007	0.0366	26.129	0.616
	E	0.393	381	25	74886.22	0.0142	0.004	0.0284	12.570	0.300
	F	0.58	228	25	74886.22	0.0142	0.004	0.0284	16.044	0.366
1.0 μm	A	0.296	635	8	748862.2	0.0045	0.125	0.0140	42.514	0.448
	B	0.344	2092	12.5	748862.2	0.0071	0.080	0.0103	111.799	0.682
	C	0.083	2553	15	748862.2	0.0085	0.067	0.0091	19.625	0.164
	D	0.355	559	15	748862.2	0.0085	0.067	0.0091	26.129	0.212
	E	0.393	381	25	748862.2	0.0142	0.040	0.0067	12.570	0.081
	F	0.58	228	25	748862.2	0.0142	0.040	0.0067	16.044	0.102

Example 6.8 Minimum Collection Efficiency Particle Size

Develop grade efficiency curves for Fabrics A and E described in Tables 6.6 and 6.7 at 70°F. Calculate the particle size that has the minimum collection efficiency for the two fabrics.

Solution:

1. Grade efficiency curve development:

Grade efficiency curves for Fabrics A and E in Table 6.6 can be determined by applying Equation [6.26] as was done in Example 6.7 for a range of particle sizes. The results of this calculation are provided in Table 6.8 and are also plotted in Figure 6.11. Figure 6.11 shows that there is a definite minimum collection efficiency near 1.0 μm diameter for both fabrics.

2. Calculation of particle size for minimum collection efficiency:

The collection efficiency for particles being removed by filtration is given by Equation [6.21].

$$\eta_{DC} = 6Sc^{-\frac{2}{3}}Re^{-\frac{1}{2}} + 3R^2Re^{\frac{1}{2}}$$

Table 6.8
Grade Efficiency for Fabrics A and E

Code	d_p, μm	α	H	D_f, μm	Sc	Re	R	η_{CD}	S_f	η_T
A	0.01	0.296	635	8	7,488.622	0.0045	0.001	0.2328	42.5141	1.000
	0.1	0.296	635	8	74,886.22	0.0045	0.013	0.0502	42.5141	0.882
	0.5	0.296	635	8	374,431.1	0.0045	0.063	0.0179	42.5141	0.534
	1	0.296	635	8	748,862.2	0.0045	0.125	0.0140	42.5141	0.448
	2	0.296	635	8	1,497,724	0.0045	0.250	0.0194	42.5141	0.562
	5	0.296	635	8	3,744,311	0.0045	0.625	0.0826	42.5141	0.970
	10	0.296	635	8	7,488,622	0.0045	1.250	0.3179	42.5141	1.000
	100	0.296	635	8	74,886,223	0.0045	12.500	31.5615	42.5141	1.000
E	0.01	0.393	381	25	7,488.622	0.0142	0.000	0.1317	12.5695	0.809
	0.1	0.393	381	25	74,886.22	0.0142	0.004	0.0284	12.5695	0.300
	0.5	0.393	381	25	374,431.1	0.0142	0.020	0.0098	12.5695	0.116
	1	0.393	381	25	748,862.2	0.0142	0.040	0.0067	12.5695	0.081
	2	0.393	381	25	1,497,724	0.0142	0.080	0.0061	12.5695	0.074
	5	0.393	381	25	3,744,311	0.0142	0.200	0.0164	12.5695	0.186
	10	0.393	381	25	7,488,622	0.0142	0.400	0.0584	12.5695	0.520
	100	0.393	381	25	74,886,223	0.0142	4.000	5.7134	12.5695	1.000

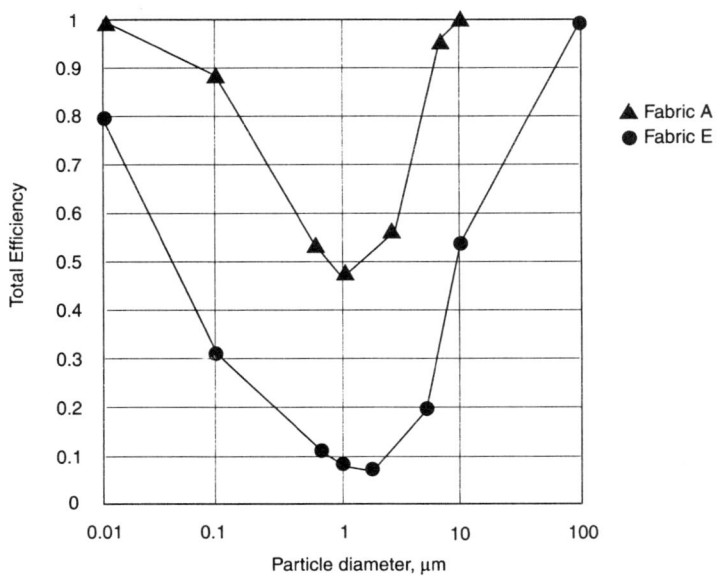

Figure 6.11
Grade Efficiency Curves for Fabrics A & E

Chapter 6: Fabric Filters

The particle size that has the minimum collection efficiency can be determined by taking the first derivative of N_{CD} and setting it equal to zero. This provides the following diameter for minimum collection efficiency.

$$d_{p \text{ (minimum efficiency)}}^{8/3} = \left(\frac{2(3\pi\mu^2/\rho kT)^{-2/3}\mu}{3\rho}\right)\frac{D_f}{v} \quad [6.33]$$

At 70°F:

μ = 1.8 • 10^{-4} g/cm sec, r = 1 • 10^{-3} g/cm^3
v = 2 fpm = 1 cm/sec
T = 70°F = 294°K
k = Boltzman's constant = 1.4 • 10^{-16} g cm^2/sec^2 °K
ρ = 1 • 10^{-3} g/cm^3

For Fabric A: D = 8 μm (8 • 10^{-4} cm)

For Fabric E: D = 25 μm (2.5 • 10^{-3} cm)

Fabric A

$$d_A^{8/3} = \left(2 \bullet \left(\frac{3 \bullet \pi \bullet (1.8 \bullet 10^{-4})^2}{(1 \bullet 10^{-3}) \bullet (1.4 \bullet 10^{-16}) \bullet (294)}\right)^{-2/3} \bullet \frac{1.8 \bullet 10^{-4}}{3 \bullet 1 \bullet 10^{-3}}\right)\frac{8 \bullet 10^{-4}}{1}$$

$$= \left(3.154 \bullet 10^{-8} \text{cm}^{8/3}/\sec\right)\frac{8 \bullet 10^{-4}}{1}$$

$d_A^{8/3} = 2.523 \bullet 10^{-11}$

$d_A = 1.06 \bullet 10^{-4}$ cm

$d_A = 1.06$ μm

Fabric E

$$d_E^{8/3} = \left(2 \bullet \left(\frac{3 \bullet \pi \bullet (1.8 \bullet 10^{-4})^2}{(1 \bullet 10^{-3}) \bullet (1.4 \bullet 10^{-16}) \bullet (294)}\right)^{-2/3} \bullet \frac{1.8 \bullet 10^{-4}}{3 \bullet 1 \bullet 10^{-3}}\right)\frac{2.5 \bullet 10^{-3}}{1}$$

$$= \left(3.154 \bullet 10^{-8} \text{cm}^{8/3}/\sec\right)\frac{2.5 \bullet 10^{-3}}{1}$$

$d_E^{8/3} = 7.885 \bullet 10^{-11}$

$d_E = 1.63 \bullet 10^{-4}$ cm

$d_E = 1.63$ μm

Example 6.8 demonstrates that the minimum collection efficiency shown in Figure 6.11 can be calculated using Equation [6.33]. This is important because the particle that is collected with minimum efficiency in fabric filters needs to be considered in the design of industrial filters.

Operating fabric filters have grade efficiency curves similar to the ones developed from theory in Figure 6.11, but also include the effect of filtration by particles collected in the filter and on its surface. Figure 6.12 shows data on fractional efficiency versus particle size for a typical baghouse using a woven fabric filtering a test dust (Parker 1977). The clean curve in the figure is for a new fabric never loaded with dust. After 10 dust collection cycles, the fabric has retained a permanent dust mat, even after cleaning. The dust mat collects particles and therefore the efficiency of collection has improved. The fully loaded fabric has additional dust cake buildup and the efficiency can be seen to further increase due to filtration by the particles in the dust cake. The figure also shows that there is a particle size (usually near 0.5 μm) that is collected with minimum efficiency. Particles larger than 1.0 μm are collected with very high efficiencies. Theoretical models for describing the filtration efficiency under these conditions are not available. Once particles are deposited, collection by the particles becomes important and collection efficiency increases with cake buildup and high overall collection efficiencies (i.e., > 99%) are assumed.

Figure 6.12
Fractional Efficiency Removal vs. Particle Size

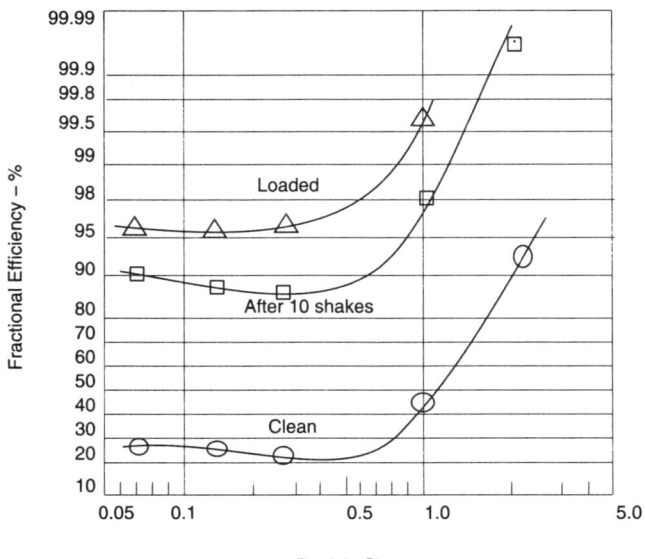

6.10 Fabric Filter Design Review

The principal design parameter for fabric filters is the air-to-cloth ratio (A/C). The air-to-cloth ratio depends on a number of variables. Before it can be determined, the fabric construction (woven or felted, including filter surface treatment and special coatings) and cleaning method (shaker, reverse air, or pulse-jet) must be selected. The type of industrial process usually dictates the type of baghouse selected. Important emission variables that need to be considered are gas temperature and moisture as well as the type of dust, including its concentration and filter cake properties. After a baghouse configuration has been selected, the type of dust and its size distribution will control the actual A/C ratio selected with reverse air the lowest, shaker next, and pulse-jet the highest. Too high of an A/C ratio results in high pressure loss, increased bag wear, blinding of filter material, and reduced collection efficiency. Lower A/C ratios result in larger baghouses and increased capital costs, but lower operating costs. There is no precise analytical method for determining the best A/C ratio. It is customary to select ratios based on similar previous experience (Tables 6.3 and 6.4). For shaker baghouses, the cleaning cycle will be set by the maximum pressure drop allowed with extra compartments provided for off-line cleaning. Collection efficiency will be high for well-designed shaker baghouses because the dust cake that is formed during the typical 30-minute to 2- hour cleaning cycle will provide efficient filtration of the emissions. For pulse-jet baghouses where the fabric controls the efficiency of filtration, important design variables include filter thickness and fiber size. These variables control the individual fiber collection efficiency and the total emissions.

6.11 Problem Set

6.1 A plant has an inlet loading into a baghouse of 10 grains/ft^3. The average filtration velocity is 10 ft/min and the gas flow rate is 25,000 acfm. What is the air-to-cloth ratio of the system?

6.2 If a plant has a volumetric flow rate of 18,000 acfm and a dust loading of 2 lb/ft^3 of gas filtered, how much filtering area would be required if the filtration velocity is 2.5 ft/min?

6.3 How many cylindrical bags, 6 inches in diameter and 25 feet long, would be needed to filter a particulate-laden gas stream; the total filtering surface area is 4,045 square feet.

6.4 Calculate K_2 for the following conditions:

    a) d_p = 10 μm, spherical particles, inlet dust concentration of 2 grains/ft^3, air-to-cloth ratio of 10 ft/min, Δp_t = 6 in. H$_2$O, t = 30 min

    b) d_p = 10 μm, spherical particles, ε = 0.8, c_i = 10 grains/ft^3, v = 3 ft/min, t = 100 min

c) Calculate the dust layer thickness in μm for the conditions given above.

6.5 It is proposed to install a pulse-jet fabric filter system to clean a 10,000 scfm air stream at 250°F, containing 4 grains/ft³ of pollutant. For a 99% efficiency, the average air-to-cloth ratio is 2.5 cfm/ft² cloth. The following information, given by filter bag manufacturers, is available at the beginning of the selection process:

Filter bag	A	B	C	D
Tensile strength	Excellent	Above average	Fair	Excellent
Recommended maximum operation temperature, °F	260	275	260	220
Resistance factor	0.9	1.0	0.5	0.9
Relative cost per bag	2.6	3.8	1.0	2.0
Standard size	8" x 16'	10" x 16'	1" x 16'	1" x 16'

a) Determine the filtering area required for this operation.

b) Based on the required area and the above information, select the most suitable filter bag and calculate the number of them that should be used. The proposal of a pulse-jet device using strong forces to clean the bags requires the selection of a fabric with at least above average tensile strength.

6.6 Given an operating baghouse with one compartment and the following conditions, what is the cleaning cycle for a total allowable pressure drop of 6 in. H_2O?

The measured minimum pressure drop is 2 in. H_2O, the face velocity = 2 ft/min, the particle mean size = 1 μm, the gas temperature = 250°F, and the dust loading = 1 grain/ft³.

6.7 A 10-compartment baghouse operates under the following conditions:

$v = 3.0$ fpm with 10 compartments on line

$S_R = 0.2$ in. H_2O/fpm

$S_T = 2.0$ in. H_2O/fpm

What is the maximum velocity and pressure drop range for the baghouse?

6.8 Estimate the combined collection efficiency η_{DC} of a single (cylindrical) fiber 1 μm in diameter when dusty air containing 0.1 μm dust particles is flowing past the fiber (perpendicular to the fiber axis) at a velocity of 10 cm/sec. Assume ambient conditions. Also, calculate the inertia parameter to verify that under these conditions, collection by an inertial impaction mechanism is negligible. Assume a particle density of $3 \cdot 10^3$ kg/m^3.

6.9 Consider the flow through a fiber filter. The average diameter of the cylindrical fibers is 100 µm. Other data are:

ρ_p = 2.58 g/cm^3

ρ = $1.2 \cdot 10^{-3}$ g/cm^3

μ = $1.8 \cdot 10^{-4}$ g/cm

The particle size distribution shows that the particle diameter ranges from 0.001 µm to 1.0 µm. The velocity in the filter varies from 200 to 1,000 cm/sec.

Calculate the target collection efficiency by interception as a function of the particle diameter for velocities of 200 and 1,000 cm/sec.

6.10 Determine the relative importance of single fiber efficiency due to interception, impaction, and diffusion for collecting unit-density particles of 0.5 µm diameter. The filter has an effective fiber diameter of 1.0 µm and a solidity of 0.02. v = 50 cm/sec.

6.11 A fibrous mat filter is made of fibers 10 µm in diameter and is to be used to filter 0.5 µm aerosol particles (density = 2.8 g/cm^3) from air at ambient conditions (assume spherical particles). A face velocity of 1 m/sec is contemplated. Estimate the individual fiber collection efficiency.

6.12 A filter having a fiber diameter of 10 µm and a solidity of 1% is used to sample 0.5 µm diameter unit-density particles. What is the face velocity when this is the minimum efficiency particle size?

6.13 What face velocity is required to achieve a 50% collection efficiency of 0.2 µm aerodynamic diameter particles by impaction? Assume the fiber diameter is 4 µm, fiber solidity is 0.05, and filter thickness is 2 mm.

6.14 What is the particle size which would be collected with minimum efficiency and what is the collection efficiency for a single element for the following conditions: collector = 10 µm diameter and velocity = 2 ft/min?

6.15 A filter is to be made from fibers with effective diameters of 10 µm and a density of 2.35 g/cm^3 with a packed porosity of 95%. The filter must collect 0.5 µm particles with 90% efficiency. The air stream velocity is 10 ft/min. How thick must the filter be? What changes in operating conditions would allow a thinner fiber?

6.12 References

Buonicore, A. J. and W. Davis (Eds.). *Air pollution Engineering Manual.* Air and Waste Management Assoc., Van Nostrand Reinhold. New York. 1992.

Danielson, J. *Air Pollution Engineering Manual*, 2nd ed., AP40, US EPA, Research Triangle Park, NC. 1973.

Davis, W. and W. Frazer. " A Laboratory Comparison of the Filtration of Eleven Different Fabric Filter Materials Filtering Resuspended Fly Ash," *Proceedings of the 75th APCA meeting*, Pittsburgh, PA. 1982.

Donovan, R. *Fabric Filtration For Combustion Sources.* Marcel Dekker, Inc. New York. 1985.

Friedlander, S. *Smoke, Dust, and Haze.* John Wiley & Sons. New York. 1977.

Joseph, G. and D. Beachler. *Control of Gaseous Emissions.* Northrop Services, Inc. EPA 450/2-81-005. Research Triangle Park, NC. 1981.

Parker, H.W. *Air Pollution*, Prentice Hall. Englewood Cliffs, NJ. 1977.

Robinson, J., R. Harrington, and P. Spaite. "A New Method for Analysis of Multicompartmented Fabric Filtration," *Atm. Envr.*, 1(4). 1967.

Sink, M. *Handbook: Control Technologies for Hazardous Air Pollutants.* Office of Research and Development. US EPA/625/6-91/014. Cincinnati, OH. 1991.

Strauss, W. *Industrial Gas Cleaning*, 2nd. ed. Pergamon Press. New York. 1975.

Chapter 7

Wet Scrubbers

7.1 Introduction

Wet scrubbers bring a contaminated gas stream into intimate contact with a liquid. Generally, in wet scrubbers the particles are collected either by liquid drops or continuum of liquid in the form of wetted walls, liquid sheets, etc. Spray droplet wet scrubbers are one of the most common air pollution control devices because they combine gaseous and particulate control and gas conditioning (both for temperature and moisture). Wet collection of particles has a number of advantages over dry methods, such as reduced dust explosion risk and being able to handle hot gases, sticky particulates, and liquids. It also has a number of disadvantages, largely associated with the problems of effluent liquid disposal and removal or recycling of the contaminants from a liquid dispersion. Applications include use with industrial boilers, lime kilns, foundry cupolas, municipal sludge incinerators, and odor control devices. The principal collection mechanisms in scrubbers are inertial impaction and interception. In addition, diffusion condensation effect and particle entrainment may also play important roles in some wet scrubbers.

There is a large variety of wet scrubbers in use (Table 7.1), the most important of which are spray towers and venturi scrubbers. The primary design parameters for wet scrubbers include: pressure drop, liquid-to-gas flow rate (L/G, gal/1,000 ft^3 of gas), and droplet-size distribution. Often, large scrubbers have a higher particle collection efficiency than small scrubbers with identical pressure drop, L/G ratio,

Table 7.1
Operating Characteristics of Particulate Water Scrubbers

Type of Scrubber	Pressure Drop ΔP	Liquid to Gas Ratio (L/G)	Liquid Inlet Pressure	Cut Diameter [d_{pc}]	Industrial Applications
Baffle spray	1-3 in. H_2O	1 gpm/1,000 acfm	<15 psig	10 μm	Mining Incineration Chemical process
Spray tower	1/2-3 in. H_2O	0.5-20 gpm/1,000 acfm (5 normal; 10 when using high pressure sprays)	10-400 psig	2-8 μm	Mining Chemical process Boilers and incinerators Iron and steel
Cyclonic	1.5-10 in. H_2O	2-10 gpm/1,000 acfm	40-400 psig	2-3 μm	Mining operations Drying operations Food process Foundries Fertilizer Asphalt production Cupolas
Venturi ejector	1/2-5 in. H_2O	50-100 gpm/1,000 acfm	15-120 psig	1 μm	Pulp and paper Chemical process
Venturi	5-100 in. H_2O	3-20 gpm/1,000 acfm	<1-15 psig	0.2 μm	Pulp and Paper Acid plants Mining Dryers Iron and steel Boilers Incinerators

1 liter/m^3 = 7.48 gallons/1,000 ft^3 (~7.5 gallons/1,000 ft^3)

etc. This is related to the longer distance the droplet travels with respect to the carrier gas. Wet scrubbers may be categorized by pressure drop (or energy consumption). Low energy scrubbers are those with a typical pressure drop less than 5 in. of water; medium energy scrubbers are those with a typical pressure drop from 5 to 15 in. of water; and high energy scrubbers are those with a typical pressure drop greater than 15 in. of water. Spray towers, for example, provide the lowest pressure drop and, correspondingly, the lowest collection efficiencies; they would be classified as low energy scrubbers. The medium pressure drop group could include centrifugal fan wet scrubbers, atomizing impingement collectors, and certain packed-bed scrubbers. The most familiar of the high pressure drop (high energy) group is the venturi type collector.

This chapter reviews the general concepts applicable to the design of wet scrubbers. Detail design methods are provided for gravity spray towers and venturi scrubbers. A general design procedure based on the relationship of pressure drop to collection efficiency is also provided. Finally, the design of mist eliminators is considered for the removal of entrained droplets before the carrier gas is released to the atmosphere.

7.2 Water Drop Formation

7.2.1 Preformed Drops (Spray Towers)

A preformed drop (spray) scrubber collects particles on liquid drops that have been atomized by spray nozzles. The properties of the drops are determined by the configuration of the nozzle, the liquid to be atomized, and the pressure at the nozzle. Horizontal and vertical gas flow paths have been used, with the spray introduced countercurrent, or cross-flow to the gas.

Particle collection in these units is a complex function of droplet size, gas velocity, L/G ratio, and droplet trajectories. There is often an optimum droplet diameter that varies with fluid flow parameters. For preformed drops, the particles fall at their terminal settling velocity, the optimum droplet diameter for particle collection is around 100 - 300 µm. Spray scrubbers that take advantage of gravitational settling can achieve a cut diameter around 5.0 µm at moderate L/G ratios. Efficiency improves with higher spray nozzle pressures and L/G ratios.

7.2.2 Gas Atomized Drops (Venturi Scrubbers)

Gas atomized drop (spray) devices use a moving gas stream to first atomize liquid into droplets, and then accelerate the droplets. Typical of these devices is the venturi scrubber. High gas velocities of 200 to 400 ft/sec between the gas and liquid drops promote particle collection. Gas atomized scrubbers have the simplest and smallest configurations of all the scrubbers. While fairly difficult to plug, they are susceptible to erosion because of their high throat velocity. They can be built with adjustable throat openings to permit variation of pressure drop and collection efficiency.

Liquid may be introduced in various places and different ways without having much effect on collection efficiency as long as it results in a uniform spray distribution. Usually it is introduced at the entrance to the throat through several straight pipe nozzles directed radially inward. Venturi scrubbers have atomized drop sizes between 10 to 20 µm and achieve cut diameters in the sub-micron size range.

7.2.3 Determination of the Drop Diameter

The water drop diameter to be used in the design of spray tower and venturi scrubbers is the *Sauter surface mean diameter*. It can be estimated from the empirical relationship developed by Nukiyama and Tanasawa (1983), for both preformed and gas atomized sprays. They studied nozzles ranging from 0.2 to 1 mm in diameter for liquid and 1 to 5 mm for air. Relative velocities ranged from 260 ft/sec to sonic, L/G from 0.85 to 15 gal/1,000 ft^3. The liquids studied were gasoline, water, alcohol, and heavy oils.

The Nukiyama-Tanaswa (empirical) equation is:

$$D_D = \frac{1,920}{v_R}\left(\frac{\sigma_L}{\rho_L}\right)^{0.5} + 597\left(\frac{\mu_L}{\sqrt{\sigma_L \rho_L}}\right)^{0.45}\left(1,000\frac{Q_L}{Q_G}\right)^{1.5} \qquad [7.1]$$

where: D_D = water droplet diameter, μm

v_R = relative velocity of the gas and liquid, ft/sec

σ_L = liquid surface tension, dynes/cm

ρ_L = liquid density, g/cm^3

μ_L = liquid viscosity, poise or g/cm • sec

Q_L = liquid flow rate

Q_G = gas flow rate

In Equation [7.1] Q_L and Q_G must be in the same volume flow rate units.

For air and water, this expression reduces to:

$$D_D = \frac{16,400}{v_R} + 1.45\left(\frac{Q_L}{Q_G}\right)^{1.5} \qquad [7.2]$$

where: Q_L/Q_G = gal/1,000 ft^3 and v_R = ft/sec

Example 7.1 Venturi Scrubber Water Drop Diameter

1. Calculate the drop diameter for a venturi scrubber with a L/G ratio of 8.5 gal/1,000 ft^3 and a gas velocity in the throat of 272 ft/sec.

Solution:

Using Equation [7.1]:

$$D_D = \frac{1,920}{v_R}\left(\frac{\sigma_L}{\rho_L}\right)^{0.5} + 597\left(\frac{\mu_L}{\sqrt{\sigma_L \rho_L}}\right)^{0.45}\left(1,000\frac{Q_L}{Q_G}\right)^{1.5}$$

and $\sigma_L = 72$ dynes/cm, $\mu_L = 0.01$ poise

$$\frac{Q_L}{Q_G} = \frac{8.5 \text{ gal}}{1,000 \text{ ft}^3} \cdot \frac{1 \text{ ft}^3}{7.48 \text{ gal}} = 1.363 \bullet 10^{-3}$$

$$D_D = \frac{1,920}{272}\left(\frac{72}{1}\right)^{0.5} + 597\left(\frac{10^{-2}}{\sqrt{1 \bullet 72}}\right)^{0.45}\left(1,000 \bullet 1.363 \bullet 10^{-3}\right)^{1.5}$$

$$= 59.896 + 45.690 \cong 106 \text{ µm}$$

2. Determine the water drop diameter when the operating parameters are $v_G = 300$ ft/sec, T = 15°C, for two L/G ratios, 5 and 10 gal/1,000 ft³.

Solution:

Using Equation [7.2]:

$$D_D = \frac{16,400}{v_R} + 1.45\left(\frac{Q_L}{Q_G}\right)^{1.5}$$

For $(Q_L/Q_G) = 5$ gal/1,000 ft³

$$D_D = \frac{16,400}{300} + 1.45(5)^{1.5}$$

$$= 71.5 \text{ µm}$$

For $(Q_L/Q_G) = 10$ gal/1,000 ft³

$$D_D = \frac{16,400}{300} + 1.43(10)^{1.5}$$

$$= 101.8 \text{ µm}$$

7.3 Design Procedure for Collection of Particles in Water Drop Scrubbers

7.3.1 Basic Design Calculations

In Chapter 3, the collection of particles on a single collection body was discussed. In wet scrubbers, particle collection is accomplished by water drops that encounter a number of particles during their passage through the device. The Kleinschmidt equation (Kleinschmidt 1939) provides the overall particle collection efficiency and follows the exponential penetration relationship developed for the lateral mixed model in Chapter 4.

$$\eta = 1 - e^{-N_t} = 1 - e^{-f\eta_{drop}} \qquad [7.3]$$

where: f = fraction of gas swept by the water drops

η_{drop} = single droplet collection efficiency (single droplet target efficiency, see Equation [3.42])

N_t = number of transfer units

The fraction of gas swept by the collecting bodies and the single droplet collection efficiency can be calculated separately. This equation assumes that the liquid drops collide with the dust particles as they travel through the gas and the droplets collect the dust particles which are in their path. For a spray tower, f is equal to the number of collecting droplets, n, encountered by the particles (Equation [3.43]). The fraction of gas swept by the drops is given by the following equation:

$$f = \frac{6.12 \bullet 10^4 \bullet H \bullet Q_L}{D_D \bullet Q_G} \qquad [7.4]$$

where: H = the distance the droplet travels with respect to gas, ft

$$H = \left(\frac{v_t}{v_t - v_G}\right) z \qquad [7.5]$$

v_G = gas velocity relative to the duct wall

v_t = drop terminal settling velocity

z = length of scrubber contact zone

Q_L = liquid flow rate, gal/min

Q_G = gas flow rate, cfm

D_D = the droplet diameter, μm

The value of f can be determined from the total liquid and gas flow rates, as shown in the following sections.

7.3.2 Single Droplet Collection Efficiency

The single droplet collection efficiency, η_{drop}, for water drops depends on three primary mechanisms shown in Figure 3.12, i.e., inertial impaction, direct interception, and diffusion. The combined collection efficiency for each droplet is given by Equation [3.41]

$$\eta_{ICD} = 1 - [1 - f(\psi)] \left\{1 - f\left(\frac{d_p}{d_o}\right)\right\} \left\{1 - f\left(\frac{1}{Pe}\right)\right\}$$

Direct interception is a mechanism that takes into account the actual particle diameter which is not contained in the theoretical derivation of inertial impaction. A scrubber uses many droplets to collect particulate matter with each drop having a combined efficiency, η_{ICD}, for collecting these particles. The collection efficiency in spray scrubbers, where the cut diameter in the 2 to 8 µm size range (see Table 7.1), is dominated by the inertial impaction term ψ in Equation [3.41]. Figure 7.1 provides a plot of experimental and calculated single particle collection efficiency based on the following equation developed by Calvert (1984) for high droplet Reynolds number.

$$\eta_I = \left(\frac{K_p}{K_p + 0.7}\right)^2 = \left(\frac{\Psi}{\Psi + 0.35}\right)^2 \quad [7.6]$$

where: $\quad K_p = 2\Psi = \dfrac{C_c \rho_p d_p^2 v}{9\mu d_d}$

v = relative velocity between the droplet and particle (terminal settling velocity of droplet for gravity spray tower or gas throat velocity in venturi scrubber

K_p = impaction parameter

Ψ = Stokes number

The overall scrubber collection efficiencies for venturi scrubbers are much higher than for spray scrubbers as shown by the 0.2 µm cut diameter in Table 7.1. This is due to the smaller droplet size and higher gas velocities in these devices. Under these conditions, particle collection efficiency of single droplets due to Brownian diffusion as well as inertial impaction need to be considered (Pilat and Prem 1976). Figure 7.2 shows grade efficiency curves for calculated collection efficiencies of a single 100 µm droplet. The figure was developed using a computer program to calculate the limiting trajectories shown in Figure 3.12. The aerodynamic diameter must be used for the particle size. This figure shows that using the terminal settling velocity for the water drops (Figure 7.8), the particle collection efficiency of water drops can be significantly enhanced for particles below 0.2 µm diameter when Brownian diffusion is taken to account. For the larger particle sizes (>1.0 µm diameter) where inertial impaction dominates, Figures 7.1 and 7.2 provide similar particle collection efficiencies. For the smaller particle sizes between 0.01 and 0.2 µm diameter where diffusion dominates, the single particle collection efficiencies are between 0.1 and 0.005. This additional collection efficiency for small particles is important if overall collection efficiencies are to exceed 98%. The computer-calculated collection efficiencies also have been shown to increase when water temperatures are colder than gas temperatures. This is due to the effect of diffusiophoresis and thermophoresis on particle collection (Pilat and Prem 1977).

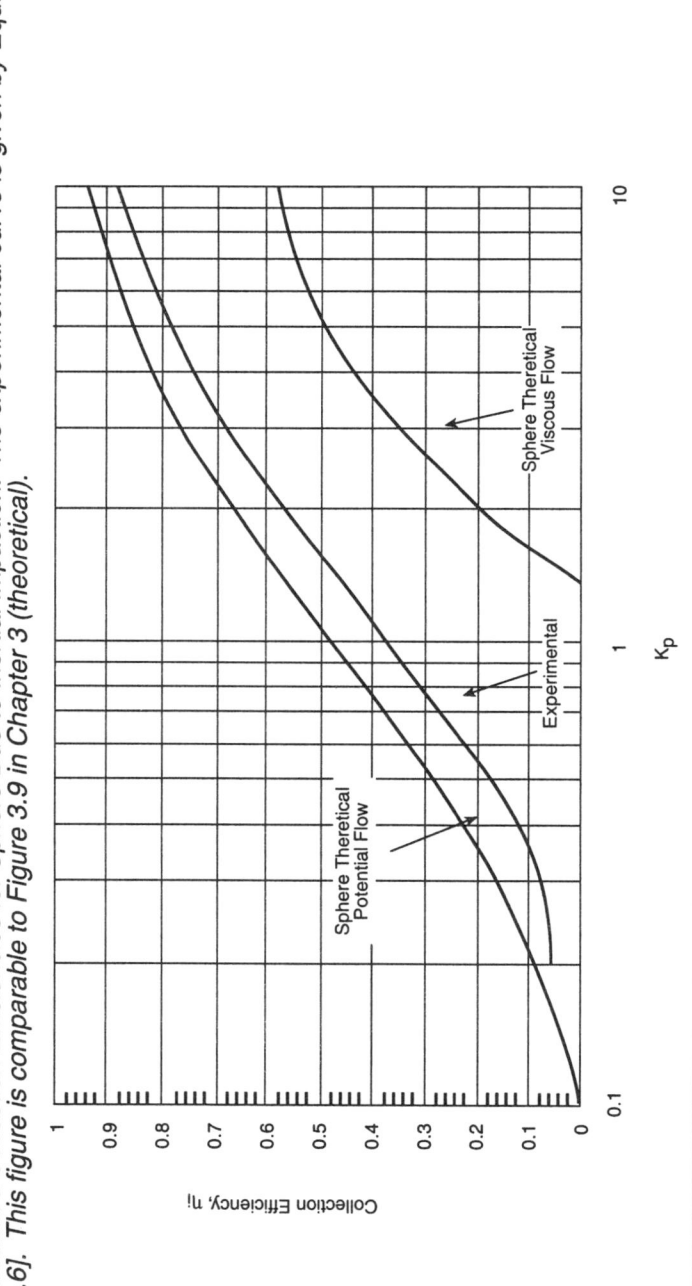

Figure 7.1
Calculated Collection Efficiencies for Sphere Due to Inertial Impaction. The experimental curve is given by Equation [7.6]. This figure is comparable to Figure 3.9 in Chapter 3 (theoretical).

Figure 7.2
Comparison of Calculated Particle Collection Efficiencies of a Single 100 μm dia. Droplet at 150° F Air Temperature and 150° F Water Droplet Temperature.

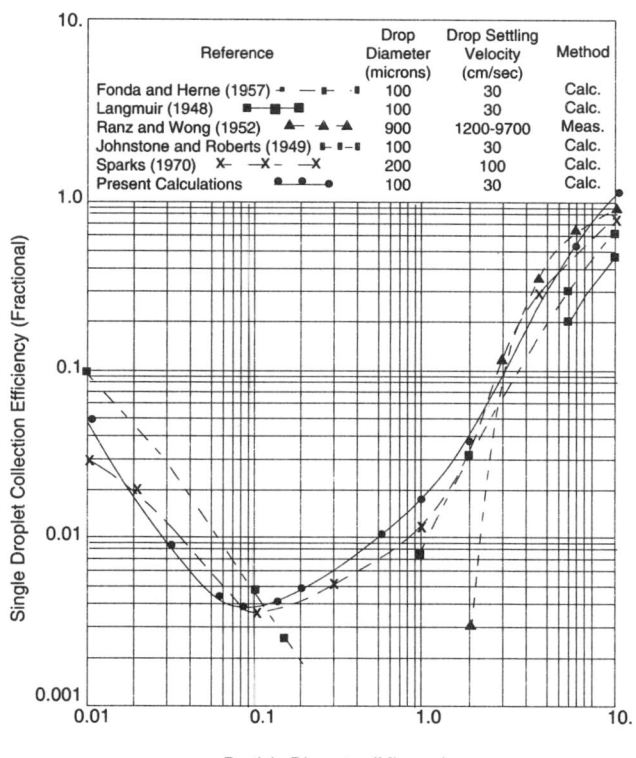

Note: The droplet terminal settling velocity is used for drop velocity.

Source: Pilat and Prem 1976

Example 7.2 Vertical Flow Gravity Spray Chamber Collection Efficiency

A vertical flow gravity spray chamber is to be designed for collection of particles of 0.01, 0.1, 1.0, and 10 μm diameters at 20°C and 1 atm. The particle density is 1 g/cm^3. Water drop sizes are 500 and 1,000 μm diameter. The gas flow rate is 1,200 lb/ft^3 hr (upward gas velocity of 4.2 ft/sec and L/G of 4.8 gal/1,000 ft^3) in a 50-ft spray section. Find the single droplet collection efficiency for each particle size, the fraction of gas swept by the water droplets, and the overall particle collection efficiency.

Figure 7.3
Calculated Single Droplet Collection Efficiency Curves (using Brownian diffusion and inertial impaction mechanisms)

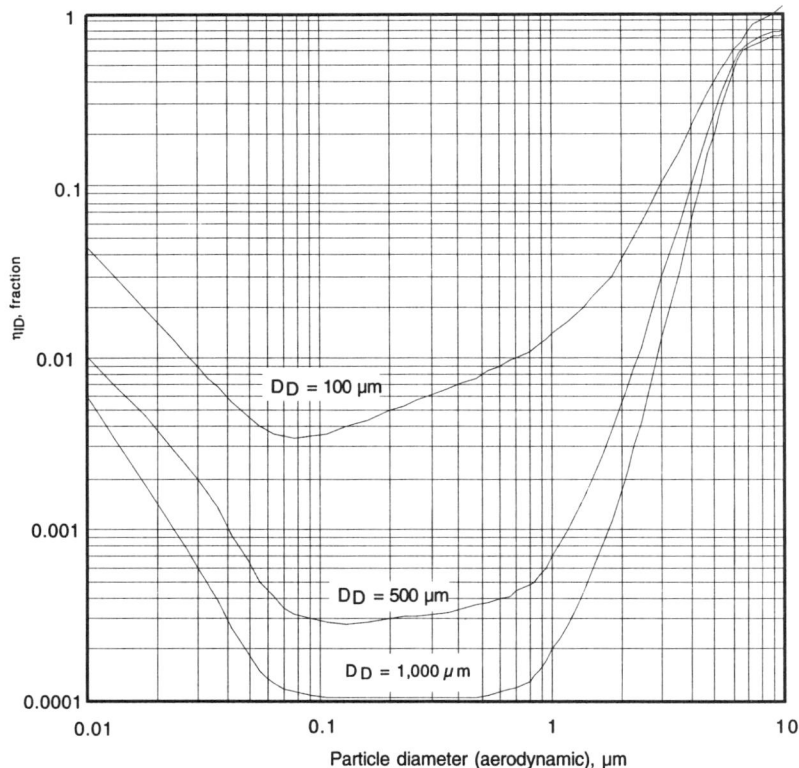

Note: Using Droplet Terminal Settling Velocity from Figure 7.8

Solution:

1. Determine particle collection efficiency of single droplet:

    From Figure 7.3:

Particle diameter	Single droplet collection efficiency, fractional	
μm	500 μm	1,000 μm
0.01	0.015	0.006
0.10	0.00027	0.00011
1.0	0.00070	0.00018
10.0	0.79	0.73

Chapter 7: Wet Scrubbers

2. Calculate fraction of gas swept:

   Using Equation [7.4]:

   $$f = \frac{6.12 \cdot 10^4 \cdot H(ft) \cdot Q_L(gal/min)}{D_D(\mu m) \cdot Q_G(acfm)}$$

   $$f = \frac{6.12 \cdot 10^4 \cdot H \cdot 4.8}{D_D \cdot 1{,}000}$$

   $$f = 293.76 \cdot \frac{H}{D_D}$$

   Using Equation [7.5]:

   $$H = \left(\frac{v_t}{v_t - v_G}\right) z = \left(\frac{v_t}{v_t - v_G}\right) 50 \text{ ft}$$

   H = distance drop travels with respect to gas

   Using Figure 3.5 or 7.8: for water drop $\rho = 1$ g/cm³

   Drop diameter = 1,000 µm, v_t = 380 cm/sec

   Drop diameter = 500 µm, v_t = 210 cm/sec

   v_G = 4.2 ft/sec • 30.48 cm/ft = 128 cm/sec

   $$H_{(500\ \mu m\ drop)} = \left(\frac{210\ cm/sec}{210\ cm/sec - 128\ cm/sec}\right) \cdot 50 \text{ ft} = 128 \text{ ft}$$

   $$H_{(1{,}000\ \mu m\ drop)} = \left(\frac{380\ cm/sec}{380\ cm/sec - 128\ cm/sec}\right) \cdot 50 \text{ ft} = 75.4 \text{ ft}$$

   $$f_{(500\ \mu m\ drop)} = 293.76 \cdot \frac{H}{D_D} = 293.76 \cdot \frac{128}{500} = 75.2$$

   $$f_{(1{,}000\ \mu m\ drop)} = 293.76 \cdot \frac{75.4}{1{,}000} = 22.15$$

3. Calculate overall particle collection efficiency:

   Using Equation [7.3]:

   $$\eta = 1 - e^{-f\eta_{drop}}$$

Particle dia.	500 μm dia. drop		1,000 μm dia. drop	
μm	$f \cdot \eta_{drop}$	Efficiency, %	$f \cdot \eta_{drop}$	Efficiency, %
0.01	1.128	67.63	0.13	12.19
0.1	0.02	1.98	0.002	0.20
1.0	0.05	0.05	0.004	0.40
10.0	59.4	100.00	16.17	100.00

7.3.3 Spray Towers

The most common low energy scrubbers are gravity spray towers in which liquid droplets are made to fall through rising gases and are drained at the bottom of the chamber (see Figure 7.4). The droplets are usually formed by liquid atomized in spray nozzles. The spray is directed into a chamber shaped to conduct the gas through the finely divided liquid. In a vertical tower, the relative velocity between the droplets and the gas is important to avoid spray droplet reentrainment. The terminal settling velocity of the droplets must be greater than the velocity of the rising gas stream. In practice, the vertical gas velocity typically ranges from 2 to 5 ft/sec and this requires that the water droplets be larger than 100 μm diameter.

Figure 7.4
A Typical Spray-Tower Configuration

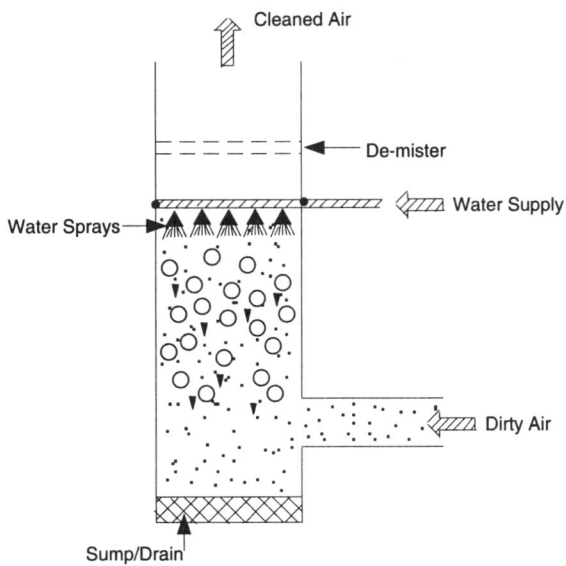

Terminal velocity of 100 μm droplet is 0.8 ft/sec. For higher velocities, a mist eliminator must be used in the top of the tower.

Although the units have a relatively large space requirement, they are inexpensive and are used in practice primarily for the collection of coarse dust (greater than approximately 25 μm) and as a precooler.

Collection of particles by the droplets is primarily by impaction and interception (Figure 7.5). The drops must be large enough to provide a rapid falling speed, yet small enough to present a small impaction target. From such considerations, it may be shown that optimum performance in counter-current scrubbers is achieved for droplets between 100 and 500 μm diameters. Collection efficiency increases with the relative velocity between the particle and the droplet, and it decreases with the droplet size. For an air stream with mean velocity of 2 to 4 ft/sec, η_{IC} is about 20% for particle diameter equal to 5 μm, falling to about 5% for a particle size of 3 μm.

Figure 7.5
Spray-Tower Wet Scrubber Operation Principle

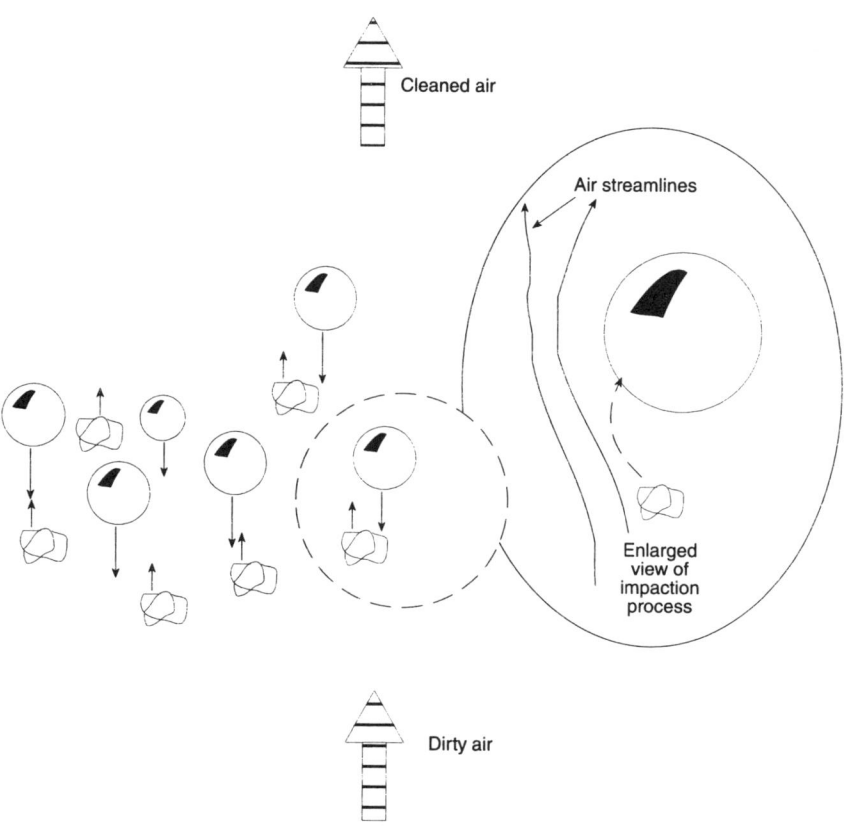

For co-current and cross-current scrubbers, smaller drop sizes are often used (see Figure 7.6) because droplet holdup by the gas does not occur and the single particle collection efficiency is larger than for bigger droplets.

Figure 7.6
Single Drop Collection Efficiency as a Function of Droplet Size by Inertial Impaction, η_I, and Interception, η_C

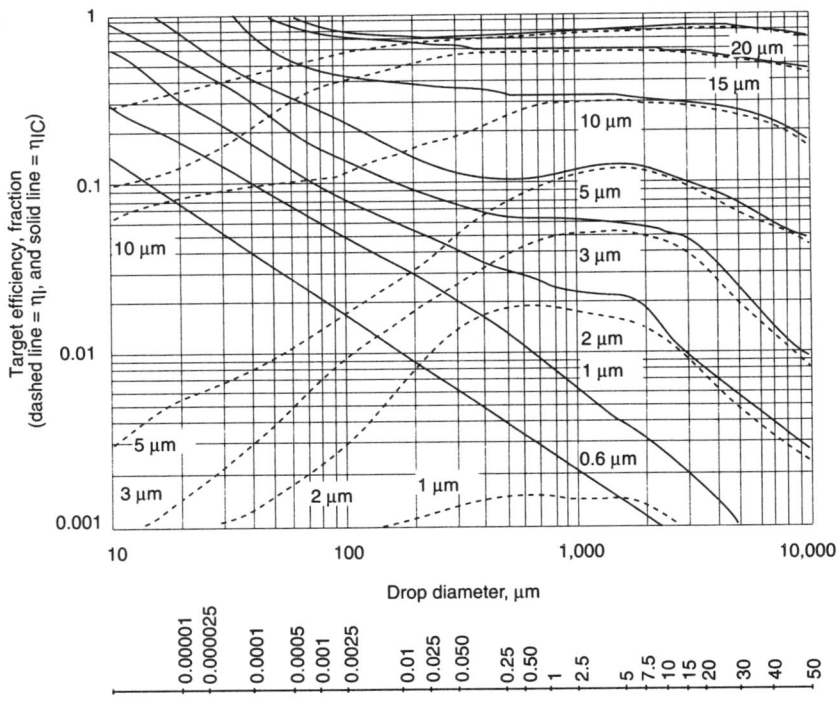

where:
$$\eta_I = \left(\frac{\Psi}{\Psi+0.35}\right) \text{(Calvert 1984)}$$

$$\eta_C = (1+R)^2 - \frac{1}{1+R} \text{ (Ranz and Wong 1952)}$$

$$\eta_{IC} = 1-(1-\eta_I)(1-\eta_C)$$

The additional collection efficiency due to interception for particles smaller than 1,000 μm is clearly shown in the difference in the single drop collection efficiency ($--\eta_I$ and $-\eta_C$) in the figure. The terminal settling velocities are from Table 7.2.

Table 7.2
Terminal Settling Velocity for Water Droplets ($\rho = 1$ g/cm³)

Diameter μm	V_{TS} cm/sec	V_{TS} ft/sec	Diameter μm	V_{TS} cm/sec	V_{TS} ft/sec	Diameter μm	V_{TS} cm/sec	V_{TS} ft/sec
0.2	0.00012	0.000004	180	61.3	2.011	600	242	7.94
0.5	0.00075	0.000025	185	63.7	2.090	650	261	8.56
1	0.003	0.000098	190	66	2.164	700	280	9.19
5	0.075	0.00246	195	68.4	2.244	750	298	9.78
10	0.3	0.00984	200	70.8	2.323	800	316	10.37
15	0.675	0.2215	210	75.5	2.48	850	334	10.96
20	1.2	0.03937	220	80.2	2.63	900	352	11.55
25	1.875	0.06152	230	84.9	2.79	950	369	12.11
30	2.7	0.08858	240	89.6	2.94	1,000	386	12.66
35	3.675	0.1206	250	94.3	3.09	1,100	419	13.75
40	4.8	0.1575	260	98.9	3.24	1,200	452	14.83
45	6.075	0.1993	270	103	3.38	1,300	484	15.88
50	7.5	0.2461	280	108	3.54	1,400	515	16.90
55	9.075	0.2977	290	112	3.67	1,500	546	17.91
60	10.8	0.3543	300	117	3.84	1,600	577	18.93
65	12.675	0.4158	310	121	3.97	1,700	606	19.88
70	14.7	0.4823	320	126	4.13	1,800	636	20.87
75	16.875	0.5536	330	130	4.27	1,900	665	21.82
80	19.2	0.6299	340	135	4.43	2,000	694	22.77
85	21.675	0.7111	350	139	4.56	2,200	738	24.21
100	24.8	0.8136	360	144	4.72	2,500	787	25.81
105	26.8	0.8793	370	148	4.86	2,800	832	27.31
110	28.9	0.9482	380	152	4.99	3,000	862	28.27
115	31.2	1.024	390	157	5.15	3,500	931	30.53
120	33.5	1.099	400	161	5.28	4,000	995	32.64
125	35.8	1.175	410	165	5.41	4,500	1,055	64.62
130	38.2	1.253	420	169	5.54	5,000	1,112	36.50
135	40.2	1.319	430	174	5.71	5,500	1,167	38.28
140	42.5	1.394	440	178	5.84	6,000	1,219	39.98
145	44.8	1.470	450	182	5.97	6,500	1,268	41.61
150	47.2	1.549	460	186	6.10	7,000	1,316	43.18
155	49.5	1.624	470	190	6.23	7,500	1,362	44.70
160	51.8	1.699	480	194	6.36	8,000	1,407	46.16
165	54.2	1.778	490	198	6.50	8,500	1,450	47.58
170	56.6	1.857	500	202	6.63	9,000	1,492	48.96
175	58.9	1.932	550	222	7.28	9,500	1,533	50.31
						10,000	1,573	51.61

(see Figure 7.8 for the equations used to calculate the values in the table)

7.3.4 Counter-Current (Counterflow) Scrubbers

Equation [7.3] shows that the collection efficiency for a spray scrubber is directly related to the number of water drops in the scrubber and to their collection efficiency. Liquid enters the top of the scrubber through a series of nozzles that distribute it uniformly. The gas enters the bottom of the scrubber and flows upward in a uniform, block-like flow. The drop is considered to be at its terminal settling velocity, v_t, relative to the gas that surrounds it. Because the surrounding gas is moving upward with velocity $v_G = Q_G/xy$, the velocity of the drop relative to a fixed coordinate of the scrubber is $v_{D\text{-fixed}} = v_t - v_G$ (Figure 7.7)(De Nevers 1995).

Figure 7.7
Counterflow Scrubber Operating Principle

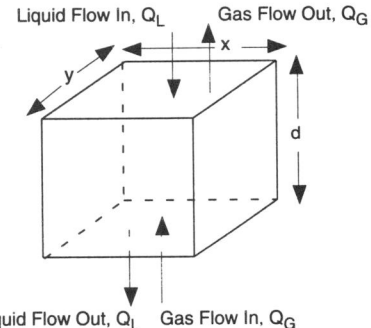

A material balance dictates that the mass of particles per unit time and volume transferred out of the gas stream should be equal to the mass of particles transferred to the water drops in the same unit time and volume.

The first step in determining a material balance is to compute the instantaneous number of drops per unit volume. The liquid flow into the system is Q_L and this makes N_D drops /time, each of volume $(\pi/6)D_D^3$. The average time each drop spends in the scrubber is:

$$\text{average time} = \frac{dz}{v_t - v_G} \qquad [7.7]$$

so at any time, the number of drops in the system is:

$$\text{drops present at anytime} = \frac{N_D dz}{v_t - v_G} \qquad [7.8]$$

The volume of gas that these drops sweep out per unit time is:

$$\frac{\text{Volume swept}}{\text{Time}} = \left(\frac{N_D dz}{v_t - v_G}\right)\left(\frac{\pi D_D^2}{4}\right)V_t = \left(\frac{Q_L}{\pi D_D^3/6}\right)\left(\frac{\pi dz}{4}\right)\left(\frac{D_D^2 v_t}{v_t - v_G}\right)$$

$$= Q_L\left(\frac{1.5}{D_D}\right)(dz)\left(\frac{v_t}{v_t - v_G}\right) \qquad [7.9]$$

Equation [7.9] represents the mass of particles swept by the liquid drops per unit volume and unit time. Mass transferred from the gas component equals that transferred to the liquid component.

$$Q_L\left(\frac{1.5}{D_D}\right)(dz)\left(\frac{v_t}{v_t - v_G}\right)\eta_{IC} c = -Q_G dc \qquad [7.10]$$

where: η_{IC} = target efficiency

c = particle concentration

Integrating provides:

$$\ln P_t = \ln\frac{c}{c_o} = -1.5\left(\frac{\eta_{IC}}{D_D}\right)\left(\frac{Q_L}{Q_G}\right)\left(\frac{v_t}{v_t - v_G}\right)z \qquad [7.11]$$

where: P_t = penetration of a given particle size (range from 0 to 1)

Q_L = liquid volumetric flow rate

Q_G = volumetric gas flow rate

v_G = gas velocity relative to the duct wall

v_t = drop terminal settling velocity

η_{IC} = collection efficiency for a single droplet for impaction and interception

D_D = droplet diameter

z = length of scrubber contact zone

The term $\left(\dfrac{v_t}{v_t - v_G}\right)z = H$ in Equation [7.4]

The droplet terminal settling velocity can be determined from Figure 7.8.

Equation [7.11] can be rewritten as:

$$Pt = \exp\left(-\frac{3Q_L v_t z \eta_d}{4Q_G r_d(v_t - v_G)}\right) = \exp\left(-0.25\frac{A_d v_t \eta_d}{Q_G}\right) \qquad [7.12]$$

Figure 7.8
Terminal Settling Velocity (v_T) for Water Sphere Drops ($\rho = 1$ g/cm³) at Standard Condition (25°C, 1 atm) (see Table 7.2)

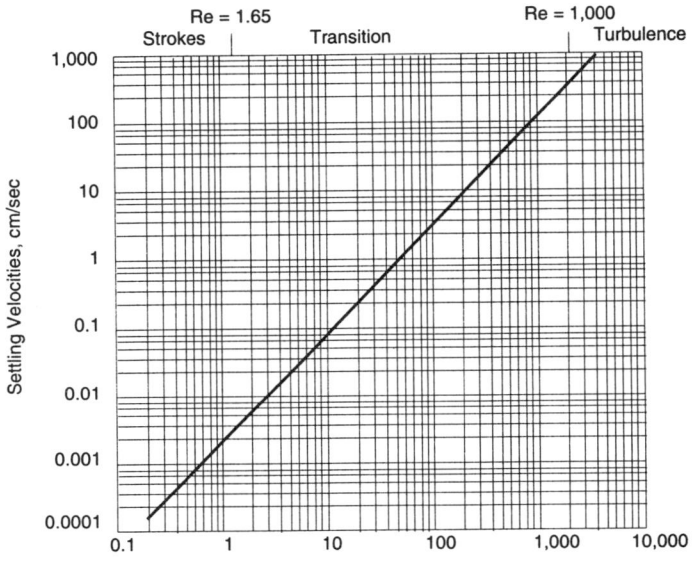

Drop Diameter, μm

Note: Settling velocity was determined as follows:

a) 1 μm < D_D < 100 μm: Stokes' Law (Re < 1.65)

b) 100 μm < D_D < 2,000 μm trial-and-error solution of 2 equations:

$$v_T = \left(\frac{4\rho_D D_D g}{3C_D \rho_g}\right)^{\frac{1}{2}}, \quad C_D = \frac{24}{Re}\left(1 + \frac{Re^{\frac{2}{3}}}{6}\right) \quad (1.65 < Re < 1,000)$$

c) D_D > 2,000 μm: $v_T = \left(\dfrac{4\rho_D D_D g}{3C_D \rho_g}\right)^{\frac{1}{2}}$, $C_D = 0.44$ (Re > 1,000)

where: r_d = droplet radius

A_d = cross sectional area of all the droplets in the scrubber

In Equation [7.12], A_d is given by:

$$A_d = \frac{3Q_L z}{r_d(v_t - v_G)} \qquad [7.13]$$

Chapter 7: Wet Scrubbers

The single droplet target efficiency is estimated from Equation [7.6] or Figures 7.1 (η_I), 7.3 (η_{ICD}), or 7.6 (η_{IC}). Equation [7.12] and Equations [7.3] and [7.4] are equivalent and can be used interchangeably.

The pressure drop through a counterflow spray scrubber is very low, around 0.4 to 0.8 in. H_2O/ft of column for gas velocities of 1 to 2 ft/sec and liquid flow rate of 4,000 to 16,000 lb/hr ft^2.

In the cross-flow case, the water is sprayed at the top of the spray chamber, while the gas flows horizontally. Comparing the cross-flow to counterflow scrubber, the only difference is the addition of a term, $v_t/(v_t - v_G)$, which accounts for the fact that each drop moves farther relative to the gas than it moves relative to the fixed geometry of the scrubber. Therefore, the cross flow scrubber efficiency is:

$$\ln P = \ln \frac{c}{c_o} = -1.5 \frac{\eta_d Q_L xyz}{D_D A Q_G} = -1.5 \left(\frac{\eta_d}{D_D}\right)\left(\frac{Q_L}{Q_G}\right) z \qquad [7.14]$$

Because drop sizes are in the 20 to 70 µm diameter size range for cross-flow and co-current spray chambers, interception becomes important and Figure 7.3 or 7.6 should be used to select the single droplet collection efficiencies. It can be seen in Figure 7.6 that there is a significant increase in target efficiency due to interception for droplets smaller than 100 µm diameter.

Example 7.3 Vertical Spray Chamber Efficiency

In a vertical spray chamber, the flow rate is 29,000 cfm at 20°C and 1 atm. The liquid flow rate is 290 gal/min. The tower height is 10 ft. The dust loading is 2.6 grains/ft^3, $\rho_p = 2.6$ g/cm^3, and the particle diameter is 5.0 µm. The water drops are 500 µm in diameter. Calculate the efficiency of the tower using Equation [7.11]. Compare this with Equation [7.3]. Select v_G at 0.2 v_T.

Solution:

1. Using Equation [7.11]:

    a) Calculate relative settling velocity

    Find drop settling velocity from Figure 7.8 or Table 7.2.

    500 µm droplet with $\rho_p = 1.0$ g/cm^3 (for water drop) $v_T = 202$ cm/sec

    $$v_T = 202 \frac{cm}{sec} \cdot \frac{ft}{30.48\ cm} = 6.6\ ft/sec$$

    $$v_G = 0.2\ v_T = 40.4\ cm/sec = 1.33\ ft/sec$$

    $$\frac{v_T}{v_T - v_G} = \frac{6.6\ ft/sec}{6.6\ ft/sec - 1.33\ ft/sec} = 1.25$$

b) Calculate tower diameter.

$$A_T = \frac{Q_G}{v_G} = \frac{29,000 \text{ ft}^3/\text{min}}{1.38 \text{ ft/sec}} \cdot \frac{1 \text{ min}}{60 \text{ sec}} = 350.24 \text{ ft}^2$$

$$D_T = (A_T \bullet 4/\pi)^{0.5} = (363.4 \bullet 4/\pi)^{0.5} = 21.5 \text{ ft}$$

c) Calculate water consumption.

$$w = \frac{290 \text{ gal}}{29,000 \text{ ft}^3} \bullet \frac{1 \text{ ft}^3}{7.48 \text{ gal}} = 1.337 \bullet 10^{-3}$$

d) Calculate single drop efficiency.

Using Equation [7.6] $\quad \eta_I = \left(\frac{\Psi}{\Psi + 0.35}\right)^2$

where $\Psi = \frac{C_c \rho_p d_p^2 v_{T(drop)}}{18 \mu D_D}$

for large particle (> 1 μm diameter) $C_c = 1$

$$\Psi = \frac{2.6 \text{ g/cm}^3 (5 \bullet 10^{-4} \text{cm})^2 \, 202 \text{ cm/sec}}{18 \bullet 1.82 \bullet 10^{-4} \text{ g/cm sec} \bullet 500 \bullet 10^{-4} \text{cm}} = 0.80$$

$$\eta_I = \left(\frac{0.80}{0.80 + 0.35}\right)^2 = 0.484$$

e) Calculate collection efficiency.

Using Equation [7.11]: $\quad \ln P_t = -1.5 \left(\frac{\eta_{IC}}{D_D}\right)\left(\frac{Q_L}{Q_G}\right)\left(\frac{v_t}{v_t - v_G}\right) z$

$$\ln P_t = -1.5 \left(\frac{0.484}{500 \text{ μm}}\right) \bullet 1.337 \bullet 10^{-3} \bullet 1.25 \bullet 10 \text{ ft} \bullet \frac{30.48 \text{ cm}}{1 \text{ ft}} \bullet \frac{1 \bullet 10^4 \text{ μm}}{1 \text{ cm}}$$

$\ln P_t \quad = \quad -7.4$

$P_t \quad = \quad 0.0004$

$\eta_T \quad = \quad 1 - P_t$

$\eta_T \quad = \quad 1 - 0.0004 = 0.9996$

$\quad \quad = \quad 99.96 \%$

Chapter 7: Wet Scrubbers

2. Using Equation [7.3]:

Equation [7.3]: $\eta = 1 - e^{-f\eta_{drop}}$

Equation [7.4]: $f = \dfrac{6.12 \cdot 10^4 \cdot H \cdot Q_L}{D_D \cdot Q_G}$

Equation [7.5]: $H = \left(\dfrac{V_t}{V_t - V_G}\right) z$

$H = 1.25 \cdot 10 \text{ ft} = 12.5 \text{ ft}$

$f = \dfrac{6.12 \cdot 10^4 \cdot 290 \cdot 12.5}{500 \cdot 29{,}000} = 15.3$

a) Calculate single drop efficiency.

Use Figure 7.1 where Ψ from preceding solution = 0.80

$K_p = 2\Psi = 2 \cdot 0.80 = 1.60$

From Figure 7.1, for $K_p = 1.60$: read $\eta_I = 0.5$

$\eta = 1 - \exp[-0.5 \cdot 15.3]$

$\eta = 0.9995$

$= 99.95\%$

b) Calculate aerodynamic diameter of the particle.

Using Figure 7.3

$d_{pa} = 5\mu m \cdot \left(\dfrac{2.6 \text{ g/cm}^3}{1 \text{ g/cm}^3}\right)^{0.5} = 8\mu m$

From Figure 7.3, for $d_{pa} = 8$ μm and $D_D = 500$ mm: read $\eta_{ID} = 0.7$

$\eta = 1 - \exp[-0.7 \cdot 15.3]$

$\eta = 0.99997 = 99.997\%$

From Figure 7.6, for $d_{pa} = 8$ μm and $D_D = 500$ μm: read $\eta_{ID} = 0.5$

$\eta = 1 - \exp[-0.5 \cdot 15.3]$

$\eta = 0.9995 = 99.95\%$

	Method			
	Equation [7.11]	Equation [7.3]		
		Figure 7.1	Figure 7.3	Figure 7.6
Efficiency, %	99.6	99.95	99.9997	99.95

The wet scrubber equations use the L/G flow ratio, which is of importance to the performance of all kinds of scrubbers. This ratio can be expressed fundamentally in terms of mass flow rate of liquid, M_L, divided by the mass flow rate of the gas, M_G. Thus:

$$\frac{M_L}{M_G} = \frac{\rho_L Q_L}{\rho_G Q_G} \qquad [7.15]$$

For an air water system at 20°C, $\rho_L/\rho_G = 830$. It is also common practice to express Q_L/Q_G in units of gallons of liquid per thousand actual cubic feet of gas, at the entrance conditions. This is commonly called the L/G ratio. Experience shows that L/G will typically fall in a range roughly 2 to 20 gal/1,000 ft³, or 0.2 to 2.5 L/m³ where 1 gal/1,000 ft³ = 0.1337 L/m³.

This ratio is directly proportional to the number of drops per unit volume of the gas: $\dfrac{6Q_L}{\pi D_D^3 Q_G} = 2.553 \bullet 10^8 \left(\dfrac{L}{G}\right)\dfrac{1}{D_D^3}$ drops/cm³, where D_D must be in microns.

Table 7.3 shows the number of drops per cm³ obtained at various values of L/G and D_D.

Table 7.3
Number of Drops per cm³ for Different Values of L/G

D_D, μm	L/G (gal/1,000 ft³)		
	2	10	20
100	511	2,553	5,106
200	64	319	638
400	8	40	80
800	1	5	10

Chapter 7: Wet Scrubbers

Example 7.4 Grade Efficiencies for a Spray Tower

Develop a set of grade efficiency curves (see section 4.3.2 for definition of grade efficiency curves) for a counter-current scrubber which is to remove particles of 2.6 gm/cm^3 density ranging in size from 0.6 μm to 20 μm. Consider L/G ratios of 5 and 10 gal/1,000 ft^3 respectively, water drop sizes of 200 and 500 microns, and tower heights of 2 ft and 10 ft. Assume the gas to be saturated with, and at the same temperature as, the liquid. Take v_G at 0.2 v_T and also at 0.1 and 0.5 v_T (see the following table).

Case	L/G	D_D	H	v_G/v_T
I	10	500	2	0.2
II	10	200	2	0.2
III	10	500	2	0.5
IV	10	500	10	0.2
V	5	500	2	0.2
VI	10	100	2	0.1

Grade Efficiencies for a Spray Tower-curves obtained using Figure 7.6 (η_{IC})

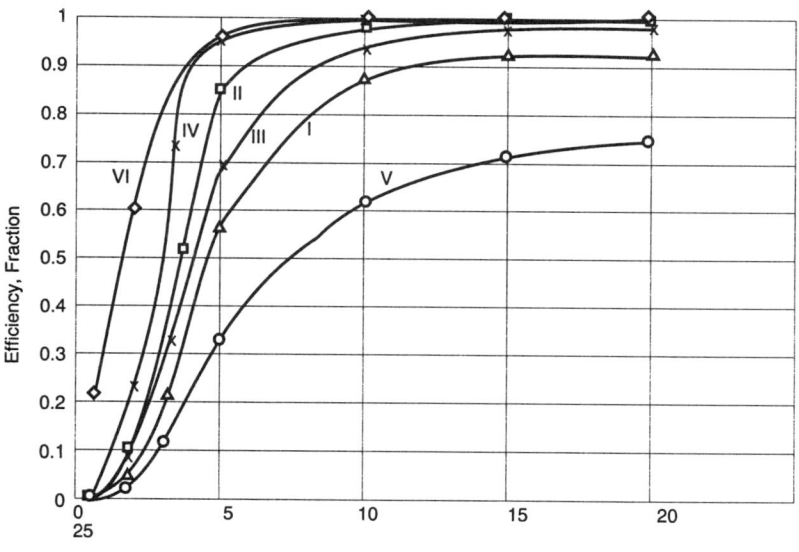

Particle Size, d_p, μm

Refering to Example 7.4, it is noted that one calculation of Case IV is shown in Example 7.3. The figure included in Example 7.4 displays the grade efficiency curves for the six cases. Case I may be regarded as the base case for which L/G = 10 gal/1,000 ft^3, D_D = 500 μm, H = 2 ft and v_G = 0.2 v_T. The other cases show that the grade efficiency is increased by smaller drop size, D_D (Case II); larger ratio v_G/v_T (Case III); and larger tower height, H (Case IV). A smaller L/G (Case V) results in a decrease in efficiency. These results are typical: the greatest efficiency would be obtained (out of these possible values) by the combination $D_D = 100$ μm, $v_G = 0.5\ v_T$, L/G = 10 gal/1,000 ft^3, and H =10 ft. Smaller drop sizes require the use of a scrubber with a larger diameter to maintain a constant v_G/v_T ratio.

The above example demonstrates important considerations arising in scrubber design. The value of Q_G is a specific requirement of the industrial operation. However, v_G, Q_L/Q_G, and D_D may be selected by the designer. Selection of v_G determines the tower cross sectional area $A = Q_G/v_G$. Obviously, $v_G < v_T$, but the larger v_G, the larger the residence time of drops in the tower and the greater the overall efficiency. v_G is usually restricted to values between 2 and 5 ft/sec to provide a reasonable tower cross sectional area for counter-current scrubbers. Figure 7.6 shows that the drop terminal settling velocities between 2 and 5 ft/sec represent droplet sizes between 800 and 1,200 μm diameter. Smaller drop sizes can be used in cross-flow scrubbers (see Equation [7.14]) to achieve higher collection efficiency.

7.3.5 Venturi Scrubbers

Venturi scrubbers are widely used as high efficiency gas cleaning devices. The heart of the system is a venturi throat, where gases pass through a contracted area, (Figures 7.9, 7.10, and 7.11) reaching velocities of 200 to 600 ft/sec, and then pass through an expansion section. From the expansion section, the gas enters a large chamber for separation of particles.

The basic particulate collection process is shown schematically in Figure 7.10 where water is injected into the region of increasing air velocity in the converging section just upstream of the throat of the venturi. Here the high fluid mechanical shear forces cause disintegration of the water stream and production of droplets (atomization). Once the liquid is atomized, it begins to collect particles from the gas impacting into the liquid as a result of the difference in velocities of the gas stream and atomized droplets.

The overall principle of particle collection is basically the same as the spray scrubbers discussed in the previous section. In the venturi scrubber, as for the spray tower, the relationships between air flow rate, the water flow rate, and the throat geometry have an important bearing on the droplet size and velocity and, hence , on the efficiency of the impaction process. Here, however, the droplet size is much smaller and the relative droplet particle velocity is much higher, so the collection efficiency for each individual droplet is much higher.

Very high gas velocities can be used in this type of scrubber, as much as 600 ft/sec. The liquid enters with zero or negligible velocity so that at the inlet the relative velocity may be as high as 600 ft/sec. This may be 100 times the maximum tolerable relative velocity in a cross-flow or counterflow scrubber. A scrubber with an

Figure 7.9
Various Venturi Scrubber Configurations

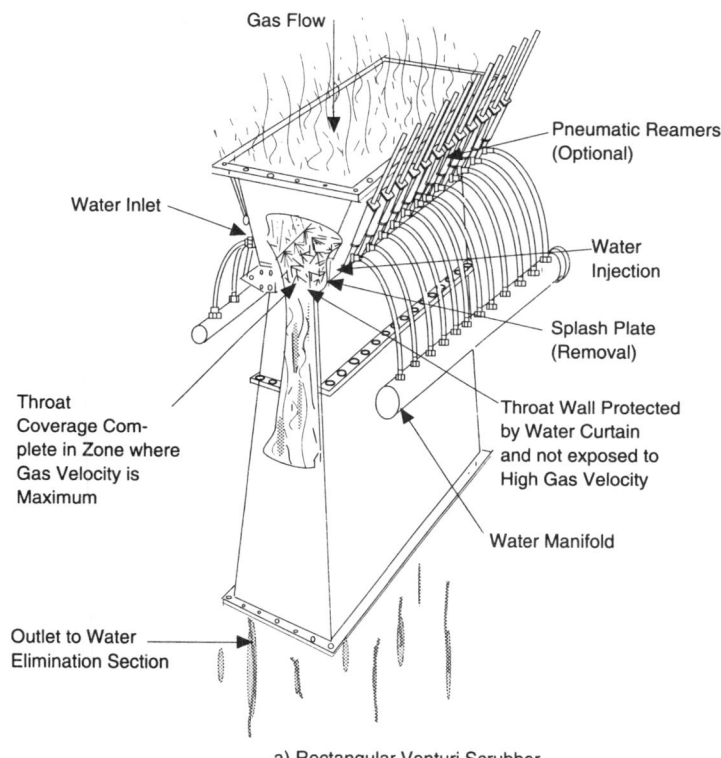

a) Rectangular Venturi Scrubber

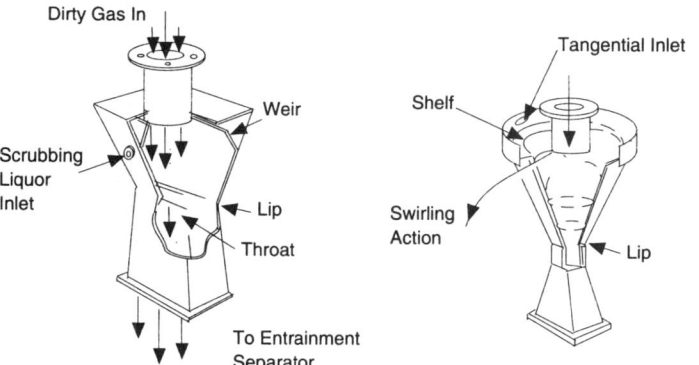

b) Weir Type Liquor Inlet

c) Tangential Liquor Inlet

Figure 7.10
Venturi Scrubber Principle of Operation

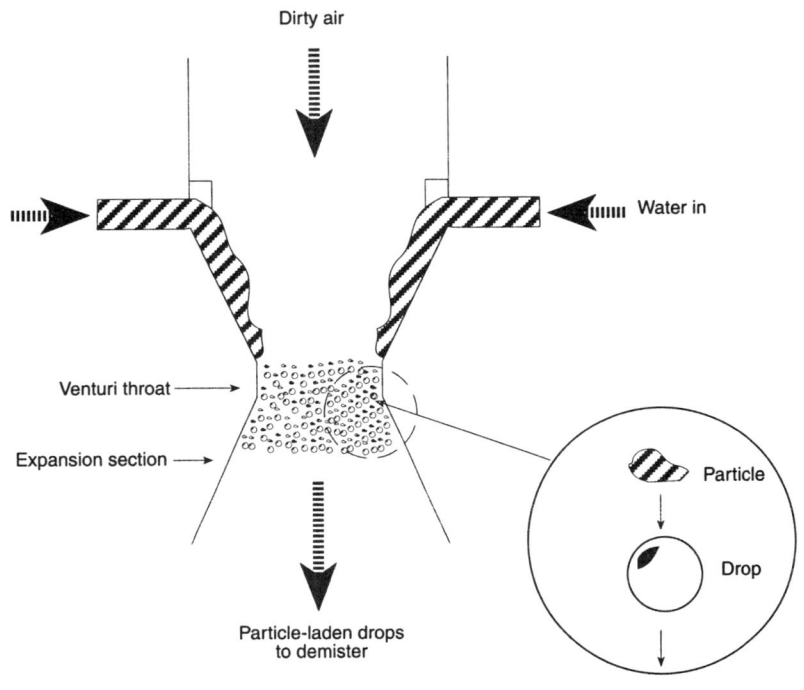

adjustable venturi throat is shown in Figure 7.11. The conical plug is positioned to increase or decrease the gas flow area. Water is injected above the venturi section, tangentially to the inlet-gas flow. A remote, automated control system is often used to move the plug.

A serious concern associated with venturi sections is the erosive effect of the gas/liquid mixture passing through the throat. As indicated, the throat velocity is extremely high, creating the potential for wear of the throat surface.

7.3.5.1 Collection Efficiency for Venturi Scrubbers — Johnstone Design Method

One of the first design equations for a venturi scrubber was developed by Johnstone et. al (1954) and is still often used for preliminary design calculations. Presently, the design procedure developed by Calvert (1975) and presented in the next section is usually used for detailed design calculations. Johnstone considered the

Figure 7.11
Flooded-Disk Scrubber. Typical liquid flow rates are 6-8 ft/sec (5-10 psig pressure), typical inlet and outlet gas velocities are 50-60 ft/sec. The calculated venturi throat velocities (over 200 ft/sec) occur in the conical plus section which is adjustable.

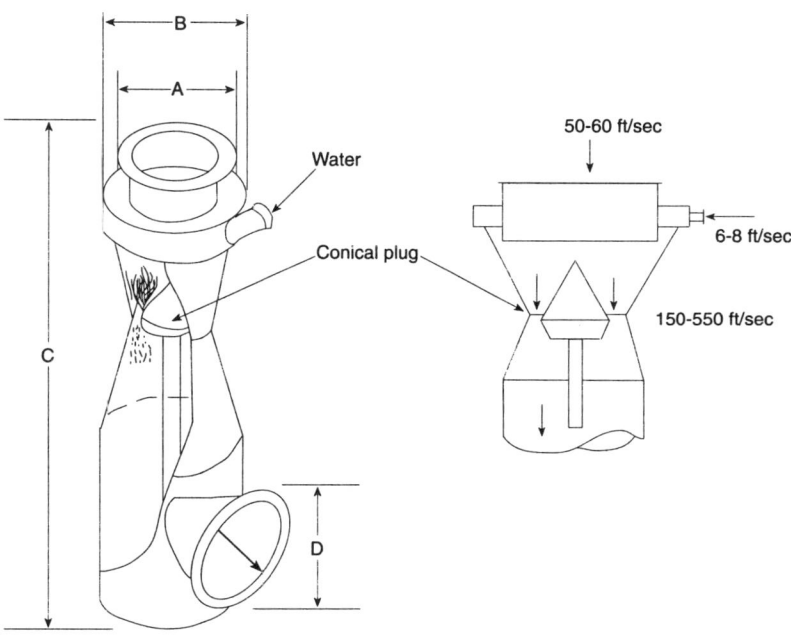

Saturated gas volume (acfm)	A	B	C	D
2,500	1'0"	3'6"	6'9"	1'1"
5,000	1'4"	3'10"	7'11"	1'6"
10,000	1'10"	4'4"	8'10"	2'0"
25,000	3'0"	5'6"	10'4"	3'3"
50,000	4'2"	7'2"	13'7"	4'5"
75,000	5'2"	8'2"	14'10"	5'7"
100,000	5'11"	8'11"	16'3"	6'0"
150,000	7'3"	10'3"	18'3"	7'6"
350,000	11'0"	11'0"	25'0"	11'0"

Source: Schiffner and Hesketh 1983

predominate particle collection mechanism to be inertial impaction and developed the following equation.

$$\eta = 1 - e^{-kLi\sqrt{\Psi}} \qquad [7.16]$$

where: k = empirical constant = 200 ft³/gal (or 0.2 ft³/gal when Q_L/Q_G = gal/ft³)

Li = L/G ratio, gal/1,000 ft³ (Q_L/Q_G = gal/ft³)

$$\Psi = \text{Stokes' number (Equation [3.28])} = \frac{C_c \rho_p d_p^2 v_t}{18\mu D_D}$$

where:

D_D = drop diameter given by Equation [7.1]

v_t = throat velocity

The equation is similar to the Kleinschmidt equation (Equation [7.3]). Figure 7.12 shows a correlation for Equation [7.16] with experimental data (Johnstone et al. 1954).

Figure 7.12
Correlation of Venturi Scrubber Efficiency -- Inertial Impaction Basis

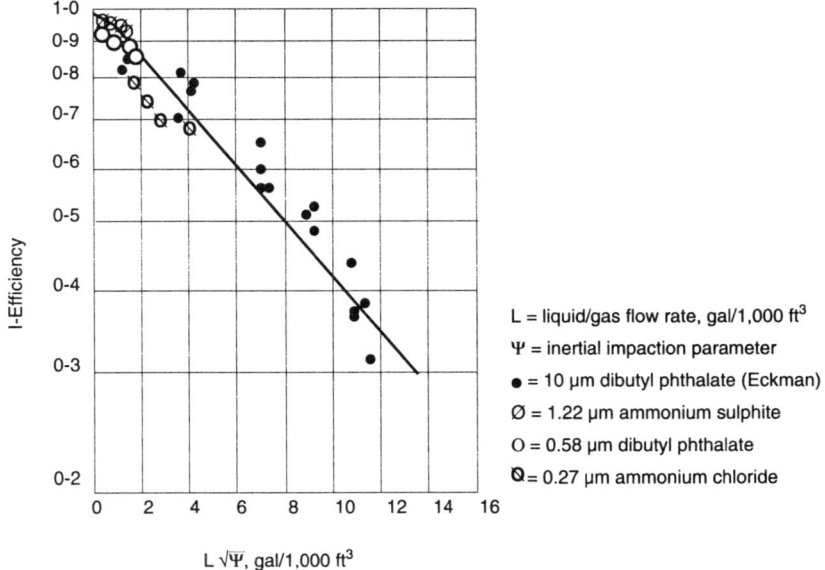

L = liquid/gas flow rate, gal/1,000 ft³
Ψ = inertial impaction parameter
● = 10 μm dibutyl phthalate (Eckman)
Ø = 1.22 μm ammonium sulphite
O = 0.58 μm dibutyl phthalate
◐ = 0.27 μm ammonium chloride

Source: Johnstone et al. 1954

Example 7.5 Venturi Scrubber Collection Efficiency Using the Johnstone Method

A fly ash-laden gas stream is to be cleaned by a venturi scrubber using a L/G ratio of 8.5 gal/1000 ft^3. The efficiency can be calculated using (Equation [7.16])

$$\eta_i = 1 - e^{-kLi\sqrt{\Psi_i}}$$

where η_i is the fractional efficiency of collection of particles of size d_{pi}. The fly ash has a particle density of 0.7 g/cm^3 and k = 200 ft^3/gal. Use a throat velocity of 272 ft/sec and a gas viscosity of 1.5 • 10^{-5} lb/ft sec. The particle size distribution is shown in the following table:

d_{pi}, μm	% by weight
<0.10	0.01
0.01-0.5	0.21
0.6-1.0	0.78
1.1-5.0	13.0
6.0-10.0	16.0
11.0-15.0	12.0
16.0-20.0	8.0
>20.0	50.0

Solution:

1. Using Equation [7.2], determine the drop size.

$$D_D = \frac{16,400}{v_R} + 1.45 \left(\frac{Q_L}{Q_G}\right)^{1.5}$$

$$= \frac{16,400}{272} + 1.45 \bullet 8.5^{1.5} = 96.23 \text{ μm}$$

2. Calculate the Stokes number.

$$\Psi = \frac{\rho_p d_p^2 v}{18 \mu D_D}$$

$$= \frac{d_p^2 \bullet 10^{-8} \text{ cm}^2 \bullet 0.7 \text{ g/cm}^3 \bullet 272 \text{ ft/sec}}{18 \bullet 1.5 \bullet 10^{-5} \text{ lb/ft sec} \bullet 96.23 \bullet 10^{-4} \text{ cm}} \bullet \frac{1 \text{ lb}}{453.6 \text{ g}} \bullet \frac{(30.48 \text{ cm})^2}{\text{ft}^2}$$

$$= 1.5 d_p^2 \quad (d_p \text{ in μm})$$

3. Determine the individual particle efficiency, η_i.

$$\eta_i = 1 - e^{-kLi\sqrt{\Psi_i}}$$
$$= 1 - e^{-0.2 \bullet 8.5\sqrt{1.5 \bullet d_p^2}}$$
$$= 1 - \exp[-2.082\, d_p]$$

4. The overall efficiency is the sum of the individual particle efficiencies calculated according to step 3 above. The results are presented in the following table:

d_p, μm	ηi	X_i, % by wt.	$\eta_i X_i$
0.05	0.0989	0.01	0.000009886
0.3	0.4645	0.21	0.0009755
0.8	0.8109	0.78	0.006325
3.0	0.9981	13.0	0.1298
8.0	1.0000	16.0	0.16
13.0	1.0000	12.0	0.12
18.0	1.0000	8.0	0.08
80.0	1.0000	50.0	0.50
		100.0	η_T=0.9971

7.3.5.2 Venturi Scrubber Collection Efficiency — Calvert Design Method

The Calvert design method employs an empirical factor to account for the variables occuring as the gas and water pass through the venturi scrubber throat. The average time it takes a drop of water to pass a section dx is:

$$\text{Average time} = \frac{dx}{v_{Dfixed}} = \frac{dx}{v_G - v_{Rel}} \qquad [7.17]$$

where: v_{Dfixed} = drop velocity referenced to fixed coordinates

v_G = local gas velocity

v_{Rel} = velocity of the drops relative to the gas = $v_G - v_{Dfixed}$

An equation analog to Equation [7.11] using v_{Rel} in place of v_t (both are the drop velocity relative to the gas for their appropriate scrubber) can be developed as follows:

Chapter 7: Wet Scrubbers

$$\frac{dc}{c} = -\frac{1.5}{D_D} \bullet \eta_d \frac{Q_L}{Q_G} \bullet \frac{v_{Rel}}{v_G - v_{Rel}} dx \qquad [7.18]$$

The drop is accelerating in the venturi throat and also decreasing in size. Initially, the relative velocity is almost equal to the gas velocity because $v_{Rel} = v_G - v_{Dfixed}$, but v_{Rel} declines rapidly as the drops accelerate. Also, the water drops are not of constant size because of atomization. The drop size decreases with distance downstream of the water injection point. Calvert (1977) introduced an empirical factor, f, to account for these variables. The modified Equation [7.11] is as follows:

$$P_t = \exp\left[\frac{Q_L v_G \rho_L d_d}{55 Q_G \mu_G} F(K_{po}, f)\right] \qquad [7.19]$$

where: K_{po} = inertial parameter evaluated for the gas velocity at the throat entrance, and

$$F(K_{po}, f) = \left[-0.7 - K_{po} f + 1.4 \ln\left(\frac{K_{po} f + 0.7}{0.7}\right) + \frac{0.49}{0.7 + K_{po} f}\right] \frac{1}{K_{po}}$$

$$K_{po} = \frac{d_{pai}^2 v_t}{9 \mu D_D} = 2\Psi \qquad [7.20]$$

where: d_{pai} = the aerodynamic diameter

v_t = throat velocity

The factor f is an empirical factor which absorbs the influence of various parameters not explicitly included in Equation [7.18]. These parameters include collection by means other than impaction, such as particle growth due to condensation or other effects, drop size other than those predicted, loss of liquid to the venturi walls, maldistribution, and other effects. Available data indicate that f = 0.25 for hydrophobic particles; f = 0.4 to 0.5 for hydrophillic particles; and f = 0.5 for large scrubber tests. For conservative design purposes, a value of 0.25 is recommended. For soluble compounds, acids, and fly ash with SO_2, f may be as high as 0.5.

Example 7.6 Venturi Scrubber Collection Efficiency Using the Calvert Method

Calculate the collection efficiency of a venturi scrubber for 1 μm (aerodynamic diameter) particles if the operating conditions are v_G = 50 m/sec at the throat entrance, and Q_L/Q_G = 1 L/m³. Assume atmospheric pressure and 25°C (at these conditions, σ_L = 72 dyne/cm and μ_L = 0.1cp); also assume f = 0.5.

Solution:

1. Determine the droplet diameter using Equation [7.1].

$$D_D = \frac{1,920}{V_R}\left(\frac{\sigma_L}{\rho_L}\right)^{0.5} + 597\left(\frac{\mu_L}{\sqrt{\sigma_L \rho_L}}\right)^{0.45}\left(1,000\frac{Q_L}{Q_G}\right)^{1.5}$$

at $\mu_L = 0.1$ cp $= 0.01$ poise, $v_G = 50$ m/sec $= 164$ ft/sec

$$D_D = \frac{1,920}{164}\left(\frac{72}{1}\right)^{0.5} + 597\left(\frac{1\bullet10^{-2}}{\sqrt{1\bullet72}}\right)^{0.45}\left(1,000\bullet1\bullet10^{-3}\right)^{1.5}$$

$$= 99.34 + 28.7 = 128.1 \; \mu m$$

2. Calculate the collection efficiency using Equation [7.19].

$$K_{po} = \frac{d_{pai}^2 v_t}{9\mu D_D} = \frac{0.0001^2 \bullet 50 \bullet 100}{9 \bullet 1.8 \bullet 10^{-4} \bullet 128 \bullet 10^{-4}} = 2.41$$

$$F(K_{po},f) = \left[-0.7 - 2.4\bullet0.5 + 1.4\ln\left(\frac{2.4\bullet0.5+0.7}{0.7}\right) + \frac{0.49}{0.7+2.4\bullet0.5}\right]\frac{1}{2.4}$$

$$= -0.1$$

$$P_t = \exp\left[\frac{0.001 \bullet 50 \bullet 100 \bullet 1.28 \bullet 10^{-4}}{55 \bullet 1.8 \bullet 10^{-4}} \bullet (-0.1)\right]$$

$$P_t = \exp[-0.464]$$

$$= 0.52$$

Collection efficiency $= (1 - 0.52) \bullet 100 = 48\%$

7.4 Pressure Drop

The design of wet scrubbers usually focuses on collection efficiency and pressure drop. In most cases, the scrubber must be designed to guarantee a specified collection efficiency, which, in turn, is strongly dependent upon pressure drop. The system pressure drop also dictates the power requirements and size of auxiliary equipment, such as the fans.

For cross-current and co-current conditions, one can estimate the pressure drop on the basis of the change in momentum of the liquid when it is accelerated from zero velocity to the gas velocity.

$$\Delta p = \rho_G \left(\frac{Q_L}{Q_G}\right) v_G^2 \qquad [7.21]$$

Chapter 7: Wet Scrubbers

The main source of pressure loss in venturi scrubbers is the mechanical energy used for accelerating the water injected into the throat; a mechanical energy balance across the throat provides the following equation for the pressure loss:

$$\Delta p = p_1 - p_2 = \frac{1}{2} v_G^2 \rho_L \frac{Q_L}{Q_G} \quad [7.22]$$

An empirical equation by Hesketh (1974) for Δp is:

$$\Delta p = 190 v_G^2 \rho_G A_t^{0.133} \left(\frac{Q_L}{Q_G}\right)^{0.78} \quad [7.23]$$

where, if consistent SI units are used, and A_t is the throat cross sectional area, Δp is obtained in Pascal.

Example 7.7 Venturi Pressure Drop

The velocity at the venturi throat entrance is 400 ft/sec, and water is introduced at a rate 7.5 gal/1,000 ft³ of gas. Estimate the pressure drop across the venturi (in inches H_2O)

Solution:

Using Equation [7.22]

$$\Delta p = \frac{1}{2} v_G^2 \rho_L \left(\frac{Q_L}{Q_G}\right)$$

$$1 \text{ gal} = 0.13368 \text{ ft}^3$$
$$7.5 \text{ gal} = 1 \text{ ft}^3$$
$$\rho_L = 62.4 \text{ lb}_m/\text{ft}^3$$
$$\Delta p = 62.4 \text{ lb}_m/\text{ft}^3 \cdot 0.001 \cdot (400 \text{ ft/sec})^2$$
$$= 9,984 \text{ lb}_m\text{ft/sec}^2$$
$$g_c = 32.2 \text{ ft lb}_m/\text{lb}_f \text{ sec}^2$$
$$\Delta p = 9,984/32.2$$
$$= 310 \text{ lb}_f/\text{ft}^2$$
$$= 2.15 \text{ psi}$$
$$1 \text{ in. } H_2O = 0.03614 \text{ psi}$$
$$\Delta p = 2.15/0.03614 = 59.58 \text{ in. } H_2O$$

234 Fundamentals of Air Quality Systems

Scrubbing systems are often categorized by system pressure drop. The effectiveness of a scrubbing system is usually directly related to the pressure drop across the scrubber. The higher the pressure drop, the greater the turbulence and mixing; therefore, the more effective the scrubbing action. This feature is illustrated by the graph in Figure 7.13 which is a plot of venturi scrubber pressure drop for a given collection efficiency and particle mean diameter for a venturi scrubber with a rectangular throat (Schiffner & Hesketh 1983). Thus, if the particle mean diameter for an emission stream and required collection efficiency are known, the pressure drop across the venturi can be estimated. This figure is applicable only for the specific venturi scrubber for which the relationship was developed. However, the figure

Figure 7.13
Venturi Scrubber Collection Efficiencies as a Function of Particle Size and Pressure Drop for a Rectangular Venturi

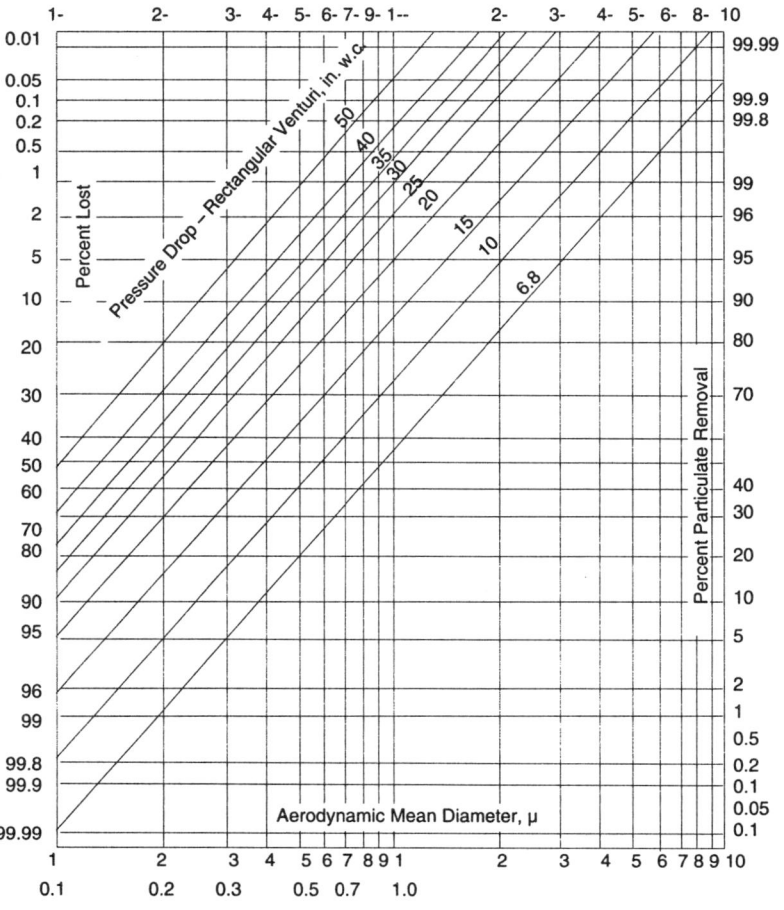

Source: Schiffner and Hesketh 1983

Chapter 7: Wet Scrubbers   235

provides a simple and easy method to obtain an initial estimate for the pressure drop required as a function of particle size and collection efficiency. The relationship between pressure drop and collection efficiency is explained in more detail in the next two sections. Once pressure drop is determined, it is possible to calculate the volume of gas and size of the venturi scrubber, using information such as that contained in Figure 7.11.

7.5 Design Optimization

Because of the wide use of scrubbers and the large number of design variables, there has been considerable attention given to optimize their design. The optimization studies generally involve the selection of the best combination of v_G and (Q_L/Q_G) for collecting a given particle size which enable calculation of the optimum-sized drops to be introduced into the scrubber. In practice, it is simplest to use the cut diameter design approach developed in Chapter 3 to find the approximate pressure drop required, and then to find the best combination of v_G and (Q_L/Q_G) for the case at hand.

Calvert (1974) published general solutions to the scrubber efficiency equations in terms of d_{pc} (see Chapter 3) which are shown in Figure 7.14 and Figure 7.15. In these figures, d_{pc}, the aerodynamic particle diameter for which the collection efficiency is 50%, is plotted against the design variables. Note that in Figure 7.15 below a certain chamber height, d_{pc} increases very rapidly, meaning that P_t approaches 100 %. These figures are valid only for air and water near standard conditions.

Figure 7.14
Performance Cut Diameter Predictions for Typical Vertical Countercurrent Spray

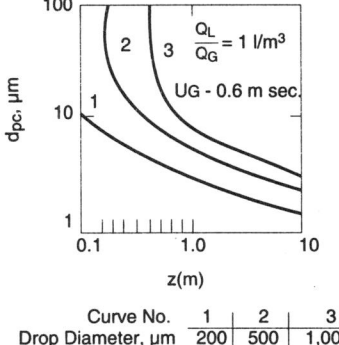

Source: Calvert and Englund 1984

Figure 7.15
Performance Cut Diameter Predictions for Typical Vertical Countercurrent Spray

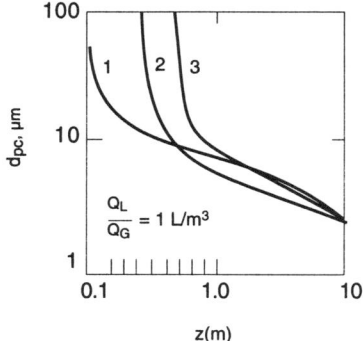

Source: Calvert and Englund 1984

Figure 7.16
Aerodynamic Cut Diameter and Pressure Drop Predictions for a Typical Venturi Scrubber

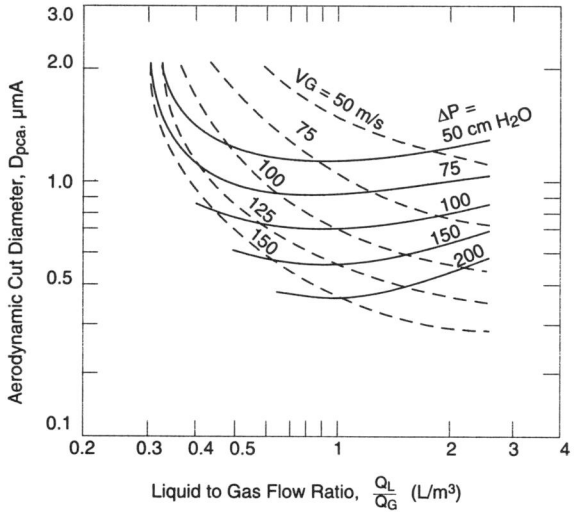

Note: v_G is the velocity at throat and f = 0.25. 1 L/m³ = 7.48 gal/1,000 ft³
Source: Calvert and Englund 1984

In Figure 7.16, combinations of L/G and v_G are determined which will produce the required cut diameter (for venturi scrubbers). As usual, there will be a range of these combinations which will work. For each, the corresponding pressure can also be determined from Figure 7.16. An optimum set of conditions may then be found, and the dimensions of the venturi throat can be calculated.

Example 7.8 Venturi Scrubber Pressure Drop

Treatment of a gas stream to remove most of the particle is required. It is judged that a venturi scrubber with a cut diameter of 0.5 µm will provide satisfactory particle removal. If $Q_L/Q_G = 1$ L/m³, what gas velocity at the throat will be needed, and what will be the expected pressure drop?

Solution:

Figure 7.16 shows cut diameter as aerodynamic diameters. For 0.5 µm, Cunningham Correction Factor of 1.24 can be computed and therefore, the aerodynamic cut diameter is as follows:

$$d_a = 0.5\mu m \left(\frac{2 \text{ g/cm}^3}{1 \text{ g/cm}^3} \bullet 1.24 \right)^{\frac{1}{2}} = 0.79 \text{ µm}$$

From Figure 7.16 it is clear that this would require a throat velocity of about 90 m/sec and a pressure drop of about 80 cm H$_2$O. This velocity can now be used in Equation [7.19] to provide a final design. If an f value of 0.5 is used for design of full scale units, then Figure 7.16 would not apply to this condition and a smaller pressure drop may be acceptable (see Figure 7.17 in the following section where f = 0.25 and f = 0.50 are both presented).

7.6 The Contact Power Theory

A general theory which avoids the details of how particles and droplets collide is the *contact power theory*. This theory is based upon a series of experimental observations made by Lapple and Kamack (1955) and Equation [7.3] where efficiency is related to the number of transfer units, N_t. The fundamental assumption of the theory is :

> "When compared at the same power consumption, all scrubbers give the same degree of collection of a given dispersed dust, regardless of the mechanism involved and regardless of whether the pressure drop is obtained by high gas flow rates or high water flow rates."

In other words, collection efficiency is a function of how much power is used and not upon details of design. Semrau (1960) developed the contact power theory from the work of Lapple and Kamack (1955). The theory is empirical in approach and relates the total pressure loss, P_T, of the system to the collection efficiency.

The total pressure loss is expressed in terms of the power expended to inject the liquid into the scrubber plus the power needed to move the process gas through the system.

$$P_T = P_G + P_L \qquad [7.24]$$

where: P_T = total contacting power, hp/1,000 acfm

P_G = power input from gas stream, hp/1,000 acfm

P_L = power input from liquid injection, hp/1,000 acfm

The power expended in moving the gas through the system, P_G, is expressed in terms of the scrubber pressure drop:

$$P_G = 0.1575 \, \Delta p \qquad [7.25]$$

where: Δp = pressure drop in inches H$_2$O

The power expended in the liquid stream, P_L, is expressed as:

$$P_L = 0.583 \rho_L \left(\frac{Q_L}{Q_G} \right) \qquad [7.26]$$

where: P_L = liquid inlet pressure, lb/in^2

Q_L = liquid feed rate, gal/min

Q_G = gas flow rate, ft^3/min

ρ_L = liquid density, lb/ft^3

The constants given in the expression for P_G and P_L incorporate conversion factors to put the terms on a consistent basis.

The total power therefore can be expressed as:

$$P_T = P_G + P_L \qquad [7.27]$$

$$P_T = 0.1575 \Delta p + 0.583 \rho_L \left(\frac{Q_L}{Q_G} \right) \qquad [7.28]$$

This power supplied can be correlated with scrubber efficiency. Semrau (1960) defined collection efficiency as a function of power using:

$$N_t = \alpha P_T^\beta \qquad [7.29]$$

where: N_t = number of transfer units

α and β = empirical constants which are determined from experiment and are dependent upon the characteristics of the particulate matter

The general efficiency equation then becomes:

$$\eta = 1 - \exp(-\alpha P_T^\beta) \qquad [7.30]$$

Table 7.4 gives values of α and β for different industries.

Table 7.4
Values of α and β for Different Industries

Aerosol	Scrubber type	α	β
Raw gas (lime dust and soda fume)	Venturi and cyclonic spray	1.47	1.05
Prewashed gas (soda fume)	Venturi, pipeline, and cyclonic spray	0.915	1.05
Talc dust	Venturi	2.97	0.362
	Orifice and pipeline	2.7	0.362
Black liquor recovery furnace fume			
Cold scrubbing water humid gases	Venturi and cyclonic spray	1.75	0.62
Hot fume solution for scrubbing (humid gases)	Venturi, pipeline, and cyclonic spray	0.740	0.861
Hot black liquor for scrubbing (dry gases)	Venturi evaporators	0.522	0.861
Phosphoric acid mist	Venturi	1.33	0.647
Foundry cupola dust	Venturi	1.35	0.621
Open hearth steel furnace fume	Venturi	1.26	0.569
Talc dust	Cyclone	1.16	0.655
Copper sulfates	Solivore		
	(A) with mechanical spray generator	0.390	1.14
	(B) with hydraulic nozzles	0.562	1.06
Ferrosilicon furnace fume	Venturi and Cyclonic spray	0.870	0.459
Odorous mist	Venturi	0.363	1.41

The contact power theory gives a relationship which is independent of the size of the scrubber. With this observation, a small pilot scrubber could first be used to determine the pressure drop needed for the required collection efficiency. The full scale scrubber design could then be scaled up from the pilot information.

The basic principle of the contact power theory can be applied in other ways. For example, Calvert (1977; 1984) has used this approach to develop a relationship between the particle cut diameter and scrubber power. Using data from scrubber installations and relationships developed in the cut power approach, Calvert shows a similar improvement in performance (i.e., lower d_{p50}) with increased power consumption (Figure 7.17). The venturi scrubber predictions are for a liquid to gas flow rate of 1 L/m^3, corresponding to about the minimum pressure drop for a given penetration (see Figure 7.16).

Figure 7.17
Cut Diameter, d_{p50}, as a Function of Gas Pressure Drop and Power Consumption

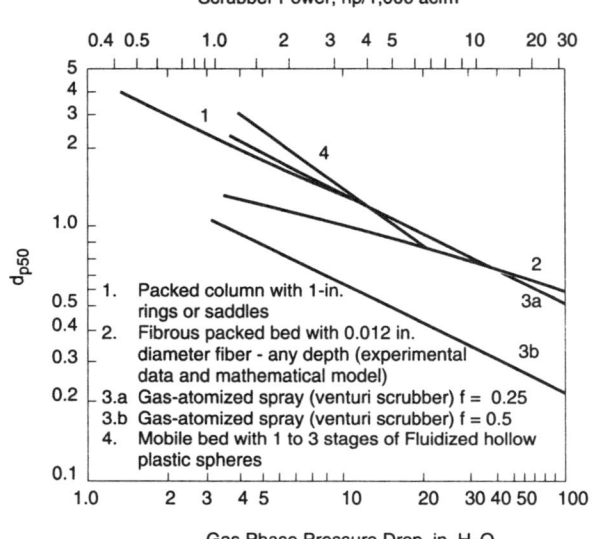

Source: Semrau 1960; Calvert 1977

The cut diameter/power plot can be used to predict scrubber performance if operating pressure drop is known. For example, consider a venturi scrubber with a pressure drop of 13 in. H_2O. From Figure 7.17, the performance cut diameter of the venturi scrubber is 0.63 μm (aerodynamic) if f = 0.5. Suppose the particle size distribution has d_{pg} = 10 μm (aerodynamic) and σ_g = 3, the overall penetration is 0.01 (from Figure 4.11).

The limit of what can be expected of a scrubber using inertial impaction is clearly indicated by Figure 7.17. If a cut diameter of 1.0 μm (aerodynamic) or smaller is required, the necessary pressure drop is in the medium-to-high energy range. High efficiency on particles smaller than 0.5 μm (aerodynamic) diameter would require extremely high pressure drop if inertial impaction was the only active mechanism.

High efficiency scrubbing of submicron particles at moderate pressure drop is possible, but it requires the application of diffusional collection as indicated by Figure 7.2. It can be seen that single particle collection efficiencies due to diffusion are high for 0.01 μm diameter particles. However, collection efficiencies for particles the size of a few tenths of a micron diameter is still poor.

Example 7.9 Pilot Scrubber Scale Up

A wet scrubber is to be used to control particulate emissions from a foundry cupola. Stack test results reveal that the particulate emissions must be reduced by 85% to meet emission standards. If a 100 acfm pilot unit is operated with a water flow rate of 0.5 gal/min at a water pressure of 80 psi, what pressure drop, Δp, would be needed across a 10,000 acfm scrubber unit?

Solution:

From Table 7.4, find $\alpha = 1.35$ and $\beta = 0.621$

Using Equation [7.3]

$$\eta = 1 - \exp(-N_t)$$

$$N_t = \ln\frac{1}{1-\eta} = \ln\frac{1}{1-0.85} = \ln 6.66 = 1.896$$

$$N_t = \alpha P_T^\beta = 1.35 P_T^{0.621} = 1.896$$

$$P_T = 1.73 \text{ hp} / 1,000 \text{ acfm}$$

From the pilot unit, it was determined the Q_L/Q_G ratio = 0.5/100 and $\rho_L = 80$ psi. Experimental data is necessary because Equation [7.28] contains three unknowns, Δp, ρ_L, Q_L/Q_G. With this data, the pressure drop can be calculated as

$$P_T = 0.1575\Delta p + 0.583\rho_L\left(\frac{Q_L}{Q_G}\right)$$

$$1.73 = 0.1575\Delta p + 0.583 \bullet 80\left(\frac{0.5}{100}\right)$$

$$\Delta p = 9.5 \text{ in. } H_2O$$

This pressure drop is the same for the pilot unit and full-scale unit.

7.7 Mist Eliminators

Wet gas cleaning equipment floods a gas with scrubbing liquid in the form of water drops. Because of this, water droplets are usually carried out of the scrubber by the exiting gas stream. A mist eliminator is passive equipment used after the wet scrubber to remove most of the entrained water droplets from the gas stream. This is necessary for the removal of particulate matter that has been collected and to prevent "local rain".

In the mist eliminator, the moisture-laden air and water droplets pass through a series of collecting baffles, screens, or packing. The gas passes through the mist eliminator while the water droplets are collected by inertial impaction. Figure 7.18 is a schematic of a mist eliminator down stream from a venturi scrubber. The collected droplets are eventually removed from the bottom of the collector.

Figure 7.18
Mist Eliminator

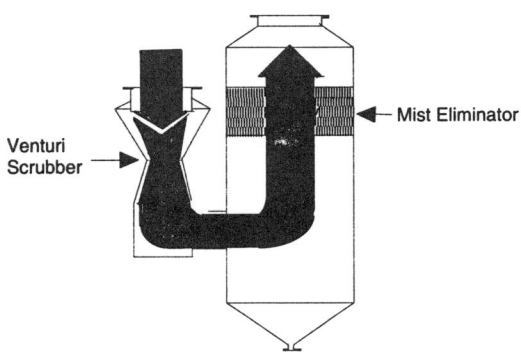

Droplets formed in the atomizing process are generally above 30 μm diameter and are called spray. A fine component is also formed with a size that varies with the atomizing energy. These droplets are classified as mist and have diameters below 30 μm. Figure 7.19 shows different types of mist eliminators. Cyclones, packed beds, wire mesh, and electrostatic precipitators are generally used for removal of mist. Wave plates or baffles are generally used to remove spray.

The velocity is the most important design factor for entrainment separators. The optimum velocity can be predicted as follows (Burkholz 1981):

$$v = k \left[\frac{\rho_L - \rho_G}{\rho_G} \right]^{\frac{1}{2}} \qquad [7.31]$$

where: k = between 0.35 and 0.45 for wire mist units and is about 0.4 for baffles

ρ_G = gas density, lb/ft^3

ρ_L = liquid density, lb/ft^3

v = superficial gas velocity, ft/sec

Figure 7.19
Methods for Droplet Separation

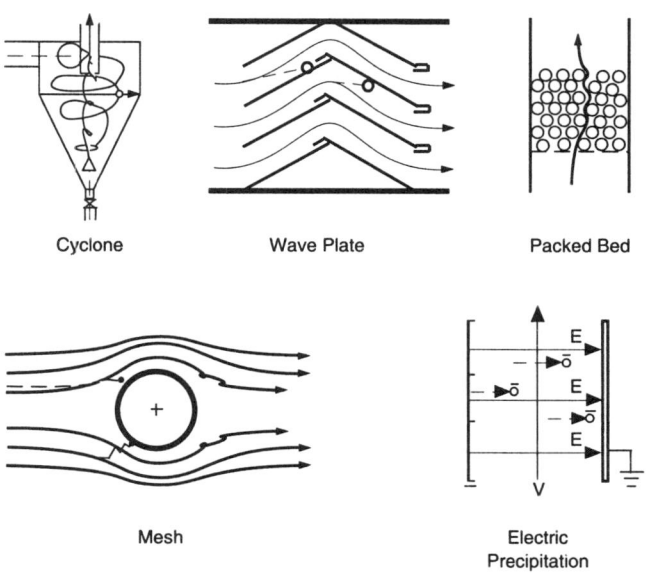

7.7.1 Baffles and Wave Plates

There are quite a few types of entrainment separators. Baffles, or wave plates, are probably the simplest of all. They are comprised of a series of inclined planes installed in a shell. The gas has to make one or more 90° turns to get through these planes. Removal efficiency for droplets as small as 30 μm can be improved by using a set of zigzag baffles so that the gas will have to follow a serpentine path, as shown in Figure 7.20. Baffle designs are used for droplets in the range of 30 to 100 μm. For those sizes, they have efficiencies of about 95%.

Collection effectiveness depends almost totally on impaction and interception for these large droplets. An average velocity of 10 ft/sec is used in demisters of this type with an upper limit of 16 ft/sec for most operations. A 50% increase in gas velocity will provide an increase in performance, but at a cost of almost doubling the pressure drop. In the usual case, the pressure drop is very low, 0.25 in. H_2O or less for velocities of 7 to 10 ft/sec.

7.7.1.1 Zigzag Baffle

With zigzag baffles, droplets are separated by repeatedly deflecting the gas stream through angles ranging from 30° to 45°, causing the droplets to impinge on the baffle surfaces. In general, the collection efficiency of zigzag baffles may be expressed as (Noll et al. 1986):

Figure 7.20
Zigzag Baffle Entrainment Separator

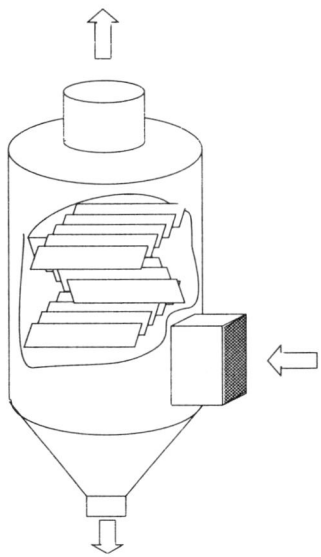

$$E = 1 - \exp\left(-\frac{A_D v_D}{Q_G}\right) \quad [7.32]$$

where: A_D = deposition surface area

v_D = deposition velocity

Q_G = gas volume flow rate

The deposition surface area is the upstream surface area of the baffles and the droplet deposition velocity is given by the terminal centrifugal velocity, where the centrifugal force acting on a droplet is due to the average radius of curvature that the gas stream follows through the baffles. Including expressions for these terms in gives Equation [7.32] yields:

$$E = 1 - \exp\left\{-\frac{w}{b}\left[\frac{v_{DI}}{v_G} + (n-1)\frac{v_D}{v_G}\right]\right\} \quad [7.33]$$

where:

$$\frac{v_{DI}}{v_G} = \left[\frac{8 \sin(\theta/2) \rho_D D_D}{3 w \cos^3(\theta/2) C_D \rho_g}\right]^{\frac{1}{2}} \quad [7.34]$$

$$\frac{v_D}{v_G} = \left[\frac{8 \sin\theta \rho_D D_D}{3 w \cos^3\theta C_D \rho_g}\right]^{\frac{1}{2}} \quad [7.35]$$

where: w = baffle width, in.

b = baffle spacing, in.

n = rows of baffles (see Figure 7.20)

θ = angle between baffle and flow direction, radians

v_D = terminal Stokes centrifugal velocity, cm/sec

v_G = superficial gas velocity, cm/sec

ρ_g = gas density, lb/ft³

ρ_D = droplet density, lb/ft³

D_D = droplet diameter, in.

C_D = droplet drag coefficient

When the droplet Reynolds number is less than 0.1, Stokes flow may be assumed and $C_D = 24/Re_D$, where:

$$Re_D = \frac{\rho_g v_D D_D}{\mu_g} \quad [7.36]$$

and μ_g is gas viscosity.

A more general drag coefficient for spheres is given by:

$$C_D = 0.22 + \frac{24}{Re_D}\left[1 + 0.15(Re_D)^{0.6}\right] \quad [7.37]$$

The pressure loss through zigzag baffles may be determined from:

$$\Delta P(\text{cm H}_2\text{O}) = 0.0102 \sum_{i=1}^{n} \frac{1}{2} f_p \rho_g (v_{Gi})^2 \quad [7.38]$$

where: f_p = dimensionless pressure loss coefficient,

ρ_g = gas density, kg/m³, and

v_{Gi} = the actual velocity between the baffles of row i, m/sec

$$v_{Gi} = \frac{v_G}{\cos(\theta/2)} \quad i = 1$$

$$v_{Gi} = \frac{v_G}{\cos\theta} \quad i > 1 \quad [7.39]$$

where v_G is the superficial gas velocity in m/sec. Experimental data indicate an average value of 1.2 for f_p.

7.7.1.2 Wave Plates

Figure 7.21 shows the profiles and dimensions of a few wave plates. Their separation performance can be calculated by approximating the gas deflection by segments having an angle of deflection α and radius of curvature R.

Figure 7.21
Wave-Plate Separator Profiles. (Dimensions in millimeters)

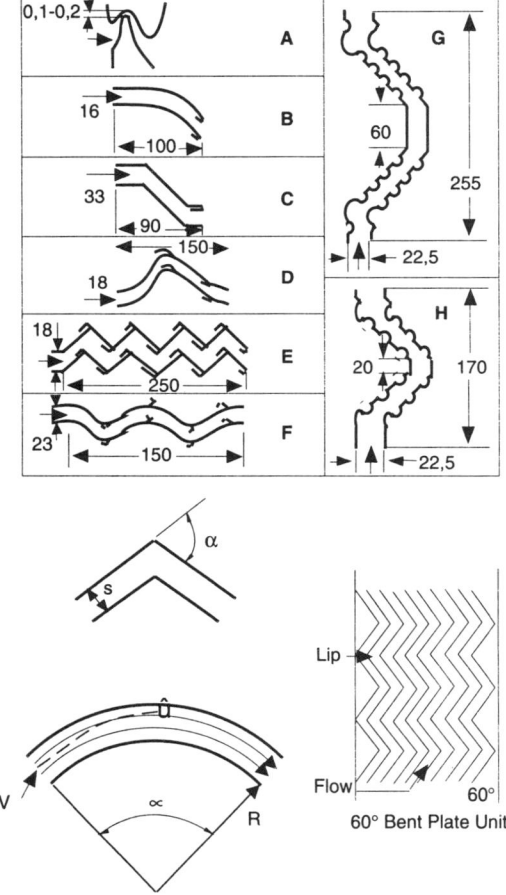

During the residence time, t, of the gas, the droplet travels along the distance:

$$s' = u_r t = u_r \alpha \frac{R}{v} \qquad [7.40]$$

radially outward (see Figure 7.21). Over this width, the outgoing stream of gas is free of droplets. If the distance between the wave plates is s, the separation efficiency is then:

$$\eta_{IC} = \frac{s'}{s} = \frac{u_r R}{v} \cdot \frac{\alpha}{s} = \frac{\rho_p v d_p^2}{18 \mu s} \alpha = \Psi \alpha \qquad [7.41]$$

where: $\quad \dfrac{u_r}{v} = \dfrac{\rho_p v d_p^2}{18 \mu R} = \Psi$

If the wave-plate separator has n bends each with the angle α, then

$$\eta_{total} = 1 - (1 - \Psi \alpha)^n \approx 1 - e^{-n\alpha \Psi} \qquad [7.42]$$

As an example, Figure 7.22 shows grade separation curves for a wave-plate separator where the arriving gas has velocities between 2.8 and 19 m/sec.

Figure 7.22
Fractional Separation Efficiencies (Grade Efficiency Curve) of a Wave-Plate Separator (Type H in Figure 7.21) at Different Gas Velocities

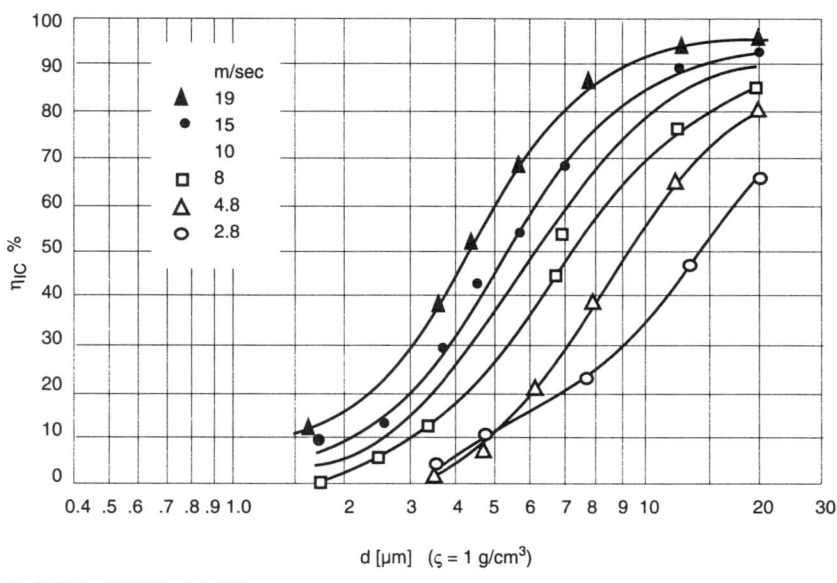

Source: Burkholz 1981; Noll et al. 1986

Example 7.10 Zigzag Baffle Separator Design

Design a zigzag baffle separator to remove particles from a spray scrubber (see Figure 7.20). The water droplets are 100 μm in diameter and the air flow rate is 100,000 acfm at 70°F and 1 atm pressure. The separator must obtain 97% collection efficiency.

Solution:

1. Flow velocity and collection efficiency in the baffle separator can be determined using Equations [7.31] and [7.33], respectively.

$$v_G = k\left[\frac{\rho_L - \rho_G}{\rho_G}\right]^{\frac{1}{2}}$$

where: $\rho_G = 0.075$ lb/ft^3

$\rho_L = 62.4$ lb/ft^3

$k = 0.4$ for baffles

$$v_G = 0.4\left[\frac{62.4 - 0.075}{0.075}\right]^{\frac{1}{2}}$$

$= 11.5$ ft/sec

2. Baffles with an angle to the flow of 45°, a width of 4 in., and spacing of 4 in. will be used.

Equation 7.33 $E = 1 - \exp\left\{-\frac{w}{b}\left[\frac{v_{DI}}{v_G} + (n-1)\frac{v_D}{v_G}\right]\right\}$

Equation 7.34 $\frac{v_{DI}}{v_G} = \left[\frac{8\sin(\theta/2)\rho_D D_D}{3w\cos^3(\theta/2)C_D\rho_g}\right]^{\frac{1}{2}}$

Using Equation [7.35]: $\frac{v_D}{v_G} = \left[\frac{8\sin\theta \rho_D D_D}{3w\cos^3\theta C_D\rho_g}\right]^{\frac{1}{2}}$

$w = 4$ in $= 4 \cdot 2.54 \cdot 10^4 = 1.016 \cdot 10^5$ μm

$b = 4$ in.

$\theta = 45°$

$D_D = 100$ μm

$C_D =$ droplet drag coefficient

3. Determine the droplet Reynolds' number using Equation [7.36].

$$Re_D = \frac{\rho_g v_d D_D}{\mu_g} = \frac{0.075 \text{ lb/ft}^3 \bullet 11.5 \text{ ft} \bullet (100 \bullet 10^{-4}/30.48)\text{ft}}{1.17 \bullet 10^{-5} \text{ lb/ft sec}}$$

$Re_D = 24.18$

Chapter 7: Wet Scrubbers

4. Determine the droplet drag coefficient.

   Using Equation [7.37]: $C_D = 0.22 + \dfrac{24}{24.18}\left[1 + 0.15(24.18)^{0.6}\right] = 2.22$

   Substituting these parameters into Equation [7.35]:

   $$\dfrac{v_{DI}}{v_G} = \left[\dfrac{8\sin(45°/2) \bullet 62.4 \text{ lb/ft}^3 \bullet 100 \text{ μm}}{3 \bullet 1.016 \bullet 10^5 \text{μm} \bullet \cos^3(45°/2) \bullet 2.22 \bullet 0.075 \text{ lb/ft}^3}\right]^{\frac{1}{2}} = 0.69$$

   $$\dfrac{v_D}{v_G} = \left[\dfrac{8\sin(45°) \bullet 62.4 \text{ lb/ft}^3 \bullet 100 \text{ μm}}{3 \bullet 1.016 \bullet 10^5 \text{μm} \bullet \cos^3(45°) \bullet 2.22 \bullet 0.075 \text{ lb/ft}^3}\right]^{\frac{1}{2}} = 1.4$$

5. Using Equation [7.33] and solving for n, the number of bends.

   $$0.97 = 1 - \exp\left\{-\dfrac{4 \text{ in.}}{4 \text{ in.}}[0.69 + (n-1) \bullet 1.4]\right\}$$

   $3.506 = 0.69 + (n-1) \bullet 1.4$

   $n = 3$

6. Determine the pressure loss through the separator with three bends or rows of baffles using Equation [7.38].

   $$\Delta P(\text{cm H}_2\text{O}) = 0.0102 \sum_{i=1}^{n} \dfrac{1}{2} f_p \rho_g (v_{Gi})^2$$

   where: $f_p \approx 1.2$

   $\rho_g = 1.275 \text{ kg/m}^3$

   $v_{GI} = \dfrac{11.5 / 3.28}{\cos(45°/2)} = 3.795 \text{ m/sec (bend one)}$

   $v_{G2.3} = \dfrac{11.5 / 3.28}{\cos 45°} = 4.96 \text{ m/sec (bends two and three)}$

   $\Delta P(\text{cm H}_2\text{O}) = 0.0102 \bullet 0.5 \bullet 1.2 \bullet 1.275(3.795^2 + 4.96^2 + 4.96^2)$

   $= 0.5 \text{ cm H}_2\text{O}$

   $= 0.2 \text{ in. H}_2\text{O}$

7. Determine the diameter of the separator.

   $$A = \dfrac{Q}{v_G} = \dfrac{100,000 \text{ cfm}}{11.5 \text{ ft/sec}} \bullet \dfrac{1 \text{ min}}{60 \text{ sec}} = 144.93 \text{ ft}^2$$

   $$d = \sqrt{\dfrac{4 \bullet 144.93}{\pi}} = 13.6 \text{ ft}$$

8. Determine the number of baffles:

$$\text{baffles} = 13.6 \text{ ft} \cdot \frac{4 \text{ in.}}{12 \text{ in./ft}} = 4.5$$

The zigzag baffle separator design is shown in the following figure.

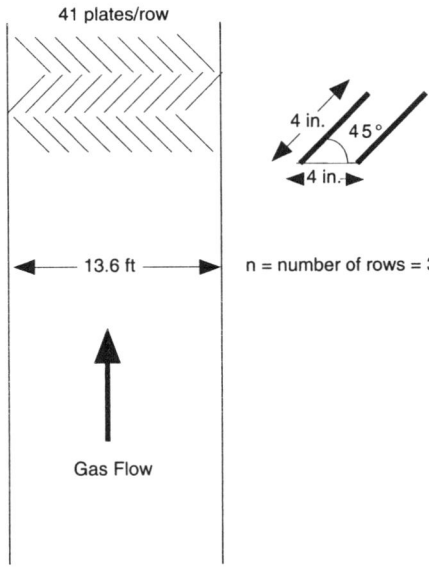

Example 7.11 Wave Plate Separator Design

Design a wave plate mist eliminator instead of a baffle chamber designed in Example 7.10 for the same conditions. Use Type H wave plate shown in Figure 7.21 with $\alpha = 60°$)

Solution:

Using Equation [7.42] $\eta_{total} = 1 - e^{-n\alpha\Psi}$ and Equation [7.41] $\eta_{IC} = \frac{\rho_p v d_p^2}{18\mu s}\alpha = \Psi\alpha$

where: $\rho_p = 1 \text{ g/cm}^3$

$v_t = 11.5 \cdot 30.48 \text{ cm/sec}$

$d_p = 100 \cdot 10^{-4} \text{ cm}$

$\mu = 1.8 \cdot 10^{-4} \text{ g/cm sec}$

From Figure 7.22 for H type separator:

$$s = 2.25 \text{ cm (opening in wave plate)}$$

$$a = 60° = \pi/3 \text{ radians}$$

$$\Psi = \frac{(1 \text{ g/cm}^3)(11.5 \bullet 30.48 \text{ cm/sec})(100 \bullet 10^{-4} \text{cm})^2}{(18 \bullet 1.8 \bullet 10^{-4} \text{ g/cm sec})(2.25 \text{ cm})}$$

$$\Psi = 4.8$$

If the wave-plate separator has n bends each with the angle $\alpha = \pi/3$, then:

$$\eta_{total} = 1 - e^{-n4.8 \bullet \pi/3}$$

$$0.97 = 1 - \exp(-5n)$$

$$5n = 3.5$$

$$n = 0.58$$

Therefore, use a 1 bend wave-plate separator to achieve the required 97% collection efficiency.

7.7.2 Knitted Wire Mesh

Knitted wire entrainment separators, shown in Figure 7.23, are used for removing droplets as small as 10 μm aerodynamic diameter from scrubber effluents. Maximum efficiencies are obtained at velocities of 8 to 18 ft/sec. As the liquid droplets are collected, they tend to coalesce throughout the entire mesh section in the loops formed by the knitting process. The collected liquid forms large drops, which can drain out against the rising air flow and fall into a sump. At higher velocities, flooding may occur. In that situation, not only are the individual wires completely covered with liquid, but the gas velocity is so high that the liquid cannot drain. The pressure drop increases rapidly under these conditions. The pressure drop for knitted wire units is generally less than 0.5 in. H_2O. Mesh units are not used if the particulate burden is greater than 100 g/m^3.

Wire-mesh mist eliminators have been widely used for many years to remove mist droplets (< 30 μm). They are made up of several layers of wire mesh spaced approximately 2 mm apart. The thickness of the wire is usually around 250 μm and the thickness of the filter 100 to 300 mm. The filter is highly porous with a large surface area. The filters can be custom-made and be used almost anywhere.

Wire-mesh and fiber filters work according to the same principle (see Chapter 6). A gas flows around a number of cylindrical elements except the predominate collection mechanisms are inertia and interception.

Figure 7.23
Knitted Wire Configurations for Mesh Entrainment Separators

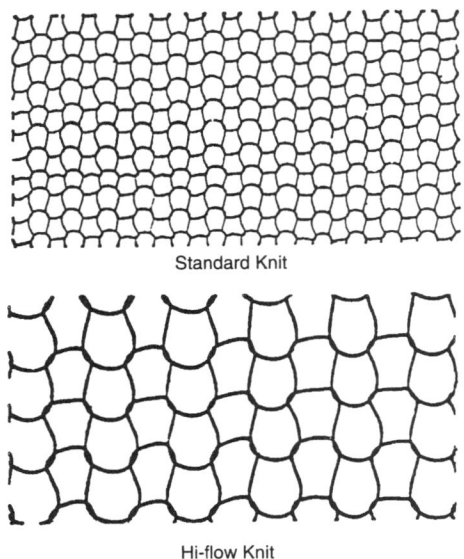

Standard Knit

Hi-flow Knit

The efficiency of a wire-mesh mist eliminator may be expressed as:

$$E = 1 - \exp\left(-f\frac{S}{\pi}\eta_{IC}\right) \qquad [7.43]$$

This equation is similar to the Kleinschmidt equation (Equation [7.3]). Here, f is a constant ranging from 0.7 and 0.85, η_{IC} is the single target efficiency of an isolated wire (cylinder), and S is the total surface area of wire per cross section of separator, given by

$$S = \frac{4Gh}{D\rho_f} \qquad [7.44]$$

where:
G = mesh bulk density
h = separator thickness
D = wire diameter
ρ_f = wire density

If we define $p = \frac{fS}{\pi}$ where p = the projected area of wire to cross-sectional area, then with p as a variable and η_{IC} values obtained from experiment, Equation [7.43] yields the grade efficiency curves shown in Figure 7.24. If multiple mesh layers are used then each layer can be treated as a separated wire-mesh. Table 7.5 (Schiffner and Hesketh 1983) is a comparison chart showing densities of various wire mesh.

Figure 7.24
Fractional Separation Efficiency Curves (Grade Efficiency Curves) of Wire-Mesh Filters, Calculated According to Equation [7.43]

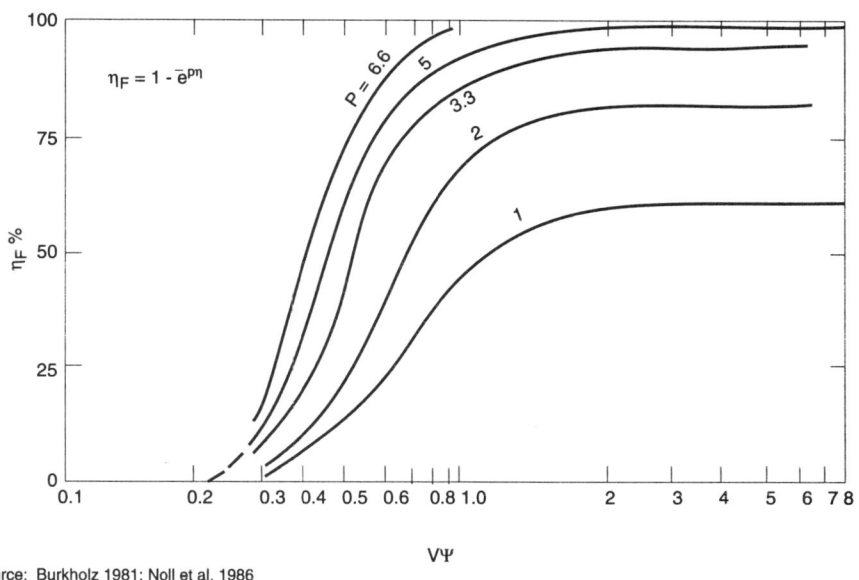

Source: Burkholz 1981; Noll et al. 1986

When wire-mesh mist eliminators become partly clogged, the disturbed gas flow around the individual wires slowly changes to a flow pattern through a porous system. This generally impairs the separating properties. There is usually a dust buildup in mist eliminators applied to scrubbers cleaning industrial emissions so that continuous spraying and periodic washing of the mist eliminator is required (typical wash rate is 3 gpm per ft^2 of surface area). Often, an entrainment separator will include baffles (to collect spray drops) followed by wire mesh (to collect mist drops). The baffles are used to remove material that could clog the wire mesh and impair collection efficiency.

Example 7.12 Wire-Mesh Mist Eliminator Design

If the emissions in Example 7.10 also contain 10 μm diameter mist droplets, design a wire-mesh (0.6 mm diameter for each wire) mist eliminator to remove 97% of the droplets. A two-part mist eliminator will be used for this example with the wire mesh following the zigzag baffles or wave plate separator.

Table 7.5
Wire-Mesh Eliminators Comparison Chart

Mesh Styles						Mesh density lb/ft^3	Mesh surface area ft^2/ft^3	Mesh voids %
KOCH	York	ACS	Vico Tex	UOP Type	Stone & Webster			
4,120		4BA	380	C	C	12.0	115	97.6
4,210	421					10.8	110	97.7
3,710	371					10.0	163	94.0
4,310	431	4CA	280	A	B	9.0	86	98.2
3,260	326	3BF	415		K	8.0	140	98.4
6,440	644					7.3	65	98.5
5,310	531	5CA				7.0	65	98.6
9,310	931	7CA	160	B	A	5.0	48	99.0
5,520		X200	611		L	20.0	450	96.0
5,540	333	X100	800		P	27.0	610	94.6
2,212	221	8T				4.0	125	97.0
2,414	241	8P				4.0	150	97.0

Solution:

1. Equation [7.44] $S = \dfrac{4Gh}{D\rho_f}$

    Use wire-mesh type KOCH 9310. From Table 7.5, the characteristics of this mesh are:

    G = 5.0 lb/ft^3

    ρ_f = wire density = 7.8•62.4 = 486.72 lb/ft^3

    D = 0.6 mm = 600 μm (typical size)

    f ~ 0.7

2. Solve Equation [7.44] for S, the total surface area.

$$S = \frac{4Gh}{D\rho_f}$$

$$= \frac{4 \bullet 5 \text{ lb/ft}^3 \bullet h(\mu m)}{600 \text{ μm} \bullet 486.72 \text{ lb/ft}^3} = 6.85 \bullet 10^{-5} h$$

Chapter 7: Wet Scrubbers

3. Use Figure 3.9 to determine h_{IC} which requires determination of K.

$$K = \frac{\rho_p v_T d_p^2 C_c}{9\mu D_c}$$

$\rho_p = 1 \text{g/cm}^3$, $v_T = 11.5 \cdot 30.48 = 350.5$ cm/sec, $d_p = 20 \cdot 10^{-4}$ cm,
$\mu = 1.8 \cdot 10^{-4}$ g/cm sec, $D_c = 600$ μm

$$K = \frac{(1 \text{ g/cm}^3)(350.5 \text{ cm/sec})(20 \cdot 10^{-4} \text{ cm})^2}{(9 \cdot 1.8 \cdot 10^{-4} \text{ g/cm sec})(600 \cdot 10^{-4} \text{ cm})}$$

$K = 14.4$

4. From Figure 3.9 $\eta_{IC} = 0.90$. Using Equation [7.43], solve for h, the separator thickness.

$$E = 1 - \exp\left(-f \frac{S}{\pi} \eta_{IC}\right)$$

$$0.97 = 1 - \exp\left(-0.7 \cdot \frac{6.85 \cdot 10^{-5} \cdot h}{\pi} \cdot 0.9\right)$$

$1.37 \cdot 10^{-5} \cdot h = 3.5$

$h = 255,474$ mm

$= 25.5$ cm

5. Check the resulting efficiency for this thickness using Figure 7.24.

$S = 6.85 \cdot 10^{-5} h$

$= 6.85 \cdot 10^{-5} \cdot 255,474$

$= 17.5$

$$p = \frac{fS}{\pi} = \frac{0.7 \cdot 17.5}{\pi}$$

$p \sim 4$

As determined: $K = 14.4$

$\Psi = K/2 = 14.4/2 = 7.2$

$\Psi^{0.5} = 2.7$

From Figure 7.24, with $p = 4$ and $\Psi^{0.5} = 2.7$, the observed $\eta_F \sim 95\%$

7.7.3 Wet Cyclones

A wet cyclone is probably the most commonly used entrainment separator for venturi scrubbers. They attain 98+ % efficiency for particles 20 to 25 µm in diameter at a pressure drop of 4 to 6 in. H_2O with velocities that are easily several hundred feet per second. A wet cyclone is just a cylindrical tank. The droplet-laden gas enters tangentially to achieve centrifugal motion, as shown in Figure 7.25. These units are simple to construct. They have no internal parts and thus do not clog.

Figure 7.25
Cyclonic Entrainment Separator

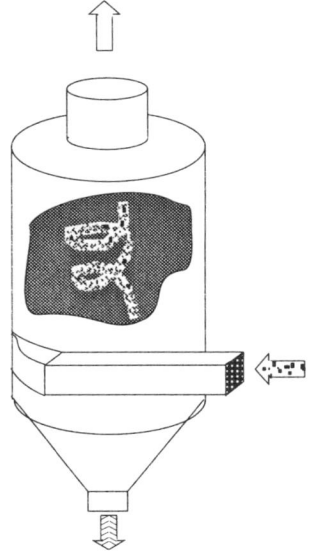

An internal wet cyclone has multiple tangential gas inlets and interior sets of vanes to impart adequate centrifugal force to the wet gas. It has the same pressure drop as a standard wet cyclone. It is useful for removal of 15 µm or larger droplets. Its primary advantage is that these units require about one third the space needed for a standard wet cyclone. Design of wet cyclones usually use the Leith and Licht model (see Chapter 5).

7.8 Design Summary

The basic design model for wet scrubbers is given by the Kleinschmidt equation, Equation [7.3]: $\eta = 1 - e^{-N_t} = 1 - e^{-f \eta_{drop}}$.

In this equation, f is a measure of the fraction of gas swept by the water drop and is a function of the liquid-to-gas flow rate, droplet size, and size of the control device. The factor η_{drop} is the collection efficiency for each drop in the scrubber and depends on three primary mechanisms: inertial impaction, interception, and diffusion. Quantification of these mechanisms relies on experimental results or computer-generated curves, such as those presented in Figures 7.1 and 7.2, where the single drop efficiency is shown to be a strong function of particle size.

Because of the large number of design variables in Equation [7.3] there has been considerable attention given to optimization of design. Figures 7.14, 7.15, 7.16 and 7.17 provide general solutions to the scrubber efficiency equation for spray and venturi scrubbers. These figures allow liquid-to-gas flow rate, the pressure drop, and collection efficiency to be determined for a specific application. One of the efficiency equations (Equations [7.3], [7.11], [7.14], [7.16], or [7.19]) can then be used to determine the size of the wet scrubber needed to meet this efficiency.

The number of mass transfer units has been related to power consumption in wet scrubbers and can be used to determine the pressure drop needed for a required collection efficiency. Figures 7.13 and 7.17 and Table 7.4 demonstrate the application of this design procedure.

7.9 Problem Set

7.1 Show that Equation [7.4] is equivalent to Equation [7.10].

7.2 A spray tower has been designed such that the fraction of area swept by the drops is equal to the volume of gas flowing in the tower. What is the collection efficiency for the tower if η_{ICD} is assumed to be equal to 1.0?

7.3 A 6-ft diameter countercurrent spray chamber is equipped with impingement nozzles that produce water droplets with an average diameter of 500 μm. To avoid excessive drop entrainment, the gas velocity will be 50% of the settling velocity of the spray. Estimate the maximum gas flow rate in ft^3/min that the spray chamber can handle.

7.4 A gravity spray tower 3 m high is operating at a L/G ratio of 1 L/m^3 with a drop diameter of 400 μm. The gas velocity is 0.1 of the drop terminal velocity which is 157 cm/sec. The operation is at 20 °C. What is the efficiency for particles of 1 μm diameter, having a density of 2.0 g/cm^3?

7.5 A vertical countercurrent spray chamber has a 2.0 m contact zone and operates with a L/G ratio of 1.0 L/m^3 and an average droplet diameter of 200 μm. During a test run, the following data were recorded:

  average scrubber temperature = 80°F
  gas velocity = 0.4 m/sec
  inlet loading = 1.5 grain/ft^3

The following particle size distribution was obtained from a cascade impactor:

Size range, μm	<4	4-8	8-16	16-30	30-50	>50
Mass, mg	25	125	100	80	20	10

Estimate the overall efficiency of the spray chamber.

7.6 Determine the collection efficiency for a cross-current and a countercurrent spray tower with the following design:

    4 ft high

    water flow = 20 gal/min

    air flow = 3,000 ft^3/min

    water drop = 500 μm diameter

    particle size = 10 μm

    velocity of gas = 2 ft/sec

7.7 Rain drops are falling at their terminal velocity in still air at 32 °F. The droplets may be assumed to be 150 μm in diameter. Assuming that the air contains suspended dust particles of density 2.2 g/cm^3, calculate and plot the collection efficiency of a single droplet as it falls, as a function of the dust particle diameter.

Data for collection efficiency as a function of the inertial parameter are:

$\sqrt{\Psi}$	η, %
0.2	0
0.35	20
0.4	40
0.55	60
0.85	80
1.0	83
1.5	85

7.8 A venturi scrubber is required to collect dust from an asphalt drier. The dust has a mass median diameter of 1.8 μm and a density of 2.6 g/cm^3. The uncontrolled emission rate is 2,310 kg/hr, but state regulations require that this be reduced to a maximum of 25 kg/hr. The air flow is 15,000 acfm at 250°F. Assuming a throat velocity of 150 ft/sec, make a preliminary determination of the necessary L/G value, and of the maximum pressure loss in the throat (use Figure 7.13). *(Note: A final design for the venturi would require information on the size distribution of the dust).*

7.9 Design a venturi scrubber for Example 7.5 using the penetration theory and Calvert's design method with f = 0.25 and f = 0.50. Compare your results.

7.10 A venturi scrubber has a throat velocity of 125 m/sec and pressure drop of 55 in. H_2O. Calculate the removal efficiency using Calvert's method for 5 μm diameter particles. The temperature is 70°F and the pressure is 1 atm. Assume that f = 0.5 and that the L/G ratio is 1 L/m³.

7.11 Provide a predicted grade efficiency curve and pressure drop for a venturi scrubber under the following conditions:

    temp = 20°C Q_L/Q_G = 1.33 L/m³

    gas velocity = 8,060 cm/sec   throat diameter = 20.3 cm

7.12 Using the contact power theory, estimate the total pressure loss in a system if the pressure drop is 5 in. H_2O and the L/G ratio is 15 gal/1,000 ft³, with a liquid inlet pressure of 1,000 psi.

7.13 A vender proposed to use a spray tower on a lime kiln to reduce the discharge of solids to the atmosphere. The inlet loading of the gas stream from the kiln is 5.0 grains/ft³ and is to be reduced to 0.05 grains/ft³ in order to meet state regulations. The vendor's design calls for a water pressure drop of 80 psi and a pressure drop across the tower of 5.0 in. H_2O. The gas flow rate is 10,000 acfm and a water rate of 50 gal/min is proposed. Assume the contact power theory to apply.

    a) Will the spray tower satisfy the state regulations?

    b) What total pressure loss is required to satisfy the state regulations?

    c) Propose a set of operating conditions that will meet the standard. The maximum gas and water pressure drop across the unit are 15 in. H_2O and 100 psi, respectively.

    d) What conclusion can be drawn concerning the use of a spray tower for this application?

7.14 The installation of a venturi scrubber is proposed to reduce the discharge of particulates from an open-hearth furnace. Preliminary design information suggests a water and gas pressure drop across the scrubber of 5.0 psi and 36 in. H_2O, respectively. A L/G ratio of 6.0 gal/1,000 ft³ is usually employed in this application. Estimate the collection efficiency of the proposed venturi scrubber using the contact power theory.

7.15 A paper mill uses a venturi scrubber to collect black liquor furnace fumes. The gas pressure drop is 30 in. H_2O and the scrubber water pressure drop is 10 psi. The gas rate is 10,000 ft³/min and the water rate is 1,000 gal/min.

a) Estimate the fume collection efficiency.

b) Estimate the electric power consumption in kilowatts, assuming that the blower and pump are 50% efficient.

c) Estimate the electricity cost for a 30-day month when the plant operates 24 hours per day and electricity costs 2¢/kWhr.

7.16 A venturi scrubber is used to collect foundry cupola dust. The gas pressure drop is 20 in. H_2O and the water pressure drop is 15 psi. The gas flow rate is 20,000 ft³/min and the water rate is 1,500 gal/min. Estimate the dust collection efficiency and the power consumption.

7.17 What would be the pressure drop required for a venturi scrubber to achieve an overall collection efficiency of 99.3% for particulate matter having a mass median diameter of 5 μm (aerodynamic) with a particle size deviation, σ_g, of 2.0 μm.

7.18 A venturi scrubber is being considered to control the emissions from an industrial operation. The following data are available from a stack sample:

Q = 14,000 acfm

C = 3 grains/acfm ρ_p = 1.5 g/cm³

MMD = 5.9 μm σ_g = 2.4

The required collection efficiency is very high, 99.5%. The initial design has been selected with a gas velocity of 150 m/sec (492 ft/sec) and a L/G ratio of 1 L/m³ (7.5 gal/1,000 ft³). Determine the collection efficiency and pressure drop for this venturi scrubber.

7.19 It has been found necessary to control particulate emissions from an asphalt plant stone dryer. The plant has a capacity of 136 metric ton/hr and a flue gas rate of 34,000 actual m³/hr. The gas temperature is 116 °C and has a dew point of 65 °C. The particle size distribution is shown on the table below. Particle density is 2.6 g/cm³. Currently, the uncontrolled emission rate is 22,310 kg/hr, compared to a maximum allowable limit of 25 kg/hr. Determine the design specifications for a wet scrubber which will meet the above performance requirements.

%<d_p	d_p, μm
2	0.9
10	2.7
50	18
80	68

7.20 A coke-producing plant has recently been cited several times for opacity violations. The regional state air pollution control office requested that the company conduct three EPA Method 5 particulate tests to determine if the plant was in compliance with the state's mass emission regulations. The average of the test results showed that the coke mass emission rate to the stack was 625 lb/hr.

Chapter 7: Wet Scrubbers 261

Air Pollution Regulations governing particulate emission standards are determined by:

$$E = 0.88 Q^{-0.1665}$$

where: E = maximum emission rate, lb/10^6 Btu

Q = total heat input, Btu/hr

Determine if the coke plant is in violation of the state's particulate emission regulations. If a control device is needed, what type of wet scrubber should be used? What would be the required scrubber pressure drop and what would be the power requirement for operating the scrubber?

Data which may be needed to solve this problem are given below:

fresh feed to coke plant	= 6,000 blb/day (1 blb = 1 • 10^9 lb)
gas fired to CO boiler	= 1,322 scf/min
oil fired to CO boiler	= 250 lb/day
heat content of gas	= 1,201 Btu/scf
heat content of oil	= 18,012 Btu/lb

The particle density is 1.5 g/cm³ and the particle size distribution leaving the CO boiler is shown in the following table:

d_p, μm	%<d_p
0.58	5
1.00	15
1.43	25
1.9	35
2.43	45
3.10	55
4.00	65
5.20	75
6.80	88

7.21 A 1/4 inch steel wire mesh with a wire diameter of 0.6 mm is used as a mist eliminator. The air velocity is 10 ft/sec. What is the collection efficiency for 1.0, 5.0, 10, and 30 μm mist droplets?

7.10 References

Burkholz, A. "Mist Elimination" *Air Pollution Control*, Part IV, Eds. Bragg and Strauss. John Wiley & Sons. New York. 1981.

Calvert, S. and H. Englund (Eds.) *Handbook of Air Pollution Technology.* John Wiley & Sons. New York. 1984.

Calvert, S. "Scrubbing" *Air Pollution* Vol. IV, Stern, A. (Ed.) Academic Press. New York. 1997.

DeNevers, N. *Air Pollution Control Engineering.* McGraw-Hill. New York. 1995.

Fonda, A. and H. Herne. *The Classical Computations of the Aerodynamic Capture of Particles by Spheres.* Int. J. Air Pollut. 3. 1960.

Friedlander, S.K. *Smoke, Dust, and Haze.* Wiley Interscience. New York. 1977.

Johnstone H., R. Field, and M. Tossler. *Gas Absorption and Aerosol Collection in a Venturi Atomizer.* Ind. & Eng. Chem. 46, 1601-1607. 1954.

Johnstone, H.F. and W.H. Roberts. Deposition of Aerosol Particles from Moving Gas Streams. Ind. Eng. 1949.

Kleinschidt, R. *Factors in Spray Scrubber Design.* Chem. & Met. Eng. Vol 46. 487. 1939.

Langmuir, I. *The Production of Rain by Chain in Cumulus Clouds at Temperatures Above Freezing.* 1948.

Lapple, C. and H. Kamack. *Performance of Wet Dust Collectors.* Chem. Eng. Prog. 51:110-121. 1955.

Licht, W. *Air Pollution Control Engineering: Basic Calculations for Particulate Collections.* Marcel Dekker. New York. 1980.

Noll, K., G. Nichols, J. Crowder, and S. Senkan. *Control Devices: Electrostatic Precipitation, Scrubbing, Mist Elimination, Adsorption, and Combustion of Toxic and Hazardous Wastes.* Chap. 7 in Air Pollution, Vol VII. Ed. A. Stern, Academic Press. New York. 1986.

Nukiyama, S. and Y. Tanasawa. *An Experiment on Atomization of Liquid By Means of Air Stream. Trans.* of Soc. Mech. Eng. Japan: 4, 86. 1983.

Pilat, M and A. Prem. "Calculated Particle Collection Efficiencies of Single Droplets Including Inertial Impaction, Brownian Diffusion, Diffusophorsis and Thermophoresis", *Atm. Envr.*, vol 10. Pergamon Press, Great Britain. 1976.

Pilat, M. J. and A. Prem. "Effect of Diffusion, Diffusophorsis and Thermophoresis on Particle Collection Efficiencies of Spray Droplets Scrubbers", *JAPCA*, vol 27. 1977.

Ranz, W. E. and J.B. Wong. *Impaction of Dust and Smoke Particles on Surface and Body Collectors.* Ind. Eng. Chem. 44. 1952.

Schiffner, K and H. Hesketh. *Wet Scrubbers.* Ann Arbor Science, Ann Arbor, MI. 1983.

Semrau, K. *Correlation of Dust Scrubber Efficiency.* J. Air Pollution Control Assoc. 10: 200-207. 1960.

Sink, M. *Handbook: Control Technologies for Hazardous Air Pollutants.* Office of Research and Development, Cincinnati, OH: US EPA/625/6-91/014. 1991.

Sparks, L. E. and M.J. Pilat. *Effect of Diffusiophoresis on Particle Collection by Wet Scrubbers.* Atmospheric Environment 4. 1970.

Stairmand, C. *The Design and Performance of Modern Gas-Cleaning Equipment.* J. Inst. Fuel, 29, 58-81. 1956.

Chapter 8

Electrostatic Precipitators

8.1 Introduction

An electrostatic precipitator (ESP) uses electric forces to separate suspended particles from gases. All precipitators operate on the same underlying principles. These principles include three major fundamental steps: electric charging of the suspended particles, collection of the charged particles in an electric field, and the removal of the precipitated material to an outside receptacle.

Electrostatic precipitators have been applied to a wide variety of dust control problems, including metallurgical operations (ore roasting, sintering, smelting, and furnaces), chemical operations (sulfuric acid and phosphate production), pulp and paper manufacturing, and large volume particulate applications, such as power plants and cement kilns. The principal advantage of electrostatic precipitators is the ability to separate small particles from large volume gas streams with relatively low energy requirements due to low pressure drop. In comparison to other dust control methods, precipitators are often higher in initial cost but lower in operating costs.

The electrostatic precipitator is composed of four essential components (Figure 8.1). The four major components are: discharge electrodes, collection electrodes, rappers, and hoppers. The *discharge electrode* is normally a wire or rod where a

Figure 8.1
Flat Surface Electrostatic Precipitator

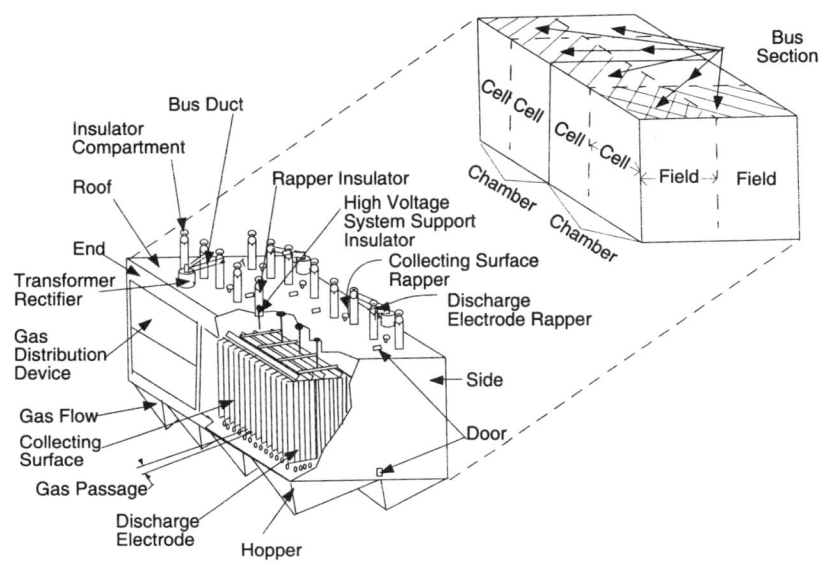

corona discharge occurs (Figure 8.2). This electrode is used to ionize the gas (which charges the particles) and creates an electric field. The *collection electrode* is a tube or flat plate which is oppositely charged (relative to the discharge electrode) and is the surface where the charged particles are collected. The *rapper* is a device used to impart a vibration or shock to dislodge the deposited dust on the electrodes. *Hoppers* are located at the bottom of the precipitator and are used to collect and store the dust removed by the rapping process.

Figure 8.2
Electric Field Generation (top view)

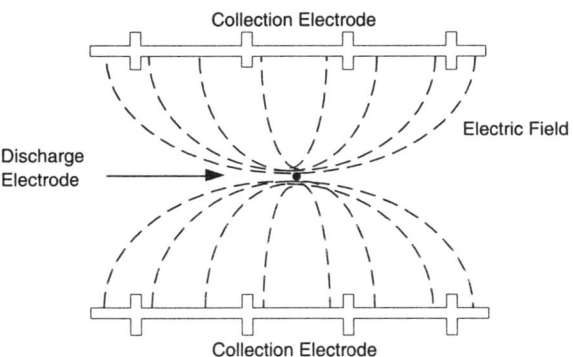

Electrostatic precipitation occurs in the space between the discharge electrode and the collection surface. A high voltage is applied to an electrode system, which is usually negatively charged, and the collecting plate electrode, which is grounded. The applied voltage is increased until it produces a corona discharge (corona), which can be seen as a luminous blue glow around the discharge electrode. The intense electric field close to the discharge electrode accelerates free electrons that are present in the gas. These electrons acquire sufficient velocity to ionize gas molecules upon collision, producing a positive ion and an additional free electron.

The additional free electrons create more positive ions and free electrons as they collide with additional gas molecules. This process is called *avalanche multiplication*. The electrons produced during the avalanche multiplication process follow the electric field toward the grounded collection electrode. When electrons impact on gas molecules, they are captured and negative gas ions are created. These negative ions serve as the principal mechanism for charging the dust particles.

Dust particles become highly charged in the first 10 or 20 cm of travel and are driven to the grounded collecting electrodes by the intense electric field of the corona. The entire process is very fast, usually taking only a few seconds (White 1963).

Increases in the applied voltage will increase the field strength and ion formation until *sparkover* occurs. Sparkover is the internal sparking between the discharge and collection electrodes. It is a sudden rush of localized electric current through the gas layer between the two electrodes. Sparking causes an immediate short-term collapse of the electric field. In general, it is very desirable to operate at voltages high enough to cause some sparking, but not at a frequency such that the electric field constantly collapses. The average sparkover rate for optimum precipitator operation is between 50 and 100 sparks per minute. For optimum efficiency, the electric field strength should be as high as possible. This is accomplished by applying a high voltage to the discharge electrode and the consequent high corona current flow from the discharge electrode to the collection electrode.

Corona characteristics depend on many factors, including electrode spacing and configuration, gas composition, pressure, temperature, dust loading and particle size, dust deposits on the corona and collection electrodes, and on the electrical resistivity of the particulates being collected. Either positive or negative corona may be used, but for industrial gas cleaning, negative polarity is preferred because of its greater stability and the higher operating voltages and currents possible. Typical corona voltage for precipitators of 10 to 15 cm electrode spacing range from 40 to 60 KV peak value. Corresponding corona current densities range from 0.1 to 1 mA/m^2, depending on the gas and dust conditions.

Resistivity is related to the ability of a particle to conduct a charge. In most industrial applications, the resistivity of the particle is such that the charge on the particle is only partially discharged upon contact with the grounded collection electrode.

The dust layer builds up on the collection plate to a thickness between 0.03 and 0.5 inches. If the dust layer becomes too thick, it is possible for the accumulated layer to act as an insulator, reducing the flow of the electric field lines.

To maintain the continuous process of precipitation, it is necessary to periodically remove collected dust particles from the discharge and collecting electrodes. Deposited particles are dislodged by mechanical impulses or rapping.

8.2 Resistivity

Particle resistivity is a condition of the particle in the gas stream that can alter the actual collection efficiency of an ESP. Resistivity is a term that describes the resistance of the collected dust layer to the flow of electric current. By definition, the resistivity is the electrical resistance of a dust sample 1.0 cm in cross-sectional area, 1-cm thick, and recorded in units of ohm-cm. It can also be described as the resistance to charge transfer by the dust. Dust resistivity values can be classified roughly into three groups:

- between 10^4 and 10^7 ohm-cm — low resistivity
- between 10^7 and 10^{10} ohm-cm — normal resistivity
- above 10^{10} ohm-cm — high resistivity

Dust resistivity of 10^9 to 10^{10} ohm-cm is desired for ESPs. As resistivity increases above this level, it becomes difficult to maintain adequate corona discharge, and back ionization can occur, freeing collected particles and stopping precipitator operation. Particles that have low resistivity are difficult to collect since they are easily charged and lose their charge upon arrival at the collection electrode; particles thus bounce off the plates and become reentrained in the gas stream.

Examples of low resistivity dust are unburned carbon in fly ash and carbon black. Particles that have normal resistivity do not rapidly lose their charge upon arrival at the collection electrode. These particles slowly leak their charge to ground and are retained on the collection plates by intermolecular adhesive and cohesive forces. This allows a particulate layer to build up, which is then dislodged into the hoppers.

Particles that exhibit high resistivity are difficult to charge. As the dust layer builds up on the collection electrode, the layer and the electrode form a high potential electric field. This causes a condition known as *back corona*. Under the influence of corona discharge, the dust layer breaks down electrically, producing small holes or craters in the layer from which back corona discharge occurs. High resistivity can generally be reduced by adjusting the temperature and moisture content of the gas stream. Particle resistivity decreases for both high and low temperatures (Figure 8.3) (White 1974, 1984). Particle volume resistivity dominates at low temperatures and surface resistivity dominates at high temperatures, yielding the resistance

Figure 8.3
Effect of Temperature and Moisture Content on the Electrical Resistivity of Dusts

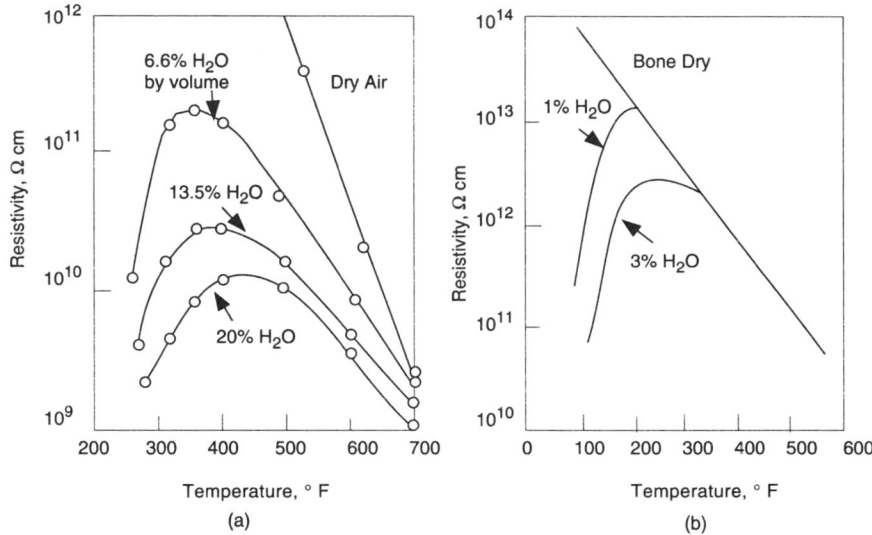

Note: (a) Moisture conditioning of cement kiln dust; (b) Effect of gas humidity in increasing the conducitivity of a typical fly ash

Source: White 1984

values seen in the figure. Fly ash resistivity is highly dependent upon SO_3, which usually comes from the combustion of sulfur. The presence of SO_3 in the gas stream has been shown to favor the electrostatic precipitation process when problems with high resistivity occur. Most of the sulfur content in the coal burned for combustion sources converts to SO_2. However, approximately 1 percent of the sulfur converts to SO_3. The amount of SO_3 in the flue gas normally increases with increasing sulfur content of the coal. The resistivity of the particles decreases as the sulfur content of the coal increases (Figure 8.4) (Seizler 1974).

The use of low sulfur western coal for boiler operations has caused fly ash resistivity problems for ESP operations. For such fly ash dusts, the resistivity can be lowered below the critical level by the injection of as little as 10 to 20 ppm SO_3 into the gas stream (Figure 8.5). The SO_3 is injected into the duct work preceding the precipitator. Other conditioning agents, such as sulfuric acid, ammonia, sodium chloride, and soda ash have been used to reduce particle resistivity.

Other substances, besides sulfur, can serve as natural conditioning agents for fly ash. The high sodium-to-silica ratio in low sulfur western coal provides good fly ash conditioning because sodium oxide is a good electrical conductor. Next in order of importance are lithium, iron, and potassium oxides. Low sulfur (0.25%)

Figure 8.4
Variation in Resistivity of Fly Ash with Sulfur Content of Coal

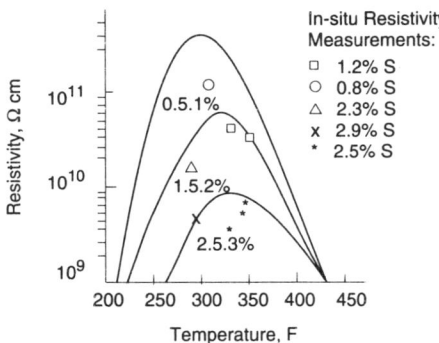

Source: White 1984

Figure 8.5
Effect of Conditioning of Fly Ash

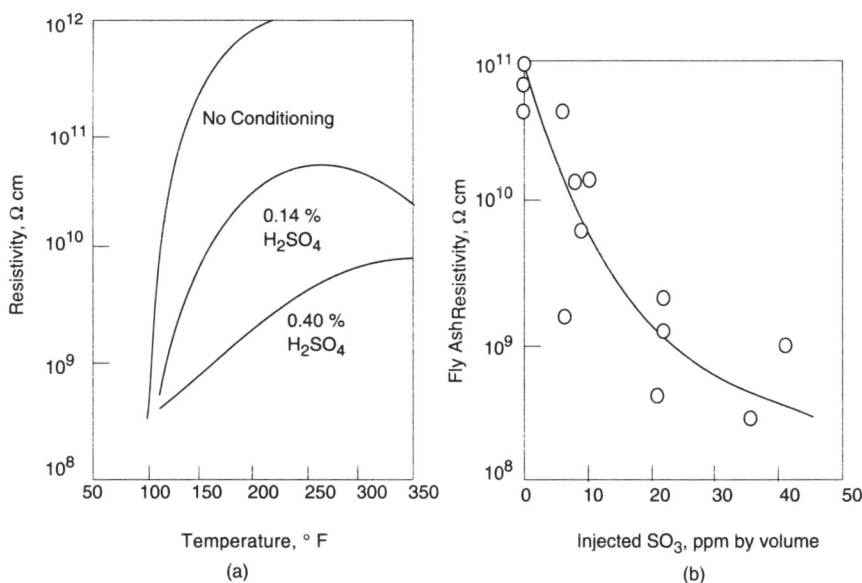

Note: (a) H2SO4 Fume and (b) SO3 Injection into Flue Gas

Source: White 1984

western coal with low sodium (0.3% Na_2O) in ash would be expected to have a fly ash resistivity of about 10^{12} ohm-cm at 175°C. The same fly ash with about 2% Na_2O would have a resistivity of about 10^{10} ohm-cm, and at 5% Na_2O, about 3 • 10^9 ohm-cm.

If the desired resistivity is not achieved by natural or added chemical conditioning, then fuel blending and/or hot precipitator operation could be considered. Hot precipitators operate at temperatures greater than 350°C. The added gas volume at high temperatures increases ESP unit size. Hot precipitators are usually located before the combustion air pre-heater section of the boiler.

8.3 Electric Fields

An electric field exists in the space around the discharge electrode and causes a charged particle in this space to experience an electrical force. The electrical force is given by:

$$F_e = qE_p \tag{8.1}$$

where: F_e = force due to the electrical field

            q = charge on particle

            E_p = electrical field strength

It is common practice to express the amount of charge, q, as n multiples of the smallest unit of charge, e, ($4.8 \cdot 10^{-4}$ stC or $1.67 \cdot 10^{-19}$ C):

$$q = ne \tag{8.2}$$

Thus, the force on a particle with n elementary units of charge is:

$$F_e = neE \tag{8.3}$$

This is the basic equation for the electrostatic force experienced by a particle.

When the smallest unit of charge, e, is defined as a single electron ($4.8 \cdot 10^{-10}$ statcoulomb [stC]), the following relationships hold:

$$\text{statampere} = \frac{\text{statcoulomb}}{\text{second}}$$

$$\text{statvolt} = \frac{\text{dyne centimeter}}{\text{stacoulomb}}$$

$$\text{dyne} = \frac{\text{statcoulomb}^2}{\text{cm}^2}$$

$$\text{statvolt} = \frac{\text{statcoulomb}}{\text{cm}}$$

In the SI system of units, with force in Newtons, N, and distance in meters, unit charge is defined independently of electrostatic force. In this system, the ampere, A, is the base unit defined as the current required to produce a specified force between two parallel wires 1 m apart. The units of charge and potential difference are derived from the ampere. The unit of charge, the coulomb, C, is defined as the amount of charge transported in 1 sec by a current of 1 A. The unit of potential difference, the volt, V, is defined as the potential difference between two points along a wire carrying 1 A and dissipating 1 watt (W) of power between the points. Table 8.1 gives conversion factors relating the two systems.

**Table 8.1
Conversion Factors for CGS and SI Electrostatic Units**

Quantity	CGS	SI
Charge	1 stC	= 3.33 • 10^{-10} C
Current	1 stA	= 3.33 • 10^{-10} A
Potential difference	1 stV	= 300 V
Charge per electron	4.8 • 10^{-10} stC	= 1.6 • 10^{-19} C

Example 8.1 Particle Electrostatic Force

A 0.6 µm particle acquires 40 excess electrons. The field strength is 2 KV/cm. What is the electrostatic force on the particle?

Solution:

$$E = \frac{2{,}000 \text{ V/cm}}{300 \text{ V /stV}} = 6.67 \text{ stV/cm}$$

$F = neE = (40 \cdot 4.8 \cdot 10^{-10} \text{ stC}) \cdot (6.67) \text{ stV/cm}$

$F = 1.28 \cdot 10^{-7}$ dyne

The central problem for the application of electrostatic theory is the determination of n and E_p, the field strength. E_p is determined by:

$$E_p = \frac{V}{X} = \frac{\text{volts}}{\text{meter}} \qquad [8.4]$$

where: V = applied voltage

X = shortest distance from the discharge wire to the collecting surface

The next section is devoted to the determination of the magnitude of particle charging.

8.4 Particle Charging Mechanisms

Particle charging is the first step in the electrostatic precipitation process. Two particle charging mechanisms occur in the corona field of a precipitator. The most important is charging by ions driven to the particles by the force of the applied electric field. A secondary charging process occurs due to the phenomenon of diffusion, which depends on the thermal energy of the ions, but not on the electric field. The field-charging process is predominant for particles larger than 0.5 µm in diameter; the diffusion process for particles smaller than about 0.2 µm; and both are important for particles in the intermediate range between 0.2 and 0.5 µm (Hinds 1982).

Particle charging by the field process under conditions existing in the unipolar corona is well understood and the theoretical predictions have been confirmed by experiment. The theory has been developed for the case of an isolated spherical particle placed in a uniform field with a uniform ion density. These conditions are met in the ordinary unipolar corona for particles in the micron size range and for particle concentrations normally occurring in practice.

Under conditions where diffusion charging can be neglected, the number of charges, n, acquired by a particle during a time, t, in an electric field, E, with an ion number concentration, N_i, is:

$$n = \left(\frac{3\varepsilon}{\varepsilon+2}\right)\left(\frac{Ed_p^2}{4e}\right)\left(\frac{\pi e Z_i N_i t}{1+\pi e Z_i N_i t}\right) \qquad [8.5]$$

where:
- ε = dielectric constant of the particle
- d_p = particle diameter
- N_i = ion concentration
- Z_i = mobility of the ions (approximately 450 cm²/stV sec)
- t = charging time

In Equation [8.5], the first two factors represent the saturation charge, n_s, that is reached after sufficient time at a given charging condition.

$$n_s = \left(\frac{3\varepsilon}{\varepsilon+2}\right)\left(\frac{Ed^2}{4e}\right) \qquad [8.6]$$

The first factor on the right in Equations [8.5] and [8.6] depends only on the material of the particle and ranges from 1 for $\varepsilon = 1$ to 3 for ε = infinity. The dielectric constant, ε, reflects the relative strength of the electrostatic field produced in different materials by a fixed potential. Most materials have $1<\varepsilon<10$; ε is 1 for vacuum, 1.00059 for air, 4.3 for quartz, 80 for pure water vapor, and infinity for conducting particles. The second factor in Equation [8.5] indicates that the saturation charge is proportional to the surface area of the particle and to the electrostatic field strength. The final factor is a time-dependent term that reaches a value of 1 when $\pi e Z_i N_i t \gg 1$. The rate of charging does not depend on the particle size or field strength, but only on the ion concentration, so charging will be 95% complete in 3 sec or less.

An approximate expression for the number of charges, n, acquired by a particle of diameter, d_p, by diffusion charging during a time, t, is:

$$n = \frac{d_p kT}{2e^2} \ln\left(1 + \frac{\pi d_p c_i e^2 N_i t}{2kT}\right) \quad [8.7]$$

where: c_i = mean thermal speed of the ions (c_i = 2.410 cm/sec)

N_i = concentration

k = Boltzmann's constant

At standard conditions, Equation [8.7] is accurate within a factor of 2 for particles 0.1 to 2 µm diameter and for $N_i t > 10^6$.

The charge acquired is proportional to d_p^2 in field charging and d_p in diffusion charging, so field charging is the dominant mechanism for particles larger than one µm and diffusion charging is the dominant mechanism for particles less than 1 µm. In between these sizes, both mechanisms are operating (Hinds 1982). Table 8.2 gives the charges acquired in 1 sec by particles of various sizes by the diffusion and field charging. For the conditions of Table 8.2, the field-charging equation is accurate for particles larger than 5 µm, and Equation [8.7] is useful from 0.2 to 2 µm. Figure 8.6 shows the particles' mobility as a function of particle size for typical charging conditions. Although particle charge decreases with particle size, mechanical mobility increases rapidly with decreasing size, and consequently there is

Table 8.2
Comparison of Calculated Methods for Charging by Field and Diffusion (Charging at $N_i t = 10^7$ ion Seconds per Cubic Centimeter)

Particle Diameter (µm)	Diffusion Charging	Field Charging E = 5 KV/cm.
	Equation [8.7]	Equation [8.5]
0.1	2.7	1.6
0.4	15.7	25.9
1.0	47	162
4	237	2,580
10	673	16,200
40	3,180	259,000

Figure 8.6
Electrical Mobility vs. Particle Size for Combined Field Diffusion Charging

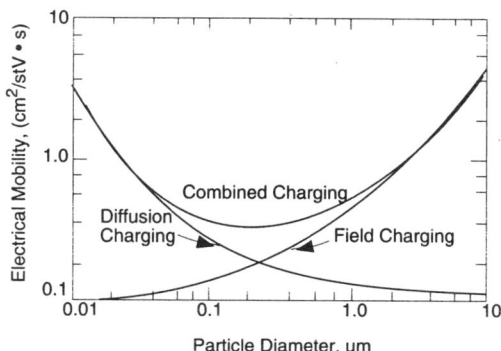

Note: E = 5 KV/cm and Nit = 10 ion sec/cm
Source: Hinds 1982

minimum mobility in the sub-micron size range. Also, mobility under typical charging conditions is a relatively weak function of particle size in the size range 0.1 to 1 μm.

Strictly speaking, both diffusion and field-charging mechanisms operate at the same time on all particles, and neither mechanism is sufficient to explain the charges measured on the particles. Empirically, it has been found that a very good approximation to the measured charge is given by the sum of the total elemental units of charge predicted by Equations [8.5] and [8.7] independent of one another (White 1984):

$$n_{tot}(t) = n_d(t) + n_f(t) \qquad [8.8]$$

where: n_{tot} = total elemental units of particle charge due to both mechanisms

n_d = elemental units of particle charge due to diffusion charging

n_f = elemental units of particle charge due to field charging

Example 8.2 ESP Charging

How many charges are acquired by a 3-μm water droplet in 1 sec by (1) diffusion charging in an ion concentration of 10/cm?, (2) field charging in an ion concentration 10/cm and a field of 6 KV/cm?, and (3) what is the total charge acquired?

Solution:

1. Calculate the number of charges acquired by a 3-μm water droplet in 1 sec by diffusion charging in an ion concentration of 10/cm.

Using Equation [8.7]:

$$n = \frac{d_p kT}{2e^2} \ln\left[1 + \frac{\pi d_p c_i e^2 N_i t}{2kT}\right]$$

$$n = \frac{(3\bullet10^{-4})\bullet(1.38\bullet10^{-16})293}{2(4.8\bullet10^{-10})^2} \ln\left[1 + \frac{\pi(3\bullet10^{-4})\bullet(2.4\bullet10^4)\bullet(4.8\bullet10^{-10})^2 \bullet 10^7 \bullet 1}{2(1.38\bullet10^{-16})\bullet293}\right]$$

$$n = n_d = 263 \ln(1+644) \quad \frac{\text{cm(dyn cm)/(K)(K)}}{\text{stC}^2}$$

$$n_d = 170$$

2. Calculate the number of charges acquired by a 3 μm water droplet in 1 sec by field charging in an ion concentration 10/cm and a field of 6 KV/cm.

Using Equation [8.5]:

$$n_f = \left(\frac{3\varepsilon}{\varepsilon+2}\right)\left(\frac{Ed^2}{4e}\right)\left(\frac{\pi e Z_i N_i t}{1+\pi e Z_i N_i t}\right)$$

$$\frac{3\varepsilon}{\varepsilon+2} = \frac{3\bullet80}{80+2} = 2.93$$

$$\frac{Ed^2}{4e} = \frac{(6,000/300)(3\bullet10^{-4})^2 (\text{stV/cm})(\text{cm}^2)}{4\bullet4.8\bullet10^{-10}\text{stC}} = 938$$

$$\pi e Z_i N_i t = \pi(4.8\bullet10^{-10})(450)(10^7)\frac{\text{stC cm}^2 \text{ sec}}{\text{stV sec cm}^3} = 6.79$$

$$n_f = (2.93)(938)(6.97/1+6.97) = 2,395$$

3. Calculate the total charge acquired

Using Equation [8.8]

$$n_{tot} = n_d + n_f = 170 + 2,395 = 2,565$$

This example demonstrates that field charging dominates for particles larger than 1 μm diameter (see Figure 8.6 and Table 8.2).

8.5 Design of Electrostatic Precipitators

8.5.1 Theoretical Determination of Particulate Migration Velocity

Once the particle is charged, it migrates toward the grounded collection electrode. An indicator of particle movement toward the collection electrode, w, is called the *particle migration velocity* or *drift velocity*. The migration velocity parameter represents the collectability of the particle within the confines of a specific collector.

The motion of particles under the influence of the electric field is opposed by the viscous drag of the gas, $3\pi\mu vd$, (Chapter 3). By equating the electric force, qE, and the drag force component due to the electric field (according to Stokes' law), the particle velocity can be obtained, which is the electrical migration velocity, w:

$$w = \frac{qE_p}{3\pi\mu d} \qquad [8.9]$$

where: q = particle charge, coulombs

E_p = strength of field in which particles are collected, V/m

μ = gas viscosity, Kg/m sec

d = particle diameter, mm

For field charging, the saturation charge, q_s, is $n_s e = \left(\frac{3\varepsilon}{\varepsilon+2}\right)\left(\frac{E_o d^2}{4}\right)$

When ε is large, migration velocity can be expressed as:

$$w = \frac{d_p E_o E_p}{4\pi\mu} \qquad [8.10]$$

where E_o is the field strength in which the particles are charged in V/m and the other parameters as defined for Equation [8.9]. For design purposes, E_p is considered to be equal to E_o. The calculation of migration velocity will be:

$$w = \frac{6.64 \cdot 10^{-18} \cdot E^2 d_p}{\mu} \qquad [8.11]$$

where: w = migration velocity, m/sec

E = average electric field, V/m

d_p = particle diameter, μm

μ = gas viscosity, kg/m-sec

Migration velocity is quite sensitive to the voltage, since the electric field appears twice in Equation [8.11]. Therefore, the precipitator should be designed using the maximum electric field for maximum collection efficiency. The migration velocity is also dependent on particle size; larger particles are collected more easily than smaller ones.

Example 8.3 ESP Drift Velocity

Calculate the theoretical drift velocity for a 2 μm diameter particle in an ESP with 40 KV and 8 in. plate spacing operating at 400°F. The wire-to-plate spacing is 4 inches.

Solution:

Calculate the theoretical migration velocity using Equation [8.11].

$$w = \frac{6.64 \cdot 10^{-18} \cdot E^2 d_p}{\mu}$$

where: E = charge per plate spacing in m

E = 40,000/(4/12 · 0.3084) = 393,700 V/m

d_p = 2 μm

μ = 2.6 · 10^{-5} kg/m sec

$$w = \frac{(6.64 \cdot 10^{-18}) \cdot (393,700)^2 (2)}{2.6 \cdot 10^{-5}}$$

= 0.079 m/sec = 0.259 ft/sec

8.5.2 Design Equations for Electrostatic Precipitators

Equation [8.10] gives particle velocity with respect to still air. In the ESP, the flow is usually very turbulent, with instantaneous gas velocities of the same magnitude as the particle velocities, but in random directions. The motion of particles toward the collecting plates is a statistical process with an average component imparted by the electric field and a fluctuating component from the gas turbulence.

This statistical motion leads to the application of the lateral mix model (Chapter 4) and to an exponential collection efficiency equation known as the Deutsch-Anderson equation:

$$\eta = 1 - \exp\frac{-wA}{Q} \qquad [8.12]$$

Chapter 8: Electrostatic Precipitators

where: η = mass collection efficiency

w = drift velocity or precipitator parameter

A = total plate area

Q = total flow rate

This equation has been used extensively for many years for theoretical collection efficiency calculations and provides the grade efficiency curve for electrostatic precipitators. A range of particle sizes will have a range of associated migration velocities.

Example 8.4 Migration Velocity

An electrostatic precipitator has two parallel 10 ft high by 16 ft long plates with corona wires positioned halfway between the plates. Find the effective migration velocity at a flow rate of 35 acfs if the required collection efficiency is 95%.

Solution:

$$\eta = 1 - \exp\frac{-wA}{Q}$$

$$\frac{-wA}{Q} = \ln(1-\eta)$$

$$w = -\frac{Q}{A}\ln(1-\eta) = \frac{35 \text{ ft/sec}}{10 \text{ ft} \bullet 6 \text{ ft} \bullet 2}\ln(1-0.95)$$

$$w = 0.328 \text{ ft/sec}$$

Example 8.5 Horizontal Flow Single-Stage Precipitator

A horizontal flow single-stage electrostatic precipitator is used to remove particulates from a dry process gas stream of a Portland cement manufacturing plant. The precipitator consists of multiple ducts formed by collecting plates 14 ft wide by 16 ft high and placed 9 inches apart. The rate of flow through each duct is estimated to be 2,400 acfm and the content of dust is 5 grains/ft^3. Calculate the collection efficiency and the amount of dust collected by a duct each day.

Solution:

1. Calculate the collection efficiency.

   Assume w = 0.19 ft/sec (see Table 8.3)

$$\eta = 1 - \exp\frac{-wA}{Q}$$

$$A = 2 \cdot 14 \cdot 16 = 448 \text{ ft}^2$$

$$Q = \frac{2,400 \text{ acfm}}{60 \text{ sec/min.}} = 60 \text{ ft}^3/\text{sec}$$

$$\eta = 1 - \exp\left[\frac{-448}{40}\right] = 0.881$$

2. Determine the amount of dust collected per day.

$$\text{lb/day} = \frac{0.881 \bullet (5 \text{ gr/ft}^3) \bullet (2,400 \text{ ft}^3\text{min}) \bullet (60 \text{ min/hr}) \bullet 24 \text{ hr/day}}{7,000 \text{ gr/lb}}$$

$$= 2,175 \text{ lb/day}$$

8.5.3 Specific Collection Area (SCA)

This is the most important design variable to determine precipitator size. Specification of the required Specified Collection Area (SCA) for a given installation is sufficient to determine the total collection surface and therefore the size of the precipitator required. There are several approaches for determining the SCA needed for a given precipitator installation. Most of these reduce to choosing an appropriate w and solving for A/Q from the Deutsch-Anderson equation which gives (Oglesby and Nichols 1975).

$$\frac{A}{Q} = \frac{1}{w} \bullet \ln\frac{1}{(1-\eta)} = \frac{1}{w} \bullet \ln\frac{1}{P_t} \quad [8.13]$$

In English units, A/Q will be in ft²/cfs and w in ft/sec. This leads to the following expression for SCA in ft²/1,000 acfm as it is commonly reported.

$$\text{SCA} = \frac{A}{Q} = \frac{16.67}{w} \bullet \ln\frac{1}{P_t} \text{ ft}^2/1,000 \text{ acfm} \quad [8.14]$$

where w is in ft/sec and P_t is the loss = 1- η

Although the Deutsch-Anderson equation is a grade efficiency equation, it has not been used in practice to design electrostatic precipitators. For design purposes, the migration velocity term is replaced by a precipitation rate parameter, w_e. This quantity is taken to represent the collection behavior for the total dust under a specific set of operating equations. It is back calculated from experimental data and, therefore, is a performance parameter. The design equation then becomes:

$$\text{SCA} = \frac{A}{Q} = -\frac{\ln(1-\eta)}{w_e} \quad [8.15]$$

where w is replaced by w_e and A/Q by SCA.

The precipitation rate parameter, w_e, is synonymous with the migration velocity, w, for uniform particle size, gas velocity, electric field, and the like. However, such uniform conditions do not exist in actual practice, although it is possible to account for some of the non-uniformity by modification of the theory. But particle reentrainment, gas sneakage around the collecting zone, sectionalization, and similar factors cannot be accounted for by theory. Moreover, for new precipitators, particle size, resistivity, and gaseous properties must be estimated. Because of these problems, values of w calculated from Equation [8.9] have been found to be several times higher than those obtained from test results on actual precipitators.

The term w_e is a semi-empirical performance parameter used for design with values for given applications derived from experimental data and experience by design engineers (Table 8.3) (Oglesby and Nichols 1975). The w_e provides a measure of how well the entire mass entering the precipitator will be collected. This quantity is obtained by replacing w by w_e and determining a single value which produces the same overall mass collection efficiency that is obtained from the collection efficiencies for all particle diameters. Although the relation of w_e to collection efficiency for values of SCA can be calculated directly from Equation [8.15], it is convenient to plot the relationship graphically, as shown in Figure 8.7.

Table 8.3
Typical Precipitation Rate Parameters, w_e, for Various Applications Determined from Field Measurements on Actual Precipitators
(Efficiency Range 90-95%)

Application	Precipitation rate Parameter (w_e)	
	ft/sec	cm/sec
Utility fly ash	0.13-0.67	4-20.4
Pulverized coal fly ash	0.33-0.44	10.1-13.4
Pulp and paper mills	0.21-0.31	6.4-9.5
Sulfuric acid mist	0.19-0.25	5.8-7.62
Cement (wet process)	0.33-0.37	10.1-11.3
Cement (dry process)	0.19-0.23	6.4-7.0
Gypsum	0.52-0.64	15.8-19.5
Smelter	0.06	1.8
Open-hearth furnace	0.16-0.19	4.9-5.8
Blast furnace	0.20-0.46	6.1-14.0
Hot phosphorous	0.09	2.7
Flash roaster	0.25	7.6
Multiple hearth roaster	0.26	7.9
Catalyst Dust	0.25	7.6
Cupola	0.10-0.12	3.0-3.7

Source: Oglesby & Nichols 1975

Figure 8.7
Chart for Finding Specific Collection Area (SCA) Based on the Precipitator Rate Parameter, w_e

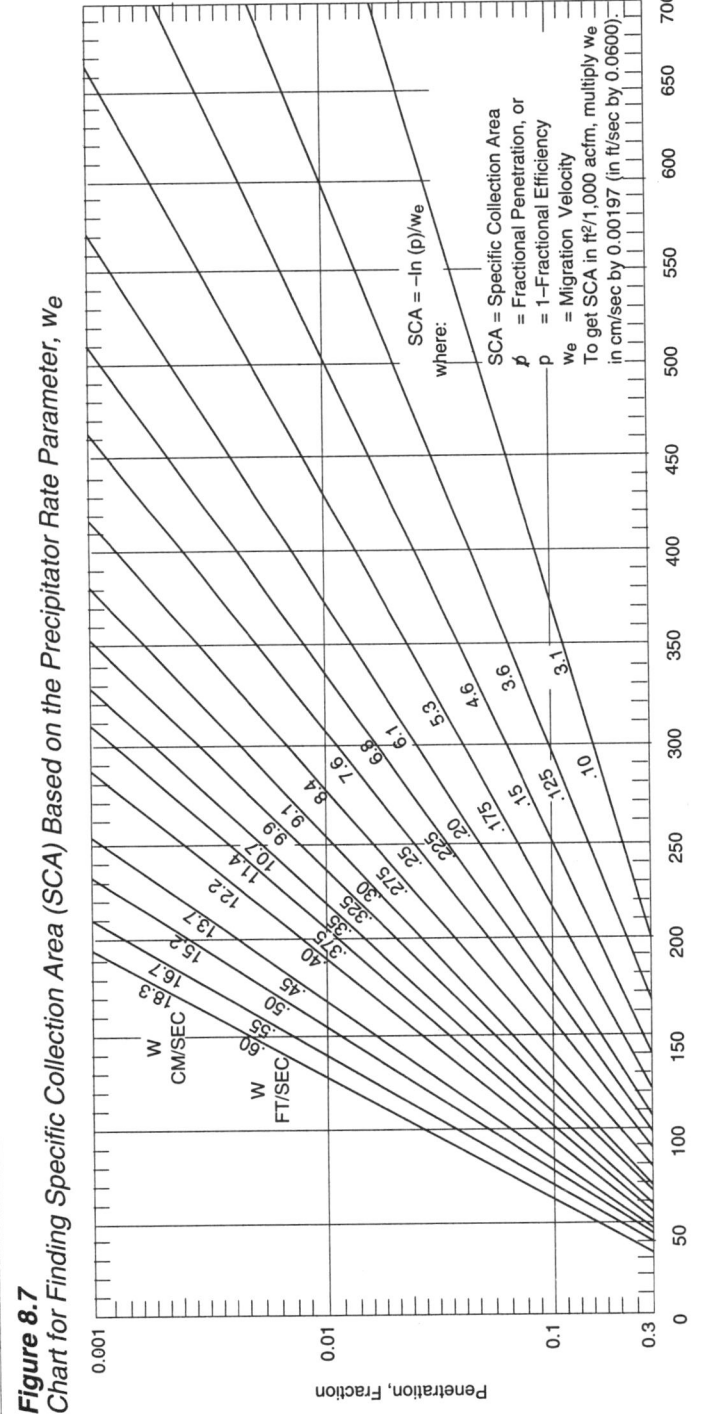

Source: Buonicore and Davis, *Air Pollution Engineering Manual*, 1992. Reprinted with permission of John Wiley & Sons, Inc.

The values of w_e shown in Figure 8.7 range from 0.10 to 0.6 ft/sec and covers most precipitator applications. The SCA required for a specific application can be read directly from the graph. For example, assume a precipitator efficiency of 95% and an application where a value of w_e of 0.2 ft/sec is judged suitable. From Figure 8.7, the required SCA is 250 ft/1,000 acfm.

The application where the greatest variation in electrical resistivity and precipitation rate parameter occurs is in the collection of fly ash from coal-fired boilers. Because of this variability, and because installations on boilers represent the largest single application of precipitators, design methods for fly ash precipitators have received the greatest attention. Several experimental relationships have been developed to show the effect of high dust resistivity. Figure 8.8 shows the relationship between the precipitation rate parameter and dust resistivity (Oglesby and Nichols 1975). It was developed for a group of fly ash precipitators of generally moderate collection efficiencies (90%-95%) and for particle size distributions of about 10 μm MMD.

Figure 8.8
Variation of Precipitation Rate Parameter with Fly Ash Resistivity

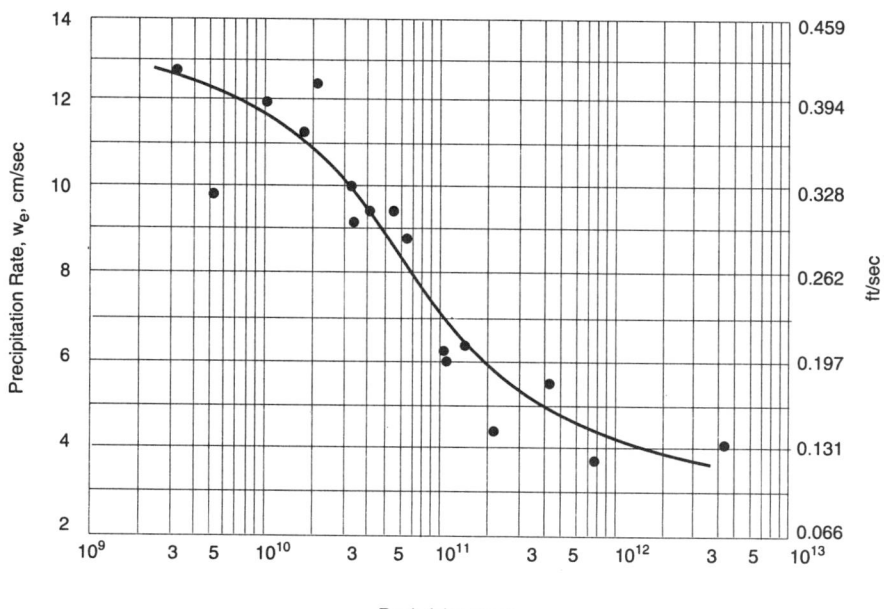

Note: The data is from large fly ash precipitators and direct field measurements of particle resistivity

Source: Calvert and Englund 1984

From the foregoing, it is apparent that factors other than resistivity can further modify the precipitation rate parameter. However, such curves are useful in establishing overall relationships which can then be modified according to specific applications.

The shape of the curve is of interest in that it is almost independent of the resistivity for values less than $5 \cdot 10^{10}$ ohm-cm. At lower resistivity, the electric field in the dust layer is too low to cause an electrical breakdown in the dust layer. Above this critical value of resistivity, the precipitation rate parameter continues to decrease, until at very high resistivity effective precipitation ceases.

A second empirical technique for sizing the precipitator is based upon the observation that the precipitation rate parameter for a given particle size varies with corona power. Figure 8.9 shows the relationship for a group of fly ash precipitators with moderate collection efficiencies.

Figure 8.9
Relationship Between Precipitation Rate Parameters and Corona Power Density for Fly Ash Precipitators

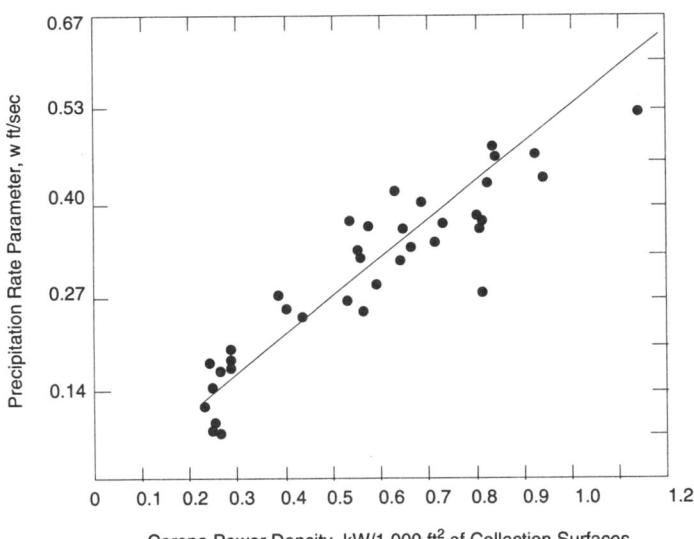

Source: Oglesby and Nichols 1975

The empirical relationship between corona power and precipitation rate parameter is useful in determining the range of performance expected; however, power input must be useful power. In the case of high dust resistivity, power input can be in the normal range, but due to back corona, the precipitator performance may be poor.

8.5.4 Electrostatic Precipitator Design for High Efficiency (>95%) Collection of Fly Ash

The data in Table 8.3 and Figure 8.8 were developed for precipitators of generally moderate collection efficiencies (90% to 95%) and represent the total average migration velocity. In practical precipitator design, collection efficiency above 95% generally requires more plate area than the area calculated by the use of the Duetsch-Anderson equation and the precipitation rate values given in Table 8.3. This is true because, contrary to the implications in the table, the precipitation rate parameter does not remain constant as collection efficiency increases. This is due to the fact that higher collection efficiencies require collection of a larger percentage of smaller particles, which have smaller migration velocities. This phenomenon can be seen in Figure 8.10 which shows variation in w_e with particle diameter and collection efficiency for fly ash (Gallaer 1983). The variation in w_e for those particles collected with maximum efficiency to those collected with minimum efficiency is about 2 to 1 (11 cm/sec vs. 5 cm/sec). The minimum efficiency, and therefore minimum w_e, is usually found for 0.5 to 1.0 µm diameter particles (Figure 8.11). For particles smaller than this size, the collection efficiency increases due to diffusion charging.

Figure 8.10
Typical Data for Effective Migration Velocity and Collection Efficiency as a Function of Particle Diameter for Fly Ash

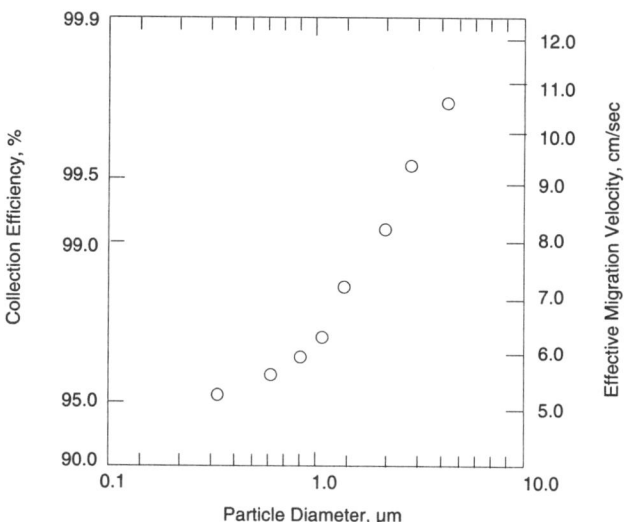

Note: Effective migration velocity varies with the MMD of the particle size distribution.
Source: Gallaer 1983

Figure 8.11
Typical Precipitator Fractional Efficiency Curves for Fly Ash

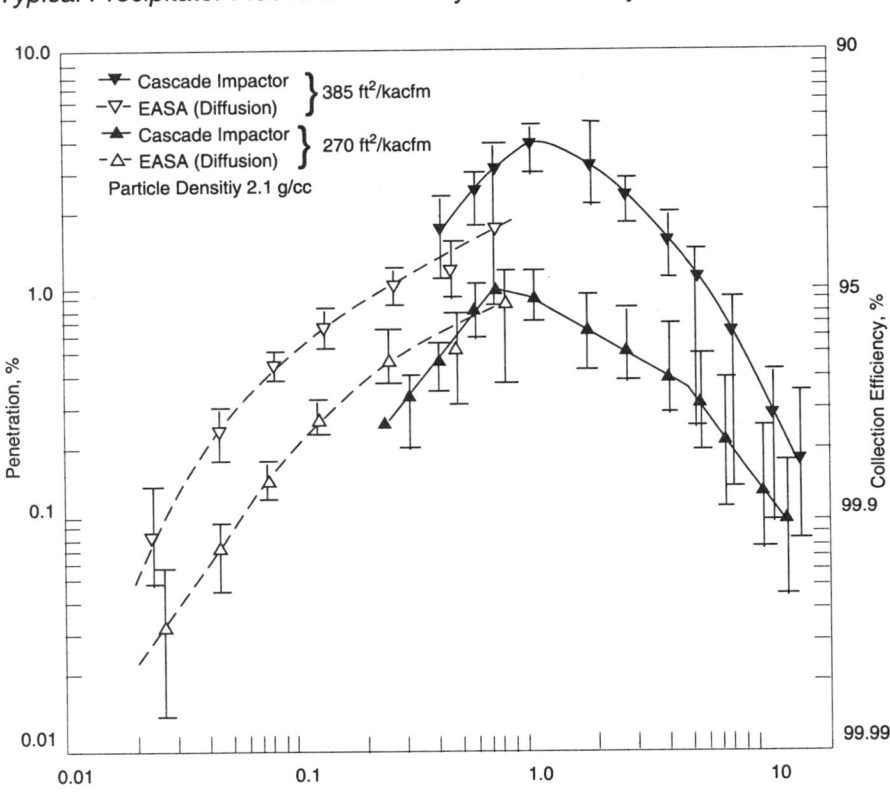

Source: Gallaer 1983

Figure 8.12 indirectly illustrates the effect of the particle size of a polydisperse dust on the plate area requirement as collection efficiency changes because w increases with particle size. The curve shows the relationship of efficiency to specific collection area (SCA) that is typical of fly ash precipitators.

The straight lines in Figure 8.12 show the specific collection area for various constant values of w_e. As indicated by the graph, the measured value of w_e varied from 19 cm/sec at the low end of the efficiency range to about 8 cm/sec at the high efficiencies. The extent of the variation depends upon the size distribution of the dust and the fractional collection efficiency of the precipitator.

To overcome variability of w_e with efficiency, a modification of the Deutsch-Anderson equation can be used (Matts and Ohnfelt 1963).

Figure 8.12
Typical Variation in Penetration with Plate Area to Gas Flow Ratio

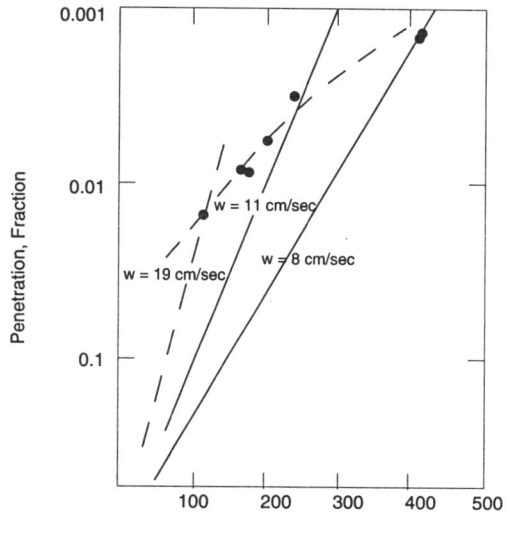

$$\eta = 1 - \exp\left[-\left(\frac{A}{Q}\right)w_e\right]^k \quad [8.16]$$

$$SCA = \frac{[-\ln(1-\eta)]^{\frac{1}{k}}}{w_e} \quad [8.17]$$

Equation [8.17] was developed in order to take into account the polydisperse nature of the fly ash particle size distribution. The w_e in this equation has historically been determined as w_k (total average migration velocity, determined experimentally from pilot plant data at the desired efficiency and varying with efficiency). Figure 8.12 shows that the Matts-Ohnfeldt equation concept using an exponential term to increase the size of the collector is important for efficiencies higher than 95%. The k is a constant, usually between 0.4 and 0.6, which depends on the standard deviation and other properties of the dust affecting efficiency. A value of 0.5 is generally used in this equation (Matts and Ohnfeldt 1963).

Figure 8.13 shows the effect of the parameter k on the efficiency-specific collection area relationship. For k =1, the expression reduces to the conventional Deutsch-Anderson equation. For values of k <1, the predicted efficiencies at high SCA are less than would be determined from Deutsch-Anderson equation with a constant precipitation rate parameter. Selection of the proper value of k is based on an empirical fit of experimental data and depends on the size distribution of the dust

Figure 8.13
Variation in Calculated Efficiency for Various Collecting Surface Areas to Gas Volumne Ratios Showing Effect of Exponent k

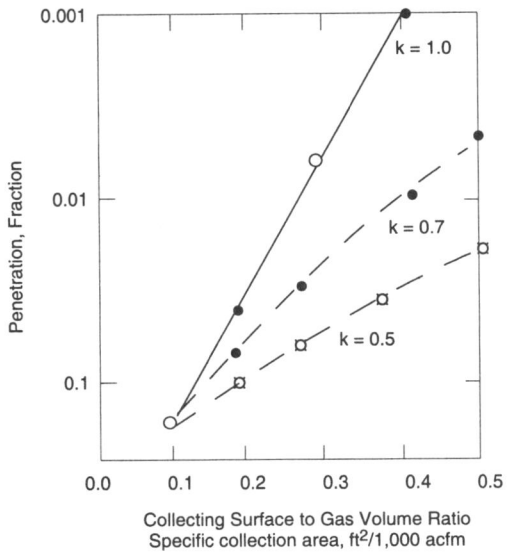

and the precipitator design parameters. Table 8.4 provides values for efficiency predicted for increased precipitator size using the Deutsch-Anderson equation with k values of 1.0 and 0.5.

Table 8.4
Efficiency Predicted for Increased Precipitator Size Using the Deutsch-Anderson Equation with k Values of 1.0 and 0.5

Relative size of precipitator, A/Q	k = 1	k = 0.5
1	90	90
2	99	96.2
3	99.9	98.1
4	99.99	99
5	99.999	99.6

The main problem with the use of the Matts-Ohnfeldt equation is that the effective migration velocity used in the Deutsch-Anderson equation cannot be used in the Matts-Ohnfeldt equation. One must determine an "equivalent effective migration velocity" in the Matts-Ohnfeldt equation by calculation of the SCA as shown in

Table 8.4. In this table, one starts with a given SCA and simply ratios the equations. The table shows that if the relative size of a 90% ESP were doubled, k = 1 would predict 99% collection efficiency, while k = 0.5 would predict 96.2%. However, to use effective migration velocities from Table 8.3 instead of w_k in the equation, it is necessary to apply the Matts-Ohnfeldt equation concept and do a statistical analysis to determine the exponent. Then, w_e will remain constant when efficiency increases.

The decision to use constant migration velocity values for 95% efficiency in Equation [8.16] is based on an evaluation of data presented in Buonicore and Davis 1992. Table 8.5 presents experimental precipitator rate parameters, w_e, for various industrial sources for different electrostatic precipitator efficiencies. These data are similar to those presented in Table 8.3 for 95% efficiency.

Table 8.5
Typical Precipitation Rate Parameters, w_e, for Various Applications

Application	Precipitation rate, ft/sec			
	95%	99%	99.5%	99.9%
Bituminous coal fly ash	0.41	0.33	0.31	0.27
Subbituminous coal fly ash in tangential-fire boiler	0.56	0.39	0.34	0.29
Other coal	0.32	0.26	0.26	0.24
Cement kiln	0.049	0.049	0.059	0.059
Glass plant	0.052	0.052	0.049	0.049
Iron/steel sinter plant with mechanical collector	0.20	0.20	0.22	0.21
Kraft-paper recovery boiler	0.085	0.082	0.10	0.095
Incinerator fly ash	0.50	0.37	0.35	0.31
Copper reverberatory furnace	0.20	0.14	0.12	0.095
Copper converter	0.18	0.14	0.13	0.12
Copper roaster	0.20	0.18	0.17	0.16

Note: These values do not necessarily agree with the values in Table 8.3 at 90 to 95% efficiency.
Source: Buonicore and Davis 1992

Table 8.5 shows decreasing values of w_e for higher efficiencies. Figures 8.14 and 8.15 provide a regression analysis using calculated SCA values based on k = 1, with k determined from the measured data in Table 8.5 (w_e and efficiency values for 99, 99.5, and 99.9%) using Equation 8.16 and constant w_e values from Table 8.5 at 95% efficiency. The k value for fly ash is 0.63 with an R^2 of 0.92 while the k value for the other industrial applications is 0.92 with an R^2 of 0.97. The values computed for k are similar to the suggested value of 0.5 for fly ash using a constant w_e. However, this is not true for the other industrial applications, and a k = 1.0 is recommended for these sources.

Figure 8.14
Calculated Value of k for Fly Ash (k=0.6259)

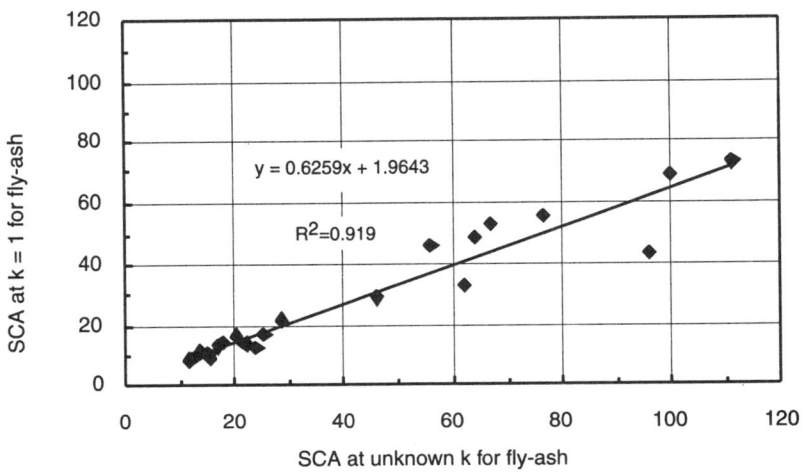

Figure 8.15
Calculated Value of k by Industrial Category (k=0.9203)

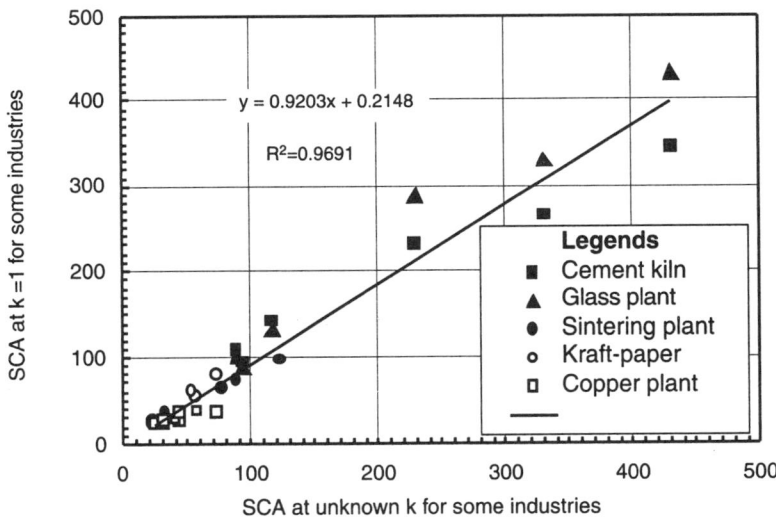

Because of the lack of firm theoretical bases, it is advisable to regard Equation [8.16] as essentially empirical, but with some fundamental elements. Figures 8.16 and 8.17 present precipitation efficiency versus SCA and w_e value when k is 0.5 and 0.6. The precipitation rate parameter values are "total average migration ve-

Figure 8.16
Precipitator Penetration vs. SCA and Precipitator Rate w_e with k Equal to 0.5

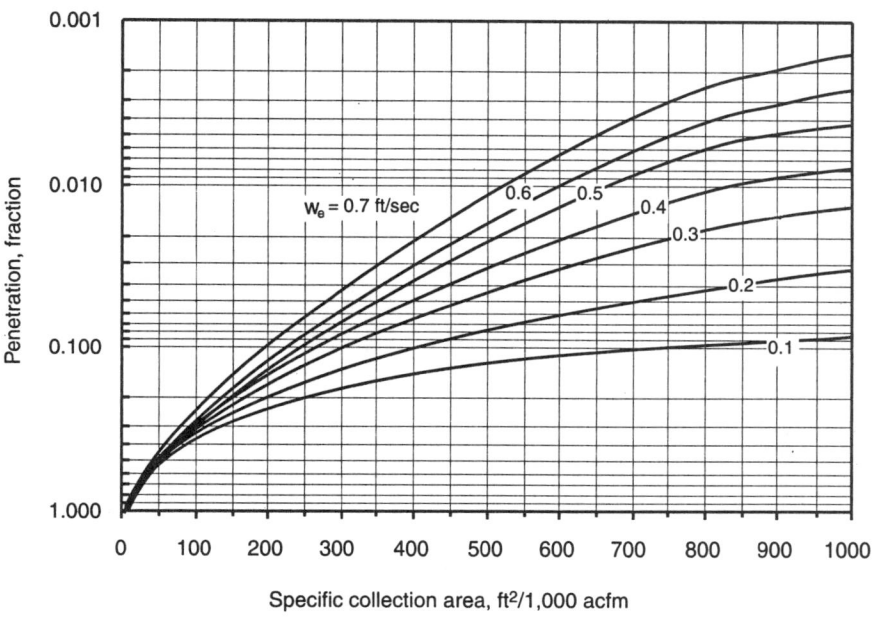

locities at 90 to 95% collection efficiency" and can be selected from Table 8.3 or 8.5 or determined from test data for an operating precipitator.

The use of the Matts-Ohnfeldt equation produces results that are different from the results from the Deutsch-Anderson equation. Therefore, the Matts-Ohnfeldt equation using the effective migration velocities should only be applied above 99% efficiency. Below 95%, the SCA from the Deutsch-Anderson equation should be used. Between these efficiencies, an average obtained from the two equations should be used.

This analysis demonstrates that the modified Matts-Ohnfeldt equation (Equation [8.16]) gives a much closer approximation of the efficiency obtained at different SCA values than using k = 1.0 for fly ash at high collection efficiencies, even though its precision might not be as high as desired. In order to increase the accuracy of Equation [8.16] and provide a final design, pilot electrostatic precipitator testing at the designed collection efficiency needs to be performed with w_e determined from the resulting data (see Example 8.7).

Figure 8.16 shows that low values of w_e, along with k= 0.5, produces an unrealistically high value of SCA (i.e., >1,000 ft²/1,000 acfm) for collection efficiencies greater than 99%. Because of this, higher collection efficiencies require a higher

Figure 8.17
Precipitator Penetration vs. SCA and Precipitator Rate w_e with k Equal to 0.6

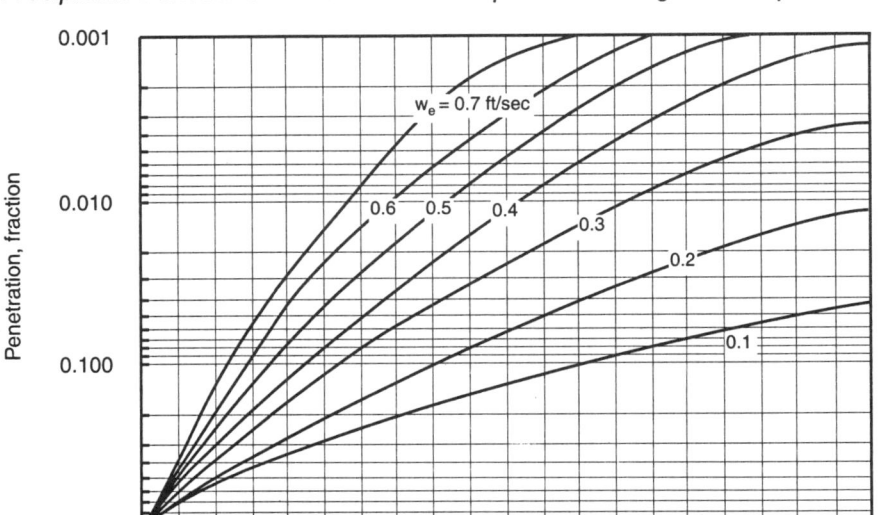

w_e value to provide an acceptable design. This can be accomplished by temperature control or gas conditioning to ensure a low resistance for the fly ash dust. In order to increase w_e, the applied voltage and current must be increased. Collection efficiency can also be increased by decreasing the gas flow rate.

Example 8.6 ESP Collection Efficiency and Plate Area Determination

1. Provide a figure showing collection efficiency for w_e values between 0.3 and 0.6 ft/sec using Equation [8.12] and [8.16]. Divide the figure into zones of low collection efficiency (<90%), moderate (90 to 99%), and high (> 99%) and SCA values.

2. Determine the plate area for an ESP for 90, 98, and 99% collection efficiency when the emissions are from a coal-fired power plant (tangential-fired boiler) with an exit gas flow of 100,000 ft^3/min.

Solution:

1. Prepare a figure as shown below:

Chapter 8: Electrostatic Precipitators

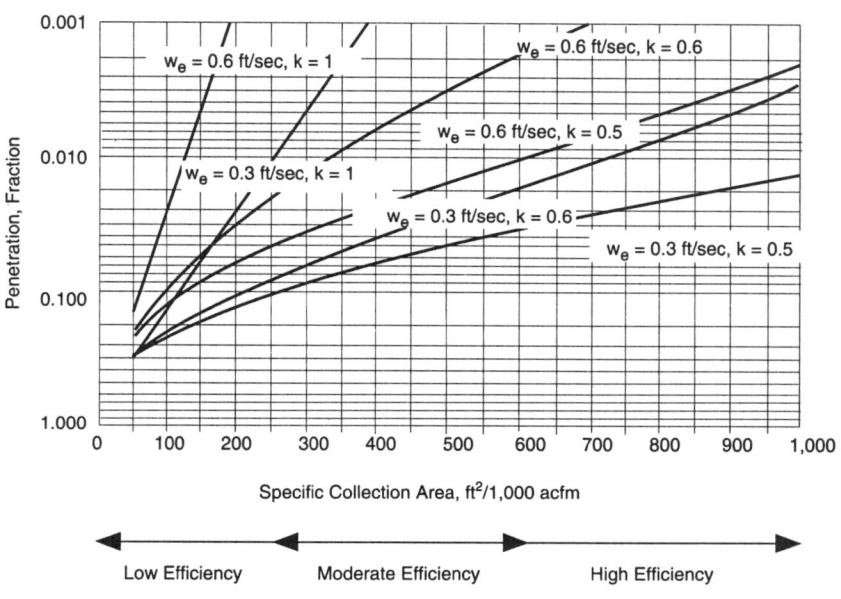

2. Determine the plate areas.

From Table 8.5: drift velocity for particles from the tangential coal-fired boiler is 0.56 ft/sec.

For 90% efficiency:

Use the Deutsch-Anderson (Equation [8.14]).

$$\eta = 1 - \exp\frac{-wA}{Q}$$

$$0.9 = 1 - \exp[-\text{SCA} \cdot 0.56 \text{ ft/sec}]$$

$$\text{SCA} = 4.11 \text{ sec}/\text{ft} \cdot \frac{1,000 \text{ cfm}}{1,000 \text{ cfm}} \cdot \frac{1 \text{ min}}{60 \text{ sec}}$$

$$= 68.5 \text{ ft}^2/1,000 \text{ acfm}$$

Plate area = $(68.5 \text{ ft}^2/1,000 \text{ acfm}) \cdot 100,000 \text{ cfm} = 6,850 \text{ ft}^2$

For 98% efficiency:

Use the average obtained from the Deutsch-Anderson equation and Matts-Ohnfeldt equation.

a) Use the Deutsch-Anderson equation

$$0.95 = 1-\exp[-SCA \cdot 0.56 \text{ ft/sec}]$$

$$SCA = 5.35 \text{ sec/ft} \cdot \frac{1,000 \text{ cfm}}{1,000 \text{ cfm}} \cdot \frac{1 \text{ min}}{60 \text{ sec}}$$

$$= 89.2 \text{ ft}^2/1,000 \text{ acfm}$$

Plate area = $(89.2 \text{ ft}^2/1,000 \text{ acfm}) \cdot 100,000 \text{ cfm} = 8,920 \text{ ft}^2$

b) Use Matts-Ohnfeldt equation, and k = 0.6.

$$0.98 = 1-\exp[-SCA \cdot 0.56 \text{ ft/sec}]^{0.6}$$

$$SCA = 17.3 \text{ sec/ft} = 288.3 \text{ ft}^2/1,000 \text{ acfm}$$

Plate area = $(288.3 \text{ ft}^2/1,000 \text{ acfm}) \cdot 100,000 \text{ cfm} = 28,830 \text{ ft}^2$

c) SCA average for 98% efficiency (k = 0.6)

$$SCA_{avg} = (89.2 + 288.3)/2 = 188.75 \text{ ft}^2/1,000 \text{ acfm}$$

Plate area = $(188.75 \text{ ft}^2/1,000 \text{ acfm}) \cdot 100,000 \text{ cfm} = 18,875 \text{ ft}^2$

For 99% efficiency, use Matts-Ohnfeldt equation and k = 0.6.

$$0.99 = 1-\exp[-SCA \cdot 0.56 \text{ ft/sec}]^{0.6}$$

$$SCA = 22.8 \text{ sec/ft} = 379.4 \text{ ft}^2/1,000 \text{ acfm}$$

Plate area = $(379.4 \text{ ft}^2/1,000 \text{ acfm}) \cdot 100,000 \text{ cfm} = 37,940 \text{ ft}^2$

The results are presented in the table below. The table demonstrates the need to use a modified form of the Deutsch-Anderson equation to provide the large SCA area required to collect fly ash at high collection efficiencies.

SCA Values for Different Efficiencies

	SCA ft²/1,000 actm					
k	Efficiency 90%		Efficiency 98%		Efficiency 99%	
	SCA ft²/1,000 acfm	Plate area, ft²	SCA ft²/1,000 acfm	Plate area, ft²	SCA ft²/1,000 acfm	Plate area, ft²
Eq [8.14] k = 1.0	68.5	6,850				
Eq [8.17] k = 0.6					379.4	37,940
Avg Eq [8.14] & Eq [8.17] k = 0.6			188.75	18,875		

Example 8.7 ESP Collection Efficiency Comparison

An electrostatic precipitator with a specific collection area (A/Q) of 300 ft^2/1,000 acfm is found to have an actual overall collection efficiency of 97.0%. If the value of A/Q is increased to 400 ft^2/1,000 acfm, estimate the anticipated overall efficiency on the basis of a constant precipitator rate parameter using (1) Equation [8.15] and (2) the Matts-Ohnfeldt Equation [8.16] with a k = 0.5 and a migration velocity, w_k, determined from the test data.

Solution:

1. By substitution of the given data into Equation [8.15], a precipitator rate parameter for the Deutsch-Anderson equation at 97% efficiency is obtained.

$$0.97 = 1 - \exp(-0.300 w_e)$$
$$w_e = 11.69 \text{ ft/min or } 0.195 \text{ ft/sec or } 5.9 \text{ cm/sec}$$

Use the precipitator rate parameter w_e with A/Q value of 400 ft^2/1,000 acfm.

$$\eta = 1 - \exp[-11.69 \text{ ft/min} \cdot 400 \text{ ft}^2/(1000 \text{ ft}^3/\text{min})]$$
$$= 0.991$$

2. Substitute the data into the Matts-Ohnfeldt equation (Equation [8.16]).

$$0.97 = 1 - \exp[-(0.300 w_e)^{0.5}]$$
$$w_k = 40.9 \text{ ft/min or } 0.68 \text{ ft/sec or } 20.7 \text{ cm/sec}$$

Using the total average migration velocity determined experimentally in the Matts-Ohnfeldt equation, the overall efficiency is estimated to be

$$\eta = 1 - \exp[(-50.9 \text{ ft/min} \cdot 400 \text{ ft}^2/(1,000 \text{ ft}^3/\text{min})^{0.5}]$$
$$= 0.982$$

Thus, if the precipitator is upgraded by increasing the specific collection area, the Matts-Ohnfeldt equation predicts a somewhat smaller increase in overall collection efficiency than given by the Deutsch-Anderson equation using a constant precipitator rate parameter.

8.6 Precipitator Design: Practical Aspects

Basic precipitator design parameters commonly used in practice for fly ash collection are summarized in Table 8.6 (Beachles and Jahnke 1981). Selection of actual values for a given design will depend on the particle and flue gas properties, total gas flow, and the required collection efficiency. The performance actually attained in practice is dependent on the mechanical properties of the precipitator. Poor

Table 8.6
Summary of Basic Design Parameters for Fly Ash Precipitators

Parameter	Range (English units)	Range (metric units)
Distance between plates (duct width)	8-12 in. (8-9 in. optimum)	20-30 cm
Gas velocity in ESP	4-8 ft/sec (5-6 ft/sec optimum)	1.2-2.4 m/sec
SCA	200-800 ft^2/1,000 cfm (300-400 ft^2/1,000 cfm optimum)	11 - 45 m^2/1,000 m^3/hr (6.5-22.0 m^2/1,000 m^3/hr)
Aspect ratio (L/H)	1-1.5 (keep plate height less than 30 ft for high efficiency)	1.1-5 (keep plate height less than 9 m for high efficiency)
Design migration velocity	0.1-0.5 ft/sec	3.05-15.2 cm/sec
Number of fields	4-8	4-8
Corona power/1000 cfm	100-500 watts/1,000 cfm	59-295 watts/1000 m^3/hr
Corona current/ft^2 plate area	10-80 microamp/ft^2	107-860 microamp/m^2
Plate area per electrical (T-R) set	5,000-80,000 ft^2/T-R set (10,000-30,000 ft^2/T-R set optimum)	465-7,430 m^2/T-R set (930-2,790 m^2/T-R set optimum)

electrode alignment, warped electrode, air inlet leakage, poor gas flow distribution, and unstable rectifier sets must be guarded against.

Although localized gas velocities in the collection zones of precipitators usually vary over a substantial range, it is useful for design purposes to use an average value calculated from the total gas flow and the flow cross section of the precipitator. The cross section is taken as the open area between collecting plates, disregarding the plates' baffles. Primary importance of the average gas velocity is its relation to rapping and reentrainment losses. These losses tend to increase rapidly above some critical velocity because of the aerodynamic forces on the particles. The critical velocity for a given type of dust depends on the quality of the gas flow, plate configuration, electrical energization, precipitator size, and other factors. For reasonably good conditions, the critical gas velocity for dust, such as fly ash, will vary between 1.5 and 2 m/sec, but may be higher in some cases. Modern requirements for very high efficiencies usually require maintaining the gas velocity on the low side of the range.

Aspect ratio is defined as the ratio of the effective duct length to duct height. Its importance in precipitator design stems from its relation to rapping loss. Collected dust falling from the plates is carried forward by the gas flow. If the ducts are short compared to their height, some of the falling dust will be carried out of the precipitator before it reaches the hopper, thereby substantially increasing the dust loss. The time required for released dust clumps to fall from the top of a 12 m high duct, for example, will be several seconds. This is sufficient time for significant amounts of dust to be carried out of the precipitation zone unless the duct length exceeds 10 or 12 m. For orientation purposes, the aspect ratio needed for high efficiencies (99%) should be at least 1 to 1.5.

The layout of an ESP is a series of ducts depicted in Figure 8.1 with a height, H, duct width, W, and a length, L. The number of ducts is calculated in a manner similar to pipe sizing, starting with flow rate equals velocity times cross sectional area. For the total number of ducts, N_d, this expands to:

$$Q = v \bullet A_v \bullet N_d \qquad [8.18]$$

or

$$N_d = \frac{Q}{v \bullet A_v} \qquad [8.19]$$

where:

$$A_v = W \bullet H = \text{cross sectional area of duct} \qquad [8.20]$$

Also important is the relationship between the collection surface area, calculated with the Deutsch-Anderson equation, and the number of ducts. Simple geometry leads to the relation:

$$A_{total} = 2 \, H \bullet L \bullet N_d = \text{area of plates} \qquad [8.21]$$

Total number of plates, N_{pl}, are calculated by noting that both sides of each plate contribute to the total area except for the two exterior plates. Therefore, N_{pl} is:

$$N_{pl} = N_d + 1 \qquad [8.22]$$

Equations [8.19] and [8.22] do not necessary produce the same answer for a given number of ducts. If given sufficient information, N_d should be calculated based on flow velocity ($v_g = Q/A_v$) (Equation [8.19]) or based on the required area using the particulate drift velocity, selecting the largest area or number of ducts to ensure that the design meets the criteria.

Example 8.8 Determination of the Number of ESP Plates

Calculate the required area to provide 99.9 % collection efficiency for 2 μm diameter particle having a migration velocity of 0.259 ft /sec. The distance between plates is 8 in. and the flow velocity is 4 ft/sec (see Table 8.5). The flow rate is 39,000 acfm.

Solution:

1. Calculate the length of each ESP plate.

   To calculate the length of the plate, it is necessary to understand that the particle must migrate to the plate in the same time the air moves from the front to the back of the ESP. Mathematically, residence time, t_r, is

   $$t_r = \frac{W}{2w_e} = \frac{L}{Q/A_v}$$

   $$L = \frac{W \bullet Q}{2 \bullet w_e \bullet A_v} = \frac{W \bullet v_g}{2 \bullet w_e}$$

   where: L = duct length

   W = duct width

   Q = volume flow rate of gas

   w_e = migration velocity of particles

   A_v = cross sectional area of duct

   v_g = gas velocity

   $$L = \frac{(8 \text{ in.} \bullet 1 \text{ ft}/12 \text{ in.}) \bullet 4 \text{ ft/sec}}{2 \bullet 0.259 \text{ ft/sec}}$$

   $L = 5.15$ ft

   select plate length of 6 ft

2. Determine the duct configuration using Table 8.5 where L/H will be selected as 1.0.

   $A_{pl} = L \bullet H = 36 \text{ ft}^2$

   If 99.9 % collection efficiency is used, then Figure 8.7 provides an SCA value near 450 ft²/1,000 cfm. Calculate the efficiency using Equation [8.12].

   $$\eta = 1 - e^{-w_e A/Q}$$

   $$.999 = 1 - e^{-(0.259 \text{fps})(A/Q)}$$

   A/Q = 26.67 sec/ft

   = 0.45 min/ft (450 ft²/1,000 cfm)

   Q = 39,000 ft³/min

   A_p = 17,550 ft² of duct required

Using Equation [8.21]

$$N_d = \frac{A_{total}}{2 \bullet H \bullet L} = \frac{17,550 \text{ ft}^2}{2 \bullet 36 \text{ ft}^2} = 244$$

$$N_{pl} = 244 + 1 = 245 \text{ plates}$$

8.7 Design Summary

Unfortunately, while the Duetsch-Anderson equation is scientifically valid, there are a number of parameters that can cause the results to be in error by a factor of two or more. The equation assumes that the gas flow rate is uniform, and that particle sneakage through the hopper section and dust reentrainment do not occur. Therefore, this equation should be used only for a preliminary estimate of the precipitation collection efficiency. Current methods of design and sizing of electrostatic precipitators are based primarily upon analogy with similar plants (Table 8.3) or empirical relationships.

One of the most significant variables in dust properties influencing precipitator design is the particle size distribution of the dust. The migration velocity decreases rather severely with decreasing particle size (Figure 8.10). However, the precipitator rate parameter is taken as a constant in the Duetsch-Anderson equation and if high efficiencies are to be achieved, additional precipitator surface area may be required.

The other important design factor is the charge on the particles which is limited by dust resistivity. Such relationships, as presented in Figure 8.8, permit a narrowing of the design precipitation rate parameters presented in Table 8.3. In applications other than fly ash, relationships between w_e and process variables have not been well-developed, and design is more dependent on analogy with existing plants. The range for many applications in Table 8.3 is great and rational design based on these values is difficult. The following guides should be considered when designing electrostatic precipitators:

   a) When considering the collection of monodispersed particles, use fundamental theory for calculation of migration velocity and the Duetsch-Anderson equation for calculation of collection efficiency. These values will be higher than test results for most applications.

   b) When considering a dust with a variation in chemical composition or with a size distribution, use the effective migration velocity, w_e, selected from Table 8.3 or 8.5 or use experimental data determined from pilot-scale testing of the source. Use the Duetsch-Anderson equation for calculation of collection efficiency.

c) High collection efficiencies of fly ash (i.e., SCA > 200, efficiency > 95%) required the use of the modified Duetsch-Anderson equation using an exponent k of 0.5 or 0.6. The effective migration value for use in the equation can be selected from Tables 8.3 and 8.5. For industrial applications other than fly ash, k = 1 is recommended, even for high collection efficiencies, based on the results presented in Figure 8.15.

8.8 Review Problem

a) An electrostatic precipitator was designed to collect fly ash under the following design conditions: Temperature of gas 300° F and sulfur fuel content of 2.5 %. A low sulfur fuel (0.5 %) has been substituted for the high sulfur fuel. If the original collection efficiency was 95%, what is the new collection efficiency?

b) Assume that you are to design a new electrostatic precipitator to replace the old one. If you want 99% collection efficiency, compare the size of a hot side precipitator working at 400° F and 0.5 % sulfur to a cold side precipitator working at 300° F and 0.5 % sulfur. Which one would you recommend and why ?

c) Consider whether chemical conditioning with SO_3 may be of benefit when low sulfur is used. This option could allow the old precipitator to be used even with the low sulfur fuel.

Solution:

a) Collection Efficiency

    1. From Figure 8.4

        2.5% S at 300°F: resistivity = $4.5 \cdot 10^9$ Ω cm

        0.5% S at 300°F: resistivity = $2.5 \cdot 10^{11}$ Ω cm

    2. From Figure 8.8

        at resistivity $4.5 \cdot 10^9$ Ω cm: w = 12.4 cm/sec = 0.407 ft/sec

        at resistivity $2.5 \cdot 10^{11}$ Ω cm: w = 5.4 cm/sec = 0.177 ft/sec

    3. Given original η = 95%

$$\eta = 1 - \exp\frac{-wA}{Q}$$

$$0.95 = 1 - \exp\left[\frac{-A}{Q} 0.407\right]$$

$$-\frac{A}{Q} = \frac{\ln(0.05)}{0.407}$$

$$\frac{A}{Q} = 7.36 \text{ ft}^2 / \left(\text{ft}^3 / \text{sec}\right)$$

4. New condition: $\eta = 1 - \exp[7.36 \cdot 0.177]$

$$= 0.728$$

$$\eta = 72.8 \%$$

By substituting the low sulfur fuel, the collection efficiency dropped from 95% to 72.8%.

b) Hotside vs. Coldside Application

   1. Hot side: T = 400°F and 0.5% sulfur

      From Figure 8.4 (resistivity = $4 \cdot 10^9$ Ω cm) and Figure 8.8: w = 12.5 cm/sec or w = 0.410 ft/sec

$$0.99 = 1 - \exp\left[\frac{-A}{Q} 0.410\right]$$

$$-\frac{A}{Q} = \frac{\ln(0.01)}{0.410}$$

$$\frac{A}{Q} = 11.23 \text{ ft}^2 / (\text{ft}^3 / \text{sec})$$

   2. Cold side: T = 300°F and 0.5% sulfur

      From review problem a)

      w = 0.177 ft/sec

$$0.99 = 1 - \exp\left[\frac{-A}{Q} 0.177\right]$$

$$-\frac{A}{Q} = \frac{\ln(0.01)}{0.177}$$

$$\frac{A}{Q} = 26.0 \text{ ft}^2 / \left(\text{ft}^3 / \text{sec}\right)$$

Alternative design using k = 0.5

1a. Hot side design:

$$0.99 = 1 - \exp\left[\frac{-A}{Q} 0.41\right]^{0.5}$$

$$\frac{A}{Q} = 51.7 \text{ft}^2 / \left(\text{ft}^3 / \sec\right)$$

2a. Cold side design:

$$0.99 = 1 - \exp\left[\frac{-A}{Q} 0.177\right]^{0.5}$$

$$\frac{A}{Q} = 120 \text{ ft}^2 / \left(\text{ft}^3 / \sec\right)$$

At first glance, it would seem that the hot side precipitator should be selected since it is the smaller. However, the Hot side design would need to be larger due to the thermal expansion of the gas at high temperatures.

$$\frac{V_1 P_1}{T_1} = \frac{V_2 P_2}{T_2} \text{ or } \frac{Q_1 P_1}{T_1} = \frac{Q_2 P_2}{T_2}$$

$T_1 = 300°F = 760°R$

$T_2 = 400°F = 860°R$

$$\frac{Q_{\text{Hot side}}}{Q_{\text{Cold side}}} = \frac{860}{760} = 1.13$$

The Hot side design precipitator should still be considered over the Cold side design precipitator based on size. However the Hot side precipitator will need to consider thermal expansion problems and require extensive insulation to maintain the high temperature in cold weather. Therefore, a detailed cost estimate of the two precipitators is required.

c) Chemical Conditioning

Chemically conditioning the gas stream by addition of SO_3 will reduce the resistivity for the Cold side precipitator and should be considered as another alternative.

If 20 ppm SO_3 is injected, the fly ash resistivity will be 10^9 Ω cm (see Figure 8.5b). Calculate the Cold side design with 10^9 Ω cm resistivity at 300°F.

1. From Figure 8.8 (resistivity = $10^9 \Omega$ cm): w = 12.9 cm/sec = 0.423 ft/sec.

$$0.99 = 1 - \exp\left[\frac{-A}{Q} 0.423\right]$$

$$-\frac{A}{Q} = \frac{\ln(0.01)}{0.423}$$

$$\frac{A}{Q} = 8.22 \text{ ft}^2 / \left(\text{ft}^3 / \text{sec}\right)$$

Alternative design using k = 0.5

1a. $\quad 0.99 = 1 - \exp\left[\frac{-A}{Q} 0.423\right]$

$$\frac{A}{Q} = 50 \text{ ft}^2 / \left(\text{ft}^3 / \text{sec}\right)$$

This calculation demonstrates that SO_3 injection is an attractive alternative and should be considered for the control of fly ash resistivity.

This problem demonstrates that when high resistivity dust is encountered, it can be overcome by several techniques. The first option exists to operate at a lower temperature which will give favorable resistivity. In many applications, such as fly ash collection, normal operating temperatures are in the vicinity of 300°F, which can be the peak resistivity for many conditions. In the case of fly ash, the minimum temperature is governed primarily by corrosion of the air heater. If low-sulfur coal is being burned, corrosion is not so much of a problem, and lower temperatures can usually be tolerated. Within the temperature range of 220 to 230°F, resistivity will generally be within acceptable values.

A second option is to operate at higher temperatures, in the range of 500 to 900°F, where volume conduction is sufficiently high to bring dust resistivity to within acceptable limits. In many applications, such as metallurgical fumes, cement kiln dusts, etc., precipitators are often operated within this temperature region. In the case of fly ash, the precipitators can be located between the economizer and air heater, where a temperature of 600 to 800°F can be maintained. These "hot precipitators" can be effective in overcoming the high dust resistivity problem.

An alternate approach to temperature control is the use of chemical conditioning agents added to the flue gas. Since the principal cause of high resistivity in fly ash is a low level of SO_3 in the flue gases, the SO_3 concentration can be increased by 5 to 20 ppm. The feasibility of SO_3 conditioning in reducing fly ash resistivity has been demonstrated on a variety of types of ash, ranging from highly basic to neutral, and at temperatures ranging from 270 to 325°F.

8.9 Problem Set

8.1 For a uniform electric field of 12,000 V/cm and an ion concentration of $10^8/cm^3$, what is the saturation charge for a 2.0 µm diameter particle? How long does it take to charge this particle to 99% of its saturation charge? ($\varepsilon = 10$)

8.2 What particle size receives equal numbers of charges in 1 sec by diffusion charging and by field charging at 20 KV/cm? Assume that each mechanism operates independently and that the ion concentration is $10^8/cm^3$ for both situations. ($\varepsilon = 1$).

8.3 Estimate the theoretical migration velocity, w, of particles 50 µm in diameter with a density of $8 \bullet 10$ kg/m^3 in still air at 1 atm pressure and 334°C ($\mu = 3.13 \bullet 10^{-5}$ Pa sec) in an electric field of strength E = 100 KV/m, if each particle carries a charge of $q = 1.6 \bullet 10^{-14}$ C.

8.4 If the saturation charge on a particle is 300 electrons, the migration velocity 0.5 ft/sec and the electric field 3 KV/cm, it is possible to calculate the particle size by two methods.

a) $w = \dfrac{qe}{6\eta rn}$

b) $q_s = 3Er^2$

Calculate r, using the two methods and explain why they are different.

8.5 Determine the migration velocity for a 1.0 µm diameter unit density sphere positioned between two parallel plates 1 cm apart that are maintained at 9,000 V potential difference. Assume that the particle is at its saturation charge. What are the number of charges and what is the ratio of the electrostatic velocity to gravitational settling velocity?

8.6 Determine the migration velocity and the ratio of this velocity to the settling velocity for a 0.6 µm particle ($\varepsilon = 10$) in a field strength of 12 KV/cm.

a) Assume the particle is charged to saturation.
b) Assume the particle is charged by field charging for 1 sec at an ion concentration of $10^8/cm^3$.

8.7 What is the maximum velocity a 2 µm particle can obtain in an electrostatic field of 1,000 V/cm? (Note: Motion may be outside the Stokes' region).

8.8 Estimate the length of plate required to obtain an overall efficiency of 90% for an electrostatic precipitator with the following:

    gas velocity   = 5 ft/sec
    wire-to-wire distance   = 0.5 ft
    the electric field intensity = 7.3 KV/in.

Chapter 8: Electrostatic Precipitators

8.9 A coal-fired power plant sends 2,400 acfm through its electrostatic precipitator. Particle migration velocity is known to be 0.35 ft/sec. What is the collection area if the overall unit efficiency is 99%? What can be done to reduce the plate area?

8.10 An electrostatic precipitator duct is 18 m high and 14 m long with 20 cm plate-to-plate spacing. A collection efficiency of 96% is obtained with a flow rate of 180 m³/min.

    a) Estimate the collection efficiency if the flow rate is increased to 200 m³/min.

    b) Estimate the collection efficiency if the plate spacing is increased to 25 cm.

8.11 An electrostatic precipitator has three ducts with plates 12 ft wide and 12 ft high. The plates are 8 in. apart.

    a) Assuming a uniform distribution of particles and a drift velocity of 0.4 ft/sec, calculate the collection efficiency at a rate of flow of 4,000 acfm at 20°C and 1 atm.

    b) Calculate the efficiency if one duct was fed 50% of the gas and the others 25% each.

8.12 A precipitator consists of two stages, each with five plates in a series (see figure below). The corona wires between any two plates are independently controlled so that the remainder of the unit can be operated in the event of a wire failure.

    The following operating conditions exist:

        gas flow rate = 10,000 acfm

      plate dimension = 10 ft x 15 ft

        drift velocity = 19.0 ft/min Section 1

                           16.3 ft/min Section 2

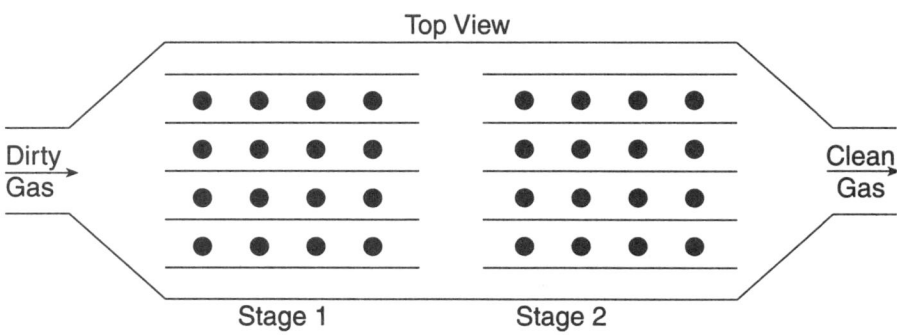

a) Determine the normal operating efficiency

b) During operation, a wire breaks in Stage 1. As a result, all of the wires in that row are shorted and ineffective, but the others function normally. Calculate the collection efficiency under these conditions.

c) Similarly, a wire breaks in Stage 2 after Stage 1 is repaired. What is the overall collection efficiency of the unit under these conditions?

8.13 An electrostatic precipitator, consisting of 5 plates (4 openings) 4 ft high and 20 ft deep with an 11 in. plate spacing, treats 50,000 acfm of air containing 4.0 grains/ft³ of particle. The particle size distribution and effective migration velocities are:

wt. % <d	d (μm)	w_e (cm/sec)
5	5	2.2
10	10	5.9
50	20	8.3
90	40	11.2
99.6	80	13.8

Calculate the overall collection efficiency.

8.14 An air flow of 100,000 acfm is produced by a cement manufacturing facility. If 99% removal efficiency is required, calculate the surface area of an electrostatic precipitator.

8.15 In the theoretical efficiency equation for electrostatic precipitators $\eta = 1 - e^{-Aw/Q}$, the drift velocity, w, can be treated as a parameter to be determined empirically. For fly ash from power plants typical values are 0.3 to 0.5 ft/sec. Plot the efficiency of such an ESP plate-type unit which is 20 ft high with 10 ft wide plates for velocities through the unit between 2 and 8 ft/sec and a plate spacing of 2 in. Why are very low bulk flow velocities not used for the air stream to maximize the efficiency?

8.16 You have been assigned to determine the size of an electrostatic precipitator for use on a power plant. The company wants to compare two different fuels. The design criteria for each one are as follows:

Eastern Coal

$SO_2 = 3\%$

Stack temperature = 350°F

Fly ash MMD = 3.7 μm, $\sigma g = 1.8$

Western Coal

SO_2 = 0.5%

Stack temperature = 350°F

Fly ash MMD = 5.0 μm, σg = 1.6

The required design efficiency is 98.94% and the electrostatic precipitator will operate at 5 KV/cm. Provide precipitator sizes based on determination of the migration velocity using a) basic particle technology theory, and b) experimental w_ϵ and empirical w_k design procedures. Discuss any differences that you obtain in the designs. Which design would you recommend?

8.17 It has been found necessary to control particulate emissions from an asphalt plant stone dryer. The plant has a capacity of 136 metric ton/hr and a flue gas rate of 34,000 actual m³/hr. The gas temperature is 116°C and has a dew point of 65°C. The particle size distribution is shown below. Particle density is 2.6 g/cm³. Currently, the uncontrolled emission rate is 2,310 kg/hr, compared to a maximum allowable limit of 25 kg/hr. Determine the design specifications for an electrostatic precipitator which will meet the above performance requirements.

Additional data: E_o = 354 KV/m

ϵ = 7.5 (for rock)

Particle size distribution: log-normal

8.18 For a large electrostatic precipitator installation on a power plant, the purchase cost is about 10 dollars per cfm for 95 to 97% efficiency. Estimate the cost of a unit for a 97% efficiency ESP for a new 1,000 megawatt plant. How much would this cost increase if the collection efficiency is increased to 99.9%?

%<dp	dp (μm)
2	0.9
10	2.7
50	18
80	68

8.10 References

Beachles, D. and J. Jahnke. Control of Particulate *Emissions*. Northrop Services, Inc. EPA 450/2-80-066. Research Triangle Park, NC. 1981.

Buonicore, T. and W. Davis (Eds). *Air Pollution Engineering Manual.* Air and Waste Management Assoc. John Wiley & Sons. New York. 1992.

Gallaer, C. *Electrostatic Precipitator Reference Manual*, EPRI Report CS-2809, Electric Power Research Institute. Palo Alto, CA. 1983.

Hesketh, H. *Air Pollution Control for Traditional and Hazardous Pollutants.* Technomic Publishers. Lancaster, PA. 1991.

Hinds, W. *Aerosol Technology: Properties, Behavior, and Measurement of Airborne Particle.* John Wiley & Sons. New York. 1982.

Matts, S. and Ohnfeldt. *Efficient Gas Cleaning with SF Electrostatic Precipitators*. Flaktfabriken, Stockholm, Sweden. 1963.

Oglesby, S. and G. Nichols. "Electrostatic Precipitators" in *Gas Cleaning for Air Quality Control*. Marchello, J. and Kelly, J.(eds). Marcel Dekker. New York. 1975.

Seizler, D. and W. Watson. "Hot Versus Enlarged Electrostatic Precipitator". J. Air Pollution Cont. Assoc. 24:115. 1974.

White, H. *Industrial Electrostatic Precipitation*. Addison-Wesley. Reading, MA. 1963.

White, H. "Resistivity Problems in Electrostatic Precipitation" J. Air Pollution Cont. Assoc. 24:314. 1974.

White, H. "Control of Particulates by Electrostatic Precipitation, " *Handbook of Air Pollution Technology*, Calvert, S., and H. Englund. Wiley-Interscience. New York. 1984.

Chapter 9

Control of Volatile Organic Compounds

9.1 Introduction

Volatile organic compounds (VOCs) are defined as organic compounds of carbon excluding carbon monoxide, carbon dioxide, carbonic acid, metallic carbides, metallic carbonates, and ammonium carbonates, having a vapor pressure of 0.02 pounds per square inch absolute or greater at standard conditions, including but not limited to petroleum fractions, petrochemicals, and solvents [*Federal Register* (CFR) 52.1596]. The vapor pressure of a pure chemical compound, central to its evaporation rate, relates only to its temperature. It has been measured for most of the pure compounds and values are readily available (Figure 9.1).

The terms VOC and hydrocarbon are not identical, but often are used interchangeably. Strictly speaking, a hydrocarbon contains only hydrogen and carbon atoms. Gasoline is normally called a "hydrocarbon fuel" because it contains mostly hydrogen and carbon atoms, but also some oxygen, nitrogen, and sulfur atoms. Table 9.1 lists some VOCs of interest in air pollution control applications.

The major anthropogenic sources of VOCs are industrial processes (46%) and automobiles (30%). VOCs in the exhaust gases from motor vehicles consist of unburned or partially burned gasoline. Gasoline contains approximately 2,000

Figure 9.1
Vapor Pressures of VOCs as a Function of Temperature

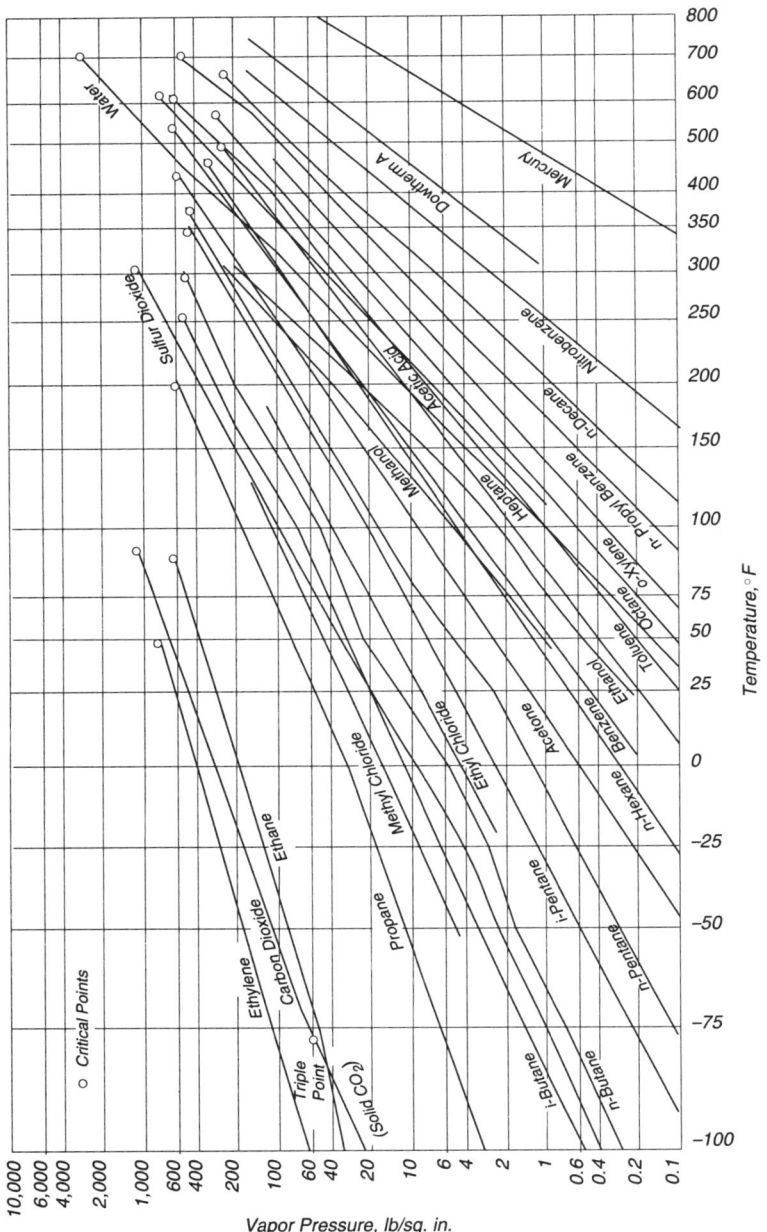

Reprinted with permission of Waveland Press, Inc. Cooper and Alley, Air Pollution Control: A Design Approach. Prospect Heights, IL: Waveland Press, Inc. 1994. All Rights Reserved.

Table 9.1
Some VOCs Identified in Ambient Air

Family	Compound	Chemical Formula
Hydrocarbons	Methane	CH_4
	Ethane	C_2H_6
	Propane	C_3H_8
	Butane	C_4H_{10}
	Pentane	C_5H_{12}
	Hexane	C_6H_{14}
	Benzene	C_6H_6
	Toluene	C_7H_8
	Ethylene	$CH_2=CH_2$
	2-Butene	$CH_3CH=CHCH_3$
Halomethanes	Methyl chloride	CH_3Cl
	Chloroform	$CHCl_3$
	Carbon tetrachloride	CCl_4
Haloethanes	1,2-Dichloroethane	CH_2ClCH_2Cl
Halopropanes	1,2-Dichloropropane	$CH_2ClCHClCH_3$
Chloroalkenes	Trichloroethylene	$CHCl=CCl_2$
	Allyl chloride	$ClCH_2CH=CH_2$
Chloroaromatics	Monochlorobenzene	C_6H_5Cl
	Dichlorobenzene	$C_6H_4Cl_2$
Oxygenated and nitrogenated species	Formaldehyde	HCHO
	Peroxyacetyl nitrate (PAN)	$CH_3COOONO_2$
	Acrylonitrile	CH_2CHCN
Chlorofluorocarbons (CFCs)	CFC-11	$CFCl_3$
	CFC-12	CF_2Cl_2

compounds. These include paraffins, olefins, and aromatics. Typical compositions vary from 4% olefins and 48% aromatics to 22% olefins and 20% aromatics (Buonicore and Davis 1992).

VOCs are widely used as solvents because they evaporate into the air, leaving either no residue (dry cleaning solvent) or a thin layer of previously dissolved solids (paints, inks). The rate of evaporation (pounds evaporated per square foot of exposed surface per hour) is roughly proportional to the vapor pressure. If quick evaporation is desired (spray paints), a solvent with a high vapor pressure at room temperature is used; if slower evaporation is desired (brushed-on paints, cleaning solvents) a solvent with a lower vapor pressure at room temperature is used.

VOCs engage in photochemical reactions in the atmosphere in the presence of sunlight and produce chemical compounds that cause visibility reduction, eye irritation, and contribute to health effects, such as emphysema.

Techniques generally used to control VOC emissions are condensation, adsorption, and combustion. The applicability of a given technique depends on the physical and chemical properties of the pollutant and the characteristics of the exhaust stream. More than one technique may be capable of controlling emissions from a given source. For example, vapors generated from loading gasoline into tank trucks at large bulk terminals are controlled by using any of these three techniques. More often, however, one control technique is used more frequently than others for a given source-pollutant combination.

Condensation is a process in which volatile gases are removed from the contaminant stream and changed to a liquid. Condensers are normally used in combination with primary control devices. Condensers can be located upstream of an incinerator, adsorber, or absorber. These condensers reduce the volume of vapors that the more expensive equipment must handle. Therefore, the size and the cost of the primary control device can be reduced. Similarly, condensers can be used to remove water vapors from a process stream with a high moisture content upstream of a control system. When used alone, refrigeration is often required to achieve the low temperatures required for condensation. Refrigeration units are used to successfully control gasoline vapors at large bulk gasoline dispensing terminals.

Combustion is defined as rapid, high temperature gas phase oxidation. Simply stated, the VOC is burned with air and converted primarily to carbon dioxide and water. Thermal incinerators are devices in which the contaminant air stream passes around or through a burner and into a refractory-lined residence chamber where oxidation occurs.

Adsorption is a mass transfer process that involves removing a gaseous contaminant by adhering it onto the surface of a solid. By far the most important adsorbent for air pollution control is activated carbon. Due to its surface, activated carbon will preferentially adsorb hydrocarbon vapors and odorous organic compounds from an air stream. Most other adsorbents (molecular sieves, silica gel, and activated aluminas) will preferentially adsorb water vapor, thereby preventing them from removing contaminants. Adsorption can be a very useful removal technique, since the vapors are not destroyed; only stored on the adsorbent surface until they can be removed by desorption. The desorbed vapor stream is highly concentrated. Depending on the nature of this material, it can be condensed and recycled or burned as the ultimate disposal technique.

9.2 Equilibrium Vapor Content

To understand which chemicals are volatile (evaporate at significant rates), the concept of vapor pressure must be considered. Figure 9.1 shows the saturated vapor pressures as a function of temperature for a variety of compounds (deNevers 1995). In a closed container, a volatile liquid will come to phase equilibrium with the vapor above it. If the container also contains a gas, like air, then at equilibrium

Chapter 9: Control of Volatile Organic Compounds

that air will be saturated with the vapor evaporated from the liquid. For the low pressures of interest for air pollution control, only small errors occur if it is assumed that the vapor mix behaves as a perfect gas, and the content of volatile liquid in the vapor mix can be estimated by:

$$y_i = x_i \frac{P_i}{P} \quad [9.1]$$

where: y_i = mole fraction = $\frac{\text{volume \%}}{100}$ (for ideal gas) of component i in the vapor

x_i = mole fraction of component i in the liquid

P_i = vapor pressure of pure component i

P = total pressure

The experimental vapor pressure data can be represented by the *Antoine equation*,

$$\log P = A - \frac{B}{(T+C)} \quad [9.2]$$

in which A, B, and C are empirical constants, determined from the experimental data. Values of Antoine equation constants for a variety of substances have been published; 23 substances are summarized in Table 9.2.

Table 9.2
Antoine Equation Constants

Compound	Chemical Formula	Range, °C	A	B	C
Acetaldehyde	C_2H_4O	-45 to +70	6.81089	992.0	230
Acetic acid	$C_2H_4O_2$	0 to +36 +36 to +170	7.80307 7.18807	1,651.1 1,416.7	225 211
Acetone	C_3H_6O	–	7.02447	1,161.0	224
Ammonia	NH_3	-83 to +60	7.55466	1,002.711	247.885
Benzene	C_6H_6	–	6.90565	1,211.033	220.790
Carbon tetrachloride	CCl_4	–	6.93390	1,242.43	230.0
Chlorobenzene	C_6H_5Cl	0 to +42 +42 to +230	7.10690 6.94504	1,500.0 1,413.12	224.0 216.0
Chloroform	CH_3Cl	-30 to +150	6.90328	1,163.03	227.4
Cyclohexane	C_6H_{12}	-50 to +200	6.84498	1,203.526	222.863
Ethyl acetate	$C_4H_8O_2$	-20 to +150	7.09808	1,238.71	217.0
Ethyl alcohol	C_2H_6O	–	8.04494	1,554.3	222.65
Ethyl benzene	C_8H_{10}	–	6.95719	1,424.255	213.206
n-Heptane	C_7H_{16}	–	6.90240	1,268.115	216.900
n-Hexane	C_6H_{14}	–	6.87776	1,171.530	224.366

Table 9.2 (cont.)
Antoine Equation Constants

Compound	Chemical Formula	Range, °C	A	B	C
Lead	Pb	525 to 1,325	7.827	9,845.4	273.15
Mercury	Hg	–	7.975756	3,255.61	281.988
Methyl alcohol	CH_4O	-20 to +140	7.87863	1,473.11	230.0
Methyl ethyl ketone	C_4H_8O	–	6.97421	1,209.6	216
n-Pentane	C_5H_{12}	–	6.85221	1,064.63	232.000
Isopentane	C_5H_{12}	–	6.78967	1,020.012	233.097
Styrene	C_8H_8	–	6.92409	1,420.0	206
Toluene	C_7H_8	–	6.95334	1,343.943	219.377
Water	H_2O	0 to 60 60 to 150	8.10765 7.96681	1,750.286 1,668.21	235 228.0

Note: $\log_{10} p = A - \dfrac{B}{T+C}$, p in mm Hg, T in °C

Example 9.1 Saturated Vapor Pressure — Water

Estimate the water content of air that is in equilibrium with pure liquid water at 68° F = 20°C.

Solution:

Using the value read from Figure 9.1:

$$y_i = x_i \frac{P_i}{P} = 1.00 \frac{0.023 \text{ atm}}{1 \text{ atm}} = 0.023$$

where $x_i = 1.00$, ignoring the small amount of air dissolved in the water.

Example 9.2 Saturated Vapor Pressure — VOC

Repeat Example 9.1 for a liquid mixture of 50 mol percent benzene and 50 mol percent toluene in equilibrium with air in a closed container.

Solution:

From Figure 9.1, the vapor pressures of benzene and toluene at 68°F are about 1.5 and 0.4 psia, respectively. Using more extensive tables, the values are 1.45 and 0.42 psia. Applying Equation [9.1]:

$$y_{benzene} = x_{benzene} \frac{P_{benzene}}{P} = 0.5 \frac{1.45 \text{ psia}}{14.7 \text{ psia}} = 0.049$$

$$y_{toluene} = x_{toluene} \frac{P_{toluene}}{P} = 0.5 \frac{0.42 \text{ psia}}{14.7 \text{ psia}} = 0.014$$

$$y_{air} = 1.0 - 0.049 - 0.014 = 0.937$$

9.3 Evaporation Loss Sources

9.3.1 Gasoline Marketing

Gasoline vapors represent a significant source of VOCs in urban areas. Air emissions ordinary occur at the time gasoline is transferred from one container to another and during storage. Figure 9.2 shows a schematic for gasoline handling. Gasoline is transferred from the refinery by ship, barge, or pipeline to large bulk-storage facilities called bulk terminals. Daily gasoline throughput at an average bulk terminal is about 250,000 gallons. Gasoline is then transferred into tank trucks (8,000 - 10,000 gallons capacity) for delivery to either bulk plants or service stations.

Figure 9.2
Gasoline Handling System

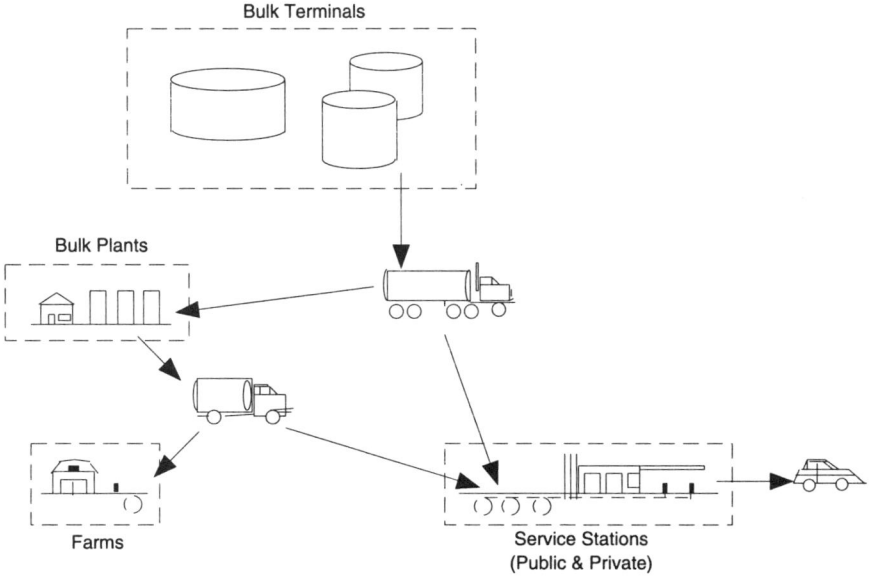

Liquid gasoline, flowing into the receiving container, displaces vapors in the vapor space of that container to the atmosphere. Displacement losses from loading operations can be estimated using the following expression (Buonicore and Davis 1992).

$$D_L = 12.46 \left[\frac{S \bullet P \bullet MW}{T} \right] \quad [9.3]$$

where: D_L = displacement loss, pounds per 10^3 gallons transferred

P = true vapor pressure of liquid loaded, psia

MW = molecular weight of vapors, lb/lb-mol

T = temperature of transferred liquid, °R or (°F + 460)

S = saturation factor (see Table 9.3)

Table 9.3
Saturation Factors for Calculating Petroleum Liquid Loading Losses

Carrier	Mode of Operation	Saturation Factor
Tank trucks and rail cars	Submerged loading:	
	Clean cargo tank	0.5
	Dedicated normal service	0.6
	Dedicated vapor balance service	1.00
	Splash loading:	
	Clean cargo tank	1.45
	Dedicated normal service	1.45
Marine vessels	Submerged loading:	
	Ships	0.2
	Barges	0.5

Equation [9.3] is derived from the ideal gas law and contains one major assumption, the degree of saturation, described by the saturation factor S. As shown in Table 9.3, the degree of saturation is dependent on the mode of operation (loading method employed, previous load, and the size of the container).

The vapor pressure of gasoline is specified by the *Reid Vapor Pressure*, RVP. Refiners adjust the RVP of their product by adjusting the ratio of low-boiling components to high-boiling components. In winter, they raise the RVP to improve the cold-starting properties of gasoline. In summer, they lower the RVP. Typical cold weather RVP values in the United States are 9 to 15 psi. Typical summer values are 8 to 10 psi. US EPA regulations limit the allowable RVP of gasoline. The limitation is only applicable in the summer months, in which VOC emissions contribute to photochemical ozone formation. The rules limit RVP to 9 psia for those areas that meet the ozone standard and to 7.8 psia for those that do not.

Following is an equation for estimating the molecular weight of gasoline vapors for use in Equation [9.3], given the RVP of the gasoline.

$$MW = 72.833 - 1.3183(P_{RVP}) + 0.15079(P_{RVP})^2 - 0.0087302(P_{RVP})^3 \quad [9.4]$$

where: MW = molecular weight of vapors, lb/lb-mole

P_{RVP} = Reid vapor pressure, psia

Following is an equation for estimating the true vapor pressure of the gasoline, given the RVP.

$$P = \exp\left[\left(0.7533 - \frac{423.0}{T}\right)S_D^{1/2}\log(P_{RVP}) - \left(1.854 - \frac{1,042}{T}\right)S_D^{1/2} \right.$$
$$\left. + \left(\frac{2,416}{T} - 2.013\right)\log(P_{RVP}) - \left(\frac{8,742}{T}\right) + 15.64\right] \quad [9.5]$$

where: P = true vapor pressure of liquid loaded, psia

P_{RVP} = Reid vapor pressure, psia

T = stock temperature, °R or (°F + 460)

S_D = slope of American Society for Testing and Material (ASTM) distillation curve at 10% evaporated (for gasoline, $S_D = 3$)

Figure 9.3 may also be used to obtain the true vapor pressure (AP-42). Using submerged-fill loading into containers (truck or storage tanks), as opposed to splash-fill, reduces the amount of vapors generated that potentially escape to the atmosphere. From the data in Table 9.3, conversion to submerged-filled loading (S = 0.6) from splash-fill (S = 1.45) reduces the amount of vapors generated by about 60%.

Vapor collection equipment is usually installed at gasoline-handling emission sources. The basic strategy is to collect vapors emitted at the end of the gasoline handling chain (service stations and bulk plants) and to transport the vapors back to the beginning of the chain (bulk terminals) for recovery or destruction.

Figure 9.4 shows a typical arrangement of vapor collection and processing at a bulk terminal. As gasoline is pumped from the storage tank into the truck tank, the air-vapor mixture is displaced through a vapor processor. Multistage refrigeration units and double-bed self-regenerating carbon adsorbers are the most common vapor recovery devices.

Figure 9.3
Vapor Pressures of Gasolines and Finished Petroleum

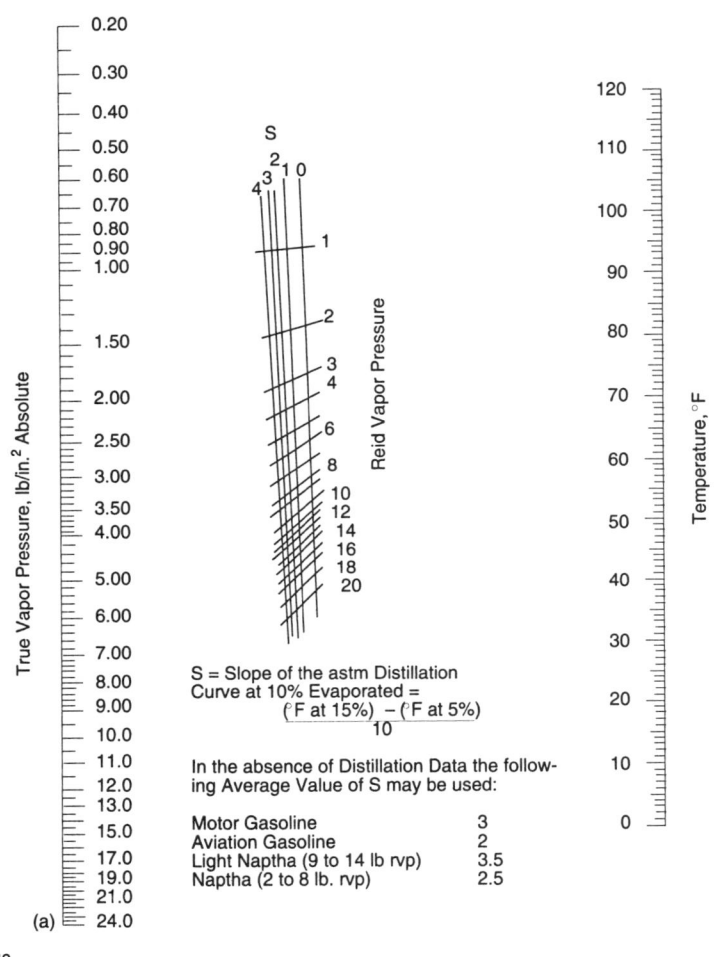

Source: AP-42

Example 9.3 Gasoline Truck Loading Vapor Loss

Calculate the loading losses, L_L, from a gasoline tank truck using vapor balance and practicing vapor recovery given the following information:

    Cargo tank volume is 8,000 gallons

    Gasoline RVP is 9 psia

    Product temperature is 80°F

    Vapor recovery efficiency is 95%

Chapter 9: Control of Volatile Organic Compounds   319

Figure 9.4
Bulk Terminal with Vapor Collection and Processor

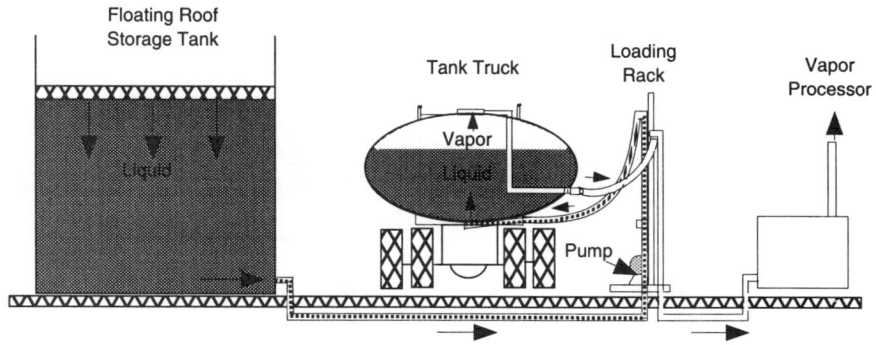

Source: AP-42

Solution:

Using loading loss equation (Equation [9.3]) with vapor recovery factor

$$L_L = 12.46 \frac{S \bullet P \bullet MW}{T}\left(1 - \frac{eff}{100}\right)$$

where: S = Saturation factor (see Table 9.3) = 1.00 (dedicated service)

P = True vapor pressure of gasoline = 6.6 psia (Figure 9.3)

MW = Molecular weight of gasoline vapors = 66

eff = Control efficiency = 95%

T = 540°R

$$L_L = 12.46\frac{(1.00)(6.6)(66)}{540}\left(1 - \frac{95}{100}\right)$$

= 0.50 lb/10³ gal

loading losses are: = (0.50 lb/10³ gal)(8.0 • 10³ gal) = 4.0 lb

Example 9.4 Service Station Vapor Emission

A gasoline service station has an average monthly throughput of approximately 30,000 gallons of gasoline. The station is equipped with a submerged vapor balance system. Based on the AP-42 table given below, estimate the monthly emissions in lb/month.

Emission source	Emission rate	
	lb/10³ gal throughput	kg/10³ liters throughput
Filling underground tank		
Submerged filling	7.3	0.88
Splash filling	11.5	1.38
Balanced submerged filling	0.3	0.036
Underground tank breathing and emptying	1	0.12
Vehicle refueling		
Displacement losses (uncontrolled)	9	4.08
Displacement losses (controlled)	0.9	0.11
Spillage	0.7	0.084
*Emissions include any vapor loss from underground tank to the gas pump.		

Solution:

1. Obtain the emission rate for balanced submerged filling in lb/10³ gal.
   Emission rate = 0.3 lb/10³ gal (from Table)

2. Calculate the emission rate for underground tank filling, E_1, in lb/month.
   E_1 = (emission rate) • (30,000 gal/month) = 9 lb/month

3. Calculate the emission rate from underground breathing and emptying, E_2.
   E_2 = (1.0 lb/10³ gal) • (30,000 gal/month) = 30 lb/month

4. Calculate the emission rate from vehicle refueling (from Table), E_3.
   E_3 = (9.0 lb/10³ gal) • (30,000 gal/month) = 270 lb/month

5. Calculate the emission rate from spillage (from Table), E_4.
   E_4 = (0.7 lb/10³ gal) • (30,000 gal/month) = 21 lb/month

6. Calculate the total emission rate.
   Total emission rate = $E_1 + E_2 + E_3 + E_4$ = 330 lb/month

9.3.2 Surface Coating

The use of solvents for a variety of surface coating operations represents a significant source of VOCs in many metropolitan areas. Table 9.4 provides a matrix of surface coating methods and the industries in which they are used.

Table 9.4
Surface Coating Application Methods by Industry

Surface coating method	Coil coating	Metal furniture	Auto and light-duty truck	Large appliance	Can	Auto refinish	Traffic marking
Dip		x		x			
Flow		x		x			
Roller	x					x	
Electrodeposition	x		x	x			
Spray							
Air atomized			x	x	x	x	x
Airless		x		x		x	
Electrostatic			x	x		x	
High volume, low pressure			x			x	
Electrostatic bell and disk				x			

VOC emissions can occur in a number of places along the production line: during atomization and application of the coating, during initial air drying of the part after it leaves the spray booth (flash-off), and in the bake oven. Table 9.5 shows the percentage of VOC emissions emitted during the various process steps for selected industries.

Table 9.5
Percentage of VOCs Emitted During Surface Coating Operations for Selected Industries

	Percentage of total VOC emissions	
Industry	Spray booth or application area and flash-off	Bake oven
Metal furniture	70	30
Automobile and light-duty truck	85-90	10-15
Large appliance	80	20
Coil coating*	8	90

*Remaining VOC emissions (2%) come from the quence section after the bake/curing oven

Air pollution control measures that can be applied to surface coating operations include thermal and catalytic incinerator, carbon adsorbers, and condensers (Table 9.6).

Table 9.6
Control Efficiencies for Pollution Control Equipment

Control device	Process being controlled	Control efficiency for control device, % by weight
Carbon adsorber	Spray booth	90
Carbon adsorber	Entire coating line	80
Thermal incinerator	Bake oven	96
Catalyst incinerator	Bake oven	90

Example 9.5 Automobile Coating Emissions

Calculate the emissions in lb/vehicle for automobile surface coating from the cathodic electrodeposited prime coat using the following table from AP-42. How much VOC is emitted per week based on a line speed of 55 automobiles/hr. There are two eight-hour shifts, 7 days per week.

Application	Area coated per vehicle, ft	Film thickness, mil	VOC content lb/gal H_2O	Volume fraction solids gal/gal H_2O	Transfer efficiency
Prime coat Solventborne spray	450 (220-570)	0.8 (0.3-2.5)	5.7 (4.2-6.0)	0.22 (0.20-0.35)	40 (35-50)
Chathodic electro-desposition	850 (660-1,060)	0.6 (0.5-0.8)	1.2 (1.2-1.5)	0.84 (0.84-0.87)	100 (85-100)
Guide coat Solventborne spray	200 (170-280)	0.8 (0.5-1.5)	5.0 (3.0-5.6)	0.30 (0.25-0.55)	40 (35-65)
Waterborne spray	200	0.8	2.8	0.62	30

(1 ft = 12,000 mils)

$$E = \frac{A_v C_1 T_f V_c C_2}{S_c e_T}$$

where: E = emission factor for VOC, mass/vehicle, lb/vehicle (exclusive of any add-on control devices)

A_v = area coated per vehicle, ft²/vehicle
C_1 = conversion factor, 1 ft/12,000 mil
T_f = thickness of the dry coating film, mil
V_c = VOC content of coating as applied less water, lb VOC/gal coating, less water
C_2 = conversion factor, 7.48 gal/ft³
S_c = solids in coating as applied, volume fraction, gal solids/gal coating
e_T = transfer efficiency fraction, fraction of total coating solids used which remains on coated parts

Solution:

The VOC emissions per automobile from a cathodic electrodeposited prime coat based on an average line speed of 55 automobiles/hr.

$$E = \frac{(850 \text{ ft}^2) \bullet (1/12,000) \bullet (0.6 \text{ mil}) \bullet (12 \text{ lb/gal H}_2\text{O}) \bullet (7.48 \text{ gal/ft}^3)}{(0.84 \text{ gal/gal}) \bullet (1.00)}$$

$= 0.45 \text{ lb VOC/vehicle } (0.21 \text{ kg VOC/vehicle})$

Total emissions per week

$$E = 0.45 \frac{\text{lb VOC}}{\text{Veh}} \bullet \frac{55 \text{ Veh}}{\text{hr}} \bullet \frac{16 \text{ hr}}{\text{day}} \bullet \frac{7 \text{ days}}{\text{week}} = 2,772 \text{ lb VOC/week}$$

9.4 Control Device Selection

9.4.1 TLV and LEL Emission Limits

There are two values of the air concentrations that are of considerable importance to the design of VOC control systems: the *threshold limit value*, TLV, and *lower explosive limit*, LEL.

- TLV relates to toxicity expressed in ppm and is an arbitrary value based on physiological considerations. It represents the conditions under which it is believed that nearly all workers may be repeatedly exposed, day after day, without adverse effects.

- LEL represents a property of the vapor. It is the lowest VOC concentration at which the mixture can sustain combustion. For insurance reasons, it is industry practice to provide enough ventilation to maintain a VOC concentration well below this limit. The usual value is set at 25 percent of the LEL. Explosive limits are usually given in percent by volume; one percent is equal to 10,000 ppm (Table 9.7) (Sax 1975).

Table 9.7
Physical Properties of Common VOCs

	Boiling Point °F	Molecular Weight	Soluble in Water	Flammable	Lower Explosive Limit[1] (vol %)	Carbon Adsorption Efficiency[2]
Acetone	133	58.1	Yes	Yes	2.15	8
Benzene	176	78.1	No	Yes	1.4	6
Butyl acetate	257	116.2	No	Yes	1.7	8
Butyl alcohol	241	74.1	Yes	Yes	1.7	8
Carbon tetrachloride	170	153.8	No	No	-	10
Ethyl acetate	171	88.1	Yes	Yes	2.2	8
Ethyl alcohol	165	46.1	Yes	Yes	3.3	8
Heptane	209	100.2	No	Yes	1.0	6
Hexane	156	86.2	No	Yes	1.36	6
Isobutyl alcohol	241	74.1	Yes	Yes	1.68	8
Isopropyl alcohol	205	60.1	Yes	Yes	2.5	8
Methyl alcohol	153	32.0	Yes	Yes	6.0	7
Methylene chloride	104	84.9	Yes	No	-	10
Methylethyl ketone	174	72.1	Yes	Yes	1.81	8
Methylisobutyl ketone	237	100.2	Yes	Yes	1.4	7
Perchloroethylene	250	165.8	No	No	-	20
Toluene	231	92.1	No	Yes	1.27	7
Trichloroethane	189	131.4	No	No	-	15
Trichlorotrifluoro ethane (113)	117.6	186.3	No	No	-	8
Naphtha	208	-	No	No	0.81	7
Xylene	292	106.2	No	Yes	1.0	10

[1] Lower explosive limit: The lowest concentration value of a vapor that will support propagation of a flame upward through a cylindrical tube.
[2] Carbon adsorption efficiency: Efficiencies are based on 200 cfm 100°F solvent-laden air per 100 lb of carbon per hour at concentrations about 15 ppm.

TLV, LEL, and 25 percent LEL values for some typical VOCs are given in Table 9.8. The TLVs for VOCs are much lower than the 25 percent LEL. This means that more dilution air is necessary to comply with the TLV than to comply with the 25-percent LEL. In general, TLV and LEL requirements demand much larger volumes of exhaust air than are necessary from a strictly operational point of view. Exhaust gas concentrations of VOCs are generally in the 1,000 to 5,000 ppm range to comply with the 25 percent of LEL requirements. Ventilation of areas where workers are exposed (i.e., paint spray booths) have much lower emission concentration (100 to 500 ppm), and require significant additional dilution air (about 10 times) in order to comply with the TLV.

Table 9.8
Comparison of TLV and LEL for Some Solvents
ppm by volume

	TLV	LEL	25% LEL
Acetone	1,000	22,000	5,500
Toluene	200	13,000	3,300
Xylene	100	10,000	2,500
Ethyl acetate	400	22,300	5,600
Isopropyl alcohol	400	25,000	6,300
Methyl ethyl ketone	200	18,100	4,500
n-Butyl acetate	150	17,000	4,300
Methylene chloride	500	None	None

Note: The TLV is not related to the LEL levels but generally have a much lower concentration

9.4.2 Selection of Control Device Based on Required Outlet Concentration

For VOCs, selection of an applicable control technique (condensation, adsorption, or combustion) is usually made on the basis of the inlet VOC concentration and the desired control efficiency. The expected emission reduction from the application of each control technique on the basis of the total VOC concentration in the emission stream is identified in Figure 9.5 (Sink 1991). Condensers cannot lower the inlet VOC concentration to levels below the saturation concentration (or vapor pressure) at the coolant temperature. When water, the most commonly used coolant, is employed, the saturation conditions represent high outlet concentrations. For example, condenser outlet VOC concentrations are often limited to 1,000 to 2,000 ppm due to the saturation conditions of most of the organic compounds at the temperature of the cooling water. This can be seen in Figure 9.5, where the efficiency of control by condensation decreases for inlet concentrations below 3,000 ppm and is less than 50% when the inlet concentration is below 500 ppm.

Figure 9.5
Approximate Percent Reduction Ranges for Different Air Pollution Control Equipment

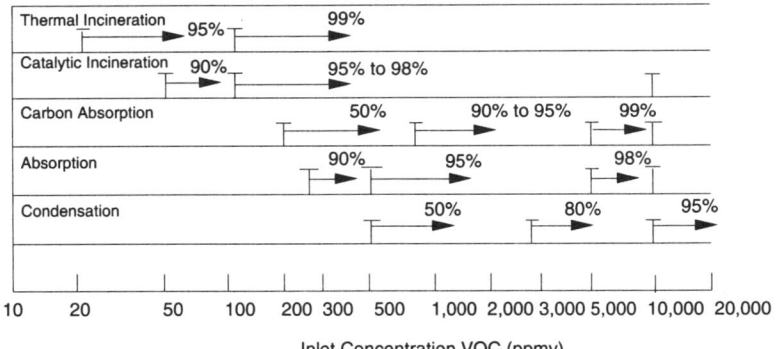

Source: AP-42

Removal efficiencies near 95% can be achieved for carbon sorption systems when the inlet concentration are 800 ppm or higher. For these inlet conditions, outlet concentrations near 50 ppm can be routinely achieved with state-of-the-art adsorption systems. Concentrations as low as 10 to 20 ppm can be achieved for some compounds. However, when the inlet concentration is below 500 ppm the collection efficiency decreases, it is below 50% for inlet concentrations below 200 ppm.

High molecular-weight compounds (Low Boiling Points) that are characterized by low volatility are strongly adsorbed on carbon. The affinity of carbon for these compounds makes it difficult to remove them during regeneration of the carbon bed. Hence, carbon adsorption is not applied to such compounds (i.e., boiling point above 400°F; molecular weight greater than about 130). Highly volatile materials (i.e., molecular weight less than about 45) do not adsorb readily on carbon; therefore, adsorption is not typically used for controlling emission streams containing such compounds. Compounds in Table 9.7 can usually be controlled by adsorption, and adsorption capacities are provided in that table.

To prevent excessive bed temperatures resulting from the exothermic adsorption process and oxidation reactions in the bed, concentrations higher than 10,000 ppm must be reduced. This is usually done by condensation ahead of the adsorption step. In contrast to incineration methods where the VOCs are destroyed, carbon adsorption provides a favorable control alternative when the VOCs in the emission stream are valuable. Low moisture content (<50%) is also required for carbon adsorption systems because higher moisture levels interfere with the adsorption capacity of the carbon for VOCs.

Thermal incineration is typically applied to emission streams that are dilute mixtures of VOCs and air. In such cases, due to safety considerations, concentrations

Chapter 9: Control of Volatile Organic Compounds

are limited to 25 percent of the LEL or to TLV levels. Additional fuel is required to support combustion which makes combustion an expensive alternative for dilute inlet streams. Catalytic incineration can be less expensive than thermal incineration in treating emission streams with low VOC concentrations due to lower auxiliary fuel requirements.

Example 9.6 Air Flow Rates

A paint-baking oven is to remove 1,000 lb/hr of toluene during a paint-drying process. For safety reasons, the concentration of toluene must be kept below 25% of the LEL. Estimate the required air flow rate. What will be the toluene concentration in ppm? Use Figure 9.5 to estimate the control efficiency for control devices that could be applied to this source to control the emissions. What control devices could be used if the concentration were reduced to 10% of the LEL? What about meeting the TLV requirement?

Solution:

From Table 9.7: for toluene, MW = 92.1

From Table 2.2: molar volume = 379 ft^3/lb mol

From Table 9.8: for toluene, TLV = 200 ppm, LEL = 13,000 ppm

$\quad$ 25% LEL = 0.25 • 13,000 = 3,250 ppm

$\quad$ 10% LEL = 0.10 • 13,000 = 1,300 ppm

$$\text{Required air flow rates (25\% LEL)} = \frac{1,000 \text{ lb/hr}}{3,250 \text{ ppm}} \bullet \frac{379 \text{ ft}^3/\text{lb mol}}{92.1 \text{ lb/lb mol}} \bullet \frac{1 \text{ ppm}}{(1/1\bullet 10^6)}$$

$$= 1.27 \bullet 10^6 \text{ ft}^3/\text{hr}$$

$$= 21,103 \text{ scfm}$$

$$\text{Required air flow rates (10\% LEL)} = \frac{1,000 \text{ lb/hr}}{1,300 \text{ ppm}} \bullet \frac{379 \text{ ft}^3/\text{lb mol}}{92.1 \text{ lb/lb mol}} \bullet \frac{1 \text{ ppm}}{(1/1\bullet 10^6)}$$

$$= 3.165 \bullet 10^7 \text{ ft}^3/\text{hr}$$

$$= 52,757 \text{ scfm}$$

$$\text{Required air flow rates (TLV)} = \frac{1,000 \text{ lb/hr}}{200 \text{ ppm}} \bullet \frac{379 \text{ ft}^3/\text{lb mol}}{92.1 \text{ lb/lb mol}} \bullet \frac{1 \text{ ppm}}{(1/1\bullet 10^6)}$$

$$= 2.057 \bullet 10^7 \text{ ft}^3/\text{hr}$$

$$= 342,924 \text{ scfm}$$

From Table 9.5, the following tabulated data can be obtained to provide an estimate of the collection efficiency for various air pollution control devices.

Outlet concentration	Flow volume, scfm	Estimated efficiency of control				
		Condensation	Absorption	Adsorption	Catalytic incineration	Thermal incineration
3,250	21,103	80	95	90-95	98	99
1,300	52,757	50	95	90-95	98	99
200	342,923	<50	<90	50	98	99

The table shows that at 3,250 and 1,300 ppm, all air pollution control devices can be used to reduce the toluene concentration by 90%, except condensation (50 to 80% efficiency). However, at 200 ppm, the only available option is combustion, but this option is usually not economically feasible due to costs associated with the large volume of air (342,924 scfm) that needs to be controlled.

9.5 Design of Condensers to Remove VOCs

9.5.1 Introduction

Condensers are widely used as preliminary air pollution control devices for removing VOC contaminants from emission streams prior to other control devices such as incinerators or adsorbers. Used in this manner, they help reduce the overall cost of control systems.

When a hot vapor stream contacts a cooler surface, heat is transferred from the hot gases to the cooler surface. Condensation occurs when the partial pressure of the pollutant in the gas stream equals its vapor pressure. Also, the volume that these vapors occupy is reduced.

Condensers are simple, relatively inexpensive devices that normally use chilled water or air to cool and condense a vapor stream. Since these devices are usually not capable of reaching low temperatures (below 80°F), high removal efficiencies of most gaseous pollutants are not obtained unless the vapors will condense at high temperatures. Condensers differ in the means of removing heat. The two different means of condensing are *direct* contact, where the cooling medium with vapors and the condensate are intimately mixed and combined, and *indirect* (or surface), where the cooling medium and vapor/condensate are separated by a surface area of some type.

Contact condensers are simpler, less expensive to install, and require less auxiliary equipment and maintenance. The condensate/coolant from a condenser has a volume 10 to 20 times that of a surface condenser. This condensate often cannot be reused and may pose a waste disposal problem. Therefore, surface condensers form the bulk of the condensers used for air pollution control. Among the applicable types of surface condensers are shell-and-tube, double pipe, spiral plate, flat plate and air-cooled. This section focuses on shell-and-tube condensers (Figure 9.6) because they are widely used in industry.

Figure 9.6
Shell-and-Tube Indirect Condensor

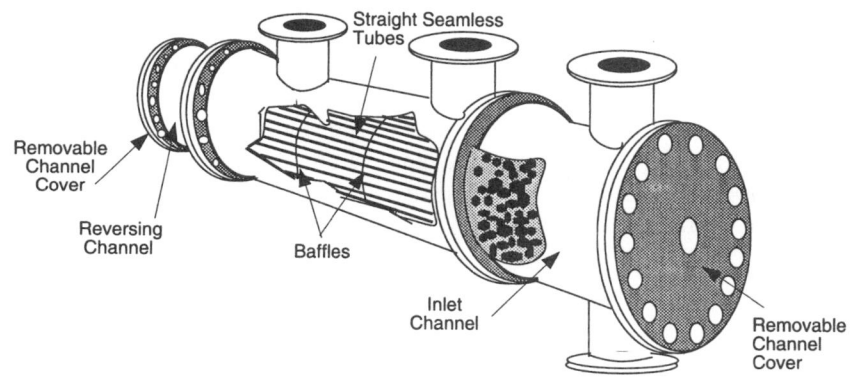

The device consists of a circular shell into which the vapor stream flows. Inside the shell are numerous small tubes through which the coolant flows. Vapors contact the cool surface of the tubes, condense, and are collected, while noncondensed vapors pass through the condenser. The cooling liquid passes through the tubes of the exchanger. By using a multipass system and shorter tube lengths, higher gas velocities through the exchanger and improved heat transfer can be achieved.

There are two widespread applications for condensation for VOC control. The first is at gasoline loading terminals. Many loading terminals treat emission discharge due to displacement with regenerable adsorbers, or destroy the VOCs in an incinerator. Others pass it through a refrigerated condenser, which usually has two stages. The first stage operates just above the freezing point of water to remove most of the water and some of the gasoline. The second, a much cooler stage, removes most of the remaining gasoline.

These condensers encounter a special fire hazard. Most VOC-contaminated air streams have VOC concentrations less than the LEL. Removing VOCs from them takes them even farther from a combustible condition. The vapors from gasoline tanks generally contain enough gasoline to be above the explosive limit. Removing VOCs from them, by condensation or adsorption, causes them to pass through the combustible range before they pass below the LEL within the control device. Inside the device, they are often combustible; therefore, special care is required to exclude all possible ignition sources and to provide flame arrestors.

The other common application of condensation for VOCs control is a condenser at the outlet of an adsorber, where the contaminated gas is steam. During regeneration, the VOC is transferred into the steam, from which it can be easily condensed.

9.5.2 Design Equations

The design of a condenser is significantly affected by the number and nature of the components present in the emission stream. In the two-component vapor system, where one of the components is noncondensible (e.g., air), condensation occurs at the dew point saturation temperature when the partial pressure of the condensible compound is equal to its vapor pressure. In most VOC control applications, the emission stream will contain large quantities of noncondensible and small quantities of condensible compounds. To separate the condensible component from the gas stream at a fixed total pressure, the temperature of the gas stream is reduced. The more volatile a compound (i.e., the lower the normal boiling point) the lower the temperature required for saturation (condensation).

Condensers are sized based on the total heat load and the overall heat transfer coefficient (McCabe and Smith 1967; Perry 1973). Condenser heat load is defined as the quantity of heat that must be extracted from the emission stream to achieve a certain level of removal. It is determined from an energy balance, taking into account the heat of condensation of the VOCs, sensible heat change of the VOCs and the sensible heat change in the emission stream (McCabe and Smith 1967; Perry 1973). For a condensation system, the heat balance can be expressed as:

Heat in = Heat out

Specific heat rate required to reduce vapor to the dew point + Latent heat rate required to condense vapors = Heat rate needed to be removed by the coolant

This heat balance is written in equation form as:

$$Q = m \bullet C_p \left(T_{G1} - T_{dew\ point}\right) + m \bullet H_v = L \bullet C_p \left(T_{L2} - T_{L1}\right) \qquad [9.6]$$

where: Q = heat transfer rate, Btu/hr

m = mass flow rate of vapor, lb/hr

L = mass flow rate of liquid coolant, lb/hr

C_p = average specific heat of a gas or liquid, Btu/lb °F

T = temperature of the streams: G for gas and L for liquid coolant, °F

H_v = heat of condensation or vaporization, Btu/lb

Chapter 9: Control of Volatile Organic Compounds

In equation [9.6], the mass flow rate, m, and inlet temperature, T_{G1}, of the vapor stream are set by the process exhaust stream. The temperature of the coolant entering the condenser, T_{L1}, is also set. The average specific heats, C_p, of both streams, the heat of condensation, H_v, and the dew point temperature can be obtained from chemistry handbooks. Therefore, only the amount of coolant, L, and its outlet temperature are left to be determined.

The enthalpy change associated with any noncondensible vapor (i.e., air) temperature change must also be included in the condenser heat load. C_p for air is the average for the temperature interval, $T_{L2} - T_{L1}$, for a dilute emission stream where the vapor concentration is in the ppm range. The condenser heat load that is due to the heat that must be extracted to cool the incoming air is significant.

The rate of heat transfer depends upon three factors: total cooling surface available, resistance to heat transfer, and mean temperature difference between condensing vapor and coolant.

$$Q = U \bullet A \bullet \Delta T_m \qquad [9.7]$$

where: U = overall heat transfer coefficient, Btu/°F ft² hr

A = heat transfer surface area, ft²

ΔT_m = Log mean temperature difference, °F

An estimate of the overall heat transfer coefficient can be used for calculations. Table 9.9 lists some typical values (Beachler and Joseph 1981).

Table 9.9
Typical Overall Heat Transfer Coefficients in Tubular Heat Exchangers

Condensing Vapor* (shell side)	Cooling Liquid (tube side)	U (Btu/°F ft² hr)
Alcohol vapor	water	100-200
High boiling hydrocarbons (vacuum)	water	20-50
Low boiling hydrocarbons	water	80-200
Organic solvents	water	100-200
Organic solvents with high percent of noncondensible present	water or brine	20-60
Naphtha	water	50-75
Stabilizer reflux vapors	water	80-120
Sulfur dioxide	water	150-200
Tall oil derivatives vegtetable oil vapors	water	20-50
Steam	feed water	400-1,000

* For water-water (liquid-liquid) heat exchanger (no phase change) the values for U range between 200-250

The choice of a coolant will depend on the required condenser efficiency. The most common coolant is usually water (Table 9.10) (Danielson 1973). In some instances, use of a chilled brine or a boiling refrigerant can achieve a collection efficiency that will be sufficient without additional control devices. It is not unusual to specify multiple-stage condensing where condensers, usually two connected in series, use cooling mediums with successively lower temperatures. For example, a condenser using cooling-tower water can be used prior to a unit using chilled water or brine, thereby achieving maximum efficiency while minimizing the use of chilled water.

Table 9.10
Coolant Selection

Required condensation temperature, T_c (°F)[a]	Coolant
T_{con}[b] 60-80	Water
$60 > T_c > 45$	Chilled water
$45 > T_c > -30$	Brine solutions (e.g., calcium chloride)
$\sim -90 > T_c > -30$	Chlorofluorocarbons (e.g., Freon-12)

[a] Also emission stream outlet temperature
[b] Summer limit

In a surface heat exchanger, the temperature difference between the hot vapor and the coolant usually varies throughout the length of the exchanger. Therefore, a mean temperature difference, ΔT_m, must be used. For the special cases, where the flow of both streams is completely cocurrent, the flow of both streams is completely countercurrent, or the temperature of one of the fluids remains constant (as in the case in condensing a pure liquid), the log mean temperature difference can be used. The log mean temperature for countercurrent flow can be expressed as:

$$\Delta T_m = \Delta T_{lm} = \frac{\Delta T_1 - \Delta T_2}{\ln\left(\frac{\Delta T_1}{\Delta T_2}\right)} \qquad [9.8]$$

where: ΔT_{lm} = log mean temperature

$\Delta T_1, \Delta T_2$ = inlet and outlet temperatures differences between the vapor and coolant, respectively

(ΔT_2 of at least 30°F is a typical design value)

The key variable in condenser system design is the required condensation temperature for a given removal efficiency or outlet concentration. Once the removal efficiency for given VOCs is specified, the required temperature for condensation can be determined from data on its vapor pressure-temperature relationship. The emission stream is usually assumed to consist of a two-component mixture: one

Chapter 9: Control of Volatile Organic Compounds

condensible component (VOCs) and one noncondensible component (air). To simplify the calculations, it is assumed that condensation occurs isothermally. It is usually assumed that the emission stream entering the condenser consists of air saturated with the VOC in question. At equilibrium between the gas and liquid phases, the partial pressure of the VOC is equal to its vapor pressure at that temperature. Therefore, by determining the temperature at which this condition occurs, the condensation temperature, T_c, can be specified. When condensers are used to control emissions, they are usually operated at the constant pressure of the emission source, which is normally close to atmosphere.

To determine cooling water requirements, use the following equation:

$$Q_{cw} = \frac{Q_{load}}{C_{pw}(T_{wo} - T_{wi})} \qquad [9.9]$$

where: T_{wo} = cooling water outlet temperature, °F

T_{wi} = cooling water inlet temperature, °F

Q_{cw} = cooling water flow rate, lb/hr

Q_{cw} can be expressed in terms of gal/min as follows:

$$Q_w = Q_{cw}[(1/60) \cdot (1/62.43) \cdot 7.48]$$
$$= 0.002 \cdot Q_{cw}$$

where the factor 62.43 is the density of water and factor 7.48 is used for converting from ft^3 to gal basis.

Example 9.7 Condenser Operation

Steam at atmospheric pressure is being condensed in a heat exchanger with water that enters and exits the unit at 80°F and 115°F, respectively. Assuming no subcooling of the condensed steam, calculate the log mean temperature driving force for the exchanger.

Solution:

Saturated steam at 1.0 atm is at 212°F. The log mean temperature difference is given by equation :

$$\Delta T_m = \Delta T_{lm} = \frac{\Delta T_1 - \Delta T_2}{\ln\left(\frac{\Delta T_1}{\Delta T_2}\right)}$$

Substituting temperatures gives

$$\Delta T_1 = 212 - 80 = 132$$
$$\Delta T_2 = 212 - 115 = 97$$
$$\Delta T_m = (132 - 97)/\ln(132/97)$$
$$= 113.6°F$$

Example 9.8 Condenser Sizing

The discharge gases from a meat-rendering plant contain essentially all steam, with a small fraction of odor-carrying gases. This steam is typically condensed (to remove the water) prior to treatment with an adsorber or incinerator. Estimate the size of a condenser to treat 60,000 lb/hr of this exhaust gas. Assume condensing occurs in a heat exchanger operating with the temperature driving force calculated in Example 9.3 and an overall heat-transfer coefficient of 135 Btu/hr ft^2 °F.

Solution:

Assuming that the discharge gas is all steam with an enthalpy of vaporization of approximately 1,000 Btu/lb, the condenser heat load becomes:

$$Q = m \cdot DH$$
$$= 60,000 \cdot 1,000$$
$$= 6.0 \cdot 10^7 \text{ Btu/hr}$$

The condenser area is as follows:

$$A = \frac{Q}{U \cdot \Delta T_m}$$
$$= \frac{6.0 \cdot 10^7}{135 \cdot 113.6}$$
$$= 3,912 \text{ ft}^2$$

Chapter 9: Control of Volatile Organic Compounds

Example 9.9 Condenser Surface Area Sizing

In an oil refinery, a stream of light hydrocarbons is to be condensed by a surface condenser, as shown in the following figure. The light hydrocarbon stream is essentially benzene. For benzene, the boiling point is 175°F, the latent heat of vaporization, H_v, is 160 Btu/lb, and the specific heat, C_p, is 0.45 Btu/lb °F. Water is used as the coolant at 60°F and $C_p = 1.0$ Btu/lb°F. The overall heat transfer coefficient, U, is 110 Btu/hr-ft^3-°F for a benzene mass flow rate, m, of 10,000 lbs/hr. What is the surface area of the tubes required for a surface condenser if the condensate temperature, T_c, can be no higher than 100°F and the outlet condenser temperature is 130°F?

Surface Condenser

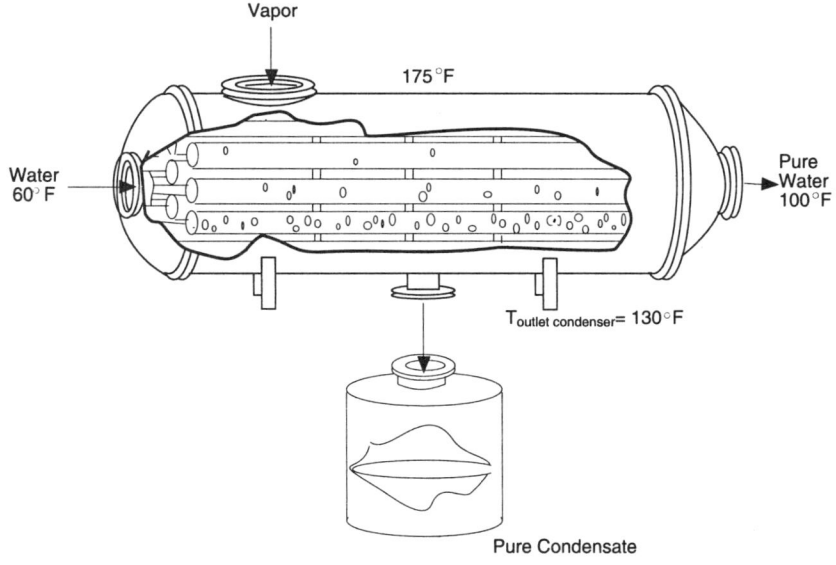

Pure Condensate

Solution:

1. Calculate the amount of cooling water required using a heat balance, i.e., Heat In = Heat Out

| Heat required to condense vapors | + | Heat required to cool vapors to outlet temperature | = | Heat needed to be removed by the coolant |

2. Calculate the heat required using:

$$Q = m \cdot H_v + m \cdot C_p (T_{G1} - T_{G2})$$

$$Q = \left(10{,}000\frac{\text{lb}}{\text{hr}}\right)\left(\frac{160\text{ Btu}}{\text{lb}}\right) + 10{,}000\frac{\text{lb}}{\text{hr}} \bullet 0.45\frac{\text{Btu}}{\text{lb °F}} \bullet (175°F - 130°F)$$

$$= 160 \bullet 10^4 \text{ Btu/hr} + 20.25 \bullet 10^4 \text{ Btu/hr}$$

$$= 180.25 \bullet 10^4 \text{ Btu/hr}$$

3. Calculate the amount of coolant water required:

$$180.25 \bullet 10^4 \text{ Btu/hr} = W \bullet C_p \bullet (T_c - T_{IN})$$

$$180.25 \bullet 10^4 \text{ Btu/hr} = W \bullet 1.0\frac{\text{Btu}}{\text{lb°F}}(100 - 60)$$

$$180.25 \bullet 10^4 \text{ Btu/hr} = W \bullet 40 \text{ Btu/hr}$$

$$W = 45{,}062.5 \text{ lb/hr} = 90.1 \text{ gal/min}$$

4. Calculate the surface area of tubes by first calculating the mean temperature change.

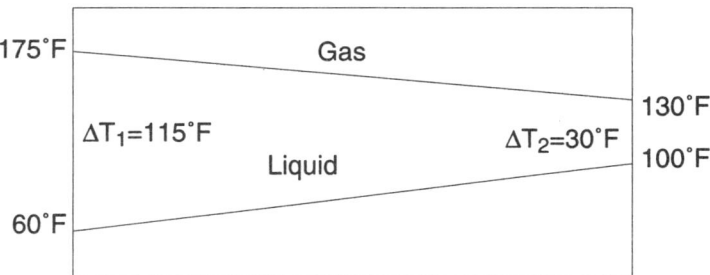

$$\Delta T_m = \frac{(T_{G1} - T_{L1}) - (T_{G2} - T_{L2})}{\ln\left(\frac{T_{G1} - T_{L1}}{T_{G2} - T_{L2}}\right)}$$

$$\Delta T_m = \frac{(175 - 60) - (130 - 100)}{\ln\left(\frac{175 - 60}{130 - 100}\right)}$$

$$\Delta T_m = \frac{(115 - 30)}{\ln\left(\frac{115}{30}\right)} = 62.2°F$$

5. Calculate the condenser surface area using:

$$A = \frac{Q}{U \cdot \Delta T_m}$$

$$A = \frac{180.25 \cdot 10^4 \text{ Btu/hr}}{\left(\frac{110 \text{ Btu/hr}}{°F \text{ ft}^2}\right) \cdot 62.2°F}$$

Surface area of the condenser tubes = 263 ft³

9.6 Chlorinated Solvents

Chlorinated solvents have properties that make them suitable for a wide range of industrial applications. Roughly a million tons are used annually in processes ranging from vapor degreasing to the fabrication of electronic components. Chlorinated solvents are non-flammable and have normal boiling points near or slightly above room temperatures. Data on the production of five of the major chlorinated solvents are given in Table 9.11 (Allen 1992). The table shows that while total production has been relatively stable over the past decade, the production rates of particular solvents have seen major fluctuations. For instance, production of 1,1,2-trichlorotrifluoroethane (CFC-113) doubled between 1979 and 1989, while production of trichloroethylene (TCE) decreased by 40 %. Some of the primary driving forces for these fluctuations have been environmental regulations. In particular, regulations related to smog formation, stratospheric ozone depletion, and global warming have influenced the patterns of solvent usage.

Three indices for ozone depletion, global warming, and smog formation potential have been developed to help measure some of the environmental impacts of chlorinated solvents. Chlorinated solvents that reach the stratospheric ozone layer can photodissociate, releasing chlorine atoms that catalyze ozone destruction. For a compound to cause stratospheric ozone depletion, it must have a lifetime in the atmosphere sufficient to reach the stratosphere and it must contain chlorine or some other halogen, such as bromine. The atmospheric lifetimes and ozone depletion potential (ODP) for the major chlorinated solvents are given in Table 9.12. The ODP provides a relative measure of the extent to which a compound contributes to ozone destruction.

The global warming potential of a chemical depends on both its ability to absorb infrared radiation and the length of time that it remains in the atmosphere. The global warming potential (GWP) of the five chlorinated solvents are listed in Table 9.12. In the case of smog formation potential, the index is based on the oxidation of the solvent by the hydroxyl radical (-OH). An approximate smog formation potential index (SFP) is also given in Table 9.12.

Table 9.11
Current and Historical Production of Chlorinated Solvents
(thousands of metric tons)

Year	Trichloro-ethylene	Tetrachloro-ethylene	Methylene chloride	1,1,1-Trichloro-ethane	CFC-113	Total
1979	145	351	287	325	47	1,155
1980	121	347	256	314	50	1,088
1981	117	313	269	279	53	1,031
1982	120	265	241	270	56	952
1983	100	248	265	266	60	939
1984	86	260	275	306	68	995
1985	73	224	263	268	73	901
1986	82	188	257	296	73	896
1987	82	215	234	315	78	924
1988	82	226	229	328	78	943
1989	82	215	213	353	78	941

Source: Allen et al. 1992

Table 9.12
Effect of Major Chlorinated Solvents

Compound	Atmospheric lifetime yrs	Ozone depletion potential index 1/g mol	Global warming potential index 1/g mol	Smog formation potential index 1/g mol
CH_2Cl_2 (methylene chloride)	0.6	$9 \cdot 10^{-3}$	$7 \cdot 10^{-3}$	$2.9 \cdot 10^{-3}$
$Cl_3C\text{-}CH_3$ (1,1,1-trichloroethane)	6.3	0.15	$5.7 \cdot 10^{-2}$	$4.4 \cdot 10^{-4}$
$HClC=CCl_2$ (trichloroethylene)	0.1	none	$1 \cdot 10^{-3}$	$5.1 \cdot 10^{-2}$
$Cl_2C=CCl_2$ (tetrachloroethylene)	0.6	none	$9 \cdot 10^{-3}$	$3.8 \cdot 10^{-3}$
$Cl_2FC\text{-}CClF_2$ (CFC-113)	90	1.3	2.5	$1.1 \cdot 10^{-5}$
$CFCl_3$ (CFC-11)	60	1.0	1.0	1.0

Source: Allen et al. 1992

The data in Table 9.12 show that CFC-113 has a high potential for ozone depletion and global warming and a low smog formation potential. Trichloroethylene has a high smog formation potential but a low ozone depletion and global warming potential. These are important results because the regulation for control of chlorinated solvents often involves substitution or replacement with less damaging material, and all aspects of environmental impact must be taken into consideration.

Chapter 9: Control of Volatile Organic Compounds   **339**

Example 9.10 Environmental Impact Analysis

Rank the chlorinated solvents in Table 9.12 by their environmental impacts, using both their production rate and potential index for ozone depletion, global warming, and smog formation.

Solution:

1. Convert the data in Table 9.11 from a mass basis to a molar basis.

$$1 \text{ metric ton } CH_2Cl_2 = \left(\frac{10^6 \text{ g}}{\text{metric ton}}\right)\left(\frac{\text{mol } CH_2Cl_2}{85 \text{ g } CH_2Cl_2}\right) = 11,765 \text{ g mol}$$

2. If it is assumed that 80% of production reported in Table 9.11 is emitted to the atmosphere, then for CH_2Cl_2:

   Atmospheric emissions in 1979 = 287,000 • 0.8 = 229,600 metric ton or $2.7 \cdot 10^9$ moles.

3. Similarly, in 1989 emissions of CH_2Cl_2 were $2.0 \cdot 10^9$ moles.

4. Combining these emissions with data in Table 9.12 gives smog formation potential (SFP) for emissions of CH_2Cl_2 for 1979 = $2.7 \cdot 10^9 \cdot 2.9 \cdot 10^{-3} = 7.83 \cdot 10^6$.

5. Repeat the foregoing calculations to produce the following table.

Atmospheric Impact of Tonnage of Solvent Emission; Ranking in Parentheses ( )						
Compound	Smog formation potential		Global warming potential		Ozone depletion potential	
	1979	1989	1979	1989	1979	1989
CH_2Cl_2 (methylene chloride)	(2) $7.8 \cdot 10^6$	(2) $5.8 \cdot 10^6$	(3) $1.9 \cdot 10^7$	(3) $1.4 \cdot 10^7$	(3) $2.4 \cdot 10^7$	(3) $1.8 \cdot 10^7$
$Cl_3C\text{-}CH_3$ (1,1,1-trichloroethane)	(4) $8.6 \cdot 10^5$	(4) $9.3 \cdot 10^5$	(2) $1.1 \cdot 10^8$	(2) $1.2 \cdot 10^8$	(1) $2.9 \cdot 10^8$	(2) $3.2 \cdot 10^8$
$HClC=CCl_2$ (trichloroethylene)	(1) $4.5 \cdot 10^7$	(1) $2.5 \cdot 10^7$	(5) $8.8 \cdot 10^5$	(5) $5.0 \cdot 10^5$	0	0
$Cl_2C=CCl_2$ (tetrachloroethylene)	(3) $6.4 \cdot 10^6$	(3) $3.9 \cdot 10^6$	(4) $1.5 \cdot 10^7$	(4) $9.3 \cdot 10^6$	0	0
$Cl_2FC\text{-}CClF_2$ (CFC-113)	(5) $2.2 \cdot 10^3$	(5) $3.7 \cdot 10^3$	(1) $5.0 \cdot 10^8$	(1) $8.3 \cdot 10^8$	(2) $2.6 \cdot 10^8$	(1) $4.3 \cdot 10^8$
Total	$6.0 \cdot 10^7$	$3.6 \cdot 10^7$	$6.5 \cdot 10^8$	$9.8 \cdot 10^8$	$5.8 \cdot 10^8$	$7.7 \cdot 10^8$

The rankings do not change between 1979 and 1989 despite the wide variation in solvent use. For example, the use of trichloroethylene dropped ~40% (from 145,000 tons to 82,000 tons) and the use of trichloroethane increased ~9% in the same 10 year period. CFC-113 is first in long term effects and trichloroethylene is first in short term effects.

The rankings do not change between 1979 and 1989 despite the wide variation in solvent use. For example, the use of trichloroethylene dropped ~40 % (from 145,000 tons to 82,000 tons) and the use of trichloroethane increased ~9 % in the same 10 year period. CFC-113 is first in long-term effects and trichloroethylene is first in short-term effects.

9.7 Problem Set

9.1 Calculate the mole fraction and gas phase equilibrium concentration in ppm of acetone in air at 100°F. The total pressure is one atmosphere.

9.2 In order to collect 40% of a toluene vapor air stream, it must be cooled to what temperature? The air stream is 100°F at 1 atm and contains 60,000 ppm of toluene.

9.3 Find the mass flow rate of organic material for an air stream at 70°F containing 20,000 ppm of heptane and flowing at a rate of 1,766 ft^3/min.

9.4 Two hundredths of a gallon of liquid propanol is vaporized in a 90,500 ft^3 closed (smog) chamber containing essentially air. Calculate the ppm of propanol in the gas contained within the chamber. Assume the density of the propanol to be 7.36 lb/gal and ambient (60°F, 1 atm) conditions to apply.

9.5 a) Use the Clapeyron equation:

$\ln p = A - (B/T)$

where: p = vapor pressure, mm Hg

T = temperature, °K

A, B = Clapeyron coefficients

Estimate the vapor pressure of acetone at 0°C. The Clapeyron coefficients have been experimentally determined to be A = 15.02 and B = 2,817.

b) Use the Antoine equation (Equation [9.2]) to estimate the vapor pressure of acetone at 0°C. The Antoine coefficients A, B, and C are 7.02447, 1,161.0 and 224, respectively.

Compare the results in part a) and b).

9.6 Determine the minimum number of cubic feet of dry air required to evaporate 20 lb of alcohol if the total pressure is maintained at 740 mm Hg. The evaporation process is isothermal at 70°F.

9.7 A vapor degreasing tank is shown in Figure 9.8. The tank is used to clean metal parts. The parts to be cleaned are placed in the vapor zone and the hot solvent and the hot solvent condenser on the cold part. Solvent loss is controlled by the use of a condenser, as shown. TCE (trichloroethylene) and TCA (1,1,1-trichloroethane) can be used as the solvent. Use the information in Table 9.12 to estimate the smog formation, global warming, and ozone depletion potential per ton of TCE and TCA used. 90 % of the solvent may eventually escape to the atmosphere. Clearly the selection process presents a dilemma. Discuss process controls that might reduce the overall emissions of both TCE and TCA.

Vapor Degreasing Tank

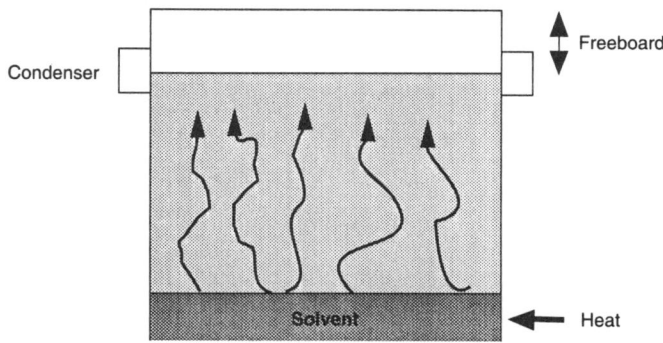

9.8 Fugitive VOC losses from an 8.0 ft degreaser used to clean machine parts are estimated at 2.5 lb/hr. However, during unloading (this occurs hourly for approximately two minutes during the eight-hour workday), the carryout loss is estimated to be 1.0 lb/min.

   a) Estimate the daily total solvent (VOC) loss from the degreaser.

   b) If the maximum and minimum operating solvent capacity of the unit are 6.0 and 5.5 ft^3, respectively, calculate the maximum period of time that should be allowed between makeup additions. Assume the VOC solvent density to be 8.83 lb/gal.

9.9 An open top, cold cleaning degreaser is presently operating at a temperature of 20°C. At this temperature the vapor pressure of the solvent is 19 mm Hg.

   a) Calculate the partial pressure of the solvent at the air solvent interface.

   b) Calculate the ppm of the solvent in the gas.

   c) A vent is employed to control emissions into the workplace. If the vent air flow rate is 220 ft^3/min (measured at 20°C), calculate the maximum emission rate of the solvent in lb/hr.

9.10 The total emissions during a coating process are affected by:
- the area to be coated,
- the thickness of the coat,
- the efficiency with which it is applied, and
- the percent of solvent in the coating.

The equation for predicting the total amount of solvent emission from any operation involving non-waterborne coating is

$$W = \frac{0.0623 \cdot A \cdot n \cdot (1 - 0.01P) D_s}{P \cdot f}$$

where:
- W = weight of solvent vapors, lb
- A = area coated, ft^2
- n = dry mils
- P = percent solids by volume
- f = transfer efficiency factor
- D_S = solvent density, lb/gal

How much VOC is emitted when painting a large 2 story house (50' • 30' • 18') with an enamel paint? The paint thickness averages 1 mm and 10% of the paint is left in the can and on the paint brushes. (Use the following table)

Typical Densities and Solids Contents of Coatings

Type of coating	Density kg/liter	Density lb/gal	Solids volume %
Enamel, air dry	0.91	7.6	39.6
Enamel, baking	1.09	9.1	42.8
Acrylic enamel	1.07	8.9	30.3
Alkyd enamel	0.96	8.0	47.2
Primer, surfacer	1.13	9.4	49.0
Primer, epoxy	1.26	10.5	57.2
Vanish, baking	0.79	6.6	35.3
Lacquer, spraying	0.95	7.9	26.1
Vinyl, roller coat	0.92	7.7	12.0
Polyurethane	1.10	9.2	31.7
Stain	0.88	7.3	21.6
Sealer	0.84	7.0	11.7
Magnet wire enamel	0.94	7.8	25.0
Paper coating	0.92	7.7	22.0
Fabric coating	0.92	7.7	22.0

9.11 Calculate the total emissions in lb/vehicle for all of the automobile surface coating operations using the table in Example 9.5. How much total VOC is emitted based on a line speed of 55 automobiles/hr. There are two eight-hour shifts, 7 days per week. (1 ft = 12,000 mils)

$$E = \frac{A_v C_1 T_f V_c C_2}{S_c e_T}$$

where: E = emission factor for VOC, mass/vehicle (lb/vehicle) (exclusive of any add on control devices)

A_v = area coated per vehicle (ft²/vehicle)

C_1 = conversion factor (1 ft/12,000 mil)

T_f = thickness of the dry coating film (mil)

V_c = VOC (organic solvent) content of coating as applied, less water (lb VOC/gal coating, less water)

C_2 = conversion factor (7.48 gal/ft³)

S_c = solids in coating as applied, volume fraction (gal solids/gal coating)

e_T = transfer efficiency fraction (fraction of total coating solids used which remains on coated parts)

9.12 a) Estimate the fraction of the gasoline that is vaporized from the gasoline tank of an auto when the contents of the tank are heated from 70°C to 100°F. The gasoline vapor in the air has a molecular weight of 65, and its vapor pressure is given approximately by ln p(psia) = 11.724 - (5,236.5 °R)/T.

b) Assume the gasoline tank holds 15 gallons and is half-full. How many gallons would this be in Los Angeles where there are 10 million automobiles?

9.13 An oil storage tank has a vent on the roof that allows the tank to breathe air.

  a) On a certain day, the tank vapor space is 1,000 gallons and the temperature ranges from 70°F at 3 a.m. to 90°F at 3 p.m. The average molecular weight of the hydrocarbon oil is 200. Its vapor pressure is 15 mm Hg at 70°F and 45 mm Hg at 90°F. Assume that the tank temperature and the oil vapor pressure vary linearly over this range so that arithmetic average values may be used. Estimate the amount of hydrocarbon vapor emitted to the atmosphere on this day by the tank.

  b) The tank's total capacity is 10,000 gallons. It is filled once a month. The average oil withdrawal from the tank is 250 gallons per day. For a

30-day month with average temperatures like those in part a) estimate the total emission of hydrocarbon vapor to the atmosphere, assuming that the tank is filled at 3 p.m.

9.14 The operating department of a refinery plans to use an existing storage tank to store a heavy gasoline product. The heavy gasoline will later be used for blending into a gasoline mixture which is to be sold at gas stations. If the tank is vented to the atmosphere, how much of this product vaporizes and at what rate does the hydrocarbon vapor enter the atmosphere?

Operation data is as follows:

heavy gasoline	Component	Equilibrium constant (30°C, 1 atm)	Molecular weight	Composition, mole %
	iso-pentane	0.83	72	5
	iso-hexane	0.27	86	10
	iso-heptane	0.089	100	20
	iso-octane	0.031	114	50
	iso-nonane	0.0088	128	15

operating temperature   30°C

maximum pump in rate   50 m^3/hr

9.15 A safety analysis for an ammonia storage tank requires knowing the source strength of ammonia in g/sec released if the tank breaks. The tank is to hold 20,000 metric tons which will spread over an enclosure of 10,000 m^2. If the boil-off rate is estimated as 1 cm/hr, calculate the emission rate and time required for all the ammonia to boil off. Would these results cause you to be concerned in the event of an accident?

9.16 A horizontal tube-and-shell surface condenser is to be used to recover toluene from a 5,000 cfm process nitrogen stream. The nitrogen is at 130°F and 1 atm. The mole fraction of toluene is 0.080. Cooling water is available at 70°F. The heat transfer coefficient is 100 Btu/hr ft^2 °F. The molecular weight of toluene is 92 and its boiling point is 110.6°C.

     a)   Estimate the toluene recovery rate.

     b)   How much cooling water is required if the exit water temperature is 90°F?

     c)   What is the heat exchanger surface area required?

9.17 A stream of dry air and carbon tetrachloride (CCl_4) has a flow rate of 2,000 acfm and contains 13% carbon tetrachloride by volume. One alternative control method is to design a condenser to remove carbon tetrachloride. The mixture enters the condenser at 80°F and atmospheric pressure. Estimate the surface area required to condense 95% of the carbon tetrachloride if the average temperature of the cooling surface is –70°F and the overall heat transfer coefficient, U, is 4 Btu/ft^2 hr °F.

9.18 A flue gas stream from a reverberatory furnace at a lead smelter has a flow rate of 20,000 acfm at 1,800°F. It contains 15% of water vapor. The gas is cooled to 275°F for filtration by a combination of water quench (1800° to 800°F) and radiation-convection in pipe-work (800° to 275°F). Water temperature is 62°F.

    a) What is the final volume of gas to be filtered?

    b) What is the moisture content of the gas to the filter?

    c) What is the area of radiation cooler required if the heat transfer coefficient is 1.00 Btu/ft^2 hr °F and the ambient temperature reaches 100°F.

9.8 References

Allen, D, B. Bakshani, and K.Rosselot. *Pollution Prevention*, American Inst. of Chemical Engineering. New York. 1992.

Buonicore, T. and W. Davis (Eds.). *Air Pollution Engineering Manual*. Air and Waste Management Assoc. Van Nostrand Reinhold. New York. 1992.

Danielson, J. *Air Pollution Engineering Manual*. US EPA. 1993.

DeNevers, N. *Air Pollution Control Engineering*. McGraw-Hill. New York. 1995.

Joseph, G. and D. Beachler. *Control of Gaseous Emissions*. Northrop Services, Inc. EPA 450/2-81-005. Research Triangle Park, NC. 1981.

McCabe, W. and C. Smith. *Unit Operations of Chemical Engineering*. McGraw Hill. New York. 1967.

Perry, J. (Ed). *Chemical Engineers Handbook* 5th ed. McGraw-Hill. New York. 1973.

Sax, N. *Dangerous Properties of Industrial Materials*, 4 Ed. Van Nostrand Reinhold. New York. 1975.

Theodore, L. and A. Buonicore (Eds.). *Air Pollution Control Equipment: Selection, Design, Operation and Maintenance*. Prentice-Hall. Englewood Cliffs, NJ. 1982.

Sink, M. *Handbook: Control Technologies for Hazardous Air Pollutants*. Office of Research and Development. US EPA/625/6-91/014. Cincinnati, OH. 1991.

Chapter 10
Adsorption

10.1 Introduction

During adsorption, gaseous components are removed from an effluent gas stream by adhering to the surface of a solid. The gas molecules being removed are referred to as the *adsorbate*, while the solid doing the adsorbing is called the *adsorbent*. Adsorbents are highly porous particles.

For air pollution control purposes, adsorption is not a final process. The contaminant gas is stored on the surface of the adsorbent. After it becomes saturated with adsorbate, the adsorbent must be regenerated. Desorbed vapors are highly concentrated and may be recovered more easily and more economically than before the adsorption step.

Adsorption technology is usually applied for pollution control of organic compounds. In general, any organic compound having a molecular weight greater than 45 is likely to be a good adsorbate. Adsorption is used to control emissions in many solvent-using operations, such as drycleaning, degreasing, surface coating, rubber processing, and flexographic and gravure printing. Adsorption systems have also been used to control toxic or odorous vapors discharged from food processing plants, rendering plants, sewage treatment plants, and many chemical manufacturing processes (such as producing fuels, fertilizers, and pharmaceutical products).

Several materials are used effectively as adsorbing agents. The most common adsorbents used in industry are activated carbon, silica gel, activated alumina (alumina oxide), and zeolites (molecular sieves) (Ruthven 1984). Adsorbents are characterized by their chemical nature, extent of their surface area, pore size distribution, pore volume, and particle size. One of the most important characteristics in distinguishing between adsorbents is *surface polarity*. Polarity determines the type of vapors for which a particular adsorbent will have the greatest affinity. Of the above adsorbents, activated carbon is the primary nonpolar adsorbent. Polar adsorbents will preferentially adsorb any water vapor that may be present in the gas stream. Since moisture is present in most pollutant air streams, the use of polar adsorbents is severely limited for air pollution systems.

Activated carbon is produced from a variety of feedstocks, such as wood, coal, and petroleum-based products. The term "activated" as applied to adsorbent materials refers to the increased internal and external surface area imparted by special treatment processes. The activation process, first, involves *carbonization*. Carbonization is the heating (in the absence of air) of the material to a temperature high enough to volatilize all volatile material. To increase the surface area, the carbon is then "activated" by using steam, air, or carbon dioxide at higher temperatures. The temperatures involved and the type of feedstock all affect the adsorption quality of the carbon.

Carbon used in gas phase adsorption systems is manufactured in granular form, usually between 4 x 6 to 4 x 10 mesh in size (Table 10.1). Bulk density of a packed carbon bed can range from 5 to 30 lb/ft^3 depending on the internal porosity of the carbon. An interval surface area of the carbon can range from $3 \cdot 10^6$ to $8 \cdot 10^6$ ft^2/lb. This is equivalent to having the surface area of 2 to 5 football fields in one gram of carbon.

Table 10.1
Properties of Activated Carbon Adsorbents

Adsorbent	Mesh size and form	Bulk density lb/ft^3	Effective diameter ft	External void fraction	External surface ft^2/ft^3	Reactivation temperature °F	Specific heat Btu/lb.°F
Activated carbon	4 x 6	30	0.0128	0.34	310	200-1,000	0.25
	6 x 8	30	0.0092	0.34	446	200-1,000	0.25
	8 x 10	30	0.0064	0.34	645	200-1,000	0.25
	4 x 10	30	0.0110	0.40	460	200-1,000	0.25
	6 x 16	30	0.0062	0.40	720	200-1,000	0.25
	4 x 10	30	0.0105	0.44	450	200-1,000	0.25

Source: Marchello 1976

In discussing the fundamentals of adsorption, it is useful to distinguish between *physical adsorption*, which involves only relatively weak intermolecular forces, and *chemisorption*, which involves essentially the formation of a chemical bond between the sorbate molecule and the surface of the adsorbent. Physical adsorption does not involve the sharing or transfer of electrons and, thus, always maintains the individuality of interacting species. The interactions are fully reversible, enabling desorption to occur. Physical adsorption is not site-specific; the adsorbed molecules are free to cover the entire surface. Chemisorption involves chemical bonding. Chemisorption is characterized mainly by large interaction potentials that lead to high heats of adsorption that approach the value of chemical bonds. Chemical adsorption is not easily reversible.

All adsorption processes are exothermic, whether adsorption occurs from chemical or physical forces. The reason for this is that in adsorption, the fast-moving gas molecules lose their kinetic energy of motion to the adsorbent in the form of heat. In chemisorption, the heat of adsorption is comparable to the heat evolved from a chemical reaction, usually over 10 kcal/mol. The heat given off by physical adsorption is much lower, approximately 100 cal/mol, which is comparable to the heat of condensation. For these reasons, chemisorption is not used extensively in air pollution control systems.

For physical adsorption processes, the capacity of an adsorbent decreases as the temperature of the system increases (Figure 10.2). As a general rule, adsorber temperatures are kept below 130°F to ensure adequate bed capacities. Temperatures above this limit can be avoided by cooling the exhaust stream that is to be treated. At low concentrations (below 100 ppm), the heat release is minimal, and is dissipated by the air flow through the bed. At higher concentrations (approximately 5,000 ppm), considerable heating of the bed can occur, which, if not removed, can cause the adsorber efficiency to rapidly decrease. In addition, granular carbon is a good insulator, which inhibits heat dissipation from the interior of the bed. In some cases, the temperature rise can cause auto-ignition of the carbon bed. Monitoring the bed temperatures and leaving the bed slightly wet after steam regeneration are techniques used to avoid bed fires.

As stated previously, activated carbon will preferentially adsorb nonpolar hydrocarbons over polar water vapor. At high relative humidities (over 50%), the number of water molecules increases such that they begin to compete with the hydrocarbon molecules for active adsorption sites. This reduces the capacity and the efficiency of the adsorption system. Exhaust streams with humidities greater than 50% may require installation of additional equipment to remove some of the moisture or increase dry bulb temperature to decrease the relative humidity (Noll 1992).

10.2 Adsorption Isotherms

Adsorption occurs by a series of steps. In the first step, the contaminant diffuses from the major body of the air stream to the external surface of the adsorbent particle. In the second step, the contaminant molecule migrates from the relatively small area of the external surface (a few m^2/g) to the pores within each adsorbent particle. The bulk of adsorption occurs in these pores because the majority of available surface area is there (hundreds of m^2/g). In the third step, the contaminant molecule adheres to the surface in the pore. Figure 10.1 illustrates this overall diffusion and adsorption process (Joseph and Beachler 1981).

Figure 10.1
Mechanism of Adsorption

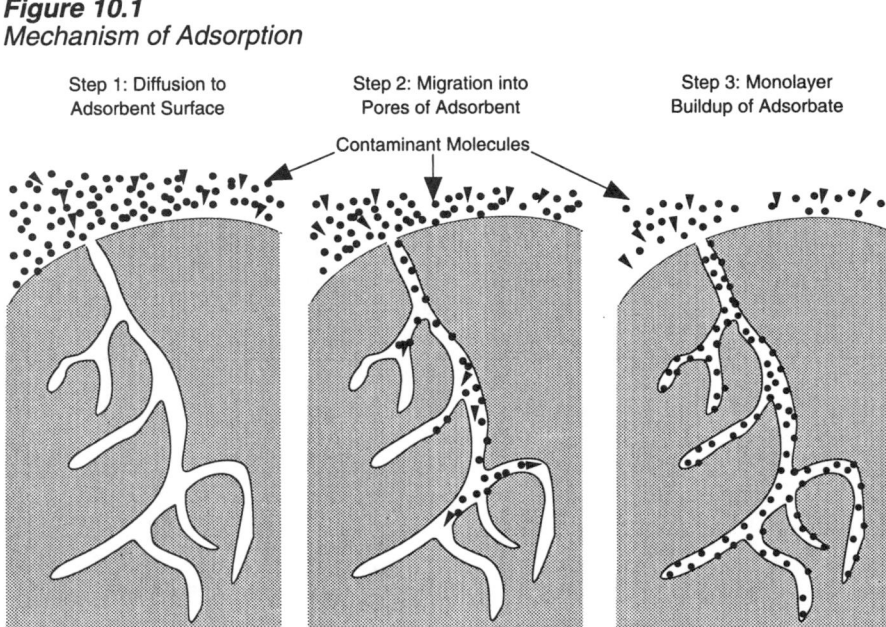

Due to these varied mechanisms, mass transfer rates in the pores are extremely difficult to predict. Unless extensive data are available concerning the specific adsorption application, determining the rate-controlling step is impossible. Therefore, empirical design procedures based on adsorption equilibrium conditions are the most common method used to predict adsorber size and performance.

Most available data on adsorption systems is determined at equilibrium conditions. Adsorption equilibrium is the set of conditions at which the number of molecules arriving on the surface of the adsorbent equals the number of molecules that are leaving. Equilibrium determines the maximum amount of vapor that may be adsorbed at a given set of operating conditions. Although a number of variables affect adsorption, the most important in determining equilibrium for an air pollution system is temperature.

The most common and useful adsorption equilibrium data is the adsorption isotherm. The isotherm is a plot of the adsorbent capacity versus the partial pressure of the adsorbate at a constant temperature. Adsorbent capacity is usually given in weight percent expressed as gram of adsorbate per 100 grams of adsorbent. Figure 10.2 is a typical example of an adsorption isotherm for carbon tetrachloride on activated carbon. Table 10.2 provides the retention on activated carbon for some common air pollutant vapors and gases at 20°C.

Figure 10.2
Adsorption Isotherm for Carbon Tetrachloride on Activated Carbon

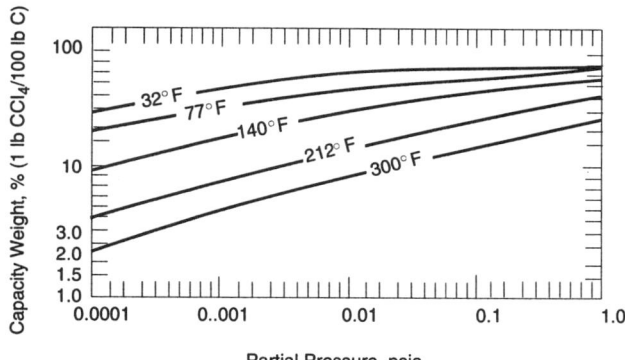

Source: Joseph and Beachler 1981

Table 10.2
Adsorption of Vapors and Gases by Activated Carbon (~ 0.01 psia)

Substance	Molecular weight	Boiling point, K	Retentivity, lb/lb C, at 20 °C
Butyric acid, $C_4H_8O_2$	88	437	0.35
Amyl acetate, $C_7H_{14}O_2$	130	421	0.34
Toluene, C_7H_8	92	384	0.29
Carbon tetrachloride, CCl_4	154	349	0.45
Ethyl mercaptan, C_2H_6S	62	307	0.23
Acetaldehyde, C_2H_4O	44	294	0.07
Sulfur dioxide, SO_2	64	263	0.10
Ammonia, NH_3	17	239	0.013
Chlorine, Cl_2	71	238	0.022
Hydrogen sulfide, H_2S	34	214	0.014

Example 10.1 Activated Carbon Saturation Capacity

A dry cleaning process exhausts a 15,000 scfm air stream containing 680 ppm carbon tetrachloride. Using Figure 10.2 and assuming the exhaust stream is at approximately 140°F and 14.7 psia, determine the saturation capacity of the carbon.

Solution:

In the gas phase, the mole fraction, y, is equal to the percent by volume.

$$y = \% \text{ volume} = 680 \text{ ppm}$$

$$= 680/10^6$$

$$= 0.00068$$

Obtaining the partial pressure:

$$p = y \cdot P$$

$$= (0.00068)(14.7 \text{ psia})$$

$$= 0.01 \text{ psia}$$

From Figure 10.2, at a partial pressure of 0.01 psia and a temperature of 140°F, the carbon capacity is 30%. This means that at saturation, 30 lb of vapor are removed per 100 lb of carbon in the adsorber.

10.3 Adsorption Equilibrium Relationships

Attempts have been made to develop generalized equations which can predict adsorption equilibrium from physical data. This is very difficult because adsorption isotherms take many shapes depending on the forces involved. Isotherms may be concave upwards, concave downwards, or "S" shaped. To date, most of the theories agree with data only for specific adsorbate systems and are valid over limited concentration ranges.

One of the most useful mathematical models to describe adsorption equilibria is the *Langmuir isotherm* (Ruthven 1984; Noll 1992). It is based on the following assumptions: (1) the adsorbed phase is a unimolecular layer and (2) at equilibrium, the rate of adsorption is equal to the rate of desorption from the surface. The quantity, f, is the fraction of the total solid surface occupied by adsorbate molecules. The rate of adsorption, r_a, is proportional to the partial pressure of the adsorbate, p, and to the fraction of the solid surface area available for adsorption, 1-f. Therefore,

$$r_a = C_a \bullet p \bullet (1-f) \qquad [10.1]$$

where C_a is a constant. Conversely, the rate of desorption, r_d, is proportional to the fraction of the surface area occupied by the adsorbate:

$$r_d = C_d \bullet f \qquad [10.2]$$

where C_d is a constant. At equilibrium, the rate of adsorption is equal to the rate of desorption. Therefore, the fraction of the surface covered is given by:

$$f = \frac{C_a \bullet p}{C_d + C_a \bullet p} \qquad [10.3]$$

Because the adsorbed phase is a unimolecular layer, the mass of adsorbate per unit mass of adsorbent, m, is also proportional to the surface covered:

$$m = C_m \bullet f \qquad [10.4]$$

where C_m is a constant. Combining Equation [10.3] and [10.4],

$$m = \frac{k_1 \bullet p}{k_2 \bullet p + 1} \qquad [10.5]$$

where $k_1 = C_a C_m / C_d$, and $k_2 = C_a / C_d$. Equation [10.5] is known as Langmuir isotherm.

Equation [10.5] can be rearranged as follows:

$$\frac{p}{m} = \frac{1}{k_1} + \frac{k_2}{k_1} p \qquad [10.6]$$

This equation is a straight line when plotted on rectangular coordinate with p/m vs. p.

At very low adsorbate equilibrium partial pressure, $k_2 \bullet p$ is near zero, and Equation [10.5] becomes

$$m = k_1 \bullet p, \text{ which corresponds to Henry's Law.} \qquad [10.7]$$

Conversely, at high equilibrium partial pressure,

$$m = \frac{k_1}{k_2} \qquad [10.8]$$

Hence, over an intermediate range of partial pressures:

$$m = k \bullet p^n \qquad [10.9]$$

where: k = constant

n = constant with a value between 0 and 1

Equation [10.9] is known as *Freundlich isotherm*. The values of k and n are obtained from experimental data. Table 10.3 gives values of the Freundlich isotherm parameters for some adsorbates on activated carbon (4 x 10 mesh). Note that these isotherms may not be extrapolated outside of the partial pressure ranges shown.

Table 10.3
Freundlich Isotherm Parameters for Some Adsorbates

Adsorbate	Temperature (K)	k • 100	n	Partial Pressure (Pa)
Acetone	311	1.234	0.389	0.69-345
Acrylonitrile	311	2.205	0.424	0.69-103
Benzene	298	12.602	0.176	0.69-345
Chlorobenzene	298	19.934	0.188	0.69-69
Cyclohexane	311	7.940	0.210	0.69-345
Dichloroethane	298	8.145	0.281	0.69-276
Phenol	313	22.116	0.153	0.69-207
Toluene	298	20.842	0.110	0.69-345
Trichloroethane	298	25.547	0.161	0.69-276
m-Xylene	298	28.313	0.0703	6.9-345

The amount of adsorbed is expresses in kg adsorbate/kg adsorbent.
The equilibrium partial pressure is expressed in Pa.
Data are for the adsorption on Calgon activated carbon (4 x10 mesh).

Source: EAB Control Cost Manual, 3rd.

Example 10.2 Equilibrium Adsorption on Activated Carbon

Calculate the equilibrium adsorptivity of toluene (in air) on activated carbon (4 x 10 mesh). The temperature is 298°K and the total pressure is 101.3 kPa. The equilibrium concentration of toluene is 1,000 ppm.

Solution:

Calculate the equilibrium partial pressure of toluene:

$$p = 1,000(101.3)/106 = 0.1013 \text{ kPa} = 101.3 \text{ Pa}$$

According to Table 10.3, Freundlich isotherm applies for this partial pressure. Substituting the appropriate parameters from that table in Equation [10.9],

$$m = 0.20842(101.3)^{0.110} = 0.346 \frac{\text{kg toluene}}{\text{kg carbon}}$$

10.4 Determination of Adsorption Capacity

The most useful theory from an engineering design viewpoint in trying to predict adsorption isotherms is the *Polanyi potential theory*. The Polanyi theory states that the adsorption potential is a function of the reversible isothermal work done by the system. The major advantage of the Polanyi theory over other models that have been employed to describe adsorption equilibrium is that the adsorption equilibrium relation for a given adsorbate-adsorbent system can be described independent of temperature. Polanyi showed that the adsorption isotherms of various vapors can be represented by the following equation (Noll 1992):

$$\ln W_e = \ln(W_0 \rho_L) - \kappa \frac{[RT \ln(P_s/P)]^2}{\beta^2} \qquad [10.10]$$

where: W_e = adsorption capacity, gram pollutant/gram carbon

W_0 = the maximum adsorption space for condensed adsorbate, cm^3/g)

κ = a structural constant of the adsorbent, (mol/cal)^2

P_s = the saturated vapor pressure of the condensed adsorbate at gas temperature T, K

P = the equilibrium pressure of adsorbate vapor at T, (i.e., the partial pressure of material in the gas phase entering the bed)

ρ_L = the liquid adsorbate density, g/cm^3

R = the ideal gas constant, 1.987 cal/g mol K

β = the dimensionless affinity coefficient

The values of κ and W_0 are parameters of a particular carbon and independent of the vapor adsorbed. Equation [10.10] can be used to determined the equilibrium adsorption isotherm of any given vapor from the adsorption isotherm of a reference vapor, provided that the value of the affinity coefficient, β, of this vapor is available. The quantity β may be determined from the ratio of the molar volume, V_m, of a test solvent to that of the reference solvent used to obtain the values of W_0 and κ for a given carbon.

The affinity coefficient can also be estimated from

$$\beta = \frac{[(n_i^2-1)/(n_i^2+2)]M_i/\rho_{Li}}{[(n_r^2-1)/(n_r^2+2)]M_r/\rho_{Lr}} \qquad [10.11]$$

where: n = the refractive index determined at the sodium D wavelength (see Tables 10.5 and 10.6)

M = molecular weight

ρ_L = liquid density

Subscripts i and r refer to a particular compound and the reference vapor, respectively.

If $\ln W_e$ is plotted for one reference vapor vs. $[RT \ln(P_s/P)]^2$, W_0 may be determined from the $\ln W_e$ intercept, and the slope of the line will be $-\kappa$ (since $\beta = 1$ for the reference material). Some typical values of W_0 and κ are given in Table 10.4. The quantity W_e for another compound is calculated using Equations [10.10] and [10.11], recognizing that W_0 and κ will remain constant.

Table 10.4
Adsorption Parameters for Various Commercial Activated Carbons

	Description	Surface area (m^2/g)	Packed density (g/cm^3)	W_0 (cm^3/g)[a]	κ $(mol/cal)^2$ $\cdot 10^8$
A.	Tsurumi 4GS-S, from coal and coconut shell	1,170	0.43	0.485	5.1
B.	Tsurumi HC-8, from coconut shell	1,270	0.44	0.525	4.7
C.	Takeda Sx, from coconut shell	1,090	0.41	0.455	4.2
D.	Hokuetsu Y-20, from coconut shell	1,098	0.45	0.440	5.1
E.	Fujisawa D-CG, from coconut shell	1,240	0.43	0.495	5.2
F.	Fujisawa A, from coal	840	0.42	0.375	4.6
G.	Kureha G-BAC, from oil pitch	1,000	0.51	0.430	4.6
H.	Columbia JXC 4/6 pellets; 25°C, 737 mm	1,194	0.461	0.404[b] 0.413[c]	1.44[b] 4.7[c]
I.	Barnebey-Cheney 6-10 mesh BC-AC from lot 0993; 23°C	–	–	0.481	1.5
J.	North American Carbon, NACAR G-352, 6-10 mesh granular	–	–	0.700	14.9

[a] Reference vapors: benzene for A-G; n-pentane, n-butanol, for H; CCl_4 for I-J.
[b] n-Heptane.
[c] n-Butanol.
Source: Wadden and Scheft 1987.

The sorption capacity for a number of carcinogenic compounds using Equations [10.10] and [10.11] are provided in Table 10.5.

Table 10.5
Predicted Values of W_e for Carcinogenic Vapors on 6 x 10 Mesh BC-AC Activated Carbon[a]

Vapor	Liquid density,[b] ρ_L (g/cm^3)	Refractive index,[c] n	W_e (kg/Kg)
Acetamide	1.159	1.4274[78]	0.494
Acrylonitrile	0.8060	1.3911	0.357
Benzene	0.8761[23]	1.5011	0.409
Carbon tetrachloride	1.5881[23]	1.4607	0.741
Chloroform	1.4832	1.4459	0.688
bis(Chloromethyl) ether	1.315	1.4346	0.608
Chloromethyl methyl ether	1.0605	1.3974	0.480
1,2-Dibromo-3-chloropropane	2.09311	1.553[11]	0.992
1,1-Dibromoethane	2.0555	1.5128	0.962
1,2-Dibromoethane	2.1792	1.5383	1.020
1,2-Dichloroethane	1.2492[33]	1.4448	0.575
Diepoxy butane (meso)	1.1157	1.4330	0.510
1,1-Dimethyl hydrazine	0.791[22]	1.4075[22,3]	0.359
1,2-Dimethyl hydrazine	0.8274	1.4204	0.375
Dimethyl sulfate	1.332	1.3874	0.615
p-Dioxane	1.0333[23]	1.4220	0.475
Ethylenimine	0.8321	1.412[25]	0.354
Hydrazine	1.0083	1.4698[22,3]	0.380
Methyl methane sulfonate	1.2943	1.4140	0.595
1-Naphthylamine	1.229[25]	1.6703[51]	0.585
2-Naphthylamine	1.0614[98]	1.6493[98]	0.506
N-Nitrosodiethylamine	0.9422	1.4386	0.442
N-Nitrosodimethylamine	1.0059	1.4368	0.458
N-Nitroso-N-methylurethane	1.133	1.4363	0.534
N-Nitrosopiperidine	1.0631[18.5]	1.4933[18.5]	0.501
N-Nitrosopropylamine	0.9163	1.4437	0.434
1,3-Propane sultone	1.39310	1.450[10]	0.646
β-Propiolactone	1.1460	1.4118	0.508
Propylenimine	0.802[25]	1.409[25]	0.361
Safrole	1.096	1.5383	0.522
Urethane	0.9862[21]	1.4144[52]	0.456
Vinyl chloride	0.9114	1.3700	0.404

[a] Barnebey-Cheney Co., Table 10.2; T = 23°C, P/P_0 = 0.0936, W_0 = 0.481 cm^3/g, $\bar{K}$ = 1.5 • 10^{-8}(mol/cal)^2.
[b] Liquid densities at 20°C relative to water unless otherwise stated.
[c] Refractive indices are for D line of the sodium spectrum and at 20°C unless otherwise stated (superscript numbers).

Source: Sansone and Jonas 1981

Table 10.6
Refractive Index for Some VOCs

Compound	Refraction Index, n
Acetone	1.357
Benzene	1.498
Butyl acetate	1.392
Butyl alcohol	1.397
Carbon tetrachloride	1.459
Chloroform	1.444
Ethyl acetate	1.412
Ethyl alcohol	1.359
Heptane	1.383
Hexane	1.498
Isobutyl alcohol	1.395
Isopropyl alcohol	1.374
Methyl alcohol	1.326
Methyl ethyl ketone	1.377
Tetrachloroethylene	1.504
Toluene	1.494
Trichloroethylene	1.475
m-Xylene	1.493
p-Xylene	1.495

Note: Measurements were made at 25°C

Example 10.3 Carbon Bed Adsorption Capacity

Estimate the adsorption capacity for a carbon sorption bed with 500 kg of Columbia JXC activated carbon. The pollutant is toluene at an air concentration of 39 mg/m³ and the air emission gas is at 25°C and 750 mm Hg.

Note: Columbia JXC Carbon (see Table 10.4) characteristics are:

Pack density, ρ_B = 0.461 g/cm³

W_0 = 0.404 cm³/g

K = 1.44 • 10⁻⁸ (mol/cal)²

For organic compounds:

	Toluene	n-Heptane
ρ_L, g/m^3	0.8669	0.6838
MW	92.13	100.21
saturated vapor pressure at 25°C	28.0	48.3
refraction index, n	1.494	1.383

Solution:

1. Determine the affinity coefficient for toluene based on n-heptane as the reference compound (see Table 10.4 and Refractive Index from Table 10.6)

$$\beta = \frac{[(1.494^2 - 1)/(1.494^2 + 2)] \bullet 92.13/0.8669}{[(1.383^2 - 1)/(1.383^2 + 2)] \bullet 100.21/0.6838}$$

$$= 0.9035$$

2. Determine the carbon capacity for toluene.

The vapor pressure equivalent of 39 mg toluene/m^3 at 25°C is

$$P = \left(\frac{39 \text{ mg}}{\text{m}^3}\right)\left(\frac{\text{mol}}{92{,}130 \text{ mg}}\right)\left(\frac{0.0245 \text{ m}^3}{\text{mol}}\right) \bullet 750 \text{ mm}$$

$$P = 0.0078 \text{ mm Hg}$$

$$\text{and } \ln W_e = \ln[(0.404)(0.8669)] - 1.44 \bullet 10^{-8} \bullet \frac{[(1.987 \bullet 298) \ln(28.0/0.0078)]^2}{(0.9035)^2}$$

$$W_e = 0.2314 \text{ kg/kg}$$

Total capacity of the carbon is: 0.2314 kg toluene/kg carbon • 500 kg carbon

$$= 116 \text{ kg of toluene}$$

10.5 Adsorption System Design

10.5.1 Adsorption Bed Capacity

The most efficient arrangement for conducting adsorption is a packed bed of adsorbent through which the stream under treatment is passed. As the polluted stream travels through this bed, adsorption of the contaminants takes place and a purified effluent exits the column (Figure 10.3).

Figure 10.3
Concentration Profile Along a Column

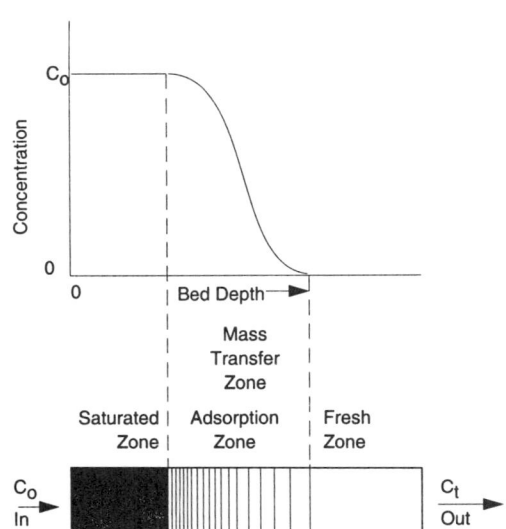

The portion of the bed close to the inlet is continuously contacted by the concentrated feed stream, whereas the subsequent portions are exposed to adsorbate not adsorbed by an earlier portion. Thus, the adsorbent becomes fully loaded at the inlet first, and then downstream.

The part of the bed that displays a gradient in solid concentration from zero to equilibrium is called the mass transfer zone (MTZ). As the name indicates, this is the active part of the bed where adsorption actually takes place. The concentration of pollutants changes continuously throughout this part of the bed, from a value close to zero at the beginning of the MTZ to the feed concentration at the end.

As the saturated part of the bed increases, the MTZ travels downstream and eventually exits the bed. This gives rise to a typical effluent concentration versus time profile, which is called the *breakthrough curve* (Figure 10.4). At breakthrough, the effluent concentration starts to rise and eventually reaches the inlet concentration. Two important terms are used to describe the capacity of the adsorbent bed. The *breakthrough capacity* is defined as the capacity of the bed at which unreacted vapors begin to be exhausted. The *saturation capacity* is the maximum amount of vapors that can be adsorbed per unit weight of carbon (this is the capacity read from the adsorption isotherm).

When the effluent quality becomes unacceptable, or after a predetermined time interval, the operation of the column is stopped and the spent adsorbent is regenerated. If the process is to be reversed by increasing the temperature, regen-

Figure 10.4
Adsorption Wave Front

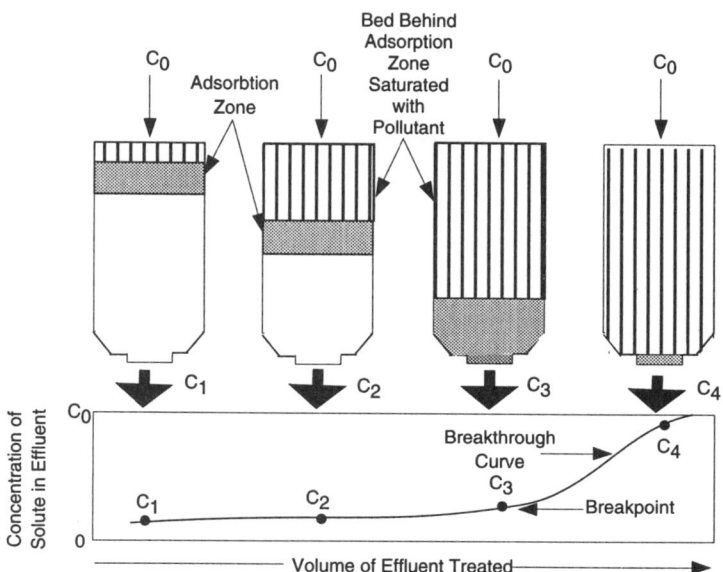

eration of an adsorbent with simultaneous recovery of the adsorbate can be achieved. This is accomplished in air adsorption systems by the use of steam. Continuous operation of an adsorption system requires the involvement of two or more beds. One of them undergoes regeneration while other(s) continue to adsorb the contaminant.

Example 10.4 Adsorber Carbon Capacity

Assume the same conditions as stated in Example 10.1. Estimate the amount of carbon that would be required if the adsorber were to operate on a 4-hour cycle. The molecular weight of CCl_4 is 154 lb/lb mole.

Solution:

From Example 10.1 we know that the carbon used will remove 30 lb of vapor for every 100 lb of carbon at saturation conditions.

1. Compute the flow rate of CCl_4.

    Q_{CCl_4} = 15,000 scfm • 0.00068

    = 10.2 scfm CCl_4

2. Convert to pounds per hour.

$$10.2 \frac{ft^3}{min} \cdot \frac{lb\ mole}{359\ ft^3} \cdot \frac{154\ lb}{lb\ mole} \cdot \frac{60\ min}{hr} = 262.5 \frac{lb\ CCl_4}{hr}$$

For a 4-hour cycle, there are: $4 \cdot 262.5 = 1,050$ lb CCl_4 adsorbed

3. The amount of carbon (at saturation) required:

$$1,050\ lb\ CCl_4 \cdot \frac{100\ lb\ carbon}{30\ lb\ CCl_4} = 3,500\ lb\ carbon$$

10.6 Adsorbent Regeneration

Because it is simple and relatively inexpensive, steam stripping is the most common desorption technique. The steam heats the bed so that the adsorption capacity is reduced to a low level. The adsorbate leaves the surface of the carbon and is removed from the vessel. During the initial heating period, no vapors are desorbed, because a fixed amount of steam is first required to raise the temperature of the cold bed to the desorption temperature. After this initial period, a substantial amount of adsorbate vapor is released. Figure 10.5 shows that the steam consumption per pound of solvent varies with time and the solvent adsorbed. The minimum points in the curves in Figure 10.5 represent the optimum steam requirements and steam time (Danielson 1973). As the steam strip time is extended, more steam per pound of solvent recovered is required and a point is reached at which steam cost exceeds recovery benefits.

Figure 10.5
Steam Consumption per Pound of Solvent Recovered

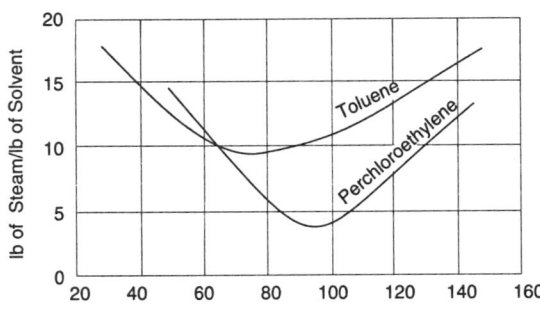

Elapsed Time, min

In Figure 10.5, the ratio of the pounds of the steam used per pound of perchloroethylene recovered is plotted for 20-minute intervals. This reaches a minimum of about 4.7 pounds after an elapsed time of 90 minutes and then rises sharply. The pounds of solvent recovered reaches a maximum at this time and then decreases. In Figure 10.5, the desorbing of toluene follows the same pattern, except that the steam consumption is higher. This is to be expected, since its latent heat is greater. It is economical to operate the stripping cycle to retrieve only part of the adsorbed solvent; that part will generally be recovered in less than 90 minutes, leaving a "heel" of solvent within the bed. A well-designed system will have steam consumption in the range of 1 to 4 pounds of steam per pound of recovered solvent supplied at pressures ranging from 3 to 15 psig.

The solvent vapors and steam that emerge from the bed during stripping are condensed in a water-cooled condenser. The cooling water will typically have an inlet temperature of 60°F. Temperature rise in the condenser will be approximately 40°F, making the outlet water temperature about 100°F. High cooling water temperatures or low water-flow rates will slow the condensation process, thereby increasing the steam-strip cycle time and adding unnecessary cost to the process.

Separators are installed following the condenser to separate the contaminant from the water. Separators work on the principle of gravitational forces where the heavier material to be separated is removed from the bottom of the decanters and the lighter material is removed at the top. If the solubility of the condensed VOC in water is low enough, decanting is sufficient; otherwise an additional separation unit, such as distillation, is required.

For the capture of vapors in a continuous operation, a minimum of two adsorption units is required. With this arrangement, one unit is adsorbing while the other is being stripped of solvent and regenerated. Sufficient time must be available to cool this unit to nearly ambient temperature before it is returned to service. After the solvent is stripped, the carbon is not only hot, but is saturated with water. Cooling and drying are usually done by blowing solvent-free air through the carbon. The ensuing evaporation of the moisture is helpful in removing the heat in the carbon.

Normally, two adsorbing units are sufficient if the regeneration and cooling of the second bed can be completed before the first unit has reached the breakpoint in the adsorbing cycle. With three units, it is possible to have one bed adsorbing, one cooling, and one regenerating. Vapor-free air from the adsorbing unit is used to cool the unit just regenerated. The air from the first bed, after being stripped of vapor, is passed through the second bed, which has been regenerated but is still hot and wet. By using the vapor-free air from the first unit to remove this moisture, the ensuing evaporation of the water effectively cools the carbon.

10.7 Maximum Adsorbent Bed Depth

Often, the adsorbent bed is sized to the maximum depth allowed by the pressure drop across the bed. Bed depth is controlled by the attrition (wear) losses of the granular activated carbon caused by excessive velocity (crushing velocity) and pressure during operation. The gas velocity is usually limited to 100 ft/min. The pressure drop across the bed is usually limited to a total of 18 to 20 in. H_2O. Data on the pressure drop per inch of bed depth for typical carbons is presented in Figure 10.6 (Turk 1977). The pressure drop per inch of bed depth is plotted versus the gas flow rate, with the carbon mesh size as a parameter. From the figure, an adsorber with a flow rate of 80 ft/min using 4 x 10 mesh carbon will have a pressure drop of approximately 6 in. H_2O per foot of bed depth. Therefore, if the pressure drop across the bed is limited to 18 in. H_2O, then the bed depth should not exceed 3 ft.

Figure 10.6
Pressure Drop vs. Flow Rate Through Granular Carbon Beds

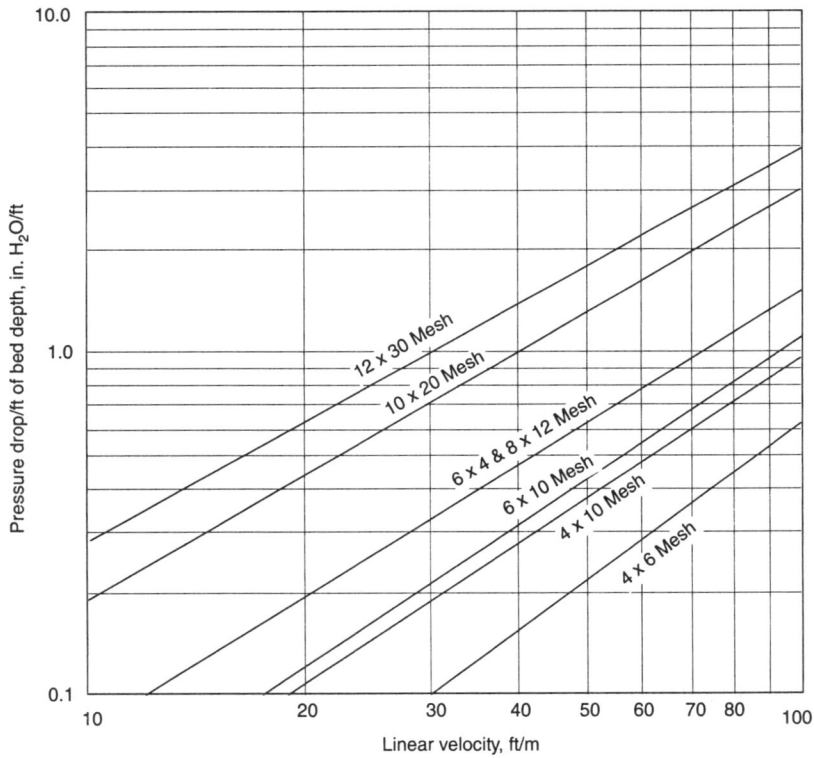

Source: Joseph and Beachler 1981

Carbon adsorption systems are designed for a maximum air flow velocity of 100 ft/min through the adsorber. A lower limit of at least 20 ft/min is maintained to avoid flow distribution problems, such as channeling. With the above stated maximum and minimum flow rates, the allowable pressure drop usually dictates the maximum bed depth. By specifying a maximum velocity through the adsorber, the minimum cross sectional area is also specified. For example, if 300 ft^3/min of contaminant gas is to be treated, and the maximum velocity through the adsorber is to be 30 ft/min, then the adsorber must have a cross sectional area of at least 10 ft^2. The restriction on maximum bed depth caused by pressure drop considerations leads to two types of fixed bed orientation, vertical and horizontal.

Vertical flow adsorbers generally have a circular bed shape. A typical three-bed adsorption system used to recover 75 lb/hr of trichloroethylene from a vapor degreasing operation is shown in Figure 10.7. These units are suited to handle flows up to 3,000 to 4,000 cfm per adsorber. Each vessel is 48 inches in diameter, contains 560 lb of carbon approximately 4 inches deep, and handles 3,500 scfm air flow. The pressure drop across the system is approximately 12 in. H$_2$O (Joseph and Beachler 1981).

Figure 10.7
Three-Bed Vertical System

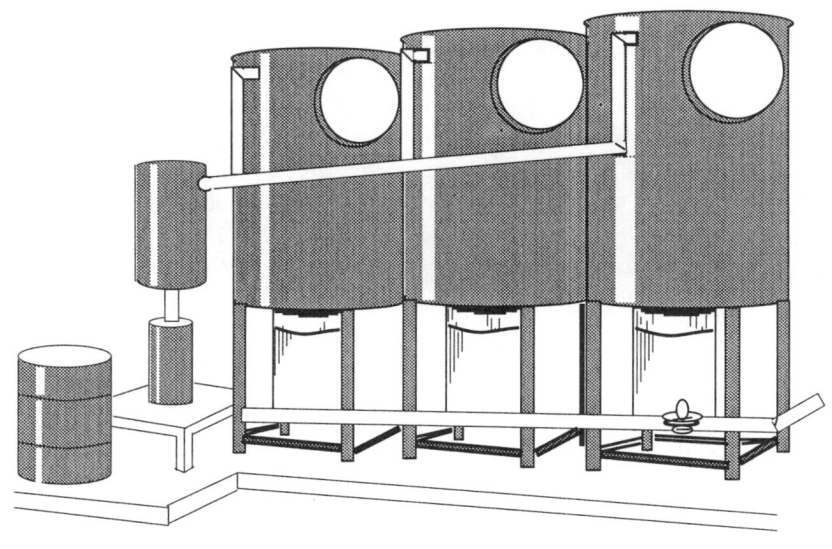

For larger flow volumes, horizontal flow adsorbers are used. In horizontal flow units, the bed length is horizontal as is the direction of the incoming air stream. The air stream flows across the bed and down. Figure 10.8 shows a horizontal system. The beds are typically rectangular with a width of 10 ft, maximum bed depths of 3 ft, and pressure drops up to 20 in. H$_2$O.

Figure 10.8
Horizontal Bed

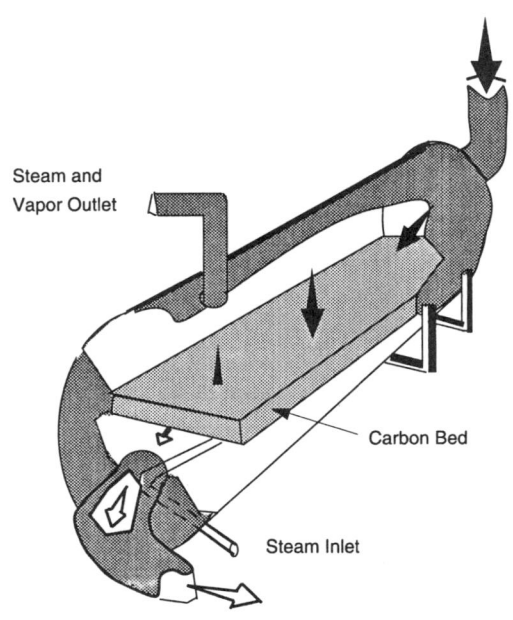

10.8 Minimum Adsorbent Bed Depth

Providing a minimum depth of adsorbent is very important in achieving efficient gas removal. If the adsorber bed depth is shorter than the required mass transfer zone, breakthrough will immediately occur rendering the system ineffective. Fortunately, the MTZ is less than 3 inches for most air pollution applications and represents only a small portion of the total bed depth (Noll 1992). Computing the length of the MTZ is very difficult since it depends upon six factors: the adsorbent particle size, gas velocity, adsorbate concentration, fluid properties of the gas stream, temperature, and pressure of the system. The MTZ can be estimated from experimental data (obtain from pilot adsorber column) using Equation 10.12 (Joseph and Beachler 1981).

$$MTZ = \frac{1}{1-X_s} D \left(1 - \frac{C_B}{C_S}\right) \quad [10.12]$$

where: D = bed depth
C_B = breakthrough capacity, %
C_S = saturation capacity, %
X_S = degree of saturation in the MTZ, % (usually assumed to be 50%)
MTZ = length of MTZ

The above equation is used mainly as a check to ensure that the proposed bed depth is longer than the MTZ. Actual bed depths are usually many times longer than the length of the MTZ. The additional bed depth allows for cycle times required to regenerate the bed. As has already been discussed, minimum regeneration times are typically 2 hours with 60 to 90 minutes allowed for steaming and 30 minutes to 1 hour for cooling and drying.

The length of the MTZ is related to the velocity; usually, the higher the velocity, the longer the unsaturated zone. An example is shown in Figure 10.9, where the effects of velocity on the length of the MTZ is demonstrated for ethanol on activated carbon (Buonicore and Davis 1992). The data in Figure 10.9 is typical of most adsorption systems for VOC control (i.e., the maximum length of the MTZ for adsorption is typically less than 3 inches). This is compared to water adsorption systems where the MTZ is often a foot or more. This difference is due to the much higher diffusivities in the air adsorption systems.

Figure 10.9
General Effect of Gas Velocity on Length of MTZ

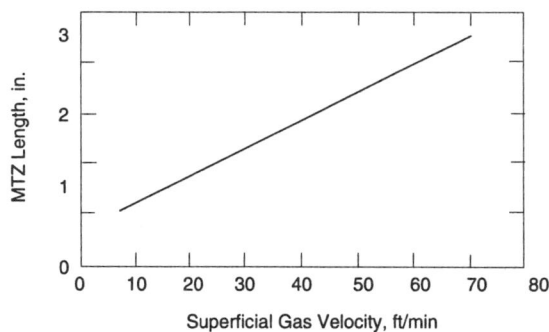

Source: Buonicore and Davis 1992

10.9 Service Time

The service time is one of the prime design parameters to be considered in sizing adsorption equipment. Service time can be calculated as follows:

$$\tau_B = \frac{C_B}{Q_0 \bullet C_0} \qquad [10.13]$$

where: τ_B = service time (adsorption time), min

Q_0 = total gas volume flow rate, ft³/min

C_0 = inlet gas concentration, lb/ft³

C_B = breakthrough capacity, lb

Equation [10.12] can be rearranged to solve for the breakthrough capacity required in Equation [10.13]. The total capacity is the sum of the capacity in the saturated bed plus the amount adsorbed in the MTZ.

(Mass Transfer Zone) + (Saturated Zone)

$$C_B = \frac{X_s C_s MTZ + C_s(D - MTZ)}{D} \qquad [10.14]$$

The total amount of adsorbate, C_S, is usually determined from the adsorption isotherm, as illustrated in Example 10.3

The usual procedure in practice is to work with a term defined as the *working charge* (or working capacity), rather than C_B. It provides a numerical value for the actual adsorbing capacity of the bed under operating conditions. If experimental data are available, the working charge, W_C, may be estimated from:

$$W_C = C_S \left[\left(\frac{D - MTZ}{D} \right) + 0.5 \left(\frac{MTZ}{D} \right) \right] - HEEL \qquad [10.15]$$

where the MTZ is assumed to be 50% saturated and HEEL is the residual adsorbate present in the bed following regeneration. This HEEL accounts for a large portion of the difference between the saturation and the working capacity.

When deep adsorbent beds are used, as in solvent recovery, the MTZ is small compared to the overall bed depth and D, the adsorption capacity of the saturation zone, essentially determines the service time. With usual solvent-recovery practices, the solvent air velocity through the bed is maintained near 100 ft/min. In these cases, the MTZ is of the order of 3 inches, and D, the adsorbent bed depth, can range from 16 to over 36 inches. The capacity of the bed can be calculated with a considerable degree of accuracy if an adsorption isotherm has been determined from:

$$W_C = C_S - HEEL \quad \text{(deep bed)} \qquad [10.16]$$

In thin beds of 2 inches and less, flow velocities near 40 ft/min are usually employed. In these applications, the MTZ and D are nearly the same, and the W_c is essentially:

$$W_C = 0.5 \bullet C_S - HEEL \quad \text{(thin bed)} \qquad [10.17]$$

The adsorption phenomenon becomes somewhat more complex if the gases or vapor to be adsorbed consist of several compounds. The adsorption of organic compounds having higher molecular weights will tend to displace those having lower molecular weights. Lighter compounds will tend to be separated or partitioned from the heavier compounds and will pass through the bed at a faster rate. This will increase the mass transfer zone and may require additional carbon bed depth or a shorter operating cycle.

10.10 Humidity Effects

The stripping of VOCs from contaminated water or soil to air often requires that the VOCs be subsequently adsorbed. The adsorption of VOCs emitted from a stripping operation can be hindered due to high relative humidity of the off-gas stream.

The presence of high humidity in a gaseous stream can interfere with the vapor adsorption process and diminish its efficiency. The loss of capacity due to high humidity is especially severe for adsorbates of low water solubility.

Water has a complicated sorptive behavior, as shown in Figure 10.10. The moisture adsorbed from zero to between 40 and 50% relative humidity is small. However, at higher relative humidity values (>40 to 50%), a sharp increase in moisture adsorption is observed. The exact point at which this transition takes place depends on the pore structure of the carbon. The adsorbed water fills the small pores in the carbon and can interfere with the adsorption capacity of contaminates, as shown in Figure 10.11.

Figure 10.10
Isotherm for Toluene Trichlorethylene and Water Vapor

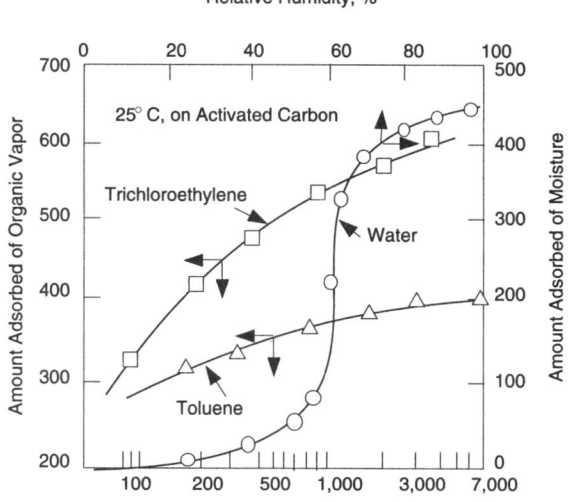

Source: Noll 1992

As the gas temperature rises, the dry adsorption capacity is reduced; however, as psychrometric charts indicate (Chapter 2, Figure 2.1), a 10 to 15°C temperature increase can reduce a 100 % relative humidity level to between 40 and 50%, at which no significant moisture effect in adsorption is expected (Noll 1992).

Figure 10.11
Amount of Trichloroethylene (TCE) Adsorbed as a Function of Relative Humidity

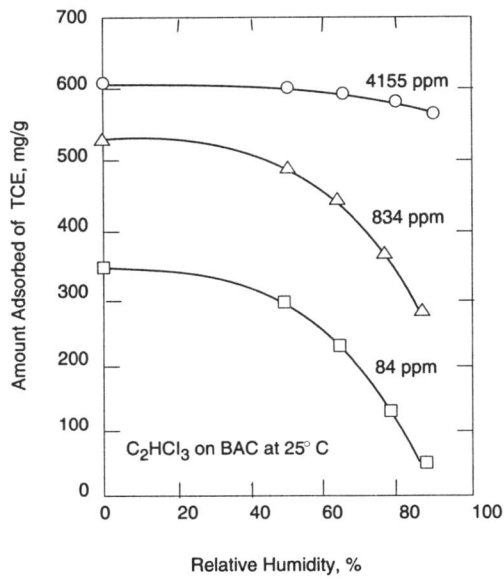

Source: Noll 1992

10.11 Design Summary

The adsorption capacity of an adsorbent for a given VOC is represented by the adsorption isotherms that relates the amount of VOC adsorbed to the equilibrium concentration at constant temperature. Typically, the adsorption capacity increases with the molecular weight of the VOC being adsorbed. Regenerable fixed carbon beds are usually 1 to 4 ft thick. The maximum adsorbent depth is based on pressure drop considerations. Superficial gas velocities through the adsorber range from 20 to 100 ft/min. Pressure drops normally range from 3 to 20 in. H_2O depending on the gas velocity, bed depth, and carbon particle size. Graphs similar to that in Figure [10.12] are used to compute the pressure drop. 60 to 90 minutes of low pressure steam is needed for regeneration. A "heel" is left on the bed because complete desorption is technically difficult to achieve and economically impractical. The amount of the heel is substracted from the adsorbent capacity to determine bed service time.

Example 10.5 Adsorbent Depth

You are asked to determine the required depth of carbon adsorbent for an adsorber that treats a degreaser ventilation stream contaminated with trichloroethylene (TCE)

Chapter 10: Adsorption 371

given the following operating and design data. The adsorption column cycle is set at four hours in the adsorption mode, two hours in heating and desorbing, one hour in cooling, and one hour in standby. The adsorber recovers 99.5% by weight of TCE. A horizontal unit with an inside width of 6 feet and length of 15 feet is to be used. Volumetric flow rate of contaminated air stream is 10,000 scfm; standard conditions are 60°F, 1 atm; operating temperature is 70°F; operating pressure is 20 psia; adsorbent is activated carbon; saturated capacity is 28 pounds TCE per 100 pounds carbon; inlet concentration of TCE is 2,000 ppm (by volume); and molecular weight of TCE is 131.5. The bulk density of the carbon is 36.

Solution:

1. Calculate the total and TCE actual volumetric flow rates.

    Q_{act} = 10,000[(70 + 460)/(60 + 460)][(14.7)/(20)]

    = 7,491 acfm

    = 4.5 • 10^5 acfh

    Q_{TCE} = $(y_{TCE})(Q_{act})$

    = (2,000 • 10^{-6})(4.5 • 10^5)

    = 900 acfh

2. Calculate the mass flow rate of the TCE, W.

    W = (Q)(ρ) = Q(PM/RT)

    = (20)(131.05)(900)/(10.73)(70 + 460)

    = 416.2 lb/hr

3. Calculate the mass of TCE adsorbed.

    M_{TCE} = (W)(0.995)(4)

    = (416.2)(0.995)(4)

    = 1,656.6 lb

4. Calculate the volume of activated carbon required.

    V_C = (TCE to be absorbed)(100 lb carbon/28 lb TCE adsorbed)/(bulk density)

    V_C = (1,656.6)(100/28)/(36)

    = 164 ft^3

5. Calculate the depth of the carbon bed.

    D = (activated-carbon volume)/(cross-sectional area)

    = (164)/(6)(15)

    = 1.83 feet

Example 10.6 Adsorber System Design

Design an adsorber system to collect 500 ppm benzene from an air-benzene mixture at 1.1 atm, 50°F, flowing at 500 acfm. Use 4 x 6 mesh activated carbon. Steam is available twice per day at the plant and a steam solvent ratio of 3.0 is to be employed. Thus, each unit is to be operated for 24 hours per total cycle (one on/one off). What is the pressure drop for this process? Also, calculate the steam flow rate per regeneration for this system. Pertinent data are as follows:

$$\rho_B, \text{ carbon } = 30 \text{ lb/ft}^3$$

$$\text{CAP} = 0.45 \text{ lb benzene per lb carbon}$$

$$\text{HEEL} = 0.03 \text{ lb benzene per lb carbon}$$

$$\text{Velocity} = 75 \text{ ft/min}$$

Solution:

Key calculations are as follows:

$$V = (500 \text{ ft}^3/\text{min})(60)(12 \text{ hours}) = 360{,}000 \text{ ft}^3 \text{ per 12 hr}$$

$$V_B = 360{,}000 \ (500/10^6) = 180 \text{ ft}^3$$

$$\rho_B = (78)(1.1)/(10.73)(510) = 0.2304 \text{ lb/ft}^3$$

$$M_B = (180)(0.2304) = 41.47 \text{ lb benzene}$$

$$W_C = \text{CAP} - \text{HEEL} = 0.45 - 0.03 = 0.42 \text{ lb benzene per lb carbon}$$

$$M_{AC} = 41.47/0.42 = 98.74 \text{ lb activated carbon}$$

$$V_{AC} = 98.74/30 = 3.291 \text{ ft}^3$$

$$A = 500/75 = 6.666 \text{ ft}^2$$

If a vertical adsorber, then

$$D = 2.91 \text{ ft}$$

$$H = 3.291/6.667 = 0.4937 \text{ ft}$$

$$= 5.9245 \text{ in.} \cong 6 \text{ in.}$$

From Figure 10.6:

$$\Delta P = 0.4 \text{ in. water/ in. bed}$$

$$\Delta P_{\text{total}} = (0.4)/(6.0) = 2.4 \text{ in.}$$

Chapter 10: Adsorption 373

$$M_{steam} = (3.0)(41.5)$$

$$= 124.5 \text{ lb steam during regeneration}$$

Extra carbon should be provided because the MTZ (2 in.) will only be 50% saturated.

Example 10.7 Adsorber Design Application

Assume the following data:

10,000 cfm air stream at 77°F

atmospheric pressure

2,000 ppm toluene

3,000 ppm carbon tetrachloride (CCl_4)

95% removal objective

isotherm Figure 10.12 and Figure 10.2

MW toluene = 92

MW CCl_4 = 154

Questions:

1. How many pounds per hour of organics must be removed?
2. Assuming steam regeneration at 212°F, what is the working capacity?
3. Design a carbon adsorption system for this application.

Solution:

1. To determine the pounds per hour of organics removed, first compute volume of organics entering the system.

$$10,000 \text{ cfm} \left(\frac{3,000 \text{ parts } CCl_4}{10^6} \right) = 30 \text{ cfm } CCl_4$$

$$10,000 \text{ cfm} \left(\frac{2,000 \text{ parts toluene}}{10^6} \right) = 20 \text{ cfm toluene}$$

Adsorption Isotherm for Toluene

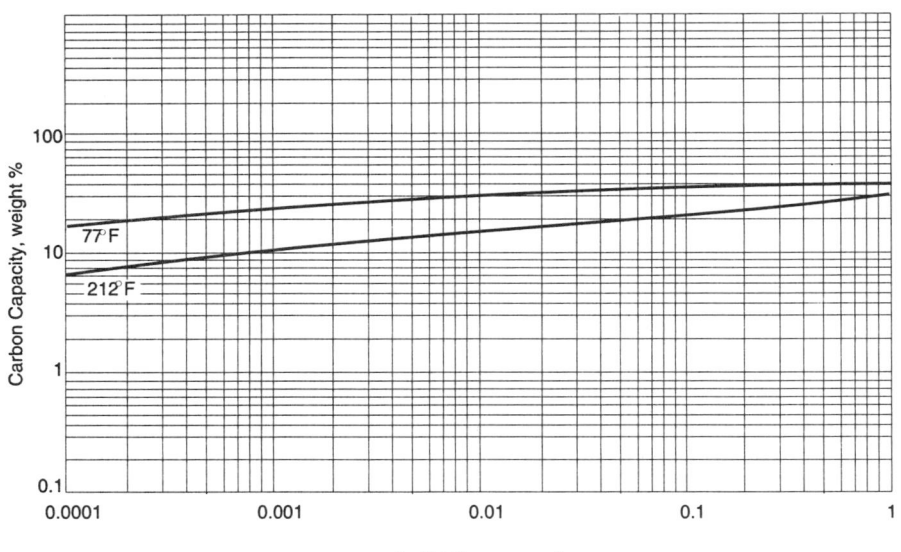

Note: *1 lb mole of ideal gas occupies 359 cf at STP (32°F and 1 atm)*

$$\text{at outlet temp (77°F)} = 359 \cdot \frac{460+77}{460+32} = \frac{391.9 \text{ cf}}{\text{lb mole}}$$

$$\text{lb/hr CCl}_4 = 30 \text{ cfm}\left(\frac{\text{lb mole}}{391.9 \text{ cf}}\right)\left(\frac{154 \text{ lb}}{\text{lb mole}}\right)\left(\frac{60 \text{ min}}{\text{hr}}\right)$$

$$= 707 \text{ lb/hr}$$

$$\text{lb/hr toluene} = 20 \text{ cfm}\left(\frac{\text{lb mole}}{391.9 \text{ cf}}\right)\left(\frac{92 \text{ lb}}{\text{lb mole}}\right)\left(\frac{60 \text{ min}}{\text{hr}}\right)$$

$$= 282 \text{ lb/hr}$$

$$\text{lb/hr of total hydrocarbons} = \text{CCl}_4 + \text{toluene}$$

$$= 707 \text{ lb/hr} + 282 \text{ lb/hr} = 989 \text{ lb/hr}$$

Calculate lb/hr of total hydrocarbons to be removed:

$$(989 \text{lb/hr})(0.95) = 939 \text{ lb/hr}$$

2. To determine the working capacity using steam regeneration at 212°F, first compute partial pressures at adsorption condition.

$p = y \cdot P_t$ = volume %

for toluene $p = (0.002)(14.7 \text{ psia}) = 0.0294 \text{ psia}$

for CCl_4 $p = (0.003)(14.7 \text{ psia}) = 0.044 \text{ psia}$

a) to compute the partial pressure at regeneration conditions, the concentration or volume is

$y_{out} = 0.05 y_{in}$

for toluene $p = (0.002)(0.05)(14.7 \text{ psia}) = 0.0015$

for CCl_4 $p = (0.003)(0.05)(14.7 \text{ psia}) = 0.0022$

1	2	3	4	5
Various Velocities Through the bed (ft/min)	Pressure drop per foot of bed depth (in. H_2O) (Fig. 10.6)	Allowable depth (18/2) (ft)	Required cross section area of bed ($Q/\underline{1}$) (ft²)	Required bed depth (carbon vol./4) (ft)
50	3	6.0	200	2.0
60	4	4.5	250	1.6
70	5	3.6	143	2.8
80	6	3.0	125	3.2
90	7	2.6	110	3.7
100	8	2.25	100	4.0

Note: To find overall working capacity, assume no co-adsorption. Treat the system as if two beds are in series, one adsorbing toluene, the other adsorbing CCl_4.

b) To compute overall working capacity at 212°F:

i. mass of carbon (lb) needed to remove toluene

$$\left(\frac{282 \text{ lb toluene}}{\text{hr}}\right)(0.95)\left(\frac{\text{lb carbon}}{0.23 \text{ lb toluene}}\right) = 1,165 \frac{\text{lb carbon}}{\text{hr}}$$

ii. mass of carbon (lb) needed to remove carbon tetrachloride

$$\left(\frac{707 \text{ lb } CCl_4}{\text{hr}}\right)(0.95)\left(\frac{\text{lb carbon}}{0.36 \text{ lb } CCl_4}\right) = 1,866 \frac{\text{lb carbon}}{\text{hr}}$$

iii. total amount of carbon = 1,165 + 1,866

$$= 3{,}031 \text{ lb/hr}$$

iv. overall W_C

$$W_C = \frac{939 \text{ lb solvent}}{\text{hr}} \bullet \frac{\text{hr}}{3{,}031 \text{ lb solvent}} = \frac{0.31 \text{ lb solvent}}{\text{lb carbon}}$$

d) Design the carbon adsorption system.

i. Determine the volume of carbon required.

If a 4-hr cycle for collection and two beds are used:

$$\frac{3{,}031 \text{ lb carbon / hr} \bullet 4 \text{ hr cycle}}{30 \text{ lb / ft}^3 \text{ of carbon}} = 404 \text{ ft}^3 \text{ of carbon in each bed}$$

ii. Various bed configurations are possible based on the mesh of the carbon selected, the maximum velocity allowed in the bed, and the velocity through the bed (see Figure 10.6). For this example, a 4 x 10 mesh carbon and a maximum bed pressure drop of 18 inches of H_2O to control carbon crushing under high pressure was selected.

The following table provides information on the design of carbon beds for various velocities through the beds.

1	*2*	*3*	*4*	*5*
Various Velocities Through the bed (ft/min)	Pressure drop per foot of bed depth (in. H_2O) (Fig. 10.6)	Allowable depth (18/*2*) (ft)	Required cross section area of bed (Q/*1*) (ft^2)	Required bed depth (carbon vol./*4*) (ft)
50	3	6.0	200	2.0
60	4	4.5	250	1.6
70	5	3.6	143	2.8
80	6	3.0	125	3.2
90	7	2.6	110	3.7
100	8	2.25	100	4.0

The following figure is a plot of columns *3* and *5* from the above table. The figure compares the required bed depths based on both pressure drop and cross sectional area. Where the two lines cross is the depth of bed that satisfies both criteria.

Bed Depths as a Function of Pressure Drop and Required Cross Section Area

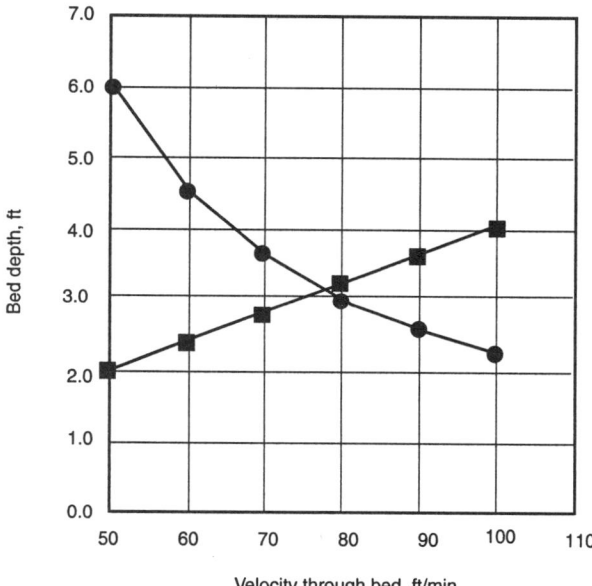

● The allowable bed depth based on 18 in. maximum pressure drop in carbon is provided in column *3* in the table.

■ The required bed depth to provide adequate cross sectional area is provided in column *5* in the table.

In the above figure, the two lines cross at a velocity of 78.5 ft/min and a bed depth of 3.2 ft.

Therefore, a bed configuration can now be selected as follows:

$$\text{Depth} = 3.2 \text{ ft}$$
$$\text{Width} = 10 \text{ ft}$$
$$\text{Length} = 12.6 \text{ ft}$$
$$\therefore \text{Volume} = 403 \text{ ft}^3$$

The velocity through the bed will be 78.5 ft/min and the pressure drop of 18 in. H_2O.

10.12 Problem Set

10.1 Carbon dioxide adsorption isotherms for activated carbon are presented in the following figure for temperature of 30°, 50°, and 95°C.

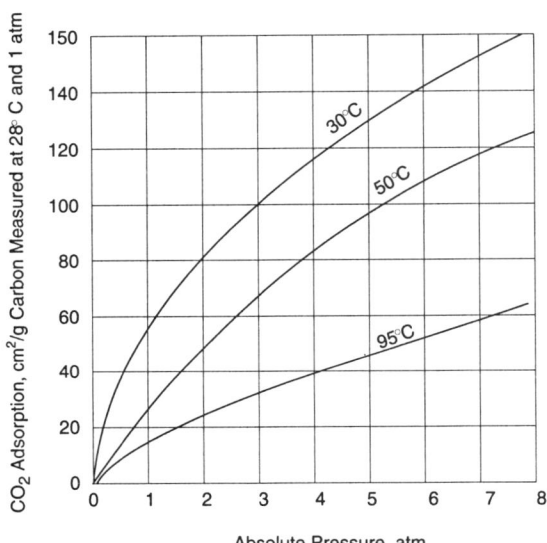

CO$_2$ Adsorption Isotherm on Activated Carbon

For a temperature of 50°C, determine the coefficients of the

    a) Freundlich equation

    b) Langmuir equation

Which equation, a or b, provides a better fit of the experimental data?

10.2 Determine the saturation capacity of a carbon bed for toluene at the following conditions:

Air flow rate = 12,000 cfm at 77°F

Concentration of toluene = 410 ppm

The system is at atmospheric pressure

Steam regeneration at 212°F

10.3 The saturated carbon capacity in a three-foot adsorption bed is 39%. The MTZ and HEEL were determined to be 4 inches and 2.5%, respectively. Calculate the working charge.

10.4 How many pounds of carbon would be needed to remove 150 lb/hr of toluene from an air stream containing 0.68% (by volume) toluene. The adsorber operates on a 2-hr cycle and the working charge can be estimated by doubling the amount of carbon needed at the saturation capacity.

10.5 A test was conducted on a carbon adsorber to determine the length of the MTZ. The carbon bed depth was 2.0 ft. The saturation capacity of the carbon at 20°C is 26 lb solvent/100 lb of carbon. The breakthrough capacity was calculated to be 24.9% from the test data. Determine the MTZ length.

10.6 A carbon adsorption system is to recover 800 lb/hr of hexane from air containing 1.8 lb hexane vapor/1,000 ft^3. The air is at 100°F and 1 atm. The adsorbent will be 6 x 8 mesh activated carbon pellets. The bed depth is 1 ft. The hexane adsorption capacity is 0.52 lb/lb carbon. The adsorption cycle is set at 2 hours to allow sufficient time for regeneration. Three adsorbing beds will be used. a) Determine the amount of carbon required; b) Select suitable dimensions for the beds; and c) Estimate the pressure drop.

10.7 A plant needs to remove acetone (MW = 58) from an air stream having a flow of 30,000 cfm, containing 0.15% (by volume) acetone. The air temperature is 20°C at 1 atm pressure. Three adsorber units, each 18 feet in diameter and 2 feet deep using 4 x 6 mesh carbon, have been proposed for this application. Two adsorbers will be operating at all times. (The vapor pressure of acetone at 20°C is 170 mm Hg). Determine the adsorption time and pressure drop for the adsorbers.

10.8 A solvent recovery system was designed to recover benzene from an air stream. The company has plans to increase production, which would result in a 75% increase in benzene that must be controlled. The following data on the present system is available.

Carbon density = 23 lb/ft^3

Carbon size = 4 x 6 mesh

Gas velocity = 100 fpm

Bed area = 120 ft^2

Bed depth = 24 in.

Concentration of benzene = 1316 ppm

Temperature = 26°C

Carbon charge = 5,200 lb

Working capacity = 36 lb benzene/100 lb carbon

Residual capacity = 2.2%

Cycle time = 10 hours adsorbing, 2 hours steaming and drying

Vapor pressure of benzene at 26°C = 100 mm Hg

MTZ = 2 in.

MW = 78.11

a) What is the new saturation capacity, breakthrough capacity, working capacity, and cycle time?

b) Can the present system handle the increased benzene load?

10.9 A printing company must reduce the amount of toluene emitted from its Rotograve printing operation. A representative from the company obtains some preliminary information on installing a carbon adsorption system. You are given the following information:

Air flow = 20,000 cfm

The adsorbers operate at 25% of LEL for toluene in the exit air

LEL for toluene = 1.2%

Toluene MW = 92.1 lb/lb mole

Carbon density = 30 lb/ft^3

Working charge = 30% of saturation capacity

Regeneration = 1 hr

Temperature = 77°F

Maximum velocity through adsorber = 100 fpm

Design a carbon adsorption system for this source.

10.10 A bulk terminal that transfers gasoline to service stations has submitted plans to install a carbon adsorption unit to control emissions from the filling of their tank trucks. You are given the following information:

Average daily throughput = 120,000 gal/day of gasoline

Maximum pumping rate = 2,000 gal/min

Air flow to adsorber = 350 cfm

Diameter of adsorber = 4 ft

Height of carbon = 3 ft

2 beds, regeneration cycle time of 1/2 hour for each bed

Emission factor = 5 lb/1,000 gal loaded

Emission limit = 0.67 lb/1,000 gal loaded

MW of gasoline = 68 lb/lb mole

Ideal gas constant, $R = 0.732 \dfrac{\text{atm ft}^3}{\text{lb mol·R}}$

T = 70°F

Carbon density = 30 lb/ft^3

Vacuum regeneration

Working capacity = 50% of saturation capacity

The Absorption Isotherm for gasoline vapors is provided in the following figure.

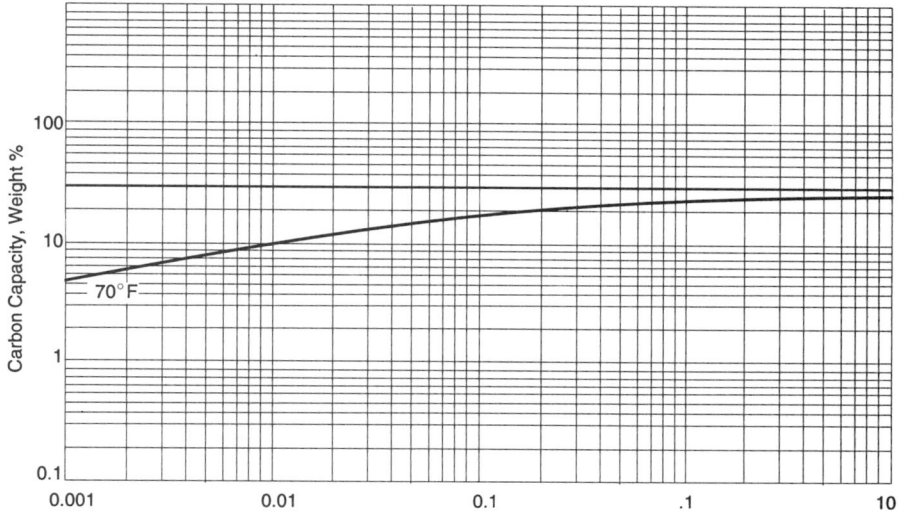

a) What is the velocity through the adsorber?

b) How much carbon is required?

c) What is the bed depth?

d) Will the unit be in compliance?

10.11 Design a carbon adsorption bed to remove 1,000 ppm toluene from a 15,000 cfm stream at 140°F. The adsorption cooling time is at least 2 hours. An adsorption cycle of 8 to 12 hours is desirable. A carbon with the following characteristics has been selected for application to this source.

W_o = Adsorption space = 0.46 cm³/g

B = Microporosity constant = 3.37 • 10⁻⁸ mol²/cal²

β = Affinity coefficient = 1.0

Determine: working capacity, gas velocity, pressure drop, bed depth.

10.12 Estimate the adsorption capacity for a carbon sorption bed with 500 kg of Columbia JXC activated carbon. The pollutant is m-xylene at an air concentration of 10 mg/m³ and the atmospheric conditions are 25°C and 750 mm Hg.

10.13 References

Buonicore, T. and W. Davis (Eds.). *Air Pollution Engineering Manual.* Air and Waste Management Assoc. Van Nostrand Reinhold. New York. 1992.

Danielson, J. (Ed.) *Air Pollution Engineering Manual.* AP40- USEPA. 1973.

EAB Control Cost Manual, 3rd ed. U.S. Environmental Protection Agency, Research Triangle Park, NC. 1987.

Joseph, G. and D. Beachler. *Control of Gaseous Emissions.* Northrop Services, Inc. EPA 450/2-81-005. Research Triangle Park, NC. 1981.

Noll, K., V. Gounaris, and W. Hou. *Adsorption Technology for Air and Water Pollution Control.* Lewis Publishers. Chelsea, MI. 1992.

Marchello, J. *Control of Air Pollution Sources*, Marcel Dekker. New York. 1976.

Ruthven, D. *Principles Of Adsorption and Adsorption Processes.* John Wiley & Sons. New York. 1984.

Sansone, E., and L. Jonas. "Prediction of Activated Carbon Performance for Carcinogenic Vapors" Amer. Ind. Hyg. Assoc. J. 42. 1981.

Turk, A. "Adsorption," *Air Pollution Vol. IV Engineering Control of Air Pollution*, Stern, A., ed. Academic Press. New York. 1977.

Wadden, R., and P. and Scheff. *Engineering Design for the Control of Workplace Hazards*, McGraw-Hill. New York. 1987.

Chapter 11

Incineration

11.1 Introduction

Incinerators refer to control devices that involve raising the waste gas temperature and sustaining it long enough for any organic waste present to combine with available oxygen. The destruction efficiency of the incinerator depends on the temperature and residence time characteristics of the unit and the organics that are burned. Combustion of organic compounds in incinerators is a widely used technique for air pollution emissions control. Table 11.1 lists some of the processes that use incineration.

The two major types of incinerators are (a) the direct flame or thermal incinerator and (b) the catalytic incinerator. Catalytic units permit the use of a lower temperature than the thermal incinerators for complete combustion, and therefore use less fuel and are made of lighter construction materials. The lower fuel cost may be offset, however, by the added cost of catalysts and higher maintenance requirements for the catalytic units.

A thermal incinerator usually consists of a refractory-lined chamber that is equipped with one or more sets of burners. A typical incinerator is shown in Figure 11.1. The contaminant-laden stream is passed through the burners where it is heated above its ignition temperature. The hot gases then pass through a residence chamber where they are held for a certain length of time to ensure complete combustion.

Table 11.1
Application of Incinerators

Adhesive tape cursing	Packing house effluent
Asphalt blowing	Paint baking ovens
Blake lining ovens	Paint removal facilities
Cat cracker regenerator off gas	Phthalic anhydride manufacturing off gas
	Plastic curing ovens
Charcoal broiler	Printing process
Coil and strip coating lines	Quench bath oil fumes
Core ovens	Resin and paint cooking
Cupola furnace stacks	Roofing paper machine hoods
Deep fat frying	Rubber curing
Fat rendering	Solvent degreasing
Fiber glass curing	solvent manufacturing off gas
Herbicide and insecticide manufacturing off gas	Sulfur plant tail gas
	Textile dryers
Lithographing ovens	Varnish burn-off
Meat smokehouses	Varnish kettles
Metal coating ovens	Vinyl sponge curing
Metal reclaiming	Wire enameling
Pulp and Paper	

Figure 11.1
Incinerator Using a Discrete Burner

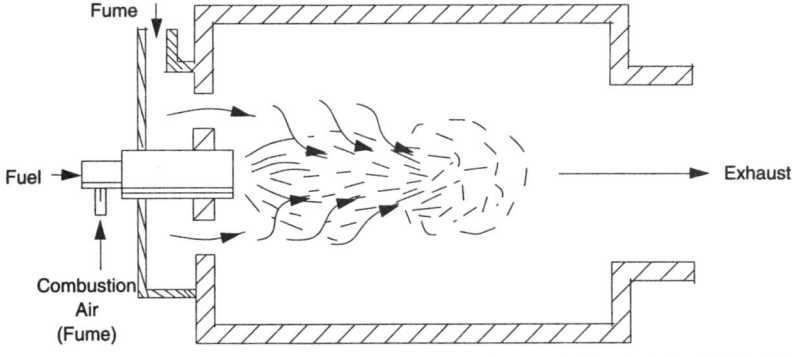

Depending on the particular needs of the system, additional fuel and/or excess air can be added through the burners. To achieve high removal efficiency in thermal incinerators, the contaminant must be held at a uniform temperature of between 1200°F and 1500°F for 0.3 to 0.5 seconds (see Figure 11.2). Time and temperature are interrelated, so that a relatively short contact period and high temperature can

Figure 11.2
Coupled Effects of Temperature and Time on Rate of Pollutant Oxidation

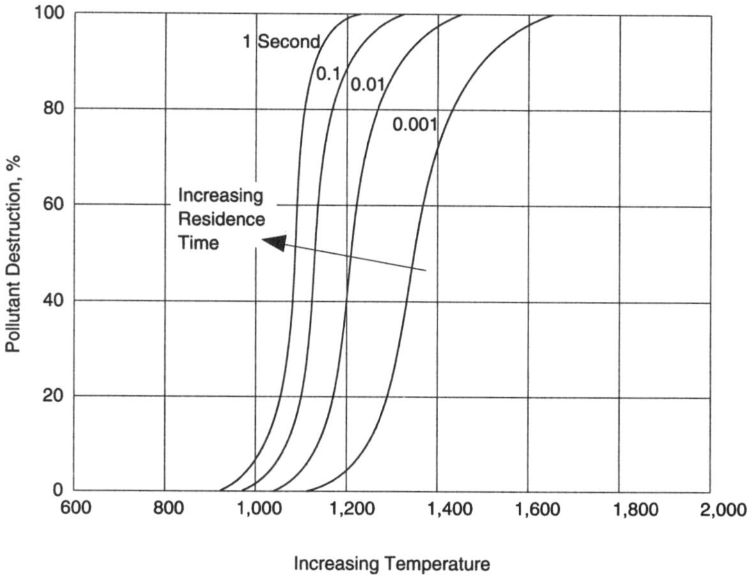

Source: Buonicore and Davis, *Air Pollution Engineering Manual*, 1992. Reprinted with permission of John Wiley & Sons, Inc.

produce an efficiency equivalent to a time/temperature unit with long contact time and low temperature. The design residence time of 0.3 or more seconds applies to the reaction zone only, with additional volume provided for initial combustion and mixing. Also, since the flue gases are discharged at elevated temperatures, a system to recover the heat may be included. The simplest method is to use the hot cleaned gases exiting the incinerator to preheat the cooler incoming gases. Design is usually for 35 to 90 percent heat recovery efficiency.

11.2 Stoichiometric Combustion Air

Oxygen is necessary for combustion. The amount of oxygen required for complete combustion is known as the *stoichiometric or theoretical oxygen* and is determined by the nature and, of course, the quantity of the combustible material to be burned.

For the case of a dilute stream of a pure hydrocarbon, HC, in air, the stoichiometry of complete combustion is:

$$C_xH_y + (b)O_2 + 3.76(b)N_2 \rightarrow xCO_2 + (y/2)H_2O + 3.76(b)N_2 \quad [11.1]$$

where: C_xH_y = the general formula for any hydrocarbon

$b = x + (y/4)$, the stoichiometric number of moles of oxygen required per mole of C_xH_y

3.76 = the number of moles of nitrogen present in air for every mole of oxygen

The equation for the theoretical stoichiometric complete combustion of the simplest hydrocarbon (methane) with air is

$$CH_4 + 2O_2 + 7.52N_2 \rightarrow CO_2 + 2H_2O + 7.52N_2 \qquad [11.2]$$

Air is considered to be 79% N_2 and 21% O_2. The mole fraction equals the volume fraction, so the moles of N_2 equals (2 moles O_2) times (79 moles N_2/21 moles O_2), or 7.52 moles. The oxidation of combustible pollutants will cause them to be converted to CO_2 and H_2O as typified by the following reaction for toluene and benzene.

$$\text{Toluene:} \quad C_7H_8 + 9O_2 \rightarrow 7CO_2 + 4H_2O \qquad [11.3]$$

$$\text{Benzene:} \quad C_6H_6 + 15/2O_2 \rightarrow 6CO_2 + 3H_2O \qquad [11.4]$$

Equation [11.1] provides not only the theoretical air requirements in terms of moles or volume, but it also permits the determination of the resulting combustion products which the flue needs to handle. Table 11.2 provides combustion constants for methane and other organic compounds (Joseph and Beachler 1981). This table also contains other useful data for combustion calculations, including molecular weights, densities, specific gravities and volumes, and the heat of combustion.

In industrial applications, more than the stoichiometric amount of air is usually used to ensure complete combustion. This extra volume is referred to as *excess air*. If ideal mixing were achievable, no excess air would be necessary. However, most combustion devices are not capable of achieving ideal mixing of the fuel and air streams. The amount of excess air is held to a minimum in order to reduce heat losses. Excess air takes no part in the reaction, but does absorb some of the heat produced. To raise the excess air to the combustion temperature, additional fuel must be used to make up for this loss of heat.

Elements other than carbon and hydrogen that may be present in the organic compound will also be released in the combustion process. Halogenated hydrocarbons, like chlorine, are generally converted to the acids. Even trace quantities of HCl require the use of corrosion-resistant materials in the incineration equipment.

11.3 Combustion Kinetics

Figure 11.2 shows that volatile organic compound (VOC) destruction rates are very sensitive to temperature; thus, sufficient time must be provided at the design

Chapter 11: Incineration

temperature to allow the reactions to reach the desired degree of destruction. Turbulence ensures sufficient mixing of oxygen and VOCs during the process. This leads to the importance of the *three T's of combustion* — temperature, time, and turbulence. The three T's relate to three characteristic times: a chemical time, a residence time, and a mixing time, given by the following equations:

$$\tau_c = 1/k \qquad [11.5]$$

$$\tau_r = V/Q = L/v \qquad [11.6]$$

$$\tau_m = L^2/\mathcal{D}_e \qquad [11.7]$$

where: τ_c, τ_r, τ_m = chemical, residence, and mixing times, respectively, sec

V = volume of the reaction zone, ft^3

Q = volumetric flow rate (at the temperature in the afterburner), ft^3/sec

L = length of the reaction zone, ft

v = gas velocity in the afterburner, ft/sec

$\mathcal{D}_e$ = effective (turbulent) diffusion coefficient, ft^2/sec

k = rate constant, sec^{-1}

The ratio of the mixing time to the residence time is called the *Peclet number*, Pe, and the ratio of the chemical time to the residence time is the inverse of the *Damkoler number*, Da. If Pe is large and Da is small, then mixing is the rate-controlling process. If Pe is small and Da is large, then the chemical kinetics are rate-controlling. At the temperatures of most incinerators, provided a reasonable flow velocity is maintained, the mixing processes will not be the limiting factor and chemical kinetics will be important.

For chemical reactions, the reaction rates are typically expressed by equations of the form:

$$\left(\text{Decrease in concentration of A per unit time}\right) = \frac{dC_A}{dt} = r = -kC_A^n \qquad [11.8]$$

where: r = reaction rate

k = a kinetic rate constant whose value is strongly dependent on the reactants

C_A = concentration of A

n = reaction order

Table 11.2
Combustion Constants

			Heat of Combustion				For 100% Total Air (mol/mol of combustible) (ft³/ft³ of combustible)							For 100% Total Air (lb/lb of combustible)							Flammable Limits % by Volume	
			(Btu/ft³)		(Btu/lb)		Required for combustion			Flue Products				Required for combustion			Flue Products					
Substance	lb/ft³	ft³/lb	Gross (high)	Net (low)	Gross (high)	Net (low)	O_2	N_2	Air	CO_2	H_2O	N_2	O_2	N_2	Air	CO_2	H_2O	N_2	Lower	Upper
Carbon, C	-	-	-	-	14,093	14,093	1.0	3.76	4.76	1.0	-	3.76	2.66	8.86	11.53	3.66	-	8.86	-	-
Hydrogen, H_2	0.0053	187.723	325	275	61,100	51,623	0.5	1.88	2.38	-	1.0	1.88	7.94	26.41	34.34	-	8.94	26.41	4.00	74.20
Oxygen, O_2	0.0846	11.819	-	-	-	-	-	-	-	-	-	-	-	-	-	-	-	-	-	-
Nitrogen, N_2	0.744	13.443	-	-	-	-	-	-	-	-	-	-	-	-	-	-	-	-	-	-
Carbon monoxide, CO	0.0740	13.506	322	322	4347	4347	0.5	1.88	2.38	1.0	-	1.88	0.57	1.90	2.47	1.57	-	1.90	12.50	74.20
Carbon dioxide, CO_2	0.1170	8.548	-	-	-	-	-	-	-	-	-	-	-	-	-	-	-	-	-	-
Paraffin series																				
Methane, CH_4	0.0424	23.565	1,013	913	23,879	21,520	2.0	7.53	9.53	1.0	2.0	7.53	3.99	13.28	17.27	2.74	2.25	13.28	5.00	15.00
Ethane, C_2H_6	0.0803	12.455	1,792	1,641	22,320	20,432	3.5	13.18	16.68	2.0	3.0	13.18	3.73	12.39	16.12	2.93	1.80	12.39	3.00	12.50
Propane, C_3H_8	0.1196	8.365	2,590	2,385	21,661	19,944	5.0	18.82	23.82	3.0	4.0	18.82	3.63	12.07	15.70	2.99	1.68	12.07	2.12	9.35
n-Butane, C_4H_{10}	0.1582	6.321	3,370	3,113	21,308	19,680	6.5	24.47	30.97	4.0	5.0	24.47	3.58	11.91	15.49	3.03	1.55	11.91	1.86	8.41
Isobutane, C_4H_{10}	0.1582	6.321	3,363	3,105	21,257	19,629	6.5	24.47	30.97	4.0	5.0	24.47	3.58	11.91	15.49	3.03	1.55	11.91	1.80	8.44
n-Pentane, C_6H_{12}	0.1904	5.252	4,016	3,709	21,091	19,517	8.0	30.11	38.11	5.0	6.0	30.11	3.55	11.81	15.35	3.05	1.50	11.81	-	-
Isopentane, C_6H_{12}	0.1904	5.252	4,008	3,716	21,052	19,478	8.0	30.11	38.11	5.0	6.0	30.11	3.55	11.81	15.35	3.05	1.50	11.81	-	-
Neopentane, C_6H_{12}	0.1904	5.252	399.	3,693	20,970	19,396	8.0	10.11	38.11	5.0	6.0	30.11	3.55	11.81	15.35	3.05	1.50	11.81	-	-
n-Hexane, C_6H_{14}	0.2274	4.398	4,762	4,412	20,940	19,403	9.5	35.76	45.26	6.0	7.0	35.76	3.53	11.74	15.27	3.06	1.46	11.74	1.18	7.40

Chapter 11: Incineration

Compound																			
Olefin series																			
Ethylene, C_2H_4	0.0746	13.412	1,614	1,513	21,644	20,295	3.0	11.29	14.29	2.0	11.29	3.42	11.39	14.81	3.14	1.29	11.39	2.75	28.60
Proplyene, C_2H_6	0.1110	9.007	23,36	2,186	21,041	19,691	4.5	16.94	21.44	3.0	16.94	3.42	11.39	14.81	3.14	1.29	11.39	2.00	11.10
n-Butene, C_4H_8	0.1480	6.756	3,084	2,885	20,840	19,496	6.0	22.59	28.59	4.0	22.59	3.42	11.39	14.81	3.14	1.29	11.39	-	-
Isobutene, C_4H_8	0.1480	6.756	3,068	2,869	20,730	19,382	6.0	22.59	28.59	4.0	22.59	3.42	11.39	14.81	3.14	1.29	11.39	-	-
n-Pentene, C_6H_{10}	0.1852	5.400	3,836	3,586	20,712	19,363	7.5	28.23	35.73	5.0	28.23	3.42	11.39	14.81	3.14	1.29	11.39	-	-
Aromatic series																			
Benzene, C_6H_6	0.2060	4.852	3,751	3,601	18,210	17,480	7.5	28.23	35.73	6.0	28.23	3.07	10.22	13.30	3.38	0.69	10.22	1.40	7.10
Toluene, C_7H_8	0.2431	4.113	4,484	4,284	18,440	17,620	9.0	33.88	42.88	7.0	33.88	3.13	10.40	13.53	3.34	0.78	10.40	1.27	6.75
Xylene, C_8H_{10}	0.2803	3.567	5,230	4,980	48,650	17,760	10.5	39.52	50.02	8.0	39.52	3.17	10.53	13.70	3.32	0.85	10.53	1.00	6.00
Miscellaneous Gas																			
Acetylene, C_2H_2	0.0697	14.344	1,499	1,488	21,500	20,776	2.5	9.41	11.91	2.0	9.41	3.07	10.22	13.30	3.38	0.69	10.22	-	-
Napthalene, $C_{10}H_8$	0.3384	2.955	5,854	5,654	17,298	16,708	12.0	45.17	57.07	10.0	45.17	3.00	9.97	12.93	3.43	0.56	9.97	-	-
Methyl alcohol, CH_3OH	0.0846	11.820	868	768	10,259	9,078	1.5	5.65	7.15	1.0	5.65	1.50	4.98	6.48	1.37	1.13	4.98	6.72	36.50
Ethyl alcohol, C_2H_5OH	0.1216	8.221	1,600	1,451	13,161	11,929	3.0	11.29	14.29	2.0	11.29	2.08	6.93	9.02	1.92	1.17	6.93	3.28	18.95
Ammonia, NH_3	0.0456	21.914	441	365	9,668	8,001	0.8	2.82	3.57	-	1.5	3.32	1.41	4.69	6.10	-	1.59	5.51	15.50	27.00
Sulfur, S	-	-	-	-	3,983	3,983	1.0	3.76	4.76	1.0	3.76	1.00	3.29	4.29	SO_2 2.00	-	3.29	-	-
Hydrogen sulfide, H_2S	0.0911	10.979	647	596	7,100	6,545	1.5	5.65	7.15	1.0	5.65	1.41	4.69	6.10	1.88	0.53	4.69	4.30	45.50
Sulfur dioxide, SO_2	0.1733	5.770	-	-	-	-	-	-	-	-	-	-	-	-	-	-	-	-	-
Water vapor, H_2O	0.0476	21.017	-	-	-	-	-	-	-	-	-	-	-	-	-	-	-	-	-
Air	0.0766	13.063	-	-	-	-	-	-	-	-	-	-	-	-	-	-	-	-	-
Gasoline	-	-	-	-	-	-	-	-	-	-	-	-	-	-	-	-	-	1.40	7.60

The overall order of a reaction is determined experimentally. If the reaction is first-order, then

$$\frac{-dC_A}{dt} = kC_A \qquad [11.9]$$

The solution to this differential equation is

$$C_A = C_{Ao}\exp(-kt) \qquad [11.10]$$

where: C_{Ao} is the initial concentration of A at time t = 0

The assumption that allows incineration to be considered first-order is that the VOC to be burned is much less than the concentration of oxygen in the contaminated air stream.

For most chemical reactions, the relation between the kinetic rate constant, k, and the temperature, T, is given by the Arrhenius equation.

$$k = Ae^{-E/RT} \qquad [11.11]$$

where: A = experimented constant, sec^{-1}
E = activation energy, cal/g-mole
R = universal gas constant, 1.987 cal/g-mole °K
T = absolute temperature, °K

Table 11.3 lists some values for A and E and Table 11.4 lists the computed value of k for three temperatures (Brunner 1993).

Example 11.1 Calculation of Kinetic Rate Constant

Show the calculation leading to the value of k in Table 11.4 for benzene at 1,000°F.

Solution:

$$k = A \exp\left(-\frac{E}{RT}\right)$$

Since: 1,000°F = 537.78°C = 810.78°K

$$k = \frac{7.43 \bullet 10^{21}}{\text{sec}} \exp-\left(\frac{95,900 \text{ cal/mol}}{1.987 \text{ cal/mol°K} \bullet 810.78°K}\right)$$

$$= 0.00011 \text{ sec}^{-1}$$

Table 11.3
Destruction Kinetics Parameters

Compounds	Frequency Factor A, sec^{-1}	Activation Energy, E (cal/g-mole)
Acetic anhydride	$1.00 \cdot 10^{12}$	34,500
Acrylonitrile	$2.13 \cdot 10^{12}$	52,100
Aniline	$9.30 \cdot 10^{15}$	71,000
Azomethane	$3.50 \cdot 10^{16}$	52,500
Benzene	$7.42 \cdot 10^{21}$	95,900
Butene	$5.00 \cdot 10^{12}$	63,000
Carbon tetrachloride	$2.80 \cdot 10^{5}$	26,000
Chloroform	$2.90 \cdot 10^{12}$	49,000
Dichlorobenzene	$3.00 \cdot 10^{8}$	39,000
Dichloromethane	$3.00 \cdot 10^{13}$	64,000
Ethyl chlorocarbonate	$9.20 \cdot 10^{8}$	29,100
Ethyle nitrite	$1.40 \cdot 10^{14}$	37,700
Ethyl peroxide	$5.10 \cdot 10^{14}$	31,500
Ethylene dibutyrate	$1.80 \cdot 10^{10}$	33,000
Ethylidene dichloride	$1.20 \cdot 10^{12}$	49,500
Hexachlorobenzene	$1.90 \cdot 10^{16}$	72,600
Hexachlorobutane	$6.30 \cdot 10^{12}$	59,000
Hexachloroethane	$1.90 \cdot 10^{7}$	29,000
Methyl iodide	$3.90 \cdot 10^{12}$	43,000
Monochlorobenzene	$8.00 \cdot 10^{4}$	23,000
Nitrobenzene	$1.40 \cdot 10^{15}$	64,000
Paracetaldehyde	$1.30 \cdot 10^{15}$	44,200
Pentachlorobiphenyl	$1.10 \cdot 10^{16}$	70,000
Propylene oxide	$1.40 \cdot 10^{14}$	58,000
Pyridine	$1.10 \cdot 10^{5}$	24,000
Tetrachlorobenzene	$1.90 \cdot 10^{6}$	30,000
Tetrachloroethylene	$2.60 \cdot 10^{6}$	33,000
Toluene	$2.28 \cdot 10^{13}$	56,500
Trichlorobenzene	$2.20 \cdot 10^{8}$	38,000
Trichloroethane	$1.90 \cdot 10^{8}$	32,000
Vinyl chloride	$3.57 \cdot 10^{14}$	63,300

Table 11.4
Thermal Oxidation Parameters, Based on First-Order Kinetics

Compounds	A, sec^{-1}	E, kcal/mol	k, sec^{-1}, at 1,000°F	k, sec^{-1}, at 1,200°F	k, sec^{-1}, at 1,400°F
Acrolein	3.30 • 10^{10}	35.9	6.99258	102.37	841.47
Acrylonitrile	2.13 • 10^{12}	52.1	0.01946	0.96	20.34
Allyl alcohol	1.75 • 10^{6}	21.4	2.99528	14.83	52.07
Allyl chloride	3.98 • 10^{7}	29.1	0.56034	4.93	27.21
Benzene	7.43 • 10^{21}	95.9	0.00011	0.14	38.59
1-Butene	3.74 • 10^{14}	58.2	0.07760	6.02	183.05
Chlorobenzene	1.34 • 10^{17}	76.6	0.00031	0.09	8.41
Cyclohexane	5.13 • 10^{12}	47.6	0.76467	26.84	438.42
1,2 -Dichloroethane	4.82 • 10^{11}	45.6	0.24851	7.51	109.11
Ethane	5.65 • 10^{14}	63.6	0.00411	0.48	19.93
Ethanol	5.37 • 10^{11}	48.1	0.05869	2.14	35.97
Ethyl acrylate	2.19 • 10^{12}	46.0	0.88094	27.44	407.99
Ethylene	1.37 • 10^{12}	50.8	0.02804	1.25	24.64
Ethyl formate	1.90 • 10^{16}	44.7	0.39562	11.18	154.04
Ethyl mercaptan	5.20 • 10^{5}	14.7	56.86353	170.64	404.29
Hexane	6.02 • 10^{8}	34.2	0.36628	4.72	35.13
Methane	1.68 • 10^{11}	52.1	0.00153	0.08	1.60
Methyl chloride	7.43 • 10^{8}	40.9	0.00708	0.15	1.66
Methyl ethyl ketone	1.45 • 10^{14}	58.4	0.02658	2.09	64.38
Natural gas	1.65 • 10^{12}	49.3	0.08565	3.41	61.61
Propane	5.25 • 10^{19}	85.2	0.00058	0.34	49.99
Propylene	4.63 • 10^{8}	34.2	0.28171	3.63	27.02
Toluene	2.28 • 10^{13}	56.5	0.01358	0.93	25.54
Triethylamine	8.10 • 10^{11}	43.2	1.85139	46.78	590.11
Vinyl acetate	2.54 • 10^{9}	35.9	0.53822	7.88	64.77
Vinyl chloride	3.57 • 10^{14}	63.3	0.00313	0.36	14.58

Source: Beard et al. 1980

Example 11.2 Estimation of Incineration Time

Estimate the time required to destroy 99.9 percent of the benzene in a waste gas stream at 1,000°F and at 1,400°F.

Solution:

From Equation [11.10] and Table 11.4:

$$k = 0.00011 \text{ sec}^{-1}$$

$$t_{1000°F} = \frac{1}{k}\ln\frac{C_{Ao}}{C_A} = \frac{1}{0.00011/\text{sec}}\ln\frac{1}{0.001} = 62,800 \text{ sec} = 17.4 \text{ hr}$$

at $1,400°F$, $k = 38.59 \text{ sec}^{-1}$

$$t_{1400°F} = \frac{1}{38.59/\text{sec}}\ln\frac{1}{0.001} = 0.2 \text{ sec}$$

11.4 Thermal Incinerator Design Principles

The design of an incinerator involves selecting the general characteristics for the unit, establishing design values for temperature and gas volume, and determining the fuel-firing rate and combustion chamber volume.

The information required for the process design calculation is:
- inlet gas flow rate, scfm;
- inlet gas temperature, °F; and
- solvent type and vapor concentration range, % or ppm.

The desired gas temperature in the incinerator is usually specified by air pollution control regulations. Frequently, air pollution regulations require the gas temperature to be above a certain minimum. This may vary from about 1,250°F for easily oxidized solvents to 1,500°F for resistant vapors. Because the final desired temperature of the gas mixture is known, the problem becomes one of finding the proper fuel-addition rate to heat the waste gas to the design temperature.

11.4.1 Material and Energy Balance

A diagram of a vapor incinerator is presented in Figure 11.3. Prediction techniques (as discussed earlier in this chapter) enable both the design temperature and the residence time to be specified to achieve a desired destruction efficiency. Therefore, what is required is the calculation of the fuel flow rate and size of the incinerator to achieve the desired residence time and temperature.

The overall material balance is:

$$\text{mass in} = \text{mass out}$$

$$m_{\text{waste gas}} + m_{\text{fuel}} + m_{\text{burner air}} = m_{\text{exhaust gas}} \qquad [11.12]$$

where: $m = Q\rho_g$, lb/min

Q = gas flow rate, scfm

ρ_g = gas density (~ 0.075 lb/ft³ for air)

Figure 11.3
Incinerator Diagram

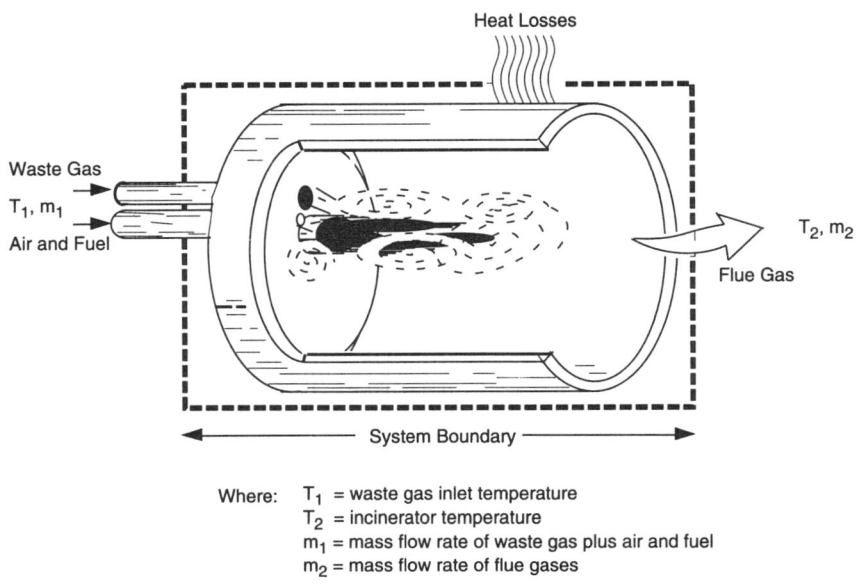

Where: T_1 = waste gas inlet temperature
T_2 = incinerator temperature
m_1 = mass flow rate of waste gas plus air and fuel
m_2 = mass flow rate of flue gases

The overall heat balance is Heat in = Heat out + Heat loss.

Heat is a relative term which is compared at a reference temperature (see Chapter 2). The heat content of a substance is arbitrarily taken as zero at a specified reference temperature. In the natural gas industry, the reference temperature is normally 60°F.

The heat content can be computed from Equation [11.13] or tables such as Table 11.5.

$$H = C_p(T - T_o) \qquad [11.13]$$

where: H = enthalpy, Btu/lb

C_p = specific heat, Btu/ lb°F

T = temperature of the substance, °F

T_o = reference temperature, °F

To determine the heat required, Equations [11.12] and [11.13] are combined:

$$Q_h = Q\rho_g H = Q\rho_g C_p(T - T_o) \qquad [11.14]$$

$$Q_h = Q \bullet \rho_g \bullet C_p \bullet \Delta T \bullet 60 \qquad [11.15]$$

Table 11.5
Heat Contents of Various Gases

Temp °C	\multicolumn{9}{c}{Relative heat content, H, in Btu/lb at atmospheric pressure}								
	O_2	N_2	Air	CO	CO_2	SO_2	H_2	CH_4	H_2O
60	0	0	0	0	0	0	0	0	0
100	8.8	9.9	9.6	10.0	8.0	5.9	137	21.0	–
200	30.9	34.8	33.6	34.9	29.3	21.4	484	76.1	–
300	53.3	59.9	57.7	59.9	52.0	37.5	832	136.4	1.165
400	76.2	85.0	81.8	85.0	75.3	54.4	1,182	202.1	1,212
500	99.4	110.3	106.0	110.6	99.8	71.8	1,532	272.6	1,259
600	123.1	136.1	130.2	136.3	125.1	89.8	1,882	347.8	1,307
700	147.2	161.7	154.5	162.4	149.6	108.2	2,233	427.4	1,355
800	171.7	187.7	178.9	188.7	177.8	127.0	2,584	511.2	1,404
900	196.6	213.9	203.4	215.6	205.6	146.1	2,935	599.2	1,454
1,000	221.7	240.7	235.0	242.7	233.6	165.5	3,291	691.1	1,505
1,200	272.5	294.7	288.5	297.8	290.9	205.1	4,007	886.2	1,609
1,400	324.3	350.8	343.0	354.3	349.7	245.4	4,729	1,094.1	1,717
1,600	377.3	407.3	298.0	407.5	416.3	286.4	5,460	1,313.0	1,829
1,800	430.7	465.0	455.0	465.3	470.9	327.8	6,198	1,542.6	–
2,000	484.0	523.8	513.0	523.8	532.8	369.1	6,952	–	–
2,200	539.3	583.2	570.7	583.3	596.1	411.1	7,717	–	–
2,400	594.4	642.3	628.5	643.0	659.2	452.7	8,490	–	–
2,600	649.0	702.8	687.3	703.2	723.2	495.2	9,272	–	–
2,800	702.8	763.1	746.6	771.3	787.4	557.5	10,060	–	–
3,000	758.6	824.1	806.3	832.6	852.0	580.0	10,870	–	–
3,200	816.4	885.8	866.0	894.0	916.7	622.5	11,680	–	–
3,400	873.4	947.6	925.9	956.0	981.6	665.0	12,510	–	–
3,600	931.0	1,010.3	986.1	1,018.3	1,047.3	707.5	13,330	–	–

Source: Joseph and Beachler 1981

where: Q_h = heat required, Btu/hr

Q = gas flow, scfm

ρ_g = gas density, lb/ft^3 (0.075 lb/ft^3 for air)

C_p = specific heat, Btu/lb °F (0.26 Btu/lb °F for air for typical temperature range in incinerators)

ΔT = temperature rise, °F

60 = constant, min/hr

11.4.2 Calculation of Required Fuel

The potential heat energy released by various fuels is termed the *gross or higher heat value*. Since water is a combustion product of most fuels, the energy available to heat combustion products must be reduced by the heat of vaporization of the water formed. The resultant heat is termed the *net or lower heat value*.

The *available heat* from a fuel is defined as the gross quantity of heat released within a combustion chamber minus the sensible heat carried away in water vapor contained in the flue gases (Figure 11.4). The available heat represents the net quantity of heat remaining for useful heating. Figure 11.5 provides the available heat from the complete combustion (no excess air) of various fuels at various flue gas temperatures (Beard, Iachetta and Lilleleht 1980).

Figure 11.4
Heat Terms

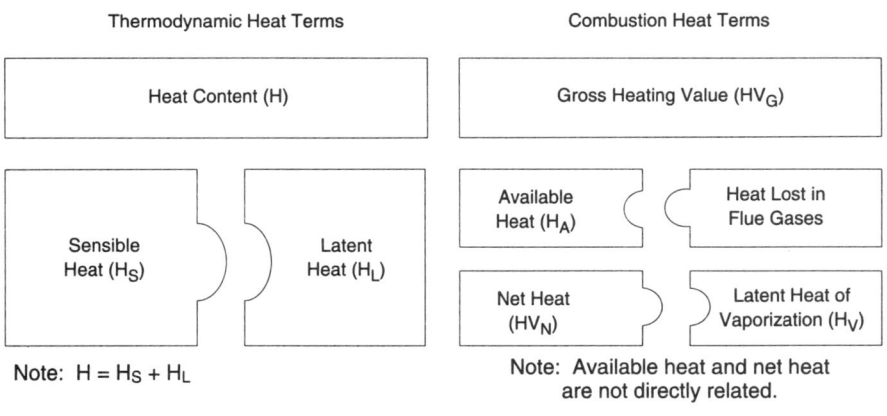

The available heat for natural gas (1,059 Btu/scf) (assuming stoichiometric air) can be calculated from the following equation:

$$Q_A = -0.237(T) + 981 \qquad [11.16]$$

where: Q_A = Available heat, Btu/scf

               T = °F (Air Temperature)

The mass flow rate of natural gas required is:

$$NG = Q_h / Q_A \qquad [11.17]$$

where: NG = natural gas required, scf/hr

              Q_h = heat required, Btu/hr (Equation [11.15])

Figure 11.5
Available Heats for Some Typical Fields

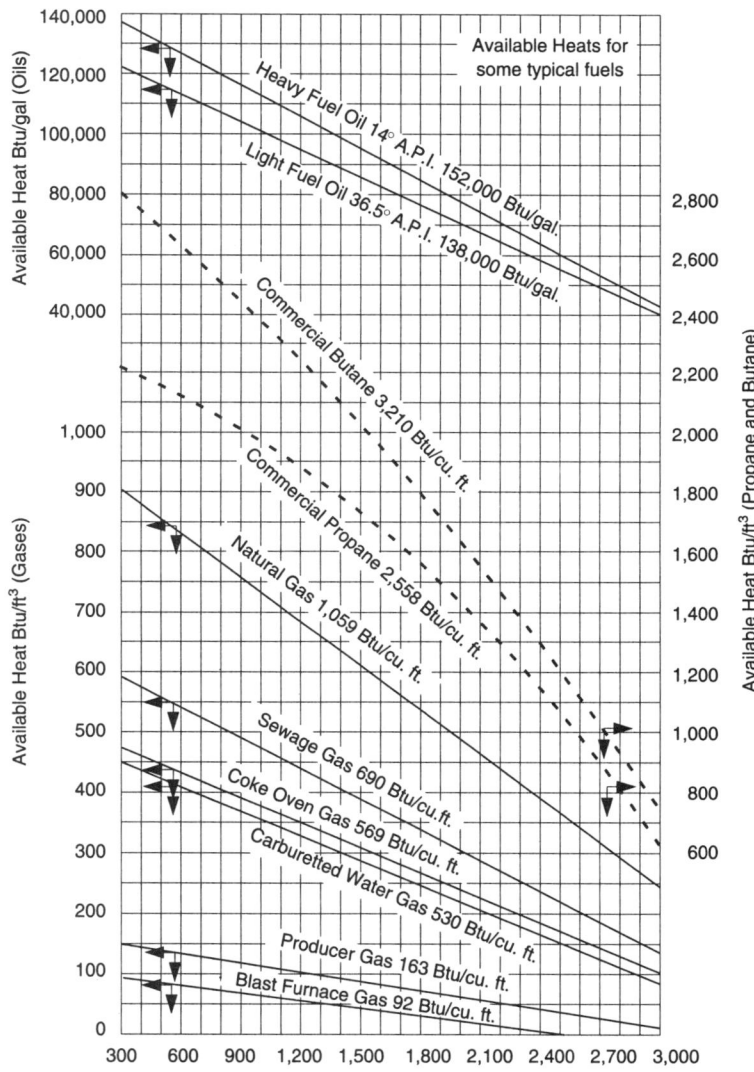

Note: Fuels listed above are identified by their gross heating values. The sum of the moisture loss and the dry flue gas loss at any particular exit gas temperature may be evaluated by subtracting the available heat from the gross heating value. Note that all available heat figures are based upon perfect combustion and a fuel input temperature of 60° F. The scales on the left side of this chart are for the solid curves. The scales on the right are for the dashed curves.

Source: Joseph and Beachler 1981

Many hydrocarbons have a high fuel value (Btu per pound of solvent) and therefore can contribute to the heat required for incineration. However, insurance regulations limit the maximum VOC concentration to 25% of the lower explosive limit (LEL). This restriction often produces concentrations of 1,000 to 3,000 ppm in the exhaust; the fuel saved will be almost negligible, but the saving could be significantly beyond the 25% LEL concentrations.

11.4.3 Sizing the Incinerator

After the heat and material balance has been completed, it is possible to size the incinerator. This is accomplished by determining the volumetric flow rate for both the process gas stream, Q_p, and the products of combustion, Q_c, of the natural gas at the operating temperature. A good estimate for the gas products is 11.5 times the natural gas required (natural gas is not listed in Table 11.2 since its chemical composition can vary).

The diameter of the combustion chamber is given by:

$$D = \sqrt{\frac{4(Q_p + Q_c)}{\pi v_T}} \qquad [11.18]$$

where: v_T = is the linear velocity of the gases in the incinerator, ft/sec

D = diameter of incinerator, ft

The linear velocity can range from 10 to 50 ft/sec but is usually between 20 and 40 ft/sec. High velocities minimize the dangers of flashback and fire hazards. The length of the reaction chamber is given by:

$$L = v_T \tau \qquad [11.19]$$

where: τ = residence time

A length-to-diameter ratio of 2.0 to 3.0 is usually employed. A minimum residence time of 0.2 to 2.0 seconds is often required by regulation and depends on the material to be incinerated (kinetic considerations).

11.4.4 Thermal Incinerator Design Summary

The design of incinerators to destroy VOCs requires that the amount of fuel and physical dimensions of the unit be specified. It is assumed that the process flow rate, inlet temperature, and combustion temperature are known and that the required residence time has been specified by the air pollution regulation or determined from kinetic considerations for the material to be destroyed.

The general design procedure is as follows:

Chapter 11: Incineration

1. Calculate the heat load required to raise the process stream from its inlet temperature to the operating temperature of the incinerator. Calculate the heat load for any radiant losses. The calculations employ stoichiometry and overall mass and heat balances.

2. Calculate the fuel (usually natural gas) required. First determine the available heat at the operating temperature. Then determine the fuel required by dividing the heat load by the available heat of the fuel.

3. The dimensions of the combustion chamber can now be determined. A linear (average) gas flow velocity of 20 to 40 ft/sec is required to prevent flashback and a length to diameter ratio of 2.0 to 3.0 is usually employed.

Example 11.3 Incinerator Design

Design an incinerator to control the emissions from a paint drying oven. Regulations require 95% removal of the toluene vapors that are present in the emissions. The emission stream is 10,000 scfm and has a temperature of 150°F. Natural gas will be used to obtain a reaction temperature of 1,350°F.

Solution:

Gas Flow Rate:

Determine the heat required using Equation [11.15]:

Q_h = 10,000 scfm • 0.076 lb/ft³ • 0.24 Btu/lb °F (1,350°F - 150°F) • 60 min/hr

= 13 • 10⁶ Btu/hr

Determine the available heat for natural gas using Equation [11.16]:

Q_A = -0.237 • 1,350 + 981 = 661.05 Btu/scf

Determine the natural gas required using Equation [11.17]:

$$NG = \frac{Q_h}{Q_A} = \frac{13 \bullet 10^6 \text{ Btu / hr}}{661.05 \text{ Btu / ft}^3} \bullet \frac{1 \text{ hr}}{60 \text{ min}} = 328 \text{ scfm}$$

For combustion of natural gas, an average value of 11.5 times air is usually used per volume of fuel. Therefore:

Combustion Air = (328)(11.5) = 3,775 scfm

The total gas flow will be = 10,000 + 328 + 3,775 = 14,103 scfm

The volumetric flow rate at the operating temperature of 1,350°F can be calculated as:

$$(14{,}103 \text{ scfm})\left(\frac{1{,}350 + 460}{60 + 460}\right) = 49{,}089 \text{ acfm}$$

Reaction Time:

For a reaction temperature of 1,350°F and 95% destruction efficiency:

$$1 - \eta = e^{-k\tau}$$

$$(100-95)/100 = e^{-k\tau}$$

$$k = Ae^{-E/RT}$$

Since: 1,350°F = 732.22°C = 1,005.2°K

$$k = 2.28 \bullet 10^{13} \bullet e^{-\left\{\frac{56{,}500 \text{ cal/mol}}{1.987 \text{ cal/mol}° \text{ K} \bullet 1005.2° \text{ K}}\right\}}$$

$$k = 11.83$$

Solving for t:

$$\frac{(100-95)}{100} = e^{-11.83t}$$

$$t = 0.25 \text{ sec}$$

Incinerator Size:

Turbulent flow is required in an afterburner to ensure adequate mixing and to approach the condition of plug flow. Therefore, a linear velocity of 20 to 40 ft/sec is required. The required reaction time is 0.25 sec, and with a linear velocity of 40 ft/sec, the chamber length will be:

40 ft/sec • 0.25 sec = 10 ft

The gas flow rate at the chosen gas velocity determines the cross-sectional area of the incinerator. The volumeric flow rate was calculated as 49,089 acfm at 1,350°F. Therefore:

$$49{,}089 \frac{\text{ft}^3}{\text{min}} \bullet \frac{\text{sec}}{40 \text{ ft}} \bullet \frac{1 \text{ min}}{60 \text{ sec}} = 20.5 \text{ ft}^2$$

The diameter of the chamber would be 5.0 ft.

11.5 Catalytic Incineration Design Principles

11.5.1 Introduction

Catalytic oxidation is similar to thermal oxidation. The main difference is that after passing through a preheater chamber, the gases pass over a catalyst bed, which promotes oxidation at a lower temperature than does thermal oxidation (see Figure 11.6). Catalysts may be porous pellets, usually cylindrical or spherical in shape, ranging from 1/16 to 1/2 inch in diameter. Other shapes include honeycombs, ribbon, and wire mesh. The catalysts are usually metals or metal salts. Platinum, palladium, cobalt, copper, chromium, molybdenum, and similar metals may be used. These can be supported on inert carriers, such as alumina or porcelain, or they can be used directly in the unsupported state. Metals such as alumina and iron readily attract and hold oxygen atoms so that a strong oxide layer is formed at the surface, thereby disqualifying these metals as catalysts. Platinum and members of the platinum family function well as catalysts because they do not form strong attachments with oxygen, but do attract and adsorb other atoms on their surfaces.

Figure 11.6
Schematic of a Catalytic Incinerator

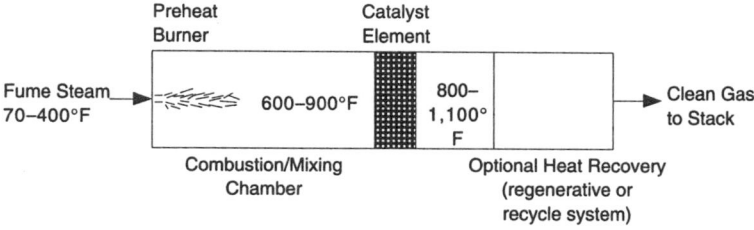

The following sequence of steps is involved in the catalytic conversion of reactants to products:

1. Transfer of reactants to the outer catalyst surface.
2. Diffusion of reactants within the pore of the catalyst.
3. Active adsorption of reactants and deposition of the products on the active centers of the catalyst.

Fouling of catalytic surfaces is the most common problem detracting from the continuous effective operation of a control device required to maintain acceptable emissions control. If the catalyst temperature is not maintained at the design level, a coating of organic material or carbon will be deposited on the surfaces and catalyst activity will suffer. Another problem involves the reaction of a contaminant in the catalyst bed to form solid oxides, which coat the surface. Generally, if the problem is one of organic solid deposits, these can be burned off by a carefully controlled increase in the bed temperature. If the contaminants are inorganic materials, the catalyst must be removed from service and subjected to an acid and/or detergent wash.

11.5.2 Design

Catalytic incineration is essentially a flameless combustion process that occurs at relatively low temperatures, usually in the range of 500 to 1,000°F. The catalyst increases the rate of reaction and permits the reaction to occur at lower temperatures. There is considerable heat released in the catalyst, which is absorbed by the gases, thereby causing a temperature rise as the gases pass through the bed.

The degree of vapor oxidation that can be expected on a catalyst is affected by the vapor composition, the reaction temperature, the volume of the catalyst, and the degree of contact between the solvent and catalyst. The basic problem in the design of a catalytic reactor is to determine the quantity of catalyst required for a given conversion and flow rate. The effect of temperature on conversion for solvent hydrocarbons is shown in Figure 11.7. To achieve destruction efficiencies in the 90 to 95% range, approximately 1.5 to 2.0 ft^3 of catalyst is required per 1,000 scfm (exhaust stream plus supplementary fuel combustion products). A simplified method to determine the total volume of a catalyst bed is to specify the *space velocity*. Space velocity is defined as the waste gas flow rate per unit volume of catalyst; its inverse is the residence time. Table 11.6 presents values for the design variables of a fixed-bed catalytic incinerator system to achieve a given destruction efficiency (Sink 1991).

Figure 11.7
Typical Temperature-Performance Curves for Various Molecular Species Being Oxidized Over Pt/Al$_2$O$_3$ Catalysts

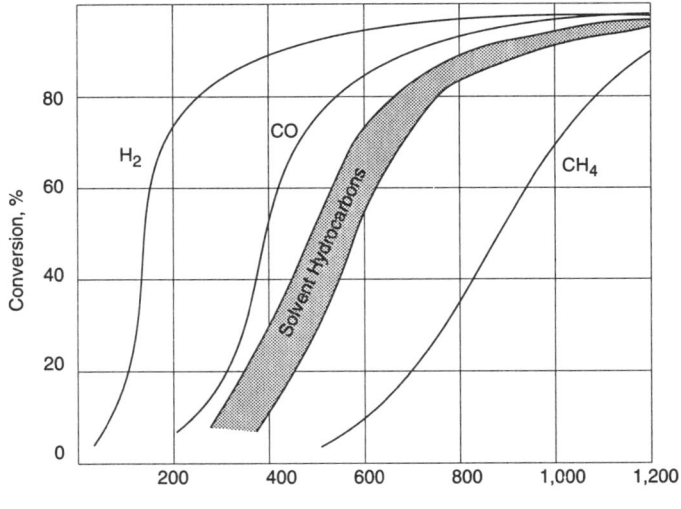

Source: Sink 1991

Table 11.6
Catalytic Incinerator System Design Variables

Required Destruction Efficiencies, DE (%)	Termperature at the Catalyst Bed Inlet, T_{ci}, °F	Temperature at the Catalyst Bed Outlet, T_{co}, °F	Space Velocity SV (hr^{-1})
90	~600	~900°F	40,000[a]
90	~600	~900°F	40,000[b]

[a] correspondends to 1.5 ft³ of catalyst per 1,000 scfm of gas stream
[b] correspondends to 2.0 ft³ of catalyst per 1,000 scfm of gas stream
Source: Modified from Sink 1991

For specific applications, other temperatures and space velocities may be appropriate, depending on the type of catalyst employed and the emission stream characteristics (i.e., composition and concentration). For example, the temperature of the flue gas leaving the catalyst bed may be lower than 1,000°F for emission streams containing easily oxidized compounds and still achieve the desired destruction efficiency.

11.5.3 Kinetics

The same reactions that occur in thermal incineration (illustrated by Equation [11.1]) can be accomplished in catalytic combustion equipment. Energy and material balances are determined similarly to those calculated for thermal incineration. The function of a catalyst is to reduce the activation energy required to cause these reactions to take place. The net result is to bring about these reactions at a reasonable rate and at temperatures lower than those that would be required without a catalyst. Thus, when the reaction rates, k, defined in Equation [11.11], are plotted against temperature for hexane, toluene, and MEK in a catalytic system, the values shown in Figure 11.8 are obtained (Tomany 1975). Comparing the data in Figure 11.8 to that in Table 11.4, it can be seen that at any given temperature, the catalytic reaction rate constant is considerably greater than for the thermal process. Thus, for toluene, Table 11.4 gives the reaction rate for thermal incineration as 0.93 sec^{-1} at a temperature of 1,200°F. As indicated by Figure 11.8, the catalytic reaction rate constant at 900°F would be 128 sec^{-1}. This latter value can be substituted directly in the reaction rate expressed by Equation [11.10], so that for a reaction conversion of 95%, a retention period of 0.024 sec would be required in the catalyst bed rather than 0.104 sec demanded in the combustion chamber of the thermal incinerator.

The retention period calculated from Equation [11.10] is the actual dwell time occupied by the total gas flow, which comprises the contaminated gases plus the combustion gases in the catalyst bed. Gas velocities in the range of 20 to 40 ft/sec are commonly used. Thus, for a specific dwell time and gas velocity, the catalyst bed depth can be computed.

Figure 11.8
Reaction Rate Constants for Catalytic Combustion of Various Solvents

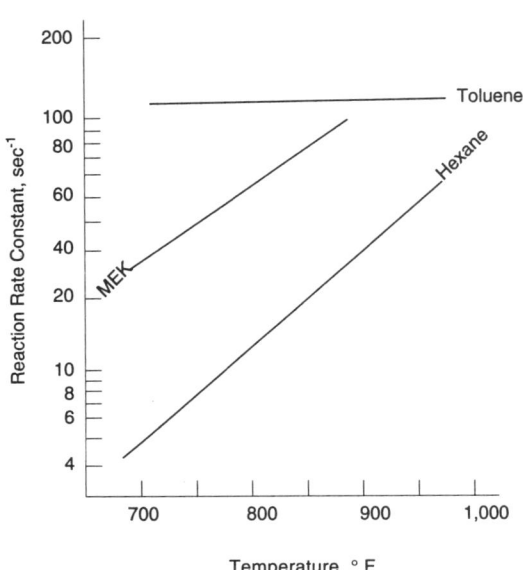

The gas flow rate at a specified gas velocity determines the cross-sectional area of the catalyst bed. Therefore, knowing the total gas flow and retention period, and assuming a reasonable gas velocity, the dimensions of the catalyst bed can be determined. For example, consider a total gas flow of 13,300 scfm. At a catalytic reaction temperature of 900°F, the volumetric flow of gas entering the catalyst bed would be 34,600 acfm. Assuming a gas velocity of 15 ft/sec, the required cross-sectional area of the catalyst would be 38.5 ft^2. Industrial catalyst elements for this type of service could be made available in standard sizes measuring 30 in. wide by 30 in. long by 4.5 in. deep. Thus, by using a configuration comprising three elements wide by two elements long, a total cross-sectional area of (2.5 • 3.0 • 6), or 45 ft^2, can be provided (see Figure 11.9).

If a 95% conversion efficiency for toluene is required, then the retention period of 0.024 sec, previously calculated, will be necessary. At a gas velocity of 15 ft/sec and dwell time value of 0.024 sec, the required catalyst bed depth would be 15 ft/sec • 0.024 sec • 12 in./ft, or 4.3 in.

Similar to thermal incinerator design concepts, the gas temperature determines the required heat input for the system and controls the reaction rate. There is considerable heat released in the catalyst, which is absorbed by the gases, thereby causing a temperature rise as they pass through the bed. Very approximately, for each gallon of solvent oxidized in the catalyst bed, the release of 120,000 Btu can be

Figure 11.9
Catalytic Incinerator Equipment Configurations

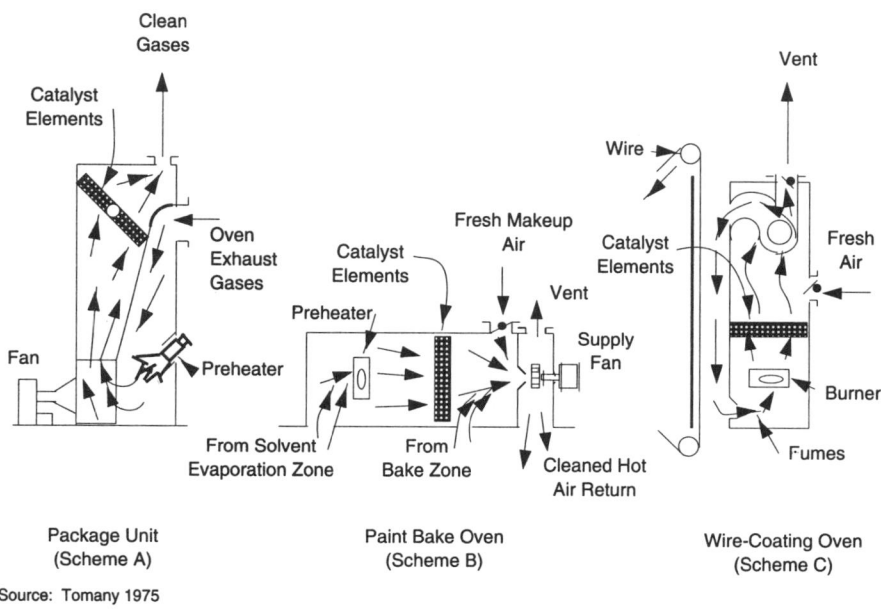

Source: Tomany 1975

expected. However, the burner is usually rated for the total heat duty without reliance on the heat content of the solvents in the contaminated gas stream. The preheat chamber volume is based on heat release rate values in the range of 40,000 to 60,000 Btu/hr ft^3. The burner may be operated at low air/fuel ratios ranging from 7 to 8:1, with the contaminated gas stream providing sufficient oxygen to complete the combustion.

Example 11.4 Catalytic Incinerator Design

The exhaust from a baking oven is 12,000 scfm. The temperature of the exhaust gas is 600°F. The exhaust contains 1.4 lb/min of hexane. Design a catalytic incinerator to meet a performance requirement of 95% converter efficiency. The required bed temperature is 950°F (see Figure 11.9).

Solution:

1. Determine heat load using Equation [11.15].

    Q_h = 12,000 scfm • 0.075 lb/ft^3 • 0.26 Btu/lb°F • (950°F - 600°F) • 60 min/hr

    = 4.91•10^6 Btu/hr

2. Calculate inlet solvent loading.

   (1.4 lb/min) • 60 min/hr • 1 gal/7 lb = 12 gal/hr

3. Calculate heat released.

   At a combustion heat value of 120,000 Btu/gal (approximate value for solvents is used in this design), the total heat released in the catalyst is:

   120,000 Btu/gal • 12 gal/hr = 1.44 • 10^6 Btu/hr

4. Calculate the temperature increase across the catalyst bed.

$$\Delta T = \frac{1.44 \bullet 10^6 \text{ Btu / hr}}{12,000 \bullet 0.075 \bullet 0.26 \bullet 60} = 102.6°F$$

   Thus, the average reaction temperature in catalyst is $950 + \frac{102.6}{2} = 1,001.3°F$

5. Calculate retention time.

   The value of k for hexane at 1,001.3°F is 90 sec^{-1} (Figure 11.8)

$$\frac{100-95}{100} = e^{-90t}$$

$$t = 0.0333 \text{ sec}$$

6. Calculate the system gas flow. Natural gas will be used to heat the bake oven gas to 950°F

   Using Equation [11.16] for Q_A

   Q_A = -0.237 • 950 + 981 = 756 Btu/ft^3

   Using Equation [11.17] to find natural gas required

$$\text{Natural gas required} = \frac{Q_h}{Q_A} = \frac{4.91 \bullet 10^6 \text{ Btu / hr}}{756 \text{ Btu / ft}^3 \bullet 60 \text{ min / hr}}$$

$$= 108 \text{ scfm}$$

7. Calculate the combustion air required.

   (108 scfm)(11.5 air/fuel) = 1,242 scfm

   Note: Use bake oven exhaust to supply the required combustion air.

   Therefore, the total air flow will be :

$$(12,000 + 1,242)\text{scfm}\left(\frac{1,001.3+460}{60+460}\right) = 37,212 \text{ acfm}$$

8. Determine the equipment size.

Based on a velocity through the bed of 20 ft/sec the cross-sectional area will be:

$$\frac{37,212 \text{ acfm}}{20 \bullet 60} = 31 \text{ ft}^2$$

The bed depth required for 0.0333 sec dwell time at 20 ft/sec can be calculated as:

(0.0333 sec) • (20 ft/sec) = 0.67 ft

The total volume of catalyst will be = 0.67 • 31 = 20.8 ft³

Using normal range of 40,000-60,000 Btu/ft³ for the heat release rate, the preheat chamber volume required for preheating the gas stream from 600°F to 950°F before it enters the catalyst bed would be:

$$\frac{4.91 \bullet 10^6 \text{ Btu / hr}}{50,000 \text{ Btu / hr} \bullet \text{ ft}^3} = 98.2 \text{ ft}^3 \quad \text{(preheat chamber volume)}$$

The space velocity (waste gas flow rate) per unit volume of catalyst will be:

$$\frac{(12,000 + 1,242) \text{ scf / min} \bullet 60 \text{ min / hr}}{20.8 \text{ ft}^3} = 38,198 / \text{hr}$$

This corresponds to $\frac{20.8 \text{ ft}^3}{12,000 \text{ scfm}} \bullet 1,000 = 1.73 \text{ ft}^3$ of catalyst per 1,000 scfm of gas stream (see Table 11.6 for comparison).

11.6 Incineration of Chlorinated Hydrocarbons

11.6.1 Introduction

In the combustion of a chlorinated hydrocarbon, hydrochloric acid will be present in the exhaust gas, as well as some free chlorine. Because hydrochloric acid and free chlorine exist in incinerators that process wastes containing chlorinated hydrocarbons, a carbon steel shell may corrode out in a few months in poorly designed incinerators. Furthermore, other combustion products may be formed which are hazardous to health. Because of this, the ultimate goal of chlorinated hydrocarbon incineration is to destroy the waste with as high a destruction efficiency (DE) as possible.

Regulations for the operation of hazardous waste incinerators specify the control of certain pollutant emissions in the stack gas. These regulations (40 CFR, part 261, Subpart O) follow:

1. For each principal organic hazardous constituent (POHC) in the waste stream, there must be at least a 99.99 % ["four nines" destruction and removal efficiency (DRE) in the incinerator and the associated air pollution control devices].

2. At least 99% of the hydrogen chloride must be removed if the emissions are more than 1.8 kg/hr.

3. Particulate emissions must not exceed 180 mg/standard m³ corrected to 7 % oxygen in the stack gas.

4. Wastes containing chlorinated dioxins, chlorinated dibenzofurans, and chlorinated phenols (RCRA codes F020-F028) require a 99.9999% (six nines) DRE of these compounds.

Incinerators that have been used for destruction of hazardous materials include rotary kilns, multiple-chamber incinerators, liquid-injection incinerators, fluidized beds, molten salt devices, plasma destructors, and pyrolysis units. Temperatures and retention times are generally higher than for hydrocarbons in conventional incinerators. Minimum requirements have been established at 2,200°F for two seconds for many applications (pesticides, PCBs). Under these conditions, destruction of greater than 99.99% can be achieved. Cement kilns are often used to dispose of hazardous materials because their temperatures are in excess of 2,500°F and they have long residence times.

11.6.2 Stoichiometry

If the organic compound is designated by $C_xH_yCl_z$ (e.g., x = 1, y = 1 and z = 3 for $CHCl_3$), the overall combustion stoichiometry can be defined for y > z by :

$$C_xH_yCl_z + \left(x - \frac{y-z}{4}\right)O_2 \rightarrow xCO_2 - zHCl - \frac{y-z}{2}H_2O \qquad [11.20]$$

Chlorinated hydrocarbons with hydrogen to chlorine molar ratios less than 5:1 are likely to produce other chlorinated products which are difficult to remove from the stack gases. For $C_xH_yCl_z$ compounds in which y < z, the formation of molecular chloride must be considered:

$$C_xH_yCl_z + xO_2 \rightarrow xCO_2 + yHCl + \frac{z-y}{2}Cl_2 \qquad [11.21]$$

In practical operations, the formation of Cl_2 is undesirable because it is highly corrosive, and it is relatively difficult to remove from the stack gases. For example, assume that trichloroethylene ($CHCl = CCl_2$) is to be incinerated; it will produce the following reaction (Santoleri 1973):

$$CHCl = CCl_2 + 2O_2 \rightarrow 2CO_2 + HCl + Cl_2$$

The HCl formed can be removed by scrubbing with water, but relatively insoluble chlorine will pass through the scrubber, unabsorbed, and into the atmosphere. By addition of natural gas or steam, the chlorine can be converted to HCl by the following:

$$CHCl = CCl_2 + 3.5O_2 + CH_4 \rightarrow 3CO_2 + 3HCl + H_2O$$

$$CHCl = CCl_2 + 3.5O_2 + H_2O + CH_4 \rightarrow 3CO_2 + 3HCl + 2H_2O$$

The use of auxiliary fuel is frequently needed not simply to suppress chlorine formation, but also to increase the heat of combustion of the mixture for effective incinerator operation. In Figure 11.10, the lower heats of combustion are shown as a function of chloride content. It is extremely difficult to stabilize flames using air when the heat of combustion of fuel mixture is less than about 4,000 kcal/kg. It can be seen in Figure 11.10 that the dichloro compound is near the stable combustion limit; thus the trichloro or higher halogenated compounds, or those with higher than 70% weight chlorine, will require auxiliary fuel.

Figure 11.10
Effect of Chlorine Content on Heat of Combustion

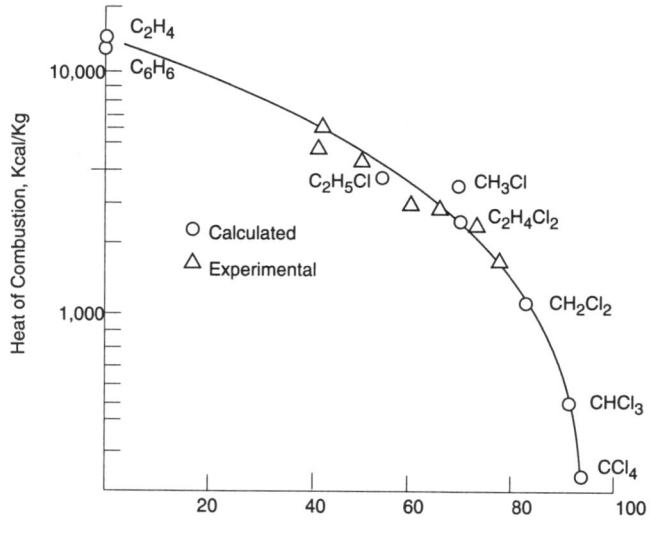

Source: Santoleri 1973

The equilibrium conditions for chlorine, steam, hydrogen chloride, and oxygen are based on the following:

$$H_2O + Cl_2 \rightarrow 2HCl + \frac{1}{2}O_2$$

$$K_p = \frac{(P_{HCl})^2 (P_{O_2})^{1/2}}{(P_{H_2O})(P_{Cl_2})}$$

The relationship between the equilibrium constant, K_p, and temperature is shown in Figure 11.11. Higher temperatures favor HCl formation and are less likely to yield other chlorinated products, which are more difficult to collect. Scrubbing of exhaust gases for acid gas control can be accomplished by packed-towers or venturi scrubbers. This illustrates the need for combustion to take place at a very high temperature, with almost theoretical air/fuel ratios required.

Figure 11.11
HCl-CL$_2$ Equilibrium Constant vs. Temperature

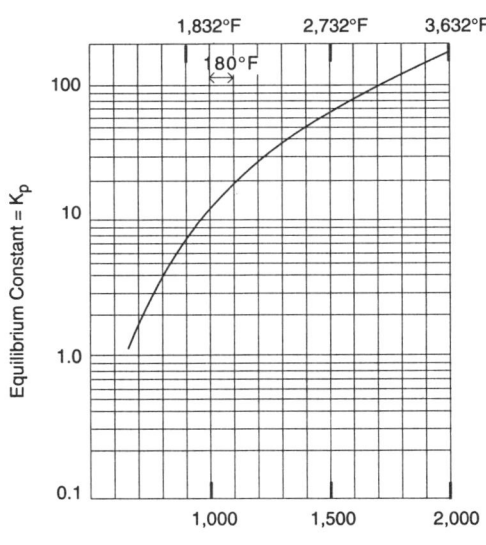

Source: Santoleri 1973

Typical design parameters for air pollution control equipment on hazardous waste incinerators are presented in Table 11.7. Either ESPs or fabric filters (dry systems) are used for particulate control combined with spray dryers for acid gas control. Venturi scrubbers (wet systems) may be used for both particulate and gas control. Commercial hazardous waste incinerators have been designed with hybrid pollution control systems with a goal of achieving the highest possible acid gas removal efficiency and the lowest possible particulate emission rate (Figure 11.12) (Buonicore and Davis 1992).

Table 11.7
Typical Design Parameters for Air Pollution Control Equipment on Hazardous Waste Incinerators

Equipment	Design Parameters
Particulate	
ESPs	SCA = 400-500 ft^2/1,000 acfm
	w = 0.2 ft/sec
Fabric filters	Pulse jet A/C = 3 - 4:1
	Reverse air A/C = 1.5 - 2:1
Venturi scrubbers	(P = 40-70 in.H$_2$O
	L/G = 8 - 15 gal/1,000 acfm
Acid gases	
Packed towers	Superficial velocity = 6 - 10 ft/sec
	Packing depth = 6-10 ft
	L/G = 20-40 gal/1,000 acfm
	Caustic scrubbing medium maintaining pH = 6.5
	Stoichiometric ratio = 1.05
Spray dryers	Low temperature:
	Retention time 15 - 20 sec
	Outlet temperature 250-320°F
	High temperature:
	Retention time 25 - 30 sec
	Outlet temperature 350-450°F
	Stoichiometric ratio (lime) = 2 - 4

11.6.3 Kinetics

From a mechanistic point of view, the destruction of chlorinated organics during incineration occurs by a complex series of reactions that may involve intermediates of greater thermal stability than the parent molecule. However, empirical evidence suggests that the degree of destruction of organic materials may be correlated by relatively simple rate expressions (Theodore and Reynolds 1987). Experimental data suggests that the first-order reaction given by Equation [11.10] can be used by incorporating a delay time as follows:

$$\ln\left(\frac{C_A}{C_{A0}}\right) = -k\theta, \quad \theta \leq \theta_1 \quad [11.22]$$

Figure 11.12
Hazardous Waste Incineration System (Hybrid)

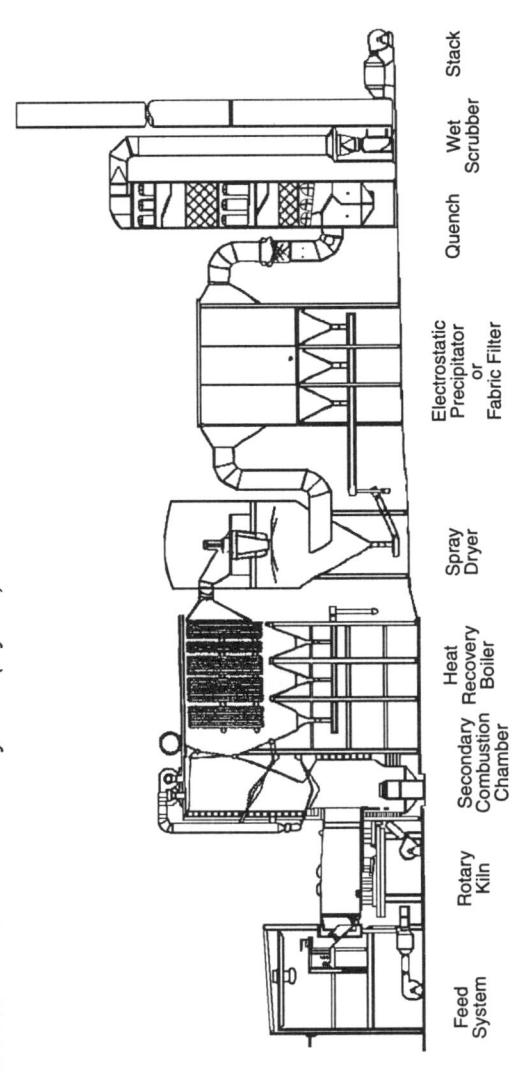

Source: Buonicore and Davis 1992

$$\ln\left(\frac{C_A}{C_{A0}}\right) = -k\theta, \quad \theta = \theta_1 \qquad [11.23]$$

$$\theta_1 = x_1 - x_2 T_f \qquad [11.24]$$

where: θ = average residence time in the reactor, hr

θ_1 = an apparent delay time (a lag prior to onset of the oxidation reaction)

x_1, x_2 = constant (hr and hr/°F, respectively)

T_f = temperature, °F

k = rate constant, hr^{-1}

$$k = Ae^{-E/RT} \qquad [11.25]$$

where: E = activation energy, Btu/lb mol

R = universal gas constant 1.987 Btu/lb mol °R

T = absolute temperature, °R

The estimated parameters and some results are shown in Table 11.8 for a variety of organic compounds. The lower limiting temperature is indicated for each compound. Use of the parameters at temperatures lower than this value or at temperatures higher than about 1,700°F to 1,800°F is not recommended.

Table 11.8
Thermal Oxidation Parameters

Compounds	A, sec^{-1}	E, kcal/mol	x_1	x_2	Lower Limiting Temp (°F)	Auto-ignition Temp (°F)	Calculated Destruction Temp from Eq. [11.22] $T_{99.99}$ at 1 s, °F	$T_{99.99}$ at 2 s, °F
Acrolein	3.30 • 10^{10}	35.9	0	0	800	453	1,020	975
Acrylonitrile	2.13 • 10^{12}	52.1	0.375	0.000250	1,250	898	1,345	1,297
Allyl alcohol	1.75 • 10^{6}	21.4	1.971	0.00146	1,050	713	1,176	1,077
Allyl chloride	3.98 • 10^{7}	29.1	0.525	0.00035	1,150	905	1,276	1,200
Benzene	7.43 • 10^{21}	95.9	2.59	0.00178	1,275	1,044	1,251	1,322
1-Butene	3.74 • 10^{14}	58.2	2.12	0.00157	1,150	723	1,232	1,195
Chlorobenzene	1.34 • 10^{17}	76.6	1.22	0.0008	1,350	1,180	1,408	1,372
1,2-Dichloroethane	4.82 • 10^{11}	45.6	0.948	0.00073	1,050	775	1,216	1,173
Ethane	5.65 • 10^{14}	63.6	7.33	0.0052	1,275	959	1,368	1,328

Table 11.8 (cont.)
Thermal Oxidation Parameters

Compounds	A, sec^{-1}	E, kcal/mol	x_1	x_2	Lower Limiting Temp (°F)	Auto-ignition Temp (°F)	Calculated Destruction Temp from Eq. [11.22] $T_{99.99}$ at 1 s, °F	$T_{99.99}$ at 2 s, °F
Ethanol	5.37 • 10^{11}	48.1	2.10	0.0015	1,250	793	1,307	1,256
Ethyl acrylate	2.19 • 10^{12}	46.0	0	0	1,000	721	1,132	1,092
Ethylene	1.37 • 10^{12}	50.8	0	0	1,200	842	1,328	1,281
Ethyl formate	1.90 • 10^{16}	44.7	0.325	0.00024	1,100	851	1,191	1,145
Ethyl mercaptan	5.20 • 10^{5}	14.7	1.87	0.0022	700	570	778	704
Methane	1.68 • 10^{11}	52.1	1.90	0.00117	1,200	999	1,545	1,486
Methyl chloride	7.43 • 10^{8}	40.9	1.518	0.00084	1,500	1,170	1,597	1,514
Methyl ethyl ketone	1.45 • 10^{14}	58.4	1.92	0.00136	1,200	960	1,290	1,247
Propane	5.25 • 10^{19}	85.2	1.02	0.000686	1,200	871	1,330	1,300
Propylene	4.63 • 10^{8}	34.2	4.54	0.00323	1,200	851	1,318	1,247
Toluene	2.28 • 10^{13}	56.5	1.35	0.000922	1,275	997	1,340	1,295
Triethylamine	8.10 • 10^{11}	43.2	1.20	0.0010	950	450	1,101	1,058
Vinyl acetate	2.54 • 10^{9}	35.9	1.076	0.00078	1,150	800	1,223	1,164
Vinyl chloride	3.57 • 10^{14}	63.3	0	0	1,250	882	1,371	1,332

Equations to predict the time-temperature kinetic rate for untested compounds applicable in the temperature ranges 1,200°F to 1,800°F were developed based on the data shown in Table 11.8. Temperatures required for three destruction levels (99.0, 99.9, and 99.99%) and four residence times (0.5, 1, 1.5 and 2 sec) were calculated (Theodore and Reynolds 1987). These temperatures were correlated with the characteristic compound structures, the autoignition temperature, and the residence times. The temperatures required are designated as T_{99}, $T_{99.9}$, $T_{99.99}$. The temperature models that appear in Table 11.9 were developed using multiple linear regression.

11.6.4 Temperature Control in Hazardous Waste Incinerators

Many incinerators burning chlorinated organics with a high heat value use dilution air (excess air) to limit the furnace temperature. This cooling is required because the material to be incinerated has a high heat content (>12,000 Btu/lb) and the resulting furnace temperatures will be considerably higher than most refractory-lined chambers can tolerate. Uncooled furnace walls constructed of refractory material normally require the furnace gas temperatures not to exceed 1,800°F to 2,200°F. Furnace temperature control, therefore, becomes of primary importance in the operation of hazardous waste incinerators burning high heat content

Table 11.9
Predictive Destruction Temperature Models

$T_{99.00} = 577 - 10.0V_1 + 110.2V_2 + 67.1V_3 + 72.6V_4 + 0.586V_5 - 23.4V_6 - 430.9V_7 + 85.2V_8 - 82.2V_9 + 65.5V_{10} - 76.1V_{11}$

$T_{99.90} = 594 - 12.2V_1 + 117.0V_2 + 71.6V_3 + 80.2V_4 + 0.592V_5 - 20.2V_6 - 420.3V_7 + 87.1V_8 - 66.8V_9 + 62.8V_{10} - 75.3V_{11}$

$T_{99.99} = 605 - 13.8V_1 + 122.5V_2 + 75.7V_3 + 85.6V_4 + 0.597V_5 - 17.9V_6 - 412.0V_7 + 89.0V_8 - 55.3V_9 + 60.7V_{10} - 75.2V_{11}$

	Variables
$T_{99.00}$ = 99.00% destruction temperature, °F	V_5 = autoignition temp, °F
$T_{99.90}$ = 99.90% destruction temperature, °F	V_6 = number of oxygens
$T_{99.99}$ = 99.99% destruction temperature, °F	V_7 = number of sulfurs
	V_8 = hydrogen/carbon ratio
V_1 = number of carbons	V_9 = allyl compound (0 = no, 1 = yes)
V_2 = aromatic (0 = no, 1 = yes)	V_{10} = carbon double bond, chlorine interaction (0 = no, 1 = yes)
V_3 = carbon-carbon double bond (0 = no, 1 = yes)	V_{11} = ln (time in s)
V_4 = number of nitrogens	

material. Cooling is usually accomplished by using sufficient excess air to produce the desired temperature. The quantity of excess air can be determined by calculating the mass of combustion gas required to absorb the heat released by the contaminated material (see Chapter 3):

When the contaminant is burned, mass must be conserved:

$$G_f = G + G_E$$

$$Q_f = G_f C_p (T_{out} - T_{in})$$

$$= (G + G_E)(C_p)(T_{out} - T_{in})$$

G_f = total mass of flue gas (lb/lb of contaminant)

G = flue gas for theoretical combustion per lb of contaminant (lb/lb of contaminant)

G_E = flue gas from excess air to maintain furnace temperature (lb/lb of contaminant)

Q_f = net heating value of contaminant (Btu/lb)

Example 11.5 Excess Air Requirement Determination

A contaminant is to be burned that has a net heating value of 14,000 Btu/lb. The furnace temperature required for 99.99% destruction has been calculated to be 1,400°F. When one lb of the contaminant is burned, the stoichiometric air required for combustion is 13 lb air/lb of contaminant. Calculate the excess air required to maintain the incinerator temperature at 1,400°F.

Fuel energy:

$$14{,}000 \text{ Btu/lb} = (G_E - 1 \text{ lb contaminant} + 13 \text{ lb air/lb contaminant}) \cdot (0.27 \text{ Btu/lb °F}) \cdot (1{,}400°F - 60°F)$$

$$G_E = \frac{14{,}000}{0.27 \bullet 1{,}340} - 14.0$$

$$= 24.7 \text{ lb/lb contaminant}$$

Excess air is $(24.7/14.0) \cdot 100 = 176\%$ (This is substantially greater than the excess air normally found necessary for proper combustion).

The temperature of the flue gas without excess air would be:

$$14{,}000 \text{ Btu/lb} = [(14 \text{ lb flue gas})/(\text{lb contaminant})] \cdot (0.27 \text{ Btu/lb°F}) (T - 60°F)$$

$$T = \frac{14{,}000}{14 \bullet 0.27} + 60 = 3{,}643°F$$

If a temperature of 2,200°F is used for combustion based on regulation requirement, then

$$G_E = \frac{14{,}000}{0.27 \bullet 2{,}140} - 14.0$$

$$= 10.2 \text{ lb/lb contaminant}$$

$$\text{excess air} = 10.2/14.0 = 72.8\%$$

Example 11.6 Incineration of VOCs (King 1996)

Design an incinerator to dispose of chlorobenzene (C_6H_5Cl) at a feed rate of 500 lb/hr. Data describing chlorobenzene is listed below. Assume chlorobenzene follows a first-order kinetic model for destruction with the rate constant in the data listed in Table 11.4. A destruction efficiency of 99.9% must be achieved. Combustion is to be accomplished at a minimum of 1,400°F with 50% excess air. Natural gas is available as a supplemental fuel, if required.

Chapter 11: Incineration

Chlorobenzene

Molecular Weight = 112.6

Vapor pressure = 12 mm Hg

Autoignition temperature = 1,245°F

Heat of combustion = 14,000 Btu/lb

k (1st order at 1,400°F) = 8.4 sec^{-1}

Natural gas

Available heat at 4,400°F = 615 Btu/ft^3

Solution:

1. Determine the size of the incinerator.

    a) Find the residence time based on first order kinetics (Equation 11.10).

    $$kt = \ln\left(\frac{C_{Ao}}{C_A}\right)$$

    $$t = \frac{1}{k}\left(\frac{C_{Ao}}{C_A}\right)$$

    Note: C_A can be in concentration or mass.

    $t = (1/8.4)\text{sec}^{-1} \cdot \ln[500/(500 \cdot 0.0001)]$

    $= 1.1$ sec

    From Equation [11.22]:

    $\theta_1 = x_1 - x_2 T_f$

    $x_1 = 1.22, \ x_2 = 0.0008$

    $k = 1.84 \text{ sec}^{-1}$

    $\theta_1 = 1.22 - 0.0008 \cdot 1,400 = 0.1$ sec

    $(\theta - \theta_1) = t$

    $(\theta - 0.1) = 1.1$

    $q = 1.2$ sec

    Choose t = 2.0 sec to provide safety factor.

b) Determine the gas volume.

Begin with a calculation of combustion gases. It is necessary to modify Equation [11.20] to account for the one hydrogen atom of benzene being replaced by one chlorine atom. Recognizing that chlorine forms HCl in the product stream, it is then possible to use the combustion formula as follows:

$$C_6H_5Cl + 7O_2 + 7(3.76)N_2 \rightarrow 6CO_2 + 2H_2O + HCl + 26.32N_2$$

$$\text{Chlorobenzene} = 500\frac{\text{lb}}{\text{hr}} \text{ or } \frac{500 \text{ lb/hr}}{112.5 \text{ lb/lb mol}}$$

$$= 4.44 \text{ lb mol/hr}$$

The products of combustion are:

$$CO_2 = 6\frac{\text{lb mol } CO_2}{\text{lb mol } C_6H_5Cl} \cdot 4.44 \text{ lb mol } C_6H_5Cl/\text{hr}$$

$$= 26.64 \text{ lb mol } CO_2/\text{hr}$$

Similarly:

$$H_2O = 2(4.44) = 8.88 \text{ lb mol/hr}$$

$$HCl = 1(4.44) = 4.44 \text{ lb mol/hr}$$

$$N_2 = 26.32(4.44) = 117 \text{ lb mol/hr}$$

The foregoing did not include the excess air necessary to achieve the desired destruction. Therefore, to include excess air:

$$N_2 = 1.5(117) = 175 \text{ lb mol/hr}$$

$$O_2 = \frac{1}{2}\left(7\frac{\text{lb mol } O_2}{\text{lb mol } C_6H_5Cl}\right) \cdot 4.44 \frac{\text{lb mol } C_6H_5Cl}{\text{hr}}$$

$$= 15.54 \text{ lb mol/hr}$$

$$\text{Total gases} = 26.64 + 8.88 + 4.44 + 175 + 15.54 \text{ lb mol/hr}$$

$$= 230.5 \text{ lb mol/hr}$$

Using the Ideal Gas Law, (PV =nRT), the volume of gas is calculated by first rearranging the equation to solve for V.

$$V = nRT/P$$

where: $P = 1$ atm

$T = 1,400 \text{ °F}$

$R = 0.7302$ atm ft^3/lb mol °R

$V = \dfrac{(230.5 \text{ lb mol/hr})(0.7302 \text{ atm ft}^3/\text{lb mol °R})(1{,}860°R)}{1 \text{ atm}}$

$V = 313{,}060$ ft^3/hr or 86.9 ft^3/sec

   c) Using a residence time in the incinerator of 2 sec, the size of the incinerator is equal to 2 sec • 86.9 ft^3/sec. or 174 ft^3.

2. Determine the requirements for supplemental fuel, if any, and estimate the temperature in the incinerator.

   a) Conduct a heat balance to determine the need for supplemental fuel to achieve 1,400°F.

$$\text{Heat}_{in} = \text{Heat}_{out} + \text{losses}$$

Heat$_{in}$ = C$_6$H$_5$Cl Heat of Combustion + Air Heat of Combustion where the Air of combustion = (O$_2$+N$_2$)1.5. From part 1., it was determined that there are 175 lb mol of N$_2$ and 15.54 lb mol of O$_2$ with 50% excess air. Therefore, the pounds of air introduced to the incinerator is:

N$_2$ = 175 lb mol/hr • 28 lb/lb mol = 4,900 lb/hr

$O_2 = 1.5\left(7\dfrac{\text{lb mol O}_2}{\text{lb mol C}_6\text{H}_5\text{Cl}}\right) \bullet 4.44 \dfrac{\text{lb mol C}_6\text{H}_5\text{Cl}}{\text{hr}} \bullet 32 \dfrac{\text{lb}}{\text{lb mol}}$

Air = 1,492 lb/hr + 4,900

     = 6,392 lb/hr (Air has specific enthalpy of 33.6 Btu/lb at 200°F)

Heat$_{in}$ = 500 lb/hr • 14,000 Btu/hr + 6,392 lb/hr (33.6 Btu/lb)

     = (7 • 10^6 + 0.215 • 10^6) Btu/hr

     = 7.215 • 10^6 Btu/hr

Heat$_{out}$ = Heat in products at 1,400°F + Losses

Mass out must equal mass in and heat losses are estimated at 25 % of heat input, therefore:

Heat$_{out}$ = 343 Btu/lb • (6,392 lb/hr + 500 lb/hr) + 0.25 (7.215 • 10^6 Btu/hr)

     = 2.36 • 10^6 Btu/hr + 1.80 •10^6 Btu/hr

     = 4.16 • 10^6 Btu/hr

The preceding calculation reveals that burning chlorobenzene will produce more than enough energy to achieve the minimum temperature, 1,400°F. In fact, the incinerator will be hotter than 1,400°F, which enables a reduction in the required residence time, and therefore the incinerator size. However, this excess energy could damage the incinerator if the temperatures in it are too high. The final temperature can be estimated as follows:

Mass • (specific enthalpy at $t_d = 2$) = Available heat

$$6{,}392 \text{ lb/hr} \cdot (\text{enthalpy}) = (7.215 \cdot 10^6) - (4.16 \cdot 10^6) \text{ Btu/hr}$$

$$\text{enthalpy} = 478 \text{ Btu/lb}$$

By interpreting an enthalpy chart for air at various temperatures, it is found that an enthalpy of 478 Btu/lb corresponds to a temperature of approximately 1,800°F. This temperature provides a safety factor to ensure destruction of the C_6H_5Cl, but it is not so high as to damage a well-designed incinerator. To maintain this temperature will require more than 50% excess air (see Example 11.8).

3. Identify any air pollution concerns for the incinerator and describe the best approach to achieve compliance.

The air pollution concern presented by the destruction of C_6H_5Cl is the emission of HCl. The rate of emission is:

$$500 \frac{\text{lb } C_6H_5Cl}{\text{hr}} \cdot \frac{36.5 \text{ lb HCl}}{112.6 \text{ lb } C_6H_5Cl} = 162 \text{ lb/hr}$$

Typically, RCRA regulations would require treatment to reduce HCl emissions to 4 lbs/hr or 99% removal. Since HCl is very soluble, removal could be easily accomplished using a wet scrubber system.

11.7 Approximate Calculations

Some assumptions can be made to simplify the rigorous approach used for the design of hazardous waste incinerators since the combustion calculations are somewhat cumbersome. A simpler set of equations can then be developed that are valid for the purpose of engineering calculations (Theodore and Reynolds 1987).

The key assumptions are as follows:

1. Although the products of combustion consist of many components, the primary components are nitrogen, carbon dioxide, and water (vapor). The average heat capacity of the combined mixture (not including the excess air) may be assigned a value of approximately 0.3 Btu/lb°F. For this condition,

$$\Delta H_p = m_p (0.3)(T - T_o) \quad [11.26]$$

where: m_p = mass of stoichiometric air, fuel, and waste entering the incinerator per unit mass of waste-fuel mixture, or equivalently, mass of products less that of excess air per unit mass of waste-fuel mixture (lb/lb mixture).

2. The average heat capacity of the (excess) air is ~ 0.27 Btu/lb°F over the temperature range 60°F to 2,000°F. If this value is rounded to 0.3,

$$\Delta H_e = m_e (0.3)(T - T_o) \quad [11.27]$$

where: m_e = mass of excess air per unit mass of waste-fuel mixture (lb/lb mol).

3. The stoichiometric air requirements for the combined waste-fuel mixture (ft³ air/lb mixture) divided by NHV (Net Heating Value), for many hydrocarbons, is approximately 0.01 ft³ air/Btu (or 100 Btu/ft³ air).

Using the density of air at 60°F, this ratio can be converted to approximately 750 lb air/10^6 Btu or $7.5 \cdot 10^{-4}$ lb air/Btu. Thus, for this condition, the stoichiometric air requirement, m_{st}, is given by:

$$m_{st} = 7.5 \cdot 10^{-4} \text{NHV} \quad [11.28]$$

The equations based on these assumptions are for 1.0 lb of the combined waste-fuel mixture, that is:

$$m_w + m_f = m_b = 1.0 \quad [11.29]$$

where: m_w = mass of w (waste) per unit mass of waste-fuel mixture (lb/lb mixture)

w, f, b = subscripts indicating waste, fuel, and burnable mixture, respectively

The approximate enthalpy balance is then:

$$\text{NHV} = (1.0 + 7.5 \cdot 10^{-4} \text{NHV})(0.3)(T - 60) + m_e (0.3)(T - 60) \quad [11.30]$$

where: T_o = 60°F, there are no heat losses, and the cooling cycle is neglected.

The fraction of excess air, EA, is defined by:

$$EA = m_e / m_{st} \quad [11.31]$$

$$m_e = (EA)(7.5 \bullet 10^{-4})(NHV) \quad [11.32]$$

and

$$NHV = [1 + (EA)(7.5 \bullet 10^{-4})(NHV) + (7.5 \bullet 10^{-4})(NHV)](0.3)(T - 60)] \quad [11.33]$$

This may be further simplified to

$$NHV = [1 + (1 + EA)(7.5 \bullet 10^{-4})(NHV)(0.3)(T - 60)] \quad [11.34]$$

Figure 11.13 provides a plot of Equation [11.34].

Three parameters or variables appear in Equation [11.34]. Specifying any two of these constitutes a complete set for this equation. Thus, three equations may be generated expressing one variable (dependent) in terms of the other two (independent),

$$T = f_1(EA, NHV)$$
$$EA = f_2(T, NHV) \quad [11.35]$$
$$NHV = f_3(T, EA)$$

The three resulting equations are:

$$T = 60 + \frac{NHV}{(0.3)\left[1 + (1 + EA)(7.5 \bullet 10^{-4})(NHV)\right]} \quad [11.36]$$

$$EA = \frac{[NHV / \{0.3(T - 60)\}] - 1}{7.5 \bullet 10^{-4} NHV} - 1 \quad [11.37]$$

$$NHV = \frac{0.3(T - 60)}{1 - (1 + EA)(7.5 \bullet 10^{-4})(0.3)(T - 60)} \quad [11.38]$$

The units of T and NHV are °F and Btu/lb, respectively — EA is a dimensionless fraction.

Example 11.7 Minimum Net Heating Value

A fuel-waste mixture is to be combusted in an incinerator at an operating temperature of 2,000°F. Calculate the minimum Net Heating Value (NHV) of the mixture

Chapter 11: Incineration   423

Figure 11.13
Operating Temperature as a Function of Excess Air and Net Heating Value According to Equation [11.34]

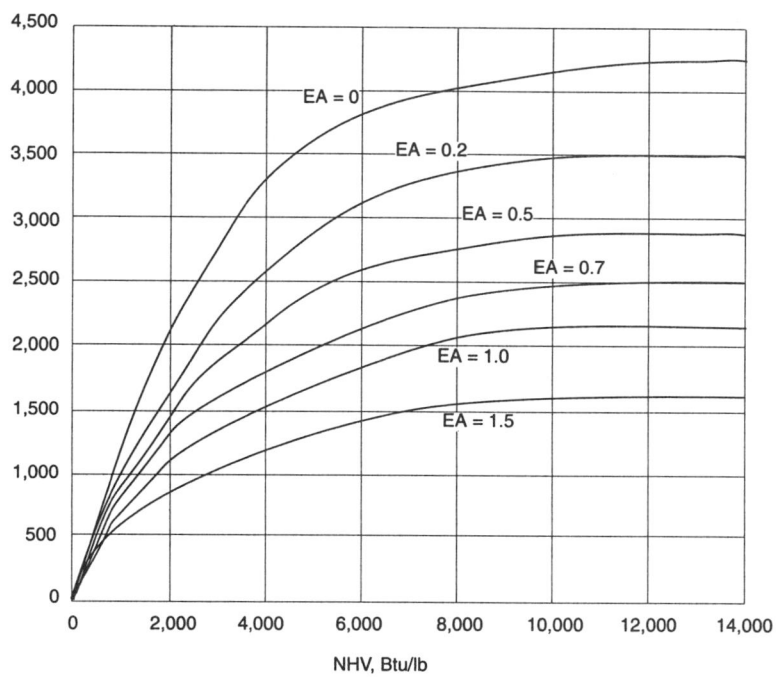

NHV, Btu/lb

in Btu/lb if 0, 20, 40, 60, 80, and 100% excess air is employed. Use Equation [11.38] to perform the calculation.

Solution:

Referring to Equation [11.38]

$$NHV = \frac{0.3(T-60)}{1-(1+EA)(7.5 \bullet 10^{-4})(0.3)(T-60)}$$

For 0% excess air,

$$NHV \text{ (0% excess air)} = \frac{0.3(2,000-60)}{1-(1)(7.5 \bullet 10^{-4})(0.3)(2,000-60)}$$

$$= 1,032.8 \text{ Btu/lb}$$

For 20% excess air,

$$NHV \text{ (20% excess air)} = \frac{0.3(2,000-60)}{1-(1.2)(7.5 \bullet 10^{-4})(0.3)(2,000-60)}$$

$$= 1,222 \text{ Btu/lb}$$

Similarly,

   NHV (40% excess air) = 1,496.5 Btu/lb

   NHV (60% excess air) = 1,929.7 Btu/lb

   NHV (80% excess air) = 2,715.8 Btu/lb

   NHV (100% excess air) = 4,582.7 Btu/lb

Example 11.8 Incineration of VOCs Approximate Calculation Procedure

Redo Example 11.6 using the approximate calculation procedure to determine the maximum allowable Btu/lb of the fuel to maintain a temperature of 1,400°F with 50% excess air. Also, calculate the excess air required to burn C_6H_5Cl with a NHV of 14,000 Btu/lb and maintain a 1,400°F temperature in the burner.

Solution:

Maximum Btu/lb allowable with 50% excess air and 1,400°F

Using Equation [11.38]
$$NHV = \frac{0.3(T-60)}{1-(1+EA)(7.5 \bullet 10^{-4})(0.3)(T-60)}$$

$$NHV = \frac{0.3(1,400-60)}{1-(1+0.5)(7.5 \bullet 10^{-4})(0.3)(1,400-60)}$$

$$NHV = 733.9 \text{ Btu / lb}$$

Air required with 14,000 Btu/lb fuel and 1,400°F

Using Equation [11.37]
$$EA = \frac{[NHV/\{0.3(T-60)\}]-1}{7.5 \bullet 10^{-4} NHV} - 1$$

$$EA = \frac{[14,000/\{0.3(1,400-60)\}]-1}{7.5 \bullet 10^{-4} 14,000} - 1$$

$$EA = \frac{34.83-1}{10.5} - 1$$

$$EA = 2.22$$

If a furnace temperature of 1,800°F was used, then the excess air (EA) would be 1.46.

11.8 Summary-Incinerator Design

The destruction of an organic compound is a function of temperature, residence time at that temperature, and the properties of the compound. The rate of destruction can be considered first order for most VOC incineration conditions (i.e., dilute emissions, 25% of LEL).

$$\frac{dC}{dt} = -kC$$

where: C = concentration at time t, sec

k = rate constant

Upon integration of this equation:

$$t = \frac{1}{k} \ln \frac{C_{Ao}}{C_A} \qquad [11.39]$$

where: $C_A = C_{Ao}$ at t = 0

Expressing the rate constant, k, in Arrhenius form:

$$k = A e^{-\frac{E}{RT}} \qquad [11.40]$$

combining Equations [11.24] and [11.25]:

$$\eta = 1 - \exp\left[-A \bullet t \bullet e^{\{-E/(RT)\}}\right] \qquad [11.41]$$

and solving for incinerator temperature:

$$T = E / \left[R \bullet \{\ln(t) + \ln[-\ln(1-\eta)] / A\}\right] \qquad [11.42]$$

where: η = destruction efficiency

A = experimental constant, sec^{-1}

E = activation energy, cal/g mol

R = universal gas constant, 1.987 cal/g mol °K

T = incinerator temperature, °K

The higher the residence time, the lower is the temperature required for destruction.

A plug-flow model can approximate furnace characteristics. In plug flow, without temperature change (isothermal) along the flow path, the mean residence time, t, is calculated as follows:

$$t = (V/Q) \cdot 60 \qquad [11.43]$$

where: t = mean residence time, sec

V = furnace volume, ft^3

Q = volumetric flow rate, acfm

11.9 Problem Set

11.1 For the following equation: $C_3H_8 + 5O_2 \rightarrow 3CO_2 + 4H_2O + Q(\text{heat})$

How many scf of air are needed for complete combustion of one scf of propane (C_3H_8)?

11.2 Calculate the flue gas composition for the theoretical combustion of natural gas (methane) with air.

11.3 Write the chemical equation for the reaction of n-butane (C_4H_{10}) with a stoichiometric amount of air. Determine the theoretical air/fuel ratio for this reaction on both a molal and mass basis.

11.4 Determine the scfm of natural gas (heating value H_a = 1,050 Btu/scf) required to heat 8500 scfm of a contaminated gas stream at 200°F to 1,400°F. Assume that there are no heat losses and that the average heat capacity of air over the temperature range is 7.5 Btu/lb mol°F.

11.5 Given the following information, determine the outlet temperature of an incinerator.

heat input = $18.7 \cdot 10^6$ Btu/hr

total heat flow = 72,000 lb/hr

air inlet temp = 200°F

average C_p = 0.26 Btu/lb°F

11.6 The heat of combustion of methane ($CH_4 + 2O_2 \rightarrow CO_2 + 2H_2O$) is 212,800 cal/g mol when the water formed is liquid. Convert this to Btu/ft^3 of CH_4 when the water is in the vapor phase.

11.7 Determine the size of an incinerator to operate with an average energy release rate of 25,000 Btu/hr ft^3 of incinerator volume. It is estimated that during operation, 5,000 lb/hr of waste with an approximate heating value of 8,000 Btu/lb is to be combusted. Assume the incinerator to be a rotary kiln with L/D ratio of 3.5.

11.8 A vertical liquid injection incinerator is operating at 2,200°F with a combustion gas flow rate of 6,000 scfm (60°F). The incinerator is rectangular in shape and is 10 ft wide (across the front), 11 ft deep, and 28 ft high. The minimum residence time required for incineration is 2 sec. Calculate a) the actual combustion gas flow rate; b) the volume required for minimum residence time; and c) the maximum residence time.

11.9 Estimate the temperature required to destroy 99.9% of an organic compound using a residence time of 0.5 sec. E (activation energy) and A (experimental constant) are 45.2 kcal/g mol and $9 \cdot 10^{10}$ sec^{-1}, respectively.

11.10 In order to meet pollution specifications for discharging hydrocarbons to the atmosphere, a gas stream must be reduced by 99.5%. It is proposed to meet this requirement by combusting the hydrocarbons in a thermal reactor operating at 1,500°F. The gas and methane (fuel) are to be fed to the reactor at 80°F and 1 atm. Design the proposed reactor using kinetic principles. The following data are available:

> flue gas flow rate from fuel combustion = 2,500 scfm
>
> process flow rate = 7,200 scfm
>
> hydrocarbon = toluene
>
> average velocity = 20 ft/sec

11.11 A combustible waste gas from a manufacturing process is burned in a natural gas boiler. The boiler burns 5,000 ft^3/hr of natural gas and 10 lb/hr of benzene (the waste gas). Determine the gross heating value per hour for the boiler. Determine the available heat if the flue gas exits at 600°F and 20% excess air is used for combustion.

11.12 An exhaust air stream from a meat smokehouse contains obnoxious odors and fumes. The exhaust is 5,000 acfm at 90°F and the fumes are to be incinerated at 1,200°F. How much gas will be required if the gross heating value of the fuel is 1,059 Btu/scf ?

11.13 Thermal incineration is to be used to remove odor and fumes from a 10,000 acfm air stream. The air enters at 200°F. Natural gas fuel is used to raise the preheated air to 1,400°F. The mean heat capacity of both the air and exhaust gases is 0.26 Btu/lb °F. The heating value of the natural gas is 950 Btu/ft^3 and 1 ft^3 of natural gas yields 11.5 ft^3 of combustion products at 60°F and 1 atm. 10% of heat is lost by radiation. Estimate the fuel requirements and combustion chamber dimensions.

11.14 Trace amounts of xylene are discharged into the atmosphere as the result of a particular chemical process. Since an economical process gas having properties similar to natural gas is available, the company has decided to incinerate the effluent in order to comply with state pollution regulations. You have been asked to perform the study. The properties of the effluent are:

flow rate = 4,000 scfm
xylene concentration = 500 ppm
temperature = 300°F

The reactor section of the incinerator is to be operated at the following conditions:

temperature = 1,400 °F
residence time = 0.3 sec
primary air = 50% excess
constant velocity = 20 ft/sec

Assume the rate of consumption of pollutant is first order with a rate constant of:

$$k = Ae^{-\left(\frac{B}{T}\right)}$$

where A and B are rate constants, which for xylene and air are given by the company as $A = 20.6 \text{ sec}^{-1}$, $B = 33,500 /°R$. The incinerator must reduce the xylene concentration to 10 ppm or less.

a) Estimate the consumption rate of fuel gas.
b) Estimate the length and cross sectional area of the incinerator.

11.15 An asphalt roofing manufacturer collects the air above the tanks and rollers in a hood-duct system at the rate of 10,000 cfm at 90°F and 1 atm. It contains 2 grains/ft^3 of micron-size oil and asphalt particles. The company has decided to investigate thermal incineration as a means of controlling both odor and particulate emissions. a) Estimate the natural gas requirement for incineration. b) Estimate the required throat and chamber dimension for the afterburner.

11.16 A 3,000 scfm air stream has an organic compound that can be incinerated at 1,800°F in a catalytic incinerator. The fumes are below the lower explosive limits. The fuel will be natural gas with a heating value of 1,050 Btu/scf. The initial temperature of the air stream is 100°F. Determine the % of natural gas saved in the catalytic incinerator.

11.17 A metal decorating plant is using a baking oven to dry painted components. The following information is available concerning the plant operation and its emissions:

exhaust volume, cfm = 6,000
exhaust temperature, °F = 400
solvents loading, lb/min = 1.4

solvents composition, wt%

$$\text{hexane} = 35$$
$$\text{toluene} = 45$$
$$\text{MEK} = 20$$
$$\text{operating period, 24 hrs per day/yr} = 365$$
$$\text{fuel type} = 0.60$$
$$\text{fuel costs, \$/1,000 scf} = 98$$

Design both a thermal and catalytic incinerator for this application. The thermal incinerator will have an exit temperature of 1,450°F. Compare the fuel costs for the two systems. The reaction rate constants for the solvents are given in Figures 11.8 and the figure below.

Reaction Rate Constants for Thermal Incineration of Various Solvents

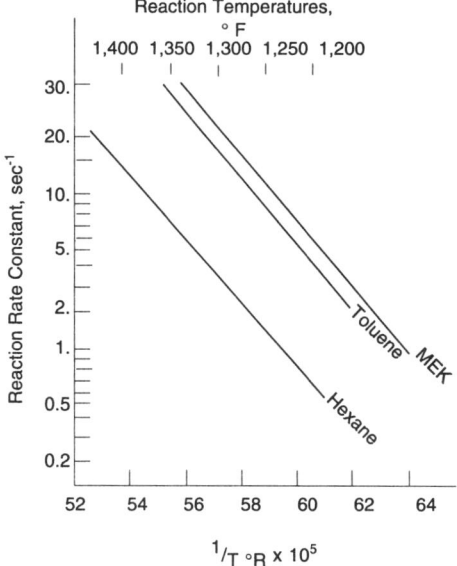

Source: Tomany 1975

11.18 A chemical plant generates chlorobenzene (C_6H_5Cl). A liquid inject incinerator is to be employed to burn the waste. Pilot studies indicate that burning with 50% excess air will satisfy the four nines (99.99% DRE) regulatory requirements. The maximum capacity of the incinerator will be 5,000 lb C_6H_5Cl/hr. Provide a preliminary design for this incinerator.

11.19 A fuel-waste mixture is to be combusted in a hazardous waste incinerator at 2,200°F for 2 sec. The net heating value of the mixture is 3,000 Btu/lb of feed. The average velocity in the incinerator is to be 20 ft/sec. Design an incinerator to burn 500 lb/hr of waste. Control the excess air to maintain the required temperature in the exit stack. What size is the incinerator (length and width)? What amount of excess air is required?

11.20 Sludge containing mercury is burned in an incinerator (mercury feed rate 9.2 lb/h). The resulting product at 500°F (40,000 lb/hr, gas, MW = 32) is filtered to remove all particulates. What happens to the mercury? Assume the process pressure is 14.7 psi and that the vapor pressure of Hg at 150°F is 0.005 psi.

11.21 A venturi scrubber manufacturer claims that large soot particles cannot be burned rapidly enough in the combustor of a power plant and that a venturi scrubber system should be used instead. The linear burning rate for combustion of soot particles is given by:

$$t = \rho_p d_p / 2q$$

where: $q = A [\exp(-E/RT)] p$

For the burning rate, $A = 8,710$ g/cm^2 sec atm, $E = 35,700$ cal/mol, and p is oxygen partial pressure of 0.1 atm. Calculate the burning time for coal soot at 1,400, 1,800, and 2,000°F for 1, 10, and 100 µm particle sizes. The density of soot particles is 0.3 g/cm^3. If a typical combustor residence time is 0.5 seconds, are the claims of the venturi scrubber manufacturer correct?

11.10 References

Beard, J., F. Lachetta, and L. Lilleleht. *Combustion Evaluation*. Associated Environmental Consultants. EPA 450/2-80-063. Charlottesville, VA. 1980.

Brunner, C. *Hazardous Waste Incineration*, 2nd ed. McGraw-Hill. New York. 1993.

Buonicore, T. and W. Davis (Eds). *Air Pollution Engineering Manual*. Air and Waste Management Assn. Van Nostrand Reinhold. New York. 1992.

Joseph, G. and D. Beachler. *Control of Gaseous Emissions*. Northrop Services, Inc. EPA 450/2-81-005. Research Triangle Park, NC. 1981.

King, W. C. *Environmental Engineering P.E. Examination Guide and Handbook*. American Academy of Environmental Engineers. Annapolis, MD. 1996.

Santoleri, J. *Chlorinated Hydrocarbon Waste Disposal and Recovery Systems*. Chemical Engineering Progress, 69/01/68. 1973.

Sink, M. *Handbook: Control Technologies for Hazardous Air Pollutants.* Office of Research and Development. US EPA/625/6-91/014. Cincinnati, OH. 1991.

Tomany, J. *Air Pollution: The Emissions, the Regulations, and the Control.* American Elsevier. New York. 1975.

Theodore, L. and J. Reynolds. *Introduction to Hazardous Waste Incineration.* Wiley-Interscience. New York. 1987.

Chapter 12

Absorption

12.1 Introduction

Absorption is the most widely used control method for inorganic gaseous emissions. Absorption is an operation in which one or more components of a gas mixture are selectively transferred into a liquid. Absorption of a gaseous component by a liquid occurs when the liquid contains less than the equilibrium concentration of the gaseous component. The difference between the actual concentration and the equilibrium concentration provides the driving force for absorption. The absorption rate depends on the physical properties of the gas/liquid system (e.g., diffusivity, viscosity, density) and the absorber operating conditions (e.g., temperature, flow rates of the gaseous and liquid streams). It is enhanced by lower temperatures, greater contacting surfaces, higher liquid-gas ratios, and higher concentrations in the gas stream.

Absorption can be physical or chemical. *Physical absorption* occurs when the absorbed compound simply dissolves in the solvent. When there is a reaction between the absorbed compound and the solvent, it is termed *chemical absorption*. If the gaseous contaminant is very soluble, then high removal efficiencies can be achieved by almost any absorption device. For a relatively insoluble contaminant, only certain systems may be able to achieve the required removal efficiency. In some cases, a chemical reagent may have to be added to the absorbing liquid to increase the solubility of the contaminant. These reagents may increase the physical solubility of the contaminant (e.g., sodium citrate added to absorb SO_2) or chemically react with the contaminant (e.g., lime scrubbing of SO_2). Selection of the proper absorbing liquids is based on the efficiency required and the liquid cost. Water is the usual choice because many gaseous contaminants are soluble in it, and it is readily available and relatively inexpensive.

The packed column (Figure 12.1) is the most common scrubber used for gas absorption (Danielson 1973). Packed columns disperse the scrubbing liquid over packing material which provides a large surface area for continuous gas-liquid contact. The gas stream being treated enters the bottom of the tower and flows upward over the packing material. Liquid is introduced at the top of the packing by sprays or weirs and flows downward over the packing material. This flow arrangement results in the highest theoretical achievable efficiency. The most dilute gas is contacted with the purest absorbing liquor, providing a constant, maximized concentration difference (driving force) for the entire length of the column.

Figure 12.1
Typical Countercurrent Pack Column

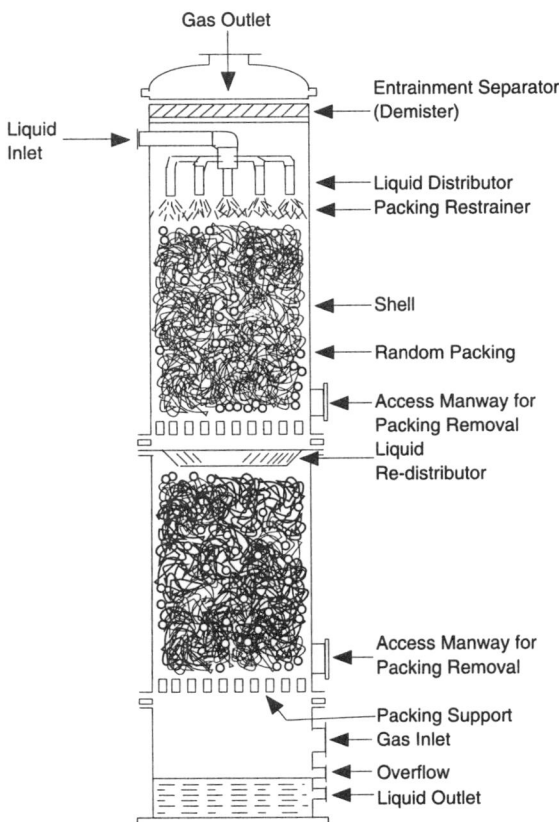

In most applications, the inlet conditions (flow rate, composition, and temperature) are usually known. The composition of the outlet gas is specified by the control requirements. The conditions of the inlet liquid are also known. Therefore, the main objectives in the design of an absorption column are the determination of the solvent flow rate and the calculation of the principal dimensions of the equipment (column diameter and height to accomplish the absorption operation) for a selected solvent.

Packed towers are most suited to applications where a high gas removal efficiency is required and the exhaust stream is relatively free from particulate matter. In the production of both sulfuric and hydrochloric acids, packed towers are used to control tail and exhaust emissions (SO_2 and HCl respectively). The scrubbing liquor for these processes can be a weak acid solution, with the spent liquor from the packed tower sent back to the process. Packed towers are also used to control HCl and H_2SO_4 fume emissions from pickling operations in the primary metals industry. They are also used to control odors in rendering plants, petroleum refineries, and in sewage treatment plants. For odor control, the scrubbing liquor usually contains an oxidizing reagent, such as potassium permanganate or sodium hypochlorite.

Packing material is the heart of the absorber. It provides the surface over which the scrubbing liquid flows. Figure 12.2 shows some of the more commonly used packings. These materials were originally made of stoneware or porcelain but presently a majority are being made of high-density thermoplastics (polyethylene and polypropylene). A specific packing is described by its trade name and overall size. For example, a column can be packed with 2-inch Raschig rings or 1-inch Tellerettes. The overall dimensions of packing materials normally range from 1/4 to 4 inches.

Figure 12.2
Common Packing Material

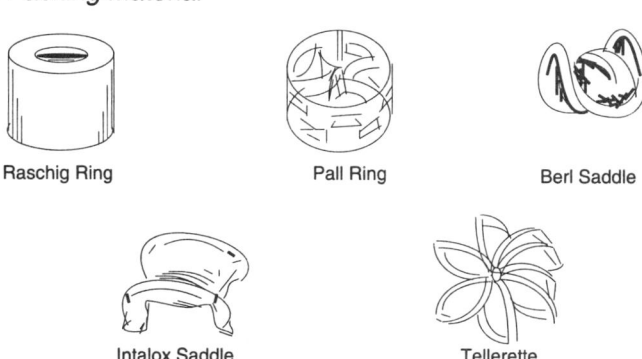

The specific packing that is selected depends on the nature of the contaminants, geometric mode of contact, size of the absorber, and scrubbing objectives. A large surface area per cubic foot of packing is desirable for mass transfer. Pressure drop is a function of the volume of void space in a tower when filled with packing. Generally, the larger the packing size, the smaller the pressure drop.

Packing material may be arranged in an absorber in either of two ways. The packing may be dumped into the column randomly, or, in certain cases, systematically stacked. Randomly packed towers provide a higher surface area, but also cause a higher pressure drop than stacked packing. In addition to the lower pressure drop, the stacked packing provides better liquid distribution over the entire surface of the packing. The large installation costs required to stack the packing material usually make it impractical unless high flow rates are required.

Once the liquid is distributed over the packing, it flows down, by force of gravity, through the packing, following the path of least resistance. The liquid tends to flow towards the tower wall where the void spaces are greater than in the center. Once the liquid hits the wall, it flows straight down the tower from that point (channels). A way must be provided to redirect the liquid from the tower wall back to the center of the column. This is usually done by using liquid redistributors which funnel the liquid back over the entire surface of packing. It is recommended that redistributors be placed at intervals of no more that 10 feet or 5 tower diameters, whichever is less.

Mobile packed beds are similar to packed towers. However, instead of being stationary they provide a zone of moving packing material. The gas and the liquid streams are mixed in this moving zone where mass transfer occurs. These absorbers are primarily used when both particulate and gaseous contaminant removal are necessary. They provide the mass transfer efficiency of a packed tower without the plugging problems. They do not, however, have the efficiency of a packed tower per energy unit consumed.

The capacity of mobile packed beds can be 5 to 6 times that of a packed or plate tower of the same diameter, due to the high gas velocities through the units. Mobile packed beds have been used to control emissions from Kraft pulp mills, cupola furnaces, and aluminum foundries. These units have also been used for SO_2 and fly ash control from large power plants.

Gas moving at high velocities that mixes with a liquid will entrain drops of that liquid. The liquid droplets must be removed from the gas stream before being exhausted to the atmosphere. Entrainment separators are used to remove the liquid droplets, although the major function of a entrainment separator is to prevent liquid loss, which saves on operating costs. Entrainment separators are, therefore, usually an integral part of any wet absorbing system.

In the design of packed towers for the control of air pollution emissions, the following assumptions are made: (1) there are no heat effects associated with the absorption operations, and (2) both the gas and liquid streams are dilute solutions (e.g., flow rates are constant throughout the absorption column and the equilibrium curve can be approximated as a straight line).

12.2 Solubility and Henry's Law

An important factor affecting the amount of a contaminant that can be absorbed is the solubility of the contaminant. Solubility is a function of both the temperature, and, to a lesser extent, the pressure of the system.

Solubility data is obtained at equilibrium conditions. This involves putting measured amounts of a gas and a liquid into a closed vessel and allowing them to sit for a period of time. Eventually, the amount of gas that is being absorbed into the liquid will equal the amount coming out of the solution. At this point, there is no net transfer of mass to either phase and the concentration of the gas in both the gaseous and liquid phases remains constant; the gas-liquid system is at equilibrium.

Henry's Law can be used to predict solubility when the solute concentrations are very dilute. In air pollution control applications this is usually the case.

$$p^* = H \cdot x \qquad [12.1]$$

where: p^* = partial pressure of solute at equilibrium, Pa

x = mole fraction of solute in the liquid

H = Henry's Law constant, Pa/mole fraction

Henry's Law can be written in a more useful form by dividing both sides of Equation by the total pressure, P_r. The left side of the equation becomes the partial pressure divided by the total pressure, which equals the mole fraction in the gas phase, y^*. Equation [12.1] now becomes:

$$y^* = H \cdot x \qquad [12.2]$$

where: y^* = mole fraction in gas phase in equilibrium with liquid

H = Henry's Law constant, mole fraction in vapor/mole fraction in liquid

Note: H is now dependent on the total pressure.

Henry's Law constants for the solubility of several gases in water are listed in Table 12.1 (Ledbetter 1974).

If equilibrium were to be reached in the actual operation of an absorption tower, the collection efficiency would fall to zero at that point, since no net mass transfer would occur. The equilibrium concentration, therefore, limits the amount of solute

Table 12.1
Henry's Law Constants for Gases in Water. ($\bullet 10^{-4}$ atm/mole fraction)

Temp °C	Air	CO_2	CO	H_2S	NO	N_2	O_2	SO_2	NH_3	COS
0	4.32	0.0728	3.52	0.0268	1.69	5.59	2.55	0.0011	0.000034	0.092
10	5.49	0.104	4.42	0.0367	2.18	6.68	3.27	0.0017	0.000054	0.148
20	6.64	0.142	5.36	0.0483	2.64	8.04	4.01	0.0024	0.000090	0.219
30	7.71	0.186	6.20	0.0609	3.10	9.24	4.75	0.0034	0.00013	0.304
40	8.70	0.233	6.96	0.0745	3.52	10.4	5.35	0.0054	0.00019	–
50	9.46	0.283	7.61	0.0884	3.90	11.3	5.88	0.0070	0.00029	–
60	10.1	0.341	8.21	0.103	4.18	12.0	6.29	–	0.00038	–
70	10.5	–	8.45	0.119	4.38	12.5	6.63	0.013	–	–
80	10.7	–	8.45	0.135	4.48	12.6	6.87	–	–	–
90	10.8	–	8.46	0.144	4.52	12.6	6.99	–	–	–
100	10.9	–	8.46	0.148	4.54	12.6	7.01	0.026	–	–

Note: The values in the table have been multiplied by 10^{-4}. The correct value for SO^2 at 0°C is 11 atm/mole fraction.

that can be removed by absorption. The most common method of analyzing solubility data is to use an *equilibrium diagram*. An equilibrium diagram is a plot of mole fraction of solute on the liquid phase, denoted as x, versus the mole fraction of solute in the gas phase, denoted as y. Equilibrium lines for the SO_2 and water system given in Table 12.2 are plotted in Figure 12.3 (Joseph and Beachler 1981). Figure 12.3 also illustrates the temperature dependence of the absorption process. At a constant mole fraction of solute in the gas, y, the mole fraction of SO_2 in the liquid, x, increases as the temperature decreases.

Example 12.1 Equilibrium Diagram

Given the following data for the solubility of SO_2 in pure water at 303 °K (30° C) and 101.3 kPa (760 mm Hg), plot the equilibrium diagram and determine if Henry's Law applies.

Equilibrium Data	
C_{SO_2}	$p^*_{SO_2}$
(g of SO_2 per 100 g of H_2O)	(partial pressure of SO_2)
0.5	6 kPa (42 mm Hg)
1.0	11.6 kPa (85 mm Hg)
1.5	18.3 kPa (129 mm Hg)
2.0	24.3 kPa (176 mm Hg)
2.5	30.0 kPa (224 mm Hg)
3.0	36.4 kPa (273 mm Hg)

Table 12.2
Partial Pressure of SO_2 in Aqueous Solution, mm Hg.

Grams SO_2 per 100 g H_2O	10°C	20°C	30°C	40°C	50°C	60°C	70°C
0.0	–	–	–	–	–	–	–
0.5	21	29	42	60	83	111	144
1.0	42	59	85	120	164	217	281
1.5	64	90	129	181	247	328	426
2.0	86	123	176	245	333	444	581
2.5	108	157	224	311	421	562	739
3.0	130	191	273	378	511	682	897
3.5	153	227	324	447	603	804	–
4.0	176	264	376	518	698	–	–
4.5	199	300	428	588	793	–	–
5.0	223	338	482	661	–	–	–

Source: Joseph and Beachler 1981

Figure 12.3
Equilibrium Lines for SO_2-H_2O System

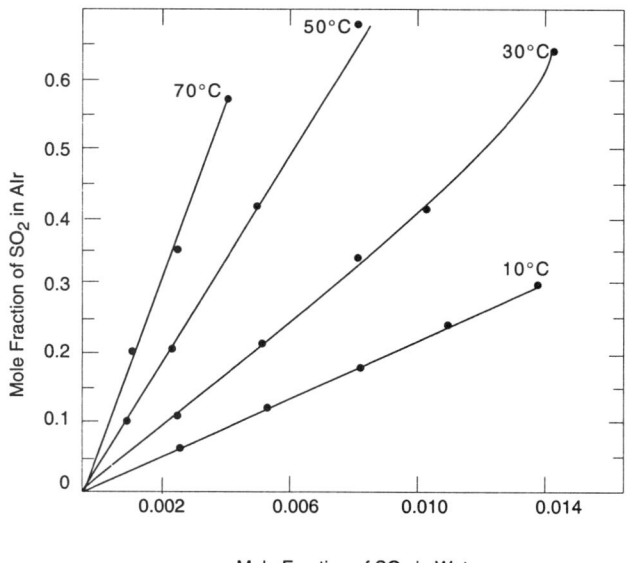

Solution:

1. The data must first be converted to mole fraction units.

   The mole fraction in the gas phase, y^*, is obtained by dividing the partial pressure of SO_2 by the total pressure of the system:

   $$y^* = \frac{p^*_{SO_2}}{P_T} = \frac{6 \text{ kPa}}{101.3 \text{ kPa}} = 0.06$$

   The mole fraction in the liquid phase, x, is obtained by dividing the moles of SO_2 in the solution by the total moles of liquid:

   $$x = \frac{\text{mol } SO_2 \text{ in solution}}{\text{mol } SO_2 \text{ in solution} + \text{mol of } H_2O}$$

   moles of SO_2 in solution = C_{SO_2}/ 64 g SO_2 per mole

   moles of H_2O = 100 g of H_2O / 18 g H_2O per mole = 5.55

   $$x = \frac{C_{SO_2}/64}{C_{SO_2}/64 + 5.55} = \frac{\frac{0.5}{64}}{\frac{0.5}{64} + 5.55} = 0.0014$$

2. Calculating additional values from Example 12.1 provides the table below.

$C_{SO_2} = \frac{\text{g of } SO_2}{100 \text{ g } H_2O}$	p* (kPa)	y* = p/103.1	$x = \frac{C_{SO_2}/64}{(C_{SO_2}/64) + 5.55}$
0.5	6.0	0.060	0.0014
1.0	11.6	0.115	0.0028
1.5	18.3	0.180	0.0042
2.0	24.3	0.239	0.0056
2.5	30.3	0.298	0.0070
3.0	36.4	0.359	0.0084

The above data is plotted in the figure below. Henry's Law applies in the given concentration range with Henry's Law constant equal to 42.7 mole fraction SO_2 in air/mole fraction SO_2 in water.

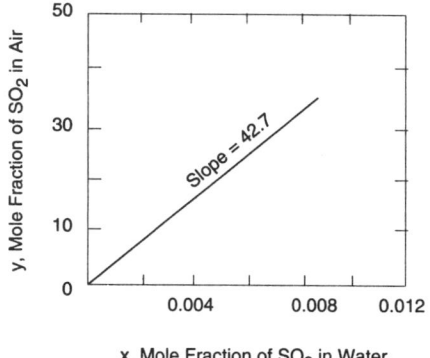

12.3 Absorption Design Theory

Absorption is a mass transfer operation. The basic model for describing the absorption process is the two-film theory — the model proposes that a mass transfer zone exists to include a small portion (film) of the gas and liquid phases on either side of the interface (Figure 12.4) (McCabe and Smith 1967; Perry 1973). The mass transfer zone is comprised of two films, a gas film and a liquid film, on their respective sides of the interface. These films are assumed to flow in a laminar or streamline motion, for which mathematical expressions that describe diffusion exist.

Figure 12.4
Visualization of Two-Film Theory

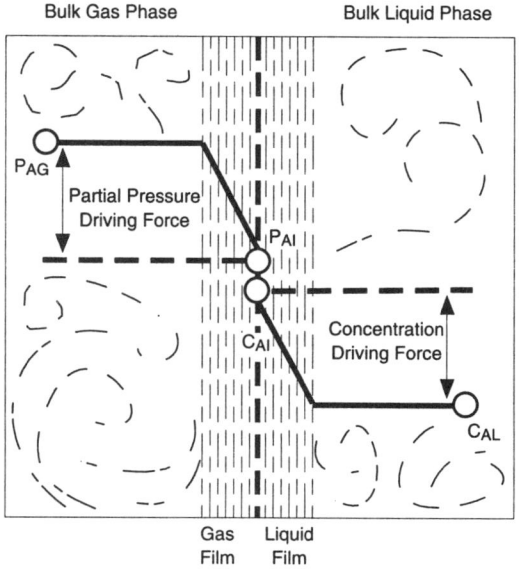

The theory assumes that there is complete mixing in both the gas and liquid bulk phases and that the interface is at equilibrium with respect to molecules transferring in or out. This implies that all resistance to diffusion occurs when the molecule is diffusing through the gas and liquid films. The concentration in the gas phase changes from P_{AG} in the bulk gas to P_{AI} at the interface.

A gas concentration is expressed by its partial pressure. Similarly, the concentration in the liquid changes from C_{AI} at the interface to C_{AL} in the bulk liquid phase as mass transfer occurs. The rate of mass transfer is then equal to the amount of molecule A transferred times the resistance molecule A encounters in diffusing through the films:

$$N_A = k_g (p_{AG} - p_{AI}) \tag{12.3}$$

$$N_A = k_l (C_{AG} - C_{AI}) \tag{12.4}$$

where: N_A = rate of transfer of component A, g mol/hr m² (lb mol/hr ft²)

k_g = mass transfer coefficient for gas film, g mol/hr m² Pa (lb mol/hr ft² atm)

k_l = mass transfer coefficient for liquid film, g mol/hr m² Pa (lb mol/hr ft² atm)

The mass transfer coefficients, k_g and k_l, represent the flow resistance the solute encounters in diffusing through each film, respectively. In practice, Equations [12.3] and [12.4] are difficult to use, since it is impossible to measure the interface concentrations p_{AI} and C_{AI}. The interface is a fictitious quantity used in the model to represent an observed phenomenon. The interface concentrations can be avoided by defining the mass transfer system at equilibrium conditions and combining the individual film resistance into an overall resistance. If the equilibrium line is straight, the rate of absorption is given by:

$$N_A = K_{OG} (p_{AG} - p_{A*}) \tag{12.5}$$

$$N_A = K_{OL} (C_{A*} - C_{AI}) \tag{12.6}$$

where: p_{A*} = equilibrium partial pressure of solute A at operating conditions

C_{A*} = equilibrium concentration of solute A at operating conditions

K_{OG} = overall mass transfer coefficient based on gas phase, g mol/hr m² Pa (lb mol/hr ft² atm)

K_{OL} = overall mass transfer coefficient based on liquid phase, g mol/hr m² Pa (lb mol/hr ft² atm)

Chapter 12: Absorption

At equilibrium, the overall mass transfer coefficients are related to the individual mass transfer coefficients by:

$$\frac{1}{K_{OG}} = \frac{1}{k_g} + \frac{H}{k_l} \quad [12.7]$$

$$\frac{1}{K_{OL}} = \frac{1}{k_l} + \frac{1}{H k_g} \quad [12.8]$$

H is Henry's Law constant (the slope of the equilibrium line). Equations [12.7] and [12.8] are useful in determining which phase controls the rate of absorption. From equation [12.8], if H is very small (which means the gas is very soluble in the liquid), then $K_{OG} \approx k_g$, and absorption is said to be gas-film-controlled. The major resistance to mass transfer is in the gas phase. Conversely, if a gas has limited solubility, H is large, and Equation [12.8] reduces to $K_{OL} \approx k_l$. The mass transfer rate is liquid-film-controlled and depends on the solute's dispersion rate in the liquid phase. Most systems in the air pollution control field are gas-film-controlled since the liquid is chosen so that the solute will have a high degree of solubility.

Mass transfer coefficients are often expressed by the symbols $K_{OG}a$, k_la, etc. where "a" represents the surface area available for absorption per unit volume of the column. This enables easy determination of the column area required to accomplish the desired separation. These mass transfer coefficients are developed from experimental data and usually reported in one of two ways: (1) as an empirical relationship based on function of the liquid flow gas or slope of the equilibrium line; or (2) correlated in terms of a dimensionless number, usually either the Reynolds or Schmidt Number. Figure 12.5 compares the effect on the mass transfer coefficient for SO_2 in water using two types of packing materials. Packing A is composed of one-inch rings and packing B of three-inch spiral tiles. Similar figures are used extensively to compare different absorbers or similar absorbers with varying operating conditions.

Figure 12.5
Comparison of Overall Absorption Coefficient for CO_2 in Water

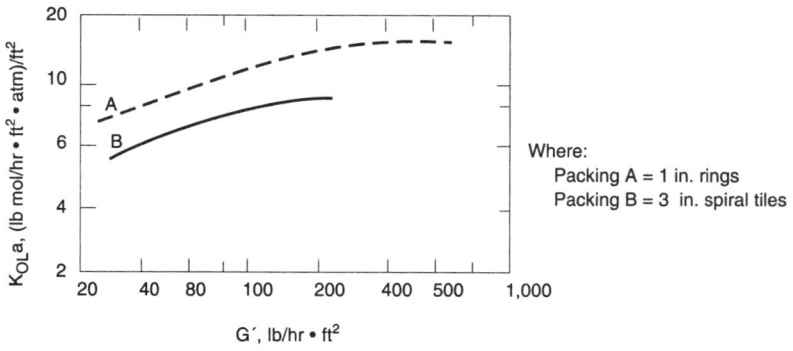

Source: Leva 1951

Although the science of absorption is considerably developed, much of the work in practical design situations is empirical in nature. The following section will apply the principles discussed to the design of gas absorption equipment. Emphasis has been placed on presenting information that can be used to estimate absorber size and liquid flow rate.

12.4 Design Procedures

12.4.1 Introduction

To design an absorption system, certain parameters are set by either operating conditions or regulations. The gas stream to be treated is the exhaust from a process. Therefore, the volume, temperature, and composition of the gas stream are given parameters. The outlet composition of the contaminant is set by the emission standard which must be met. The temperature and inlet composition of the absorbing liquid are also known. The main unknowns in designing the absorption system are: (1) the flow rates required, (2) the diameter of the vessel needed to accommodate the gas and liquid flow, and (3) the height of absorber required to achieve the needed removal.

12.4.2 Liquid-to-Gas Ratio

In design, the first task is to determine the flow rates and composition of each stream entering the system. A material balance is used to determine flow rates and composition of individual streams. Figure 12.6 presents a typical countercurrent flow absorber in which a material balance is illustrated.

An overall mass balance across the absorber in Figure 12.6 yields:

$$\text{lb mole in} = \text{lb mole out}$$
$$G_m(\text{in}) + L_m(\text{in}) = G_m(\text{out}) + L_m(\text{out}) \qquad [12.9]$$
$$G_{m1} + L_{m2} = G_{m2} + L_{m1}$$

where: L_m = liquid flow rate, g mol/hr

G_m = gas flow rate, g mol/hr

The numerals 1 and 2 are the bottom and top of the absorber, respectively.

A simplified equation can be obtained by assuming that as the gas and liquid streams flow through the absorber, their total mass does not change appreciably (e.g., $G_{m1} = G_{m2}$ and $L_{m1} = L_{m2}$). This is justifiable for most air pollution control systems, since the mass flow rates of contaminant are very small compared to the liquid and gas mass flow rates. For example, a 10,000 cfm exhaust stream containing 1,000 ppm SO_2 would be only 0.1% SO_2 (by volume) of 1.0 cfm. If the scrubber were 100% efficient, the gas mass flow rate would change from 10,000 cfm at G_{m1} to 9,999 cfm at G_{m2}.

Figure 12.6
Typical Operating Line Diagram

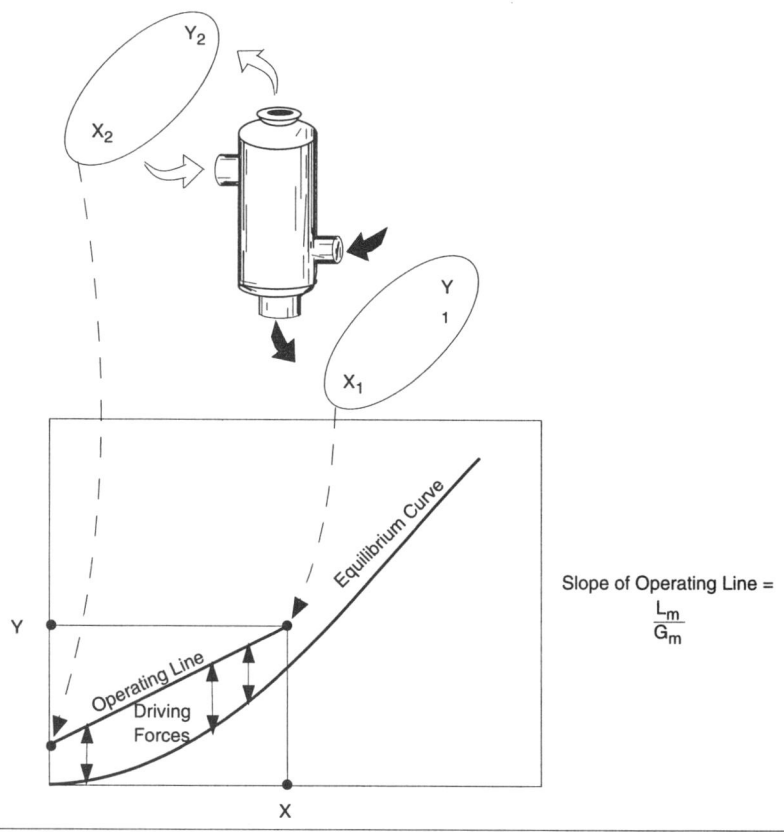

$$G_m(Y_1 - Y_2) = L_m(X_2 - X_1) \quad [12.10]$$

$$Y_1 - Y_2 = \frac{L_m}{G_m}(X_2 - X_1) \quad [12.11]$$

where: X = mole fraction of solute in pure liquid

Y = mole fraction of solute in inert gas

Equation [12.11] is the equation of a straight line. When this line is plotted on an equilibrium diagram, it is referred to as an *operating line*. This line defines operating conditions within the absorber. An equilibrium diagram with a typical operating line plotted on it is shown in Figure 12.6. The slope of the operating line is the liquid mass flow rate divided by the gas mass flow rate, which is the liquid-to-gas-ratio, or L_m/G_m.

12.4.3 Liquid Requirements

In the design of most absorption columns, the quantity of gas to be treated, G_m, and the inlet solute concentrations, Y_1, are set by process conditions. Minimum acceptable standards specify the outlet solute concentrations, Y_2. The composition of the liquid into the absorber, X_2, is also generally known or can be assumed to be zero if there is no recycling of liquid. By plotting these data on an equilibrium diagram, the minimum amount of liquid required to achieve the requires outlet concentrations, Y_2, can be determined.

Figure 12.7a is a typical equilibrium diagram with operating points plotted for a countercurrent flow adsorber. At the minimum liquid rate, the inlet gas concentration of solute Y_1 is in equilibrium with the outlet liquid concentration of solute, X_{max}. The liquid leaving the absorber is saturated with solute and can no longer dissolve any more solute unless additional liquid is added. This condition is represented by point B on the equilibrium curve.

The slope of the line drawn between point A and point B represents the operating conditions at minimum flow rate in Figure 12.7b. Note how the driving force decreases to zero at point B. The slope of the line AB is (L_m/G_m) min, and may be determined graphically or from the equation for a straight line. By knowing the slope of the minimum operating line, the minimum liquid rate can be determined by substituting the known gas flow rate.

Determining the minimum liquid flow rate, L_m/G_mmin, is important since absorber operation is usually specified as some factor of it. Generally, liquid flow rates are specified at 25 to 100% greater than the required minimum. Typical absorber operation would be at 50% greater than the minimum liquid flow rate (i.e., 1.5 times the minimum liquid-to-gas-ratio) (Joseph and Beachler 1981). Setting the liquid rate in this way assumes that the gas flow rate by the process does not change appreciably. Line AC in Figure 12.7c is drawn at a slope of 1.5 times the minimum L_m/G_m. Line AC is referred to as the operating line since it describes absorber operating conditions.

Example 12.2 SO₂ Removal by Absorption in Water

Using the data given in the sketch, compute the minimum liquid rate of pure water required to remove 90% of the SO_2 from a gas stream of 84.9 m³/min (3,000acfm) containing 3% SO_2 by volume. The temperature is 293°K and the pressure is 101.3 kPa.

Figure 12.7
Graphic Determination of Liquid Flow Rate

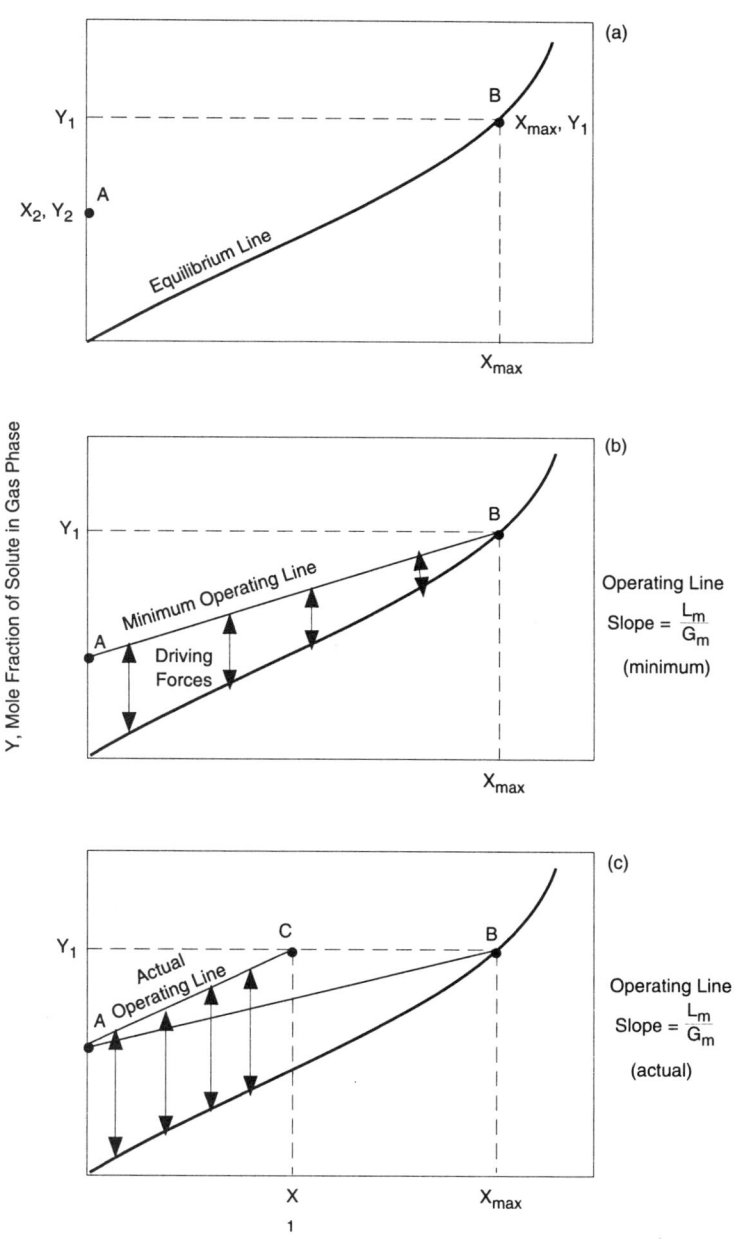

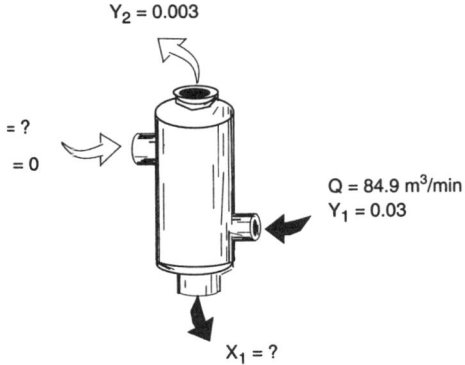

Y_1 = 3% by volume = 0.03

Y_2 = 90% reduction from Y_1 or only 10% of Y_1; therefore

Y_2 = (0.10)(0.03) = 0.003

Solution:

1. Compute minimum liquid-to-gas ratio.

    At the minimum liquid rate, Y_1 and X_1 will be in equilibrium; the liquid will be saturated with SO_2.

    At equilibrium:

    $Y_1 = H \bullet X_1$

    $H = 42.7 \dfrac{\text{mole fraction } SO_2 \text{ in air}}{\text{mole fraction } SO_2 \text{ in water}}$ (from Example 12.1)

    $0.3 = 42.7 \bullet X_1$

    X_1 = 0.000703 mole fraction

    The minimum liquid-to-gas ratio is:

    $$Y_1 - Y_2 = \dfrac{L_m}{G_m}(X_1 - X_2)$$

    $$\dfrac{L_m}{G_m} = \dfrac{Y_1 - Y_2}{X_1 - X_2}$$

    $$\dfrac{L_m}{G_m} = \dfrac{0.03 - 0.003}{0.000703 - 0} = 38.4 \dfrac{\text{g mol water}}{\text{g mol of air}}$$

Chapter 12: Absorption

2. Compute the minimum required liquid flow rate.

First, convert m³ of air to g moles. At 0°C and 101.3 kPa there are 0.0224 m³/g mole (359 ft³/mole) of an ideal gas.

At 30°C: $0.0224 \dfrac{m^3}{min} \left(\dfrac{293°K}{273°K} \right) = 0.024 \, m^3 / g \, mol$

$G_m = 84.9 \dfrac{m^3}{min} \left(\dfrac{g \, mol \, air}{0.024 \, m^3} \right) = 3,538 \dfrac{g \, mol \, air}{min}$

$\dfrac{L_m}{G_m} = 38.4 \left(\dfrac{g \, mol \, water}{g \, mol \, air} \right)$ at minimum conditions

$L_m = (38.4)(3,538) = 136.0 \, kg \, mol \, water / min$

In mass units:

$L = \dfrac{136 \, kg \, mol}{min} \left(\dfrac{18 \, kg}{kg \, mol} \right) = 2,488 \dfrac{kg}{min} (647 \, gal / min)$

3. Multiply the slope of the minimum operating line times 1.5 to get the slope of the actual operating line (line AC).

38.4 • 1.5 = 57.6

12.4.4 Packed Tower Diameter

Once all the streams entering and leaving the absorber are identified, flow rates calculated, and operating conditions determined, the physical dimensions of the packed tower can be calculated. The tower must be of sufficient diameter to accommodate the gas and liquid and be of sufficient height to ensure that the required amount of mass is transferred with the existing driving force.

The tower diameter is selected to give a gas velocity that is satisfactory. The gas velocity in countercurrent towers must be low enough to prevent excessive holdup and/or carryover of the liquid, yet high enough for economical design.

The main parameter which affects the diameter of a packed column is the gas velocity at which liquid droplets become entrained in the existing gas stream. If the gas flow rate through the column is too high, the liquid flowing down over the packing will be held in the void spaces between the packing. This gas-to-liquid flow ratio is termed the *loading point*. A further increase in gas velocity will cause the liquid to completely fill the void spaces in the packing. The liquid forms a layer over the top of the packing and no more liquid can flow down through the tower. The pressure drop increases substantially and mixing between the phases is minimal. This condition is referred to as *flooding* and the gas velocity at which it occurs is the *flooding velocity*.

Normal practice is to set the diameter of the packed column so that the gas velocity is a certain percent of the flooding velocity. A typical operating range for the gas velocity through the column is 50% to 75% of the flooding velocity. It is assumed that by operating in this range, the gas velocity will also be below the loading point.

One of the more commonly used correlations to determine the flooding velocity is U.S. Stoneware's generalized pressure drop correlation, as presented in Figure 12.8 (Leva 1951). Pressure drops for packed absorption towers are usually in the range of 0.2 to 0.4 in. H_2O/ft for water as the absorbent and are lower or higher for lighter or heavier absorbents.

Figure 12.8
Generalized Flooding and Pressure Drop Correlation

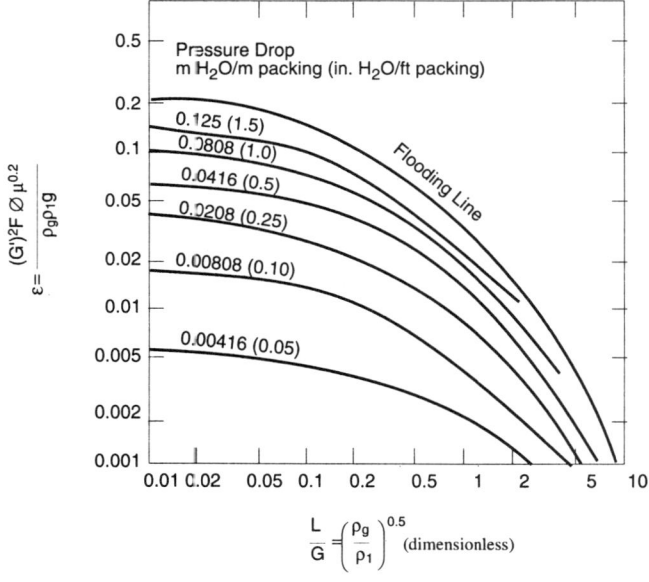

Source: Leva 1951

The procedure to determine the column diameter is as follows:

1. Calculate the value of the abscissa.

$$\frac{L}{G} = \left(\frac{\rho_g}{\rho_l}\right)^{0.5}$$
[12.12]

where: L and G = mass flow rates (any constant set of units may be used as long as the abscissa is dimensionless)

Table 12.3
Packing Data

Packing	Size (in.)	Weight (lb/ft^3)	Surface area, a (ft^2/ft^3 packing volume)	Void fraction (%)	Packing factor, F (ft^2/ft^3)
Raschig rings (ceramic and porcelain)	1/2	52	114	65	580
	1	44	58	70	155
	1 1/2	42	36	72	95
	2	38	28	75	65
	3	34	19	77	37
Raschig rings (steel)	1/2 • 1/32	77	128	84	300
	1 • 1/32	40	63	92	115
	2 • 1/16	38	31	92	57
Berl saddles (ceramic)	1/4	55	274	63	900
	1/2	54	155	64	240
	2	48	79	68	110
		38	32	75	45
Intalox saddles (ceramic)	1/4	54	300	75	725
	1/2	45	190	78	200
	1	44	78	77	98
	2	42	36	79	40
Intalox saddles (plastic)	1	6.00	63	91	30
	2	3.75	33	93	20
	3	3.25	27	94	15
Pall rings (plastic)	5/8	7.0	104	87	97
	1	5.5	63	90	52
	2	4.5	31	92	25
Pall rings (metal)	5/8 • 0.018 thick	38	104	93	73
	1 1/2 • 0.03 thick	24	39	95	28
Tellerettes	1	7.5	55	87	40
	2	3.9	38	93	20
	3	5.0	30	92	15

Source: Joseph and Beachler 1981

ρ_g = density of the gas stream

ρ_1 = density of the absorbing liquid

2. From the point calculated in Equation [12.12], proceed up the graph to the flooding line and read the ordinate, ε.

3. Rearrange the equation of the ordinate and solve for G'.

$$G' = \left[\frac{(\varepsilon)(\rho_g)(\rho_1)(g)}{F\emptyset\mu_1^{0.2}}\right]^{0.5} \qquad [12.13]$$

where: ε = pressure drop

F = packing factor given in Table 12.3 for different types of packing

$\emptyset$ = ratio of specific gravity of the scrubbing liquid to that of water

μ_1 = viscosity of liquid (for water = 0.8 centipoise)

G' = mass flow rate of gas per unit cross sectional area of column at flooding, g/sec m^2 (lb/sec ft^2)

ρ_1 = density of the absorbing liquid, kg/m^3 (lb/ft^3)

ρ_g = density of the gas stream, kg/m^3 (lb/ft^3)

g = gravitational constant = 9.82 m/sec^2 (32.2 ft/sec^2)

4. G' at the operating point is a fraction of G' at flooding.

$$G'_{operating} = (f)(G'_{flooding}) \qquad [12.14]$$

5. The cross sectional area of column A is calculated from:

$$A = \frac{G}{G'_{operating}} \qquad [12.15]$$

6. The diameter of the column is obtained from:

$$d_r = \left(\frac{4A}{\pi}\right)^{0.5} = 1.13\ A^{0.5} \qquad [12.16]$$

Chapter 12: Absorption 453

The following example illustrates the use of U.S. Stoneware's pressure drop correlation (Figure 12.8) to compute the minimum allowable diameter of a packed tower.

Example 12.3 Packed Tower Diameter Determination

For the absorber in Example 12.2, determine the packed tower diameter if the operating liquid rate is 1.5 times the minimum. The gas velocity should be no greater than 75% of the flooding velocity and the packing material is two-inch ceramic Intalox saddles.

Solution:

From Example 12.2:

$$G_m = 3{,}538 \text{ g mole/min}$$

$$L_{minimum} = 2{,}448 \text{ kg/min}$$

Convert gas molar flow to a mass flow, assuming molecular weight of the gas to be 29 kg/kg mole.

$$G = 3.538 \frac{\text{kg mol}}{\text{min}} \bullet \frac{29 \text{ kg}}{\text{kg mol}} = 102.6 \text{ kg / min}$$

Adjusting the liquid flow to 1.5 times the minimum:

$$L = 1.5\,(2{,}448) = 3{,}672 \text{ kg/min}$$

The densities of air and water at 30°C are:

$$\rho_l = 1{,}000 \text{ kg/m}^3$$

$$\rho_g = 1.17 \text{ kg/m}^3$$

1. Using the relationship in Equation [12.12], compute the abscissa.

$$\left(\frac{L}{G}\right)\left(\frac{\rho_g}{\rho_l}\right)^{0.5}$$

$$\left(\frac{3{,}672}{102.6}\right)\left(\frac{1.17}{1{,}000}\right)^{0.5} = 1.22$$

2. From Figure 12.8, proceed up to the flooding line from 1.22. The ordinate is 0.019.

3. Use equation 12.13 to solve for the superficial flooding velocity, G'.

Note: A superficial velocity is a flow rate per unit cross sectional area.

$$G' = \left[\frac{(\varepsilon)(\rho_g)(\rho_1)(g)}{F\emptyset\mu_1^{0.2}}\right]^{0.5}$$

For water, $\rho_1 = 1.0$ and $\mu_1 = 0.0008$ Pa sec

From Table 12.3 for two-inch Intalox saddles, $F = 40$ ft^2/ft^3 or 131 m^2/m^3, $g = 9.82$ m/sec^2

$$G' = \left[\frac{(0.019)(1.17)(1000)(9.82)}{(1)(131)(0.0008)^{0.2}}\right]^{0.5}$$

$= 2.63$ kg / m^2 sec at flooding

4. The superficial gas velocity at the operating point is obtained from Equation [12.14].

$$G'_{operating} = f \, G'_{flooding} = (0.75)(2.65) = 1.97 \text{ kg/m}^2 \text{ sec}$$

5. From Equation [12.15], the cross-sectional area of the tower is:

$$A = \frac{G}{G'_{operating}}$$

$$= \frac{(102.6 \text{ kg / min})(\text{min / 60 sec})}{1.97 \text{ kg / m}^2 \text{ sec}} = 0.87 \text{ m}^2$$

6. The diameter of the tower, from equation [12.16] is:

$$d_r = \left(\frac{4A}{\pi}\right)^{0.5}$$

$$= \left(\frac{(4)(0.87 \text{ m}^2)}{\pi}\right)^{0.5}$$

$= 1.05$ m or at least 1.1 m (3.5 ft)

Figure 12.8 may also be used to estimate the pressure drop of the tower once $G'_{operating}$ is set.

This is done by plugging G' back into Equation [12.13] and solving for ε.

$$\varepsilon = \frac{G'^2 \emptyset F\mu_1^{0.2}}{\rho_g \rho_1 g}$$

$$= \frac{(1.97)^2 (1)(131)(0.0008)^{0.2}}{(1.17)(1,000)(9.82)}$$

$= 0.0106$

The abscissa remains unchanged and equals 1.22.

The pressure drop through the column is the point at which these two lines cross.

From Figure 12.8:

$$\Delta p \approx \frac{0.0416 \text{ meter of } H_2O}{\text{meter of packing}} \text{ or } \left(\frac{0.5 \text{ inches of } H_2O}{\text{foot of packing}}\right)$$

12.4.5 Packed Tower Height

Once the diameter of the tower has been set, the height of the tower must be determined to obtain the desired efficiency of absorption. The height of a packed column refers to the depth of packing material needed to accomplish the required removal efficiency. The more difficult the separation, the larger the packing height required. For example, a much larger packing height would be required to remove SO_2 than to remove Cl from an exhaust stream using water as the absorbent. This is because Cl_2 is more soluble in water than SO_2.

A number of theoretical equations are used to predict the required packing height. These equations are based on diffusion principles. Depending on which phase is controlling the absorption process, either Equation 12.5 or 12.6 is used as the starting point to derive an equation to predict column height. A material balance is then set up over a small differential section of the column. Derivations of the equations used can be found in a number of chemical engineering texts or other books dealing with mass transfer and will not be covered here (Perry 1973).

The general form of the design equation for a gas-phase controlled resistance is given by:

$$Z = \frac{G'}{K_g a P} \int_{Y_2}^{Y_1} \frac{dY}{(1-Y)(1-Y^*)} \qquad [12.17]$$

where: Z = height of packing, m

a = interfacial contact area, m^2

P = pressure of the system, kPa

In analyzing Equation [12.17], the term $G'/K_g aP$ has the dimension of meters and is defined as the height of a transfer unit. The term inside the integral is dimensionless and represents the number of transfer units needed to make up the total packing height. Using the concept of transfer units, Equation [12.17] can be simplified to:

$$Z = HTU \bullet NTU \qquad [12.18]$$

where: HTU = height of a transfer unit, m

NTU = number of transfer units

The number and the height of a transfer unit are based on either the gas or liquid phase. Equation [12.18] now becomes:

$$Z = N_{OG} \bullet H_{OG} = N_{OL} \bullet H_{OL} \quad [12.19]$$

where: N_{OG} = number of transfer units based on overall gas-film coefficient

N_{OL} = number of transfer units based on overall liquid-film coefficient

H_{OG} = height of transfer units based on overall gas-film coefficient, m

H_{OL} = height of transfer units based on overall liquid-film coefficient, m

The height of a transfer unit, H_{OG}, is usually determined experimentally for the system under consideration. Information on many different systems using various types of packings has been compiled by the manufacturers of gas absorption equipment. The data are usually in the form of graphs depicting, for a specific system and packing, the H_{OG} versus the gas rate (lb/hr ft^2) with the liquid rate (lb/hr ft^2) as a parameter (Figure 12.9) (Leva 1951). Manufacturers' H_{OG} data are summarized in Table 12.4 for plastic packings (Buonicore and Davis 1992)

Figure 12.9
Column Packing Comparison for Ammonia and Water System

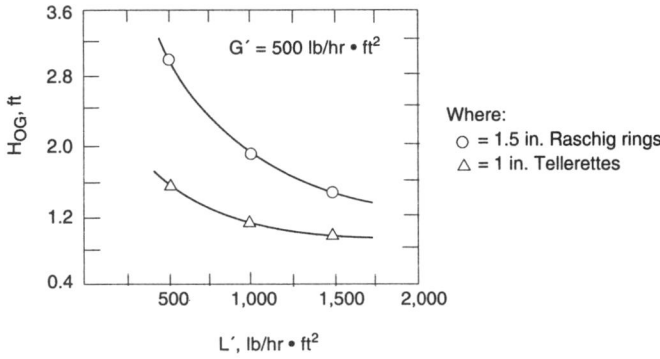

Source: Leva 1951

When no experimental data are available or if only a preliminary estimate of absorber efficiency is needed, there are generalized correlations available to predict the height of a transfer unit. The correlations for predicting the H_{OG} or the H_{OL} are empirical in nature and are a function of type of packing, liquid and gas flow rates, concentration and solubility of the contaminant, liquid properties, and system temperature (Danielson 1973).

Table 12.4
Range of Manufacturers' H_{OG} Data for Plastic Packings

Chemical System (in air)	H_{OG}, feet
$HCl - H_2O$	0.6-1.1
$HCl - NaOH$	0.5-0.7
$Cl_2 - NaOH$	0.8-1.2
$NH_3 - H_2SO_4$	0.3-0.5
$NH_3 - H_2O$	0.3-0.7
$SO_2 - NaOH$	0.7-2.0
$HF - H_2O$	0.4-0.7
$CH_3COCH_3 - H_2O$	0.8-1.3
$H_2S - NaOH$	0.8-1.6

The number of transfer units, NTU, can be calculated by a variety of methods (Leva 1951). For the case where the solute concentration is very low and the equilibrium line is straight, Equation [12.20] can be used to determine the number of transfer units, N_{OG}, based on the gas phase resistance. Equation [12.20] is derived from the integral portion of Equation [12.17].

$$N_{OG} = \frac{\ln\left[\frac{Y_1 - mX_2}{Y_2 - mX_2}\left(1 + \frac{mG_m}{L_m}\right) + \frac{mG_m}{L_m}\right]}{1 - \frac{mG_m}{L_m}} \qquad [12.20]$$

where:
    m = slope of equilibrium line
    G_m = molar flow rate of gas, kg mol/hr
    L_m = molar flow rate of liquid, kg mol/hr
    X_2 = mole fraction of solute entering the column
    Y_1 = mole fraction of solute in entering gas
    Y_2 = mole fraction of solute in exiting gas

Equation [12.20] may be solved directly or graphically by using the Colburn Diagram which is presented in Figure 12.10. The Colburn Diagram is a plot of the N_{OG} versus $\ln[Y_1 - mX_2/Y_2 - mX_2]$ at various values of mG_m/L_m. The term mG_m/L_m is referred to as the *absorbtion factor*. Figure 12.10 is used by first computing the value of $\ln[Y_1 - mX_2/Y_2 - mX_2]$, reading up the graph to the line corresponding to mG_m/L_m, and then reading across to obtain the N_{OG}. The absorption factor usually has values between 1.25 to 2.0.

Figure 12.10
Colburn Diagram

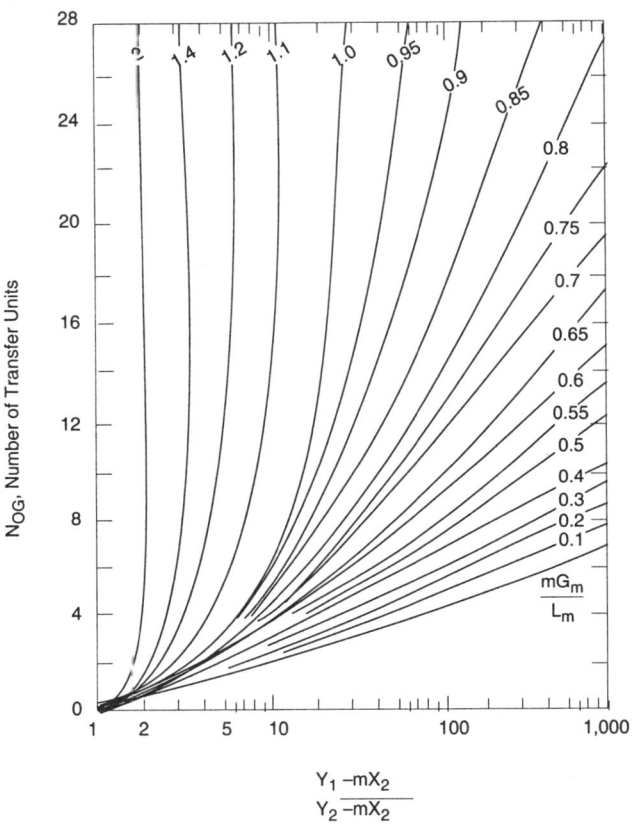

The equation [12.20] can be further simplified for situations where a chemical reaction occurs or if the solute is extremely soluble. In these cases, the solute exhibits almost no partial pressure and, therefore, the slope of the equilibrium line approaches zero (m → 0). For either of these cases, Equation [12.20] reduces to:

$$N_{OG} = \ln \frac{Y_1}{Y_2} \qquad [12.21]$$

The number of transfer units depends only on the inlet and outlet concentration of the solute (contaminant). For example, if the conditions of Equation [12.21] are met, to achieve 90% removal of any pollutant requires 2.3 transfer units. Equation [12.21] applies only when the equilibrium line is straight and approaches zero (for very soluble or reactive gases). The following example illustrates the use of Equation [12.20].

Example 12.4 Packed Tower Packing Height Determination

From pilot plant studies of the absorption system in Example 12.2, it was determined that the H_{OG} for the SO_2 - water system is 0.829 m (2.72 ft). Calculate the total height of packing required to achieve the 90% removal. The following data were taken from Example 12.2:

$$m = 42.7 \frac{\text{kg mol H}_2\text{O}}{\text{kg mol air}}$$

$$G_m = 3.5 \frac{\text{kg mol}}{\text{min}}$$

$$L_m = 3,672 \frac{\text{kg}}{\text{min}} \cdot \frac{\text{kg mol}}{18 \text{ kg}} = 204 \frac{\text{kg mol}}{\text{min}}$$

$X_2 = 0$ (no recycle liquid)

$Y_1 = 0.03$

$Y_2 = 0.003$

Solution:

$$N_{OG} = \frac{\ln\left[\frac{Y_1 - mX_2}{Y_2 - mX_2}\left(1 + \frac{mG_m}{L_m}\right) + \frac{mG_m}{L_m}\right]}{1 - \frac{mG_m}{L_m}}$$

$$= \frac{\ln\left[\left(\frac{0.03}{0.003}\right)\left(1 - \frac{(42.7)(3.5)}{204}\right) + \frac{(42.7)(3.5)}{204}\right]}{1 - \frac{(42.7)(3.5)}{204}}$$

$$= 4.58$$

The total packing height is thus:

$$Z = H_{OG} \cdot N_{OG}$$

$Z = (0.829)(4.58) = 3.79$ m packing height.

12.5 Review of Design Procedure

Typical superficial velocities for packed column range from 3 to 4 ft/sec, with the lower value applying to ceramic packing and the higher value to plastic packing. Pressure drop in packed scrubbers for air pollution control is near 0.25 in. H_2O per foot of packings for ceramic and 0.2 in. H_2O per foot of packing for plastic. For larger diameter towers, larger packing is generally used and vice versa.

For plastic packing, the liquid and gas flow rates are both typically in the range of 1,500 to 2,000 lb/hr ft^2. For ceramic packing, the range of both flow rates is 500 to 1,000 lb/hr ft^2. For difficult-to-absorb gases, the gas flow rate is usually lower and the liquid flow rate higher.

The quantity of liquid to be used is typically estimated from the minimum liquid-to-gas ratio as determined from material balances and equilibrium considerations. As a rule-of-thumb for purposes of rapid estimates, it has frequently been found that the most economical value for the absorption factor will be in the range from 1.25 to 2.0.

Once the gas and liquid streams entering and leaving the absorber column and their concentrations are identified, flow rates calculated, and operating conditions (type of packing) determined, the diameter of the column can be calculated. The column must be of sufficient diameter to accommodate the gas and liquid streams. The calculation of the column diameter is based on flooding considerations; the usual operating range being taken as 50 to 60 percent of the flooding rate.

The height of a packed column is calculated by determining the required number of theoretical transfer units and multiplying by the height of a transfer unit. In emission control applications, gas film resistance will typically be the controlling factor. In most air pollution designs, the number of transfer units, N_{OG}, is obtained graphically for dilution solution (Figure 12.4). The height of a transfer unit, H_{OG}, is also usually determined experimentally and is available from the literature (Table 12.4). The packing height, Z, is then simply the product of the H_{OG} and the N_{OG}.

Example 12.5 Packed Tower Design Using English Units

Examples 12.2 through 12.4 are repeated in English units.

An exhaust stream of 3,000 acfm is known to contain 3% SO_x by volume. It is determined that the SO_x content be reduced 90% by scrubbing with water.

Henry's Law constant = 42.7 mole fraction SO_2 in gas/mole fraction SO_2 in liquid.

$$\rho_l = 62.4 \text{ lb/ft}^3$$

ρ_g = 0.0732 lb/ft³

packing = 2-in. Intalox saddle

temperature = 86°F

1. What is the liquid requirement in gal/min at 1.5 times the minimum L/G ratio?
2. What is the column diameter at 75% of flooding?
3. Determine the number of transfer units.
4. Based on a H_{OG} of 2.72 ft, what is the total height of the tower?

Solution:

Material balance diagram:

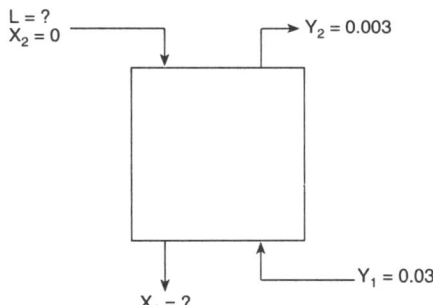

Exhaust gas flow rate = 3,000 acfm

1. Determine the Liquid Requirement.

    L_m minimum may be completed two ways:

    a) graphically by plotting the inlet and outlet concentrations on an equilibrium diagram.

    b) by using Henry's Law and the equation for a straight operating line as follows:

    Using Henry's Law: $Y = H \cdot X$

    the bottom of tower Y_1 and X_1 are in equilibrium

    $$H = 42.7 \frac{\text{mol fraction SO}_2 \text{ in gas}}{\text{mol fraction SO}_2 \text{ in liquid}}$$

    $Y_1 = X_1 \cdot H$

$$X_1 = Y_1/H$$

$$X_1 = (0.03)/42.7 = 0.000703 \frac{\text{mol of } SO_2}{\text{mol } H_2O}$$

Minimum L_m/G_m ratio:

$$L_m/G_m = \frac{Y_1 - Y_2}{X_1 - X_2} = \frac{0.03 - 0.003}{0.000703 - 0.0}$$

$$L_m/G_m = 38.4 \frac{\text{lb mol water}}{\text{lb mol air}}$$

Operating L_m/G_m ratio (1.5 times minimum)

(L_m/G_m)operating $= (1.5)(38.4) = 57.6$

Determine the operating liquid requirement:

$$G_m = \left(\frac{3,000 \text{ cf}}{\text{min}}\right)\left(\frac{\text{lb mol}}{398 \text{ cf}}\right) = 7.54 \frac{\text{lb mol air}}{\text{min}}$$

$$L_m = \left(57.6 \frac{\text{lb mol } H_2O}{\text{lb mol air}}\right)\left(7.54 \frac{\text{lb mol air}}{\text{min}}\right)$$

$L_m = 434.3$ lb mol H_2O / min

In gallons per minute that is:

$$\text{gal/min} = 434.3 \frac{\text{lb mol}}{\text{min}}\left(\frac{18 \text{ mol}}{\text{lb mol}}\right)\left(\frac{\text{gal}}{8.34 \text{ lb}}\right)$$

gal / min = 937.34

2. Determine the column diameter.

   Given:

   $L_m = 434$ lb mole H_2O/min $\rho_l = 62.4$ lb/ft^3

   $G_m = 7.54$ lb mole air/min $\rho_g = 0.0732$ lb/ft^3

   Compute the abscissa $(L/G)\left(\frac{\rho_g}{\rho_l}\right)^{0.5}$:

   $$(L/G)\left(\frac{\rho_g}{\rho_l}\right)^{0.5} = \frac{(7,817.4)}{(218.7)}\left(\frac{0.0732}{62.4}\right)^{0.5} = 1.225$$

Chapter 12: Absorption

Gas flow rate at flooding conditions from graph (Figure 12.8):

$$\frac{G'^2 \varnothing F \mu_l^{0.2}}{\rho_s \rho_l g} = \text{ordinate} = \varepsilon$$

for water $\varnothing = 1.0$ and $\mu_l = 0.8$ centipoise

if 2-in. ceramic Intalox saddles are used:

F = 40

$$\frac{G'^2 (40)(1.0)(0.8)^{0.2}}{(0.0732)(62.4)(32.2)} = 0.019$$

$$G' = 0.27 \text{ lb / ft}^2 \text{ sec}$$

$$G_{operating} = (f)(G'_{flooding})$$

$$G_{operating} = (0.75)(0.27) = 0.20 \text{ lb / ft}^2 \text{ sec}$$

Calculate the cross sectional area of the column:

$$A = \frac{\text{mass flow rate(lb / sec)}}{\text{operating flow rate lb / sec ft}^2}$$

$$A = \frac{(218.7 \text{ lb / min})(\text{min / 60 sec})}{0.20 \text{ lb / sec ft}^2}$$

$$A = 18.23 \text{ ft}^2$$

Calculate the column diameter:

$d_r = (4A/\pi)^{1/2}$

$d_r = 1.13 \, A^{1/2}$

$d_r = 1.13 \, (18.23 \text{ ft}^2)^{1/2} = 4.8 \text{ ft}$

3. Determine the number of Transfer Units.

$X_2 = 0$ $m = 42.7 \dfrac{\text{mole fraction SO}_2 \text{ in air}}{\text{mole fraction SO}_2 \text{ in liquid}}$

$Y_1 = 0.03$ $L_m = 434.4$ lb mole H_2O/min

$Y_2 = 0.003 \quad G_m = 7.54$ lb mole air/min

Solve for N_{OG}:

$$N_{OG} = \frac{\ln\left[\frac{Y_1 - mX_2}{Y_2 - mX_2}\left(1 + \frac{mG_m}{L_m}\right) + \frac{mG_m}{L_m}\right]}{1 - \frac{mG_m}{L_m}}$$

$$\frac{mG_m}{L_m} = \frac{(42.7)(7.54)}{434.4} = 0.741$$

$$N_{OG} = \frac{\ln\left[\frac{0.03 - 0}{0.003 - 0}(1 - 0.741) + 0.741\right]}{1 - 0.741}$$

$$N_{OG} = 4.65$$

Compute the total height of the packed section:

given $H_{OG} = 2.72$ ft

$$Z = N_{OG} \cdot H_{OG} = (4.65)(2.72)$$

$$Z = 12.65 \text{ ft of packing}$$

Graphic solution:

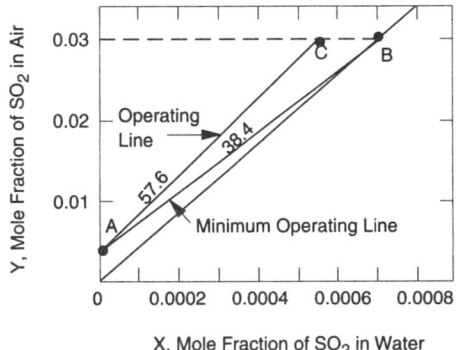

12.6 Alkaline Absorption for SO₂

Example 12.5 demonstrates that unacceptably large amounts of water are required to remove SO_2 from lean waste gas streams such as power plants, where the SO_2 concentration is typically less than 3,000 ppm. This happens because of the large Henry's Law constant ($\cong 50$) for the SO_2/H_2O system. The amount of scrubbing water can be substantially reduced if a reagent is added to the water to increase the solubility of SO₂ gas being removed. For example, the L/G ratio for a dilute alkaline solution is 5 compared to 57 for water.

The important concepts for the design of an alkaline scrubber system are: (1) quantification of the raw chemicals required and waste volumes generated, (2) volumetric flow rates in the scrubber, and (3) operational considerations, such as controlling scaling in the systems as precipitates are formed in the reactions (King 1996). The reactions for the most common absorption chemicals are as follows:

Limestone absorption:

$$2CaCO_3 + H_2O + 2SO_2 \rightarrow 2CaSO_3 + 2CO_2 + H_2O \qquad [12.22]$$

Lime [as $Ca(OH)_2$)] absorption:

$$Ca(OH)_2 + SO_2 + H_2O + \frac{1}{2}O_2 \rightarrow CaSO_4 \bullet 2H_2O \qquad [12.23]$$

Trona (natural sodium carbonate) absorption:

$$NaCO_3 + SO_2 \rightarrow Na_2SO_3 + CO_2 \qquad [12.24]$$

Double Alkali (lime or limestone plus sodium carbonate) absorption:

1) $NaCO_3 + SO_2 \rightarrow NaSO_3 + CO_2$ \qquad [12.25]

2) $NaSO_3 + CaCO_3 + \frac{1}{2}O_2 \rightarrow CaSO_4 + NaCO_3$ \qquad [12.26]

Example 12.6 Countercurrent Wet Scrubber with Lime Addition

A counter-current wet scrubber with lime treatment, as shown, is used to control SO₂.

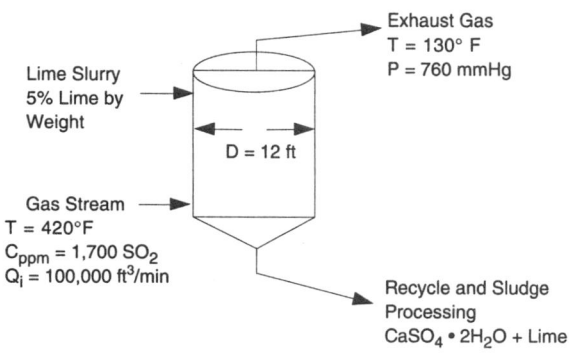

1. Determine if the gas velocity in the scrubber is within design standards (7 to 20 ft/sec at entrance).

$$v = \frac{Q_1}{Area}$$

$$v = \frac{100,000 \text{ ft}^3 / \text{min}}{3.14(6)^2 \text{ ft}^2}$$

$$= 884.6 \text{ ft} / \text{min}$$

$$= 14.74 \text{ ft} / \text{sec}$$

2. Determine the concentration of SO_2 in the exhaust gas in both mg/m³ and ppm, if the scrubber achieves a mass removal efficiency of 92%.

Mass of SO_2 in the gas stream.

$$C(mg/m^3) = 4.09 \cdot 10^{-2} (C_{ppm}) MW_{SO_2}$$

$$= 4.09 \cdot 10^{-2} (1,700) 64$$

$$= 4,450 \text{ mg} / m^3$$

$$= 4.45 \text{ g} / m^3$$

Mass SO_2 = 4.45 g/m³ • 100,000 ft³/min • 0.0283 m³/ft³

$$= 12,593 \text{ g/min}$$

$$V_2 = V_1 \cdot \frac{P_1}{P_2} \cdot \frac{T_2}{T_1}$$

$$= 100,000 \cdot \frac{780}{760} \cdot \frac{(130 + 460)}{(420 + 460)}$$

$$= 68,810 \text{ ft}^3 / \text{min} \cdot 0.0283 \text{ m}^3 / \text{ft}^3$$

$$= 1,947 \text{ m}^3 / \text{min}$$

$$C_{mass} = \frac{(12,593 \text{ g} / \text{min}) \cdot (1 - 0.92)}{1,947 \text{ m}^3 / \text{min}} = 0.517 \text{ g} / m^3$$

$$C_{ppm} = \left(\frac{mg}{m^3}\right) = \frac{C_{mass}}{4.09 \cdot 10^{-2} \cdot MW_{SO_2}}$$

$$C_{mass} = \frac{517 \text{ mg} / m^3}{4.09 \cdot 10^{-2} \cdot 64}$$

$$C_{ppm} = 198 \text{ ppm}$$

Chapter 12: Absorption

3. Determine the stoichiometric lime consumption rate and calcium sulfate generation rate. *(Note: $CaO + H_2O \rightarrow Ca(OH)_2$)*

$$Ca(OH)_2 + SO_2 + H_2O + \frac{1}{2}O_2 \rightarrow CaSO_4 \cdot 2H_2O$$

$$56 \quad + 64 + 36 + \quad 16 \quad \rightarrow \quad 172$$

$$0.875 + 1 + 0.56 + 0.25 \rightarrow \quad 2.69$$

$$SO_2 \text{ removed } = 92\% \text{ of } 12{,}593 \text{ g/min}$$

$$= 11{,}586 \text{ g/min} \bullet 1 \text{ lb}/453.6 \text{ g}$$

$$= 25.54 \text{ lb/min}$$

$$\text{Lime} = 25.54 \text{ lb/min} \bullet 0.875 \frac{\text{lb lime}}{\text{lb SO}_2} = 22.35 \frac{\text{lb lime}}{\text{min}}$$

$$CaSO_4 = 25.54 \text{ lb/min} \bullet 2.69 \frac{\text{lb CaSO}_4}{\text{lb SO}_2} = 68.65 \text{ lb/min}$$

4. Calculate the lime slurry feed rate based on either five times the stoichiometric requirement or a liquid-gas ratio of 6.0 lb liquid/lb gas. Specify the largest required liquid feed rate.

Lime slurry is 5% by weight CaO and 22.35 lb lime/min is required.

$$\frac{22.35 \text{ lb/min}}{0.05} = 447 \text{ lb/min slurry weight}$$

Assuming the specific gravity of the slurry is equal to that of water:

$$Q_{liq} = \frac{447 \text{ lb/min}}{8.33 \text{ lb/gal}} \bullet 5 \text{ safety factor}$$

$$= 268 \text{ gal/min}$$

Based on the liquid-to-gas ratio (by weight) approach:

$$\text{weight of air} = 100{,}000 \text{ ft}^3/\text{min} \bullet 0.045 \text{ lb/ft}^3$$

$$= 4{,}500 \text{ lb air/min}$$

$$Q_{liq} = 4{,}500 \text{ lb air/min} \bullet 6 \text{ lb liquid/lb air}$$

$$= 27{,}000 \text{ lb/min slurry} \bullet 0.12 \text{ gal/lb}$$

$$= 3{,}237 \text{ gal/min}$$

Since 3,237 gal/min > 268 gal/min, liquid-to-gas ratio controls the liquid flow rate.

12.7 Problem Set

12.1 Air containing 2,000 ppm of SO_2 is contacted with pure water. What is the highest concentration of SO_2 in the water that can be achieved at 20°C?

12.2 A gas stream at 20°C and 700 mm Hg (0.921 atm) contains 1.5% H_2S. Determine the equilibrium concentration of H_2S in water at these conditions and determine what could be done with pressure and temperature to increase this concentration.

12.3 Air at 30°C containing 3,000 ppm of SO_2 is equilibrated with a solution containing 5 moles of sodium hydroxide per 100 mole of water. What mole fraction of SO_2 will the liquid have?

12.4 If Henry's Law Constant for CO_2 in a CO_2-water system is 1,420 atm/mole fraction at 78°F, what is the partial pressure of CO_2 in a 1,000 cfm exhaust if the mole fraction of CO_2 in the liquid is 0.005?

12.5 Given the following information for a packed countercurrent gas scrubber, determine the liquid flow rate in lb mol/hr ft^2.

    gas flow rate = 18 lb mol/hr ft^2

    the mole fractions of pollutant in inlet and outlet gas are 0.08 and 0.002, respectively

    the mole fractions of pollutant in inlet and outlet liquid are 0.001 and 0.05, respectively

12.6 Given the following information for a packed countercurrent gas scrubber, determine the outlet liquid concentration, X_1.

    gas flow rate: 10 lb mol/hr ft^2

    liquid flow rate: 40 lb mol/hr ft^2

    mole fraction of contaminant in inlet gas: $Y_1 = 0.02$

    mole fraction of contaminant in outlet gas: $Y_2 = 0.004$

    inlet liquid is contaminant free: $X_2 = 0$

12.7 An exhaust stream of 9,000 cfm contains 0.2% by volume SO_x and is to be scrubbed with pure water. Henry's Law Constant for this SO_x in water solution is 42.7 mole fraction/mole fraction. What is the maximum theoretical mole fraction of SO_x in the outlet liquid slurry?

12.8 Determine the packing height of a countercurrent scrubber required to reduce the SO_x concentration by 90% if dilute NaOH is used as absorbing solution, given the following information.

liquid rate = 1,000 lb/hr ft²

gas rate = 750 lb/hr ft²

mole fraction of SO_x in inlet gas = 0.02

assume no SO_x in inlet liquid

the H_{OG} = 1.63

12.9 Determine the packing height of a countercurrent scrubber required to reduce the outlet ammonia concentration by 90%, given the following information:

liquid rate (water) = 1,000 lb mol/hr ft²

gas flow rate (air) = 700 lb mol/hr ft²

mole fraction of NH_4 in inlet gas = 0.023

mole fraction of NH_4 in outlet liquid = 0.015

assume no NH_4 in inlet water stream

slope of equilibrium line = 0.93

the H_{OG} = 1.5

12.10 The vapor - liquid equilibrium relationship for ammonia (NH_3) at 30°C is given by Y = 1.32X. A gas stream flowing at 50 kg mol/hr contains 5% V/V ammonia. It is desired to reduce the concentration to 0.5% V/V. For the design of an absorber, determine:

a) the minimum liquid flow rate required

b) the liquid flow rate required at 1.5 times the minimum liquid-to-gas ratio, L/G_{min}

12.11 Calculate the tower diameter for a gas (air) flow rate of 1,750 acfm at 1 atm pressure and 20°C temperature. A pressure drop of 0.5 in. H_2O/ft of packing height is required. Plastic Pall rings, 1-inch in diameter, are to be used. The scrubbing liquid is water. The liquid-to-gas flow rate is 1.5 mol liquid/mol gas.

12.12 A gas stream containing 2.0% by volume of HCl is to be scrubbed countercurrently. The scrubber is to remove 99.5% of HCl with a liquid flow 150% above the minimum. Henry's Law Constant for HCl = 0.2.

a) Calculate the molar liquid/gas ratio if clean water is used in the scrubber.

b) Calculate the molar liquid/gas ratio if 50% of the used scrubbing water is recycled with clean water.

12.13 A packed column was designed to remove 99% of the HCl from a gas stream containing 2% HCl. It has been discovered that inlet HCl concentrations may reach 5%. Will it be possible to use the column previously designed (Problem 12.12) to reduce the concentrations to the same level as planned for the 2% HCl stream? The original column was designed with an L/G ratio equal to 500% of the minimum. What will be the relationship between the actual and the minimum L/G ratios for the new conditions?

Operation data and assumptions:

    liquid flow rate = 3,690 lb/hr

    gas mass flow rate = 5,000 lb/hr

    scrubbing liquid = pure H_2O

    packing = 1-in. Raschig rings

    H_{OG} = 2.5 ft at the operating conditions

    Henry's Law constant = 0.2

    Column height = 15 ft

    Column diameter = 2.6 ft

Use the following equations for the packed scrubber design:

$$y_2 = (0.01 y_1)/[(1-y_1)+(0.01 y_1)]$$

$$z = N_{OG} H_{OG}$$

$$N_{OG} = \ln\{[(y_1 - mx_2)/(y_2 - mx_2)](1 - 1/A) + 1/A\}/(1 - 1/A)$$

$$A = L/(mG)$$

where:
- y_2 = mole fraction of gas exiting column
- y_1 = mole fraction of gas entering column
- x_2 = mole fraction of liquid exiting column
- x_1 = mole fraction of liquid entering column
- x_1^* = equilibrium mole fraction in liquid, $x_1^* = y_1/m$
- z = height of column, ft
- N_{OC} = number of transfer units, dimensionless
- H_{OC} = height of a transfer, 2.5 ft
- m = slope of equilibrium line, mole fraction/mole fraction
- G = gas flow rate, lb/hr

12.14 An incinerator burning chlorobenzene has a HCl mass flow rate of 1,598 lb/hr in the exit gas stream. Regulations require that the HCl emissions over 4 lb/hr be controlled at an efficiency of 99%. HCl is easily absorbed in water. For this condition, the overall gas transfer units (N_{OG}) can be calculated from Equation [12.21]: $N_{OG} = \ln(y_1/y_2)$. The absorber operating temperature and pressure are 150°F and 1 atm, and the packing is 3.5-in. ceramic Raschig rings. The approximate H_{OG} is 5.5 ft. The superficial gas velocity is 8 ft/sec. Determine the tower height and diameter.

12.15 A packed tower is to be used to remove 90% of the ammonia from an air stream containing of 5% of ammonia by volume. The feed gas rate is 100 lb mol/hr at 20°C and 1 atm. Water at 90 lb mol/hr containing no ammonia is used as a solvent and the packing is 1-inch Raschig rings. Assume that the tower is isothermal at 20°C. It is to operate at 60% of flooding. Determine a) the tower diameter, b) the packed height, and c) the pressure drop. For this system m = 0.75, μ_l = 2.4 lb/hr ft, Sc_G = 0.66 and Sc_L = 570.

12.16 For the packed bed absorber shown, calculate:

a) The mass flow rate of NaOH required in gal/hr and

b) The purge flow rate in gal/hr. Assume that the purge flow (blow-down) is 5% Total Dissolved Solids, with a specific gravity = 1.0.

12.17 A company has submitted plans for a H_2S scrubber. Hydrogen sulfide is to be removed from a waste air discharge by scrubbing with a triethanolamine-water solution in a packed tower at atmospheric pressure. Specifications indicate that the gas flow rate is 10,000 acfm at 70°F and contains 15,000 ppm H_2S. The inlet liquid is to be solute-free. The H_2S must be reduced to 500 ppm (250 on SO_2 basis). The pilot plant data indicate that Henry's Law applies (m = 2.0) and the H_{OG} = 1.94 ft. The company proposes to install a 5-ft diameter, 10-ft high packed tower.

a) What is the minimum L/G (gal/1,000 ft³) that you would approve?

b) If the tower is operated at 80.65 lb moles of scrubbing liquid per minute, is it adequate?

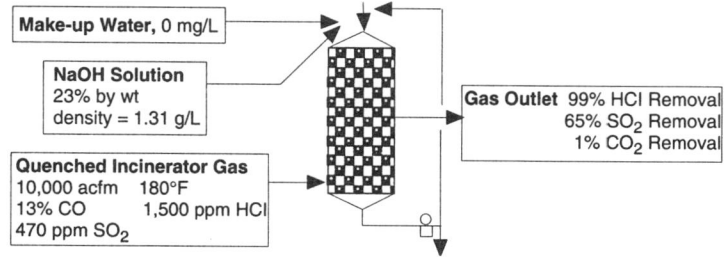

12.18 A company has submitted plans for a packed ammonia scrubber on a 1,575 cfm air stream containing 2% by volume NH_3. The emission regulation is 0.1% by volume NH_3. The company's permit supplies the following information:

tower diameter is 3.57 ft

packing height is 8 ft

operating temp is 75°F

gas flow rate is 1,575 cfm

ammonia free liquid flow rate is 1,000 lb/hr ft^2

packing is 1 1/2" Raschig rings

air density is 0.0743 lb/ft^3

Henry's Law Constant, H = 0.972 (slope of equilibrium line, m)

From Perry's Chemical Engineering Handbook, you obtained the ammonia-water absorption system figure presented below. Find a packing height, z, that would satisfy the regulation.

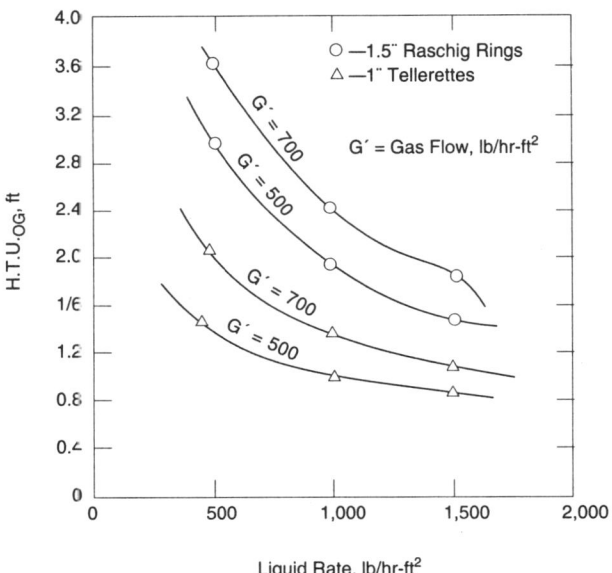

12.8 References

Bethea, R. *Air Pollution Control Technology*. Van Nostrand Reinhold. New York. 1978.

Buonicore, T. and W. Davis (Eds.). *Air Pollution Engineering Manual*. Air and Waste Management Assn. Van Nostrand Reinhold. New York. 1992.

Danielson, J. *Air Pollution Engineering Manual (AP40)*. USEPA, Research Triangle Park, NC. 1973.

Joseph, G. and D. Beachler. *Control of Gaseous Emissions*. EPA 450/2-81-005. Northrop Services, Inc. Research Triangle Park, NC. 1981.

King, W.C. *Environmental Engineering P.E. Examination Guide and Handbook.* American Academy of Environmental Engineers. Annapolis, MD. 1996.

Ledbetter, J. Air Pollution, Part B Prevention and Control. Marcel Dekker. New York. 1994.

Leva, M. *Tower Packing and Packed Tower Design.* The United States Stoneware Company. Akron, OH. 1951.

McCabe, W. and C. Smith. *Unit Operations of Chemical Engineering.* McGraw-Hill. New York. 1967.

Perry, J. (Ed). *Chemical Engineers Handbook* 5th ed. McGraw-Hill. New York. 1973.

Chapter 13

Control of Gaseous Emissions from Motor Vehicles

13.1 Introduction

Motor vehicles are a major source of air pollutants that contribute to environmental health hazards and other adverse environmental effects. Table 13.1 shows that 63% of the carbon monoxide, 34% of nitrogen oxide, and 38% of the volatile organic pollutant emissions were from mobile sources in 1990 in the United States. Sulfur dioxide and particulate matter from vehicles are of less concern because of the relative low percentage of diesel vehicles in the United States. In other countries, mobile sources of SO_2 and particulate matter may be significant.

Table 13.2 (Cooper & Alley 1994) shows the number of vehicles in use in the United States and other countries. The Organization for Economic Cooperation and Development (OECD) countries have more vehicles than non-OECD countries. The OECD has 80% of all cars, 70% of all trucks and buses, and 53% of two- and three-wheelers. Asian countries have a much greater percent of trucks and buses and two- and three-wheelers. The air pollution problems in cities in non-OECD countries are significant because of the two- and three-wheelers. Two- and

Table 13.1
Mobile Source Emissions of Air Pollutants in the U.S. in 1980 and 1990

Source Category	Pollutant emissions, millions of metric tons/year									
	CO		VOC		NO_x		TPM*		SO_x	
	1980	1990	1980	1990	1980	1990	1980	1990	1980	1990
Highway vehicles	48.7	30.3	7.7	5.1	7.9	5.6	1.1	1.3	0.4	0.6
Aircraft	1.0	1.1	0.2	0.2	0.1	0.1	0.1	0.1	0.0	0.0
Railroads	0.3	0.2	0.2	0.1	0.8	0.5	0.1	0.0	0.1	0.1
Vessels	1.4	1.7	0.4	0.5	0.2	0.2	0.0	0.0	0.3	0.2
Other off-highway	4.7	4.4	0.5	0.5	1.0	1.1	0.1	0.1	0.1	0.1
Total mobile sources	56.1	37.7	9.0	6.4	10.0	7.5	1.4	1.5	0.9	1.0
Total all sources	79.6	60.1	72.6	18.7	20.9	19.6	8.5	7.5	23.4	21.2
% Mobile sources	70	63	40	34	48	38	16	20	4	5

* Total Particulate Matter
Source: U.S. Environmental Protection Agency, EPA-450/4-91-026, 1991.

Table 13.2
Types and Numbers of Vehicles in Use, 1988

	Millions of Vehicles			
	Cars	Trucks and Buses	2- and 3- Wheelers	Total
OECD* Countries				
USA	141.3	43.1	7.1	191.5
Canada	11.9	4.0	0.4	16.3
Europe	138.5	18.0	22.0	178.5
Japan	30.8	21.7	18.2	70.7
Australia	8.8	2.3	0.4	11.5
Sub-total OECD	331.3	89.1	48.1	468.5
Non-OECD countries				
Eastern Europe	30.0	11.7	3.9	45.6
Latin America	27.3	7.9	4.3	39.5
Africa	8.0	4.4	1.2	13.6
Middle East	6.6	3.7	0.6	10.9
Asia	10.2	10.0	33.3	53.5
Sub-total non-OECD	82.1	37.7	43.3	163.1
Total	413.4	126.8	91.4	631.6

*OECD stands for Organization for Economic Cooperation and Development

Source: Reprinted by permission of Waveland Press, Inc. Cooper and Alley, Air Pollution Control: A Design Approach. Prospect Heights, IL: Waveland Press, Inc. 1994. All rights reserved

Chapter 13: Control of Gaseous Emissions from Motor Vehicles

three-wheeler vehicles use two-stroke engines with higher emissions and many of the trucks and buses are diesels that produce more particulate matter.

Figure 13.1 (Griffin 1994) indicates the severity of air pollution problems experienced by cities around the world. Most of the Pb, CO, NO_2, and O_3 is directly or indirectly due to emissions from motor vehicles.

Figure 13.1
Relative Air Quality Worldwide

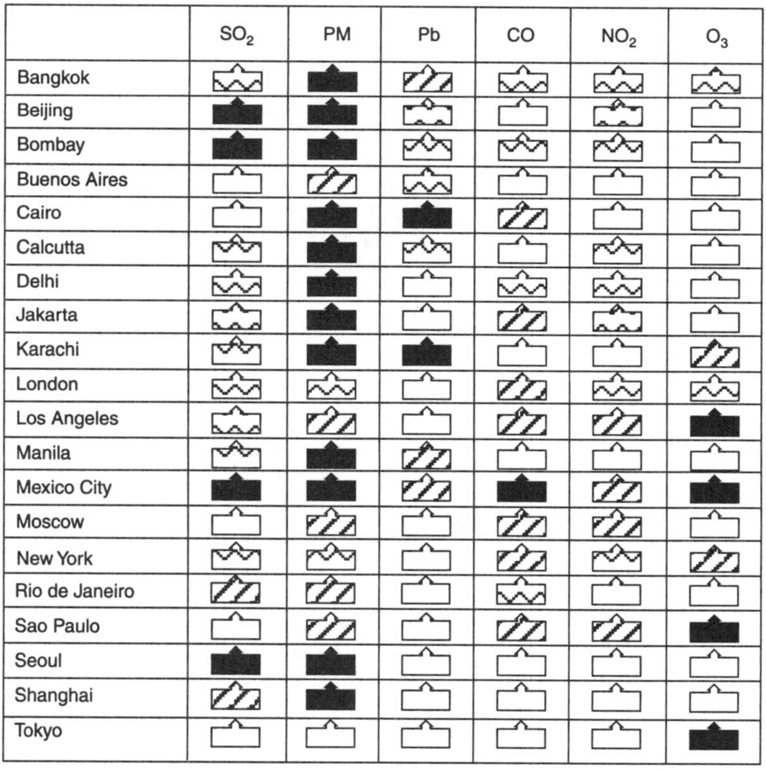

	SO_2	PM	Pb	CO	NO_2	O_3
Bangkok	∿	●	▨	∿	∿	∿
Beijing	●	●	∿		∿	
Bombay	●	●	∿	∿	∿	
Buenos Aires		▨	∿			
Cairo		●	●	▨		
Calcutta	∿	●	∿		∿	
Delhi	∿	●		∿	∿	
Jakarta	∿	●		▨	∿	
Karachi	∿	●	●			▨
London	∿	∿		▨	∿	∿
Los Angeles	∿	▨		▨	▨	●
Manila	∿	●	▨			
Mexico City	●	●	▨	●	▨	●
Moscow		▨		▨	▨	
New York	∿	∿		▨	∿	▨
Rio de Janeiro	▨	▨		∿		
Sao Paulo		▨		▨	▨	●
Seoul	●	●				
Shanghai	▨	●				
Tokyo						●

■ Serious problem, World Health Organization (WHO) guidelines exceeded by more than a factor of two

▨ Moderate to heave pollution, WHO guidelines exceeded by up to a factor of two (short-term guidelines exceeded on a regular basis at certain locations)

∿ Low pollution, WHO guidelines are normally met (short-term guidelines may be exceeded occasionally)

☐ No data available or insufficient data for assessment

13.2 Reactions in the Atmosphere

Nitrogen oxides (NO_x) and volatile organic compounds (VOCs) emissions from automobiles are responsible for the formation of photochemical air pollution that causes significant adverse effects in urban areas. Photochemically derived air pollutants are called *secondary pollutants*. Figure 13.2 (Marchello 1976) illustrates the chemical reactions that occur due to auto exhaust emissions in the atmosphere. Nitrogen oxides react with sunlight, UV, and oxygen in the air to produce ozone, O_3, and nitrogen dioxide, NO_2. The NO_2 and O_3 then react with hydrocarbons, HC, to produce aldehydes, organic oxidants, and suspended particulate matter.

Figure 13.2
Photochemical Smog Formation

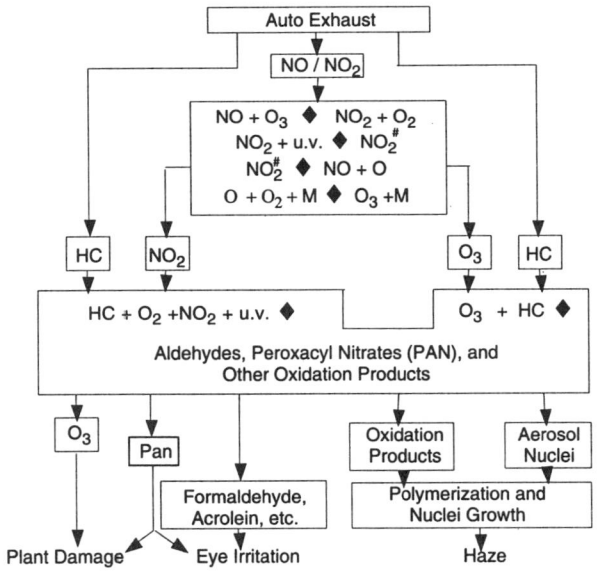

The overall reaction is given below as Equation [13.1]:

$$NO_x + \text{hydrocarbons} + \text{sunlight} + O_2 \rightarrow O_3 + \text{other irritating smog components}$$

NO and NO_2 are often treated together as one problem or as a quasi species, and written NO_x. NO is a colorless gas that has some harmful effects on health, but these effects are substantially less than those of an equivalent amount of NO_2. In the atmosphere, NO reacts with O_2 to form NO_2, a brown-colored gas that is a serious respiratory irritant. The principal concern with NO_x is that nitrogen oxides contribute to the formation of ozone, O_3, which is a strong respiratory irritant and one of the principal constituents of urban smog. Hydrocarbons will react directly in the atmosphere with oxides of nitrogen and be removed by photochemical activity to form ozone.

Table 13.3 (Griffin 1994) provides information on the sources and sinks for a number of contaminants and their estimated "half-life" in the atmosphere. The mechanism responsible for the greatest removal of air pollutants in the atmosphere is atmospheric processes themselves. These are primarily oxidation reactions and are photochemically induced. CO may react to form carbon dioxide, and thus remain in the carbon cycle. VOCs are oxidized to CO and CO_2, where they remain in the gaseous state pending further reactions and/or removals. Oxidation products, such as ozone and peroxyacyl nitrates, PAN, will remain until they are either reduced back to oxygen or further oxidized. NO_x may be transformed to nitric acid and then nitrate, with subsequent removal by mechanical action. Nitrates may be further reacted in aerosols to form ammonia and calcium nitrate salts, which may be removed by either dry or wet deposition processes. Sulfur dioxide is ultimately converted to sulfate ions, which are subject to mechanical removal.

Table 13.3
Gaseous Air Pollutants, Sources/Sinks/Residence Times

Pollutant	Anthropogenic Source	Natural Source	Rural Atmospheric Concentration	Residence Time	Principal Sinks
CH_4	Combustion	Biogenic	1.5 ppm	8 years	Reaction with OH
CO	Auto exhaust	Forest fires +photochem	0.25 ppm	1-3 months	Oxidation
CH_3Cl	Combustion	Oceanic	600 ppt	1-2 years	Stratospheric reactions
CO_2	Combustion	Biological	345 ppm	2-4 years	Biogenic
HCl, Cl_2	Combustion, manufacturing	Volcanoes	0.5 ppb	7 days	Precipitation
H_2S+sulfides	Sewage + chemicals	Volcanoes + biogenics	0.15 ppb	1-2 days	Oxidation
NO, NO_2	Mobiles & fossil fuel combustion	Biogenic & lighting	0.1 ppb	2-5 days	• Oxidation • Precipitation • Dry deposition
NH_3	Waste treatment	Biogenic	1-10 ppb	1-7 days	• Reaction with SO_2 • Oxidation
N_2O	Combustion	Biogenic	330 ppb	20-100 years	Photochemical in stratosphere
SO_2	Fossil fuel (combustion)	Volcanoes	0.05-20 ppb	2-4 days	Oxidation

13.3 Engine Operation and Air Pollution Emissions

Emissions from motor vehicles originate from the crankcase, the carburetor, the fuel tank, and the engine exhaust (Table 13.4) (Marchello 1976). Crankcase emissions are due to blowby past the piston rings which escape through the crankcase ventilation cap and draft tube. Hydrocarbon emissions from the carburetor result from fuel evaporation during the hot-soak period after shutdown. Fuel vapor loss from the tank occurs primarily when the temperature increases and during filling. Hydrocarbons and carbon monoxide in the exhaust are the products of incomplete combustion, while nitrogen oxides result from the reaction of oxygen and nitrogen contained in the combustion air.

Table 13.4
Analysis of the Automobile with No Emission Controls Showing the Sources of Emissions

Source	Pollutant %			
	CO	HC	NO_x	Particles
Exhaust	100	62	100	90
Crankcase emission		20		10
Fuel tank evaporation		9		
Carburetor evaporation		9		

Exhaust emissions from uncontrolled engines average 12 to 18 grams/mile of hydrocarbons, 80 to 125 grams/mile of carbon monoxide, and 3 to 6 grams/mile of nitrogen oxides. Pollutant formation in the combustion chamber is influenced by such factors as air/fuel ratio, ignition timing and quality, intake-manifold vacuum, engine compression ratio, engine speed and load, fuel distribution between cylinders and within a cylinder, coolant temperature, and combustion chamber configuration, flame-propagation, and deposits. The range of engine operation and the exhaust composition of each operating mode for uncontrolled engines is given in Table 13.5.

13.4 Combustion of Gasoline in Motor Vehicles

The theoretical quantity of air required to burn gasoline may be obtained by employing a theoretical fuel to represent the actual gasoline blend of hydrocarbons (see Equation [11.1]). The stoichiometric combustion of gasoline can be represented by (on a mole basis):

$$C_7H_{13} + 10.25\ O_2 + 38.54\ N_2 \rightarrow 7\ CO_2 + 6.5\ H_2O + 38.54\ N_2 \qquad [13.2]$$

And on a mass basis:

Table 13.5
Gasoline Engine Operation

	Idle	Acceleration	Cruising	Deceleration
Engine speed, rpm	400-500	400-3,000	1,000-3,000	3,000-400
Cylinder vacuum, in.Hg	16-20	0-7	7-19	20-25
Airflow, cfm	6-8	30-35	15-35	6-8
Air-fuel ratio	11:1-12.5:1	11:1-13:1	13:1-15:1	11:1-12.5:1
Exhaust analysis Precontrol:				
CO, %	4-6	0-6	1-4	2-4
NO, ppm	10-50	1,000-4,000	1,000-3,000	10-50
HC, ppm	500-1,000	50-500	200-300	4,000-12,000

$$97 \text{lb } C_7H_{13} + 328 \text{lb } O_2 + 1{,}080 \text{lb } N_2 \rightarrow 308 \text{lb } CO_2 + 1{,}170 \text{lb } H_2O + 1{,}080 \text{lb } N_2 \quad [13.3]$$

It is useful to define a mixture ratio, or air/fuel ratio (A/F). This is the ratio of the mass of air required per unit mass of fuel for combustion. That is,

$$\text{mixture ratio} = A/F = \frac{\text{mass of air}}{\text{mass of fuel}} \quad [13.4]$$

For comparative purposes, the stoichiometric air/fuel ratio is quite useful. In terms of Equation [13.2] this is

$$(A/F)_{\text{stoich}} = \frac{328 + 1{,}080}{97} = 14.5$$

If gasoline is represented by octane C_8H_{18} rather than C_7H_{13}, the air/fuel ratio is 15.1. These values are typical of the stoichiometric air/fuel ratios for many individual hydrocarbons or hydrocarbon mixtures (see Table 11.2). The ratio of the mass of fuel actually supplied to the mass of air may be larger or smaller than the stoichiometric ratio. The equivalence ratio, ER, expresses the actual fuel/air ratio.

$$ER = \frac{F/A_{\text{actual}}}{F/A_{\text{stoich}}} = \frac{A/F_{\text{stoich}}}{A/F_{\text{actual}}} = \lambda^{-1} \quad [13.5]$$

The inverse of the equivalence ratio is often called *lambda*, and is referred to as the *relative air/fuel ratio*.

The combustion reaction in the internal combustion engine happens very rapidly and is not complete so that CO and VOC's (unburned fuel) are emitted in significant quantities. The difficulties involved in eliminating exhaust emissions are illustrated in Figure 13.3 (Wark et al. 1998). Pollutant levels vary with the air/fuel ratio used in the engine. At low air/fuel ratios, the nitrogen oxides are reduced, but high carbon monoxide and hydrocarbon levels are produced. At high air/fuel ratios, all three pollutant levels are low, but operating difficulties, such as misfire and stalling, are encountered.

Figure 13.3
The Effects of the Air/Fuel Ratio on Hydrocarbon, Carbon Monoxide, and Nitric Oxide Exhaust Emissions

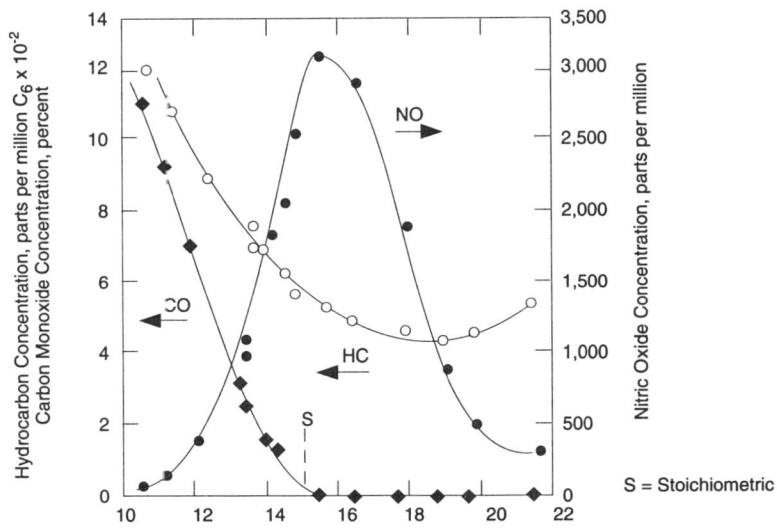

Source: Agnew 1968. General Motors Corporation

The values presented in Equation [13.2] indicate that 52.04 moles of products are formed by the combustion of 1 mole of fuel. If 0.10% of the fuel is unburned and is exhausted with the combustion products as unburned hydrocarbon, the exhaust will contain approximately 20 ppm of unburned hydrocarbons. Should the quantity of unburned fuel be 1%, the exhaust would contain roughly 200 ppm.

Figure 13.4 illustrates the variety of air pollutants (organic compounds) in the exhaust of a spark-ignited, gasoline-powered four-stroke engine. A comparison of the fuel composition with exhaust hydrocarbon composition demonstrates the strong correlation between the exhaust species distribution and the fuel species.

Carbon monoxide, like unburned hydrocarbons, results from the incomplete combustion of the fuel. Consequently, those conditions that promote or enhance complete combustion are intended to reduce the quantity of carbon monoxide in the exhaust gas of the engine. Figure 13.3 shows that as the air/fuel ratio increases from 10 to 16, the CO in the exhaust gas decreases from 12% to nearly zero. Thus, one method of reducing the CO emission is to operate the engine with lean A/F

Figure 13.4
Variety of Air Pollutant in the Exhaust vs. Gasoline

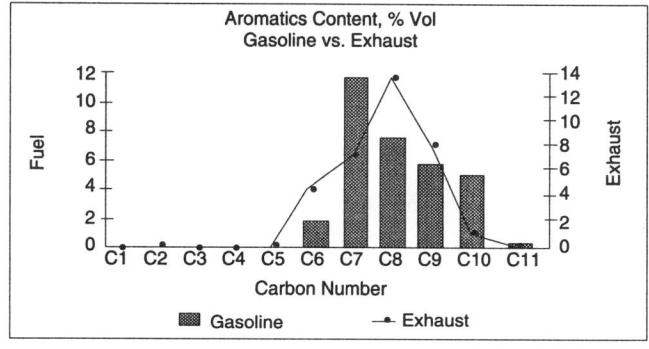

Source: Griffin 1994

mixture ratios. The value of the air/fuel ratio also exerts a major influence upon the quantity of unburned hydrocarbons emitted by a given engine. Figure 13.3 shows that the value of unburned hydrocarbons in the exhaust gas first decreases as the air/fuel ratio increases from 11 to approximately 16, and then increases as the air/fuel ratio increases further to 22.5. The increase in unburned hydrocarbons is attributed to what is known as *misfire*. The mixture is so lean that combustion does not always proceed from the ignition spark.

The production of NO is maximized at high temperatures (see next section) in the presence of extra oxygen and Figure 13.3 shows high values for this pollutant near stoichiometric combustion conditions that produce the maximum temperatures.

Example 13.1 CO Mole Fraction Computation (DeNevers, Air Pollution Control Engineering, 1995. Reproduced with permission of the McGraw-Hill Companies)

Calculate the expected CO mole fraction for combustion of a gasoline with C_8H_{17}, and with the air supplied being 90% of that required for complete combustion.

Solution:

For complete combustion, use the general Equation [11.1]

$$C_xH_y + \left(x + \frac{y}{4}\right)O_2 \rightarrow xCO_2 + \left[\frac{y}{2}\right]H_2O$$

If there is not enough oxygen to complete the reaction, then the combustion products will contain CO and unburned hydrocarbons. If we assume that the oxygen fed is less than the stoichiometric amount with an oxygen deficit of z moles per mole of fuel, and further assumed that all of the oxygen deficit causes CO formation, Equation [11.1] can be rewritten as:

$$C_xH_y + \left(x + \frac{y}{4} - z\right)O_2 \rightarrow (x - 2z)CO_2 + \left[\frac{y}{2}\right]H_2O + (2z)CO$$

Each mole of oxygen from air brings with it (0.79/0.21 = 3.76) moles of nitrogen, so that the total moles of combustion products will be:

$$n_{total\ out} = 3.76\left(x + \frac{y}{4} - z\right) + (x - 2z) + \left(\frac{y}{2}\right) + 2z$$

and the mole fraction of CO will be:

$$y_{CO} = \frac{2z}{3.76[x + (y/4) - z] + z + (y/2)}$$

For this example, x = 8 and y = 17

$$n_{oxygen} = 0.9 n_{stoich\ oxygen} = \left(x + \frac{y}{4} - z\right) = 0.9\left(x + \frac{y}{4}\right)$$

from which

$$z = 0.1\left(x + \frac{y}{4}\right) = 0.1\left(8 + \frac{17}{4}\right) = 1.225$$

and

$$y_{CO} = \frac{2 \bullet 1.225}{3.76[8 + (17/4) - 1.225] + 8 + (17/2)} = 0.042 = 4.2\%$$

13.5 Nitrogen Oxide Emissions

The nitrogen oxides are formed principally by reactions between atmospheric oxygen and nitrogen inducted into the engine. The quantities of oxides of nitrogen formed are complicated functions of temperature, pressure, time of reaction, and the quantities of the reactants present. Nitric oxide is formed within the combustion chamber during the time period that maximum temperature exists, and it persists in above equilibrium quantities during expansion and exhaust.

Emissions of oxides of nitrogen result from the fixation of atmospheric nitrogen in the flame zone. The principal high-temperature reaction is the formation of nitric oxide:

$$N_2 + O_2 \leftrightarrow 2NO \qquad [13.6]$$

Some of the nitric oxide formed is further oxidized to nitrogen dioxide:

$$NO + \tfrac{1}{2}O_2 \leftrightarrow NO_2 \qquad [13.7]$$

Nitrogen oxides are produced in all fossil-fuel combustion processes using air as the oxidant. The thermodynamic equilibrium of Equation [13.6] is given by

$$K_e = \frac{(NO)^2}{(N_2)(O_2)} = 21.9 \exp\left(\frac{-43,400}{RT}\right) \qquad [13.8]$$

where NO, N_2, and O_2 are in mole fractions, R is the gas constant (1.987 cal/g mol °K), and T is the absolute temperature in °K. Table 13.6 provides the equilibrium constants as a function of temperature.

Table 13.6
Equilibrium constants for the formation of NO by Equation [13.6] and NO_2 by Equation [13.7]

Temperature		K_e of Formation	
K	°F	NO	NO_2
300	80	$7 \cdot 10^{-31}$	$1.4 \cdot 10^6$
500	440	$2.7 \cdot 10^{-18}$	4.9
1,000	1,340	$7.5 \cdot 10^{-9}$	0.11
1,500	2,240	$1.07 \cdot 10^{-5}$	0.011
2,000	3,140	0.00040	0.0035
2,500	4,040	0.0035	0.0018

Source: DeNevers, Air Pollution Control Engineering, 1995. Reproduced with permission of the McGraw-Hill Companies.

The rate of NO formation in g mol/cm^3 sec is given by Equation [13.9] below:

$$\frac{d(NO)}{dt} = 9 \cdot 10^{14} \left[\exp\left(\frac{-135,000}{RT}\right)\right](N_2)(O_2)^{\tfrac{1}{2}} - 4.1 \cdot 10^{13} \left[\exp\left(\frac{91,600}{RT}\right)\right]\frac{(NO)^2}{(O_2)^{\tfrac{1}{2}}}$$

assuming the presence of excess air and equilibrium for oxygen atom formation from oxygen molecules.

For most stationary combustion processes, the residence time available is too short for the oxidation of nitric oxide to nitrogen dioxide, the thermodynamically favored species at lower temperatures. As a result, no more than 5 to 10% of this reaction takes place prior to emission. Heat release rates from combustion and heat removal rates to heat transfer surfaces may be counterbalanced to control NO_x formation. High heat release increases NO_x levels because of the sensitivity of the reaction equilibrium and kinetics to peak temperature (Equation [13.8]).

Example 13.2 NO_x Equilibrium Concentrations

Calculate the equilibrium concentrations of NO and NO_2 for air that is held at 2,000°K (3,140°F) long enough to reach chemical equilibrium.

Solution:

The definition of the equilibrium constant is:

$$K_e = \frac{[NO]^2}{[O_2][N_2]} \qquad [13.8]$$

Solving Equation [13.8] for equilibrium concentration of NO:

$$[NO] = (K_e[N_2][O_2])^{1/2}$$

Substituting values, including the value of K_e from Table 13.6 into this equation:

$$[NO] = (0.0004[0.78][0.21])^{1/2} = 0.0081 = 8,100 \text{ ppm}$$

Solving Equation [13.7] for $[NO_2]$:

$$[NO_2] = (K_e[NO][O_2])^{1/2}$$

Substituting values, including the value of K_e from Table 13.6 into this equation:

$$[NO_2] = (0.0035[0.0081][0.21])^{1/2} = 2.44 \cdot 10^{-3} = 2,440 \text{ ppm}$$

Example 13.2 demonstrates that the equilibrium concentrations of nitrogen oxides in combustion gases at 3,140°F are 8,100 ppm of NO and 13 ppm of NO_2. However, because actual residence times are short and mixing is incomplete, Table 13.7 shows that these levels are not reached in operating combustion units. How much thermal NO is actually formed in a flame as well as how much is then converted back to N_2 and O_2 as the gases cool is a strong function of how fast the gases heat and cool in the flames. Table 13.7 provides estimated peak temperature and combustion times for various kinds of burners (DeNevers 1995) and provides typical NO_x emission levels for some sources based on measurement of exit gases.

Table 13.7
Estimated Peak Temperatures, Combustion Times, and NO_x Production in Various Kinds of Burners

Type of Flame or Burner	Estimated Approximate Peak Temperature, °F	Estimated Approximate Combustion Time, sec	Typical NO_x Levels, ppm
Kitchen gas stove, hot water heater, furnace	2,100	0.005	25-75
Propane torch	2,500	0.001	300-1,000
Medium size industrial furnace without preheating	2,900	1-2	200-700
Large coal-fired furnace with preheating	3,500	2-4	300-400
Automobile engine (at 2000 RPM)	4,000	0.0025	1,000

Source: DeNevers, Air Pollution Control Engineering, 1995. Reproduced with permission of the McGraw-Hill Companies.

The abatement of nitrogen oxide emissions from stationary combustion sources depends primarily upon equipment designs which avoid their formation. Stationary combustion sources are able to meet emission regulations through modifications such as staged or overfired combustion and the use of partial recycle of stack gas. Without control, power-plant boilers using natural gas release gases containing from 300 to 1400 ppm of nitrogen oxides (Table 13.7). Units using oil range from about 200 ppm for tangentially fired units up to 700 ppm for horizontal firing. Coal-fired units emit between 300 and 600 ppm. Domestic heaters and incinerators give lower emissions, on the order of 25 to 75 ppm.

Example 13.3 Auto Combustion Products (DeNevers, Air Pollution Control Engineering, 1995. Reproduced with permission of the McGraw-Hill Companies)

Estimate the time that combustion products are at high temperature in (a) a large coal-fired boiler and (b) in an auto engine.

Solution:

(a) In a typical 500 MW coal-fired power plant that uses pulverized coal, the gas that flows through the system is about 1.3 Mscfm, which is equivalent to 5.3 Macfm at the firebox outlet temperature of 1,800°F. The main firebox is roughly 46 ft • 46 ft • 165 ft high. The flow is vertically upward with the average velocity given by:

$$V_{avg} = \frac{Q}{A} = \frac{5.3 \cdot 10^6 \text{ ft/min}}{(46 \text{ ft})^2} = 2,505 \text{ ft/min} = 42 \text{ ft/sec} = 13 \text{ m/sec}$$

and the time to traverse the 165-ft high firebox is 165/42 = 4 sec. This value somewhat overstates the time the gas spends in the hottest part of the furnace, because 1) much of the gas is admitted above the bottom of the furnace and 2) this value is based on the outlet temperature. At an average temperature of 3,000°F, one would calculate a time of 2.6 sec.

(b) In an automobile engine at 2,000 RPM, the time for one revolution is 0.03 sec. The combustion takes place in perhaps one-twelfth of this time, so the estimated time of the combustion is 0.0025 sec = 2.5 m sec. Here the distance traveled by the flame front is about 2 inches, so that the flame speed must be:

$$V_{flame} \approx \frac{2 \text{ in.}}{0.0025 \text{ sec}} = 800 \text{ in./sec} = 66.7 \text{ ft/sec} = 20.3 \text{ m/sec}$$

This speed is reasonable for a premixed flame, but would not be plausible for a diffusion flame. The high NO_x emission levels (1,000 ppm) for automobiles shown in Table 13.7 is due to the high combustion temperatures present in the combustion zone and the short residence time that prevents slow enough gas cooling for conversion of NO_x back to N_2 and O_2.

13.6 Emission Standards

In the 1950s, typical new cars were emitting nearly 13 grams/mile of hydrocarbons, 3.6 grams/mile of NO_x and 87 grams/mile of CO. By 1996, standards set by USEPA and applied to automobiles had reduced emissions for new vehicles to less than 0.25 grams/mile of hydrocarbons, 0.4 grams/mile of NO_x and 3.4 grams/mile of CO, representing 98, 90, and 96% reduction in HC, NO_x, and CO, respectively. Despite controls, the emissions from highway vehicles have continued to be a serious problem, primarily due to the increased number of vehicles. This has tended to offset emission reductions achieved on each vehicle.

Table 13.8 presents the emission standards for tailpipe emissions in the United States. The standards are reported as emission factors (average rate emission for a defined activity). Emission factors are given in units of grams of pollutant per mile driven and are calculated based on a pre-set driving cycle that includes various time periods of vehicle idle, acceleration, cruise, and deceleration (see Table 13.5).

The 1990 Clean Air Act contained proposed emission limits, known as Tier 2 Standards, for Light Duty Vehicles, of 0.2 grams per mile for NO_x and 1.7 grams per mile for CO that could go in to effect in 2004 if needed to reduce air pollution levels in urban areas. The USEPA Tier 2 draft report suggests that the lower standards have both technical and economic feasibility.

Table 13.8
Light Duty Vehicles (LDV) Exhaust Emission Standards, grams/mile[a]

Year	Carbon Monoxide	Hydrocarbons	Oxides of Nitrogen	Particulate Matter
Prior to Controls	3.4% [b]	850 ppm	1,000 ppm	-
1980	7.0	0.41	2.0	-
1981	3.4	0.41	1.0	-
1982-86	3.4	0.41	1.0	-
1987-93	3.4	0.41	1.0	0.60 [c]
1994+	3.4	0.41	0.4	0.20 [c]
1996	3.4	0.25 (non-methane)	0.4	0.08 [d]

[a] The federal test procedure since 1975, has been CVS-75, a constant volume sample test that includes hot and cold starts
[b] % = percent by volume
[c] Applied to diesels only
[d] All LDVs

To reach these high levels of control for automobiles requires substantial modifications to internal combustion engines. Add-on air pollution control devices have been developed to meet the emission standards. These include catalytic converters, carbon adsorption systems, and emission recycle to the combustion chamber.

Control of fuel composition plays an important role in the control of air pollution emissions in the United States. Gasoline volatility is measured by the Reid Vapor Pressure (RVP) (see Chapter 9). The higher the RVP, the greater the tendency for the fuel to evaporate. Higher RVPs produce higher VOC emissions from cars, from refueling operations, and from fuel storage. USEPA mandates have resulted in gasolines with lower RVPs, especially during the summer months (the peak ozone season). The Clean Air Act of 1990 specifies that the Reid Vapor Pressure (RVP) shall not exceed 9.0 pounds per square inch (psi) unless the fuel blend is for gasoline with 10% ethanol, in which case the RVP requirement is 10.0 psi.

Recent emphasis has been to lower the emissions of air toxics, such as benzene, and to lower the percentages of those components of gasoline which are photochemically reactive (those which promote greater ozone production per unit mass). Figure 13.5 (Griffin 1994) indicates the photochemical oxidation potentials for different classes of organic compounds which contribute to photochemical ozone production. Reducing the amount of class 5 and 6 compounds in gasoline reduces the gasoline vaporization into the atmosphere.

Fuel modification is required for the nine worst ozone nonattainment areas in the United States with a 1980 population greater than 250,000. These requirements include reformulated and oxygenated fuels. A reformulated fuel must meet certain general requirements for NO_x, oxygen, benzene, and heavy metal content and must

Figure 13.5
Hydrocarbon Reactivity as Indicated by the Rate at which NO is Oxidized

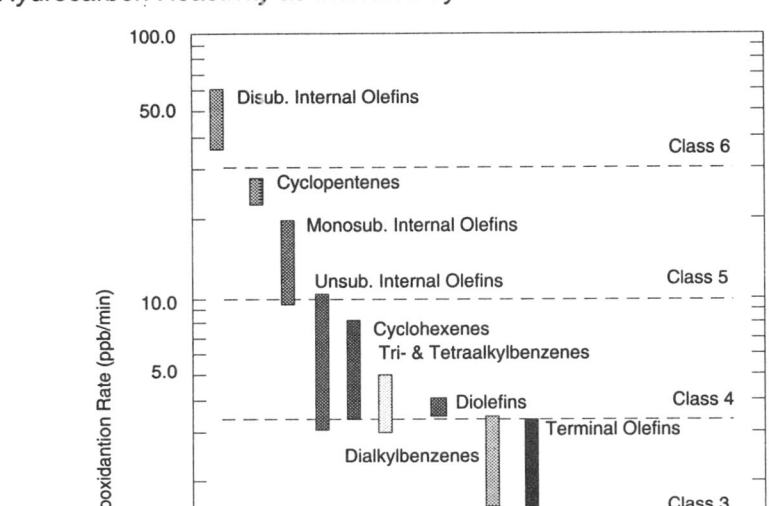

also achieve reductions in ozone-forming VOCs and toxic air pollutants. Table 13.9 details the minimum composition requirements for reformulated fuels. These include a minimum 2.0% by weight oxygen content, a benzene level of 1.0% or less by volume, and a prohibition on any heavy metals, including lead or manganese. The aggregate limit of aromatic hydrocarbons is 25% by volume, and the use of additives is required as needed to meet VOC and toxic emission standards. In addition, the emission of ground level ozone-forming compounds and toxic air pollutants must be reduced by 25% and 20%, respectively, from the aggregated baseline levels.

CO nonattainment areas must use oxygenated fuels that have a minimum content of 2.7% oxygen by weight. These are required to be sold during the high CO portion of the year (typically wintertime conditions). Since the combustion of

Table 13.9
Composition of Baseline and Reformulated Gasolines

	Baseline gasoline	Reformulated fuel
Constituent, %		
Oxygen	0	2.0 (min)[b]
Benzene	1.53	1.0 (max)
Lead	[a]	0.0
Aromatics	32	25 (max)
Sulfur, ppm	339	-
NOx	-	No increase
Reductions (2000 A.D.), %:		
Ozone VOCs	-	25
Toxics[c]	-	20 (min)

[a] Not applicable
[b] 2.7% (min) in CO nonattainment areas for Oxygenated Fuels
[c] Total emissions of: benzene, formaldehyde, acetaldehyde, 1,3 butadiene, and polycyclic organic matter (POM)

these reformulated gasolines can cause increased emissions of NO_x, the new law also prohibits vehicles which are using reformulated gasoline from increasing NO_x emissions beyond the levels associated with vehicles burning conventional fuels.

As a result of the Clean Air Act Amendments, the state of California has mandated the gasoline composition requirements in Table 13.10. The lower sulfur levels in gasoline are required because current sulfur content of gasoline is high enough to impair the function of catalytic converters by poisoning the catalyst. Sulfur atoms can bond with the catalyst surface preventing the breakdown of NOx and hydrocarbons.

Table 13.10
1996 California Gasoline Composition Limits

Component	Before 1996	After 1996
Aramotics[a]	32%	25%
Olefin[a]	9.9%	6%
Benzene[a]	2%	1%
Sulfur, ppm	150	40
Volatility, psi	7.8	7
90% Boiling point, °F	350	300
50% Boiling point, °D	220	210

[a] % by volume

13.7 Control Devices

13.7.1 Crankcase and Evaporative Emissions

Since the flow rate of pollutants (HC, CO) from the crankcase is at least an order of magnitude lower than the exhaust flow rate, incineration can be employed as a common means of treating these combustible pollutants when the engine is running. Crankcase blowby occurs only during engine operation; thus, direct incineration of all blowby is the control method used. The general features of the system are shown in Figure 13.6. The combustion gases that leak past the piston rings into the crankcase are metered through a positive crankcase ventilation (PCV) valve into the intake manifold of the engine and burned in the combustion chamber. The fresh air for crankcase ventilation is normally drawn from the air cleaner so that if the blowby exceeds the capacity of the PCV valve, the backflow out the crankcase air intake will be entrained in the air flow to the carburetor or intake manifold.

Figure 13.6
Crankcase Blowby Recycle System

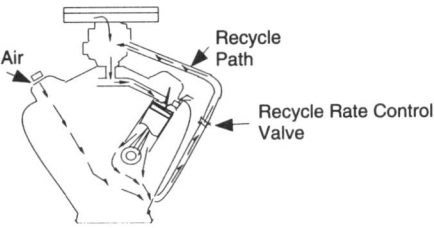

Fuel-evaporation emissions from the carburetor and fuel tank are controlled by a canister filled with activated carbon that adsorbs the hydrocarbon vapors (see Figure 13.7). The adsorbed vapors are desorbed by an air purge during power operation and sent to the carburetor. Without control, fuel tank losses are approximately 30 grams/day and carburetor losses are roughly 40 grams/day. The canister system reduces these losses to less than 6 grams/day. The Clean Air Act of 1990 requires that all vehicles must have on-board vapor recovery systems that are at least 95% efficient.

When the engine is running at sufficient speed, air is drawn through the charcoal canister in the opposite direction, stripping the adsorbed fuel off the charcoal and returning it to the engine air inlet, thus regenerating the adsorbent and making it ready for its next adsorption task. Suitable valves are used to guarantee that the flows are always in the correct direction at the right times. Activated charcoal can store 30 to 40 grams fuel/100 grams charcoal (Figure 13.11). Typically, vehicles may emit 100 to 200 grams of fuel/day depending primarily on ambient and vehicle temperature changes, fuel volatility, fuel tank vapor volume, and carburetor fuel bowl volume. A typical vehicle canister may employ 1,000 grams of charcoal.

Figure 13.7
Evaporative Emission Control System

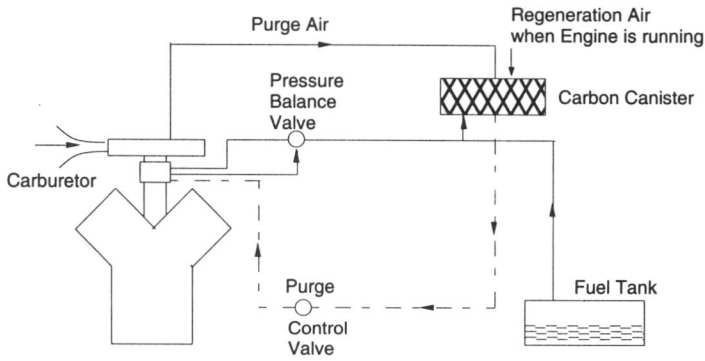

The charcoal canister is large enough to handle the emissions due to thermal expansion in the fuel tank and hot soak emission. It is not used to deal with the displacement vapor emissions that occur when the tank is filled (see Example 9.4).

Example 13.4 Auto Evaporative Emissions Control

The charcoal in a canister for recovery of auto evaporation losses contains 800 grams of charcoal and can hold 0.3 grams HC/gram of charcoal. If a typical gas tank holds 15 gallons, calculate the displaced HC vapor in the tank that could be recovered by the carbon system.

Solution:

Find Emission Factor from Table in Example 9.4.

The emission rate for uncontrolled displacement losses = 9 lb/10^3 gal.

$$\frac{9 \text{ lb}}{10^3 \text{ gal}} \bullet 15 \text{ gal} = 135 \bullet 10^{-3} = 0.135 \text{ lb / refuel} = 61.3 \text{ g / refuel}$$

$$\frac{61.3 \text{ g HC}}{\text{refuel}} \bullet \frac{\text{g carbon}}{0.3 \text{ g HC}} = 204 \text{ grams of carbon required}$$

The canister would be able to adsorb the emission amount if it is not already saturated due to other emissions. However, because of the rapid rate of emission (~10 gal/min fuel rate), a fan would be required to overcome the pressure drop in the carbon bed. The present system does not require a fan because of the low rate of emission of the evaporative losses.

13.7.2 Catalytic Converters for Control of Exhaust Emissions

Catalytic converters are used to treat the engine exhaust emissions in order to meet the standards listed in Table 13.8. The catalyst provides for the oxidation of CO and VOCs to end products CO_2 and H_2O, and the chemical reduction of NO_x to N_2 and O_2. A three-way catalytic converter performs both oxidation and reduction simultaneously in one unit (Figure 13.8). This three-way catalyst requires precise computer control of the engine A/F ratio driven by a real-time oxygen sensor in the exhaust system because the A/F ratio strongly influences emissions (Figure 13.3). Table 13.11 lists the emissions from a typical test on a three-way catalyst-equipped vehicle.

Figure 13.8
Three-Way Catalytic Converter Systems

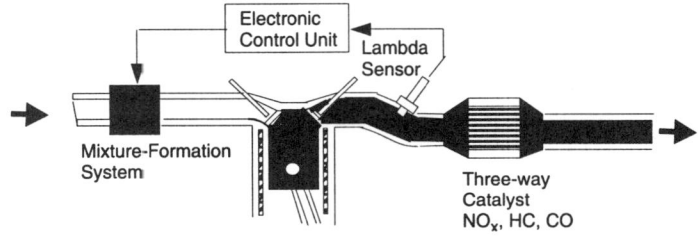

Table 13.11
Gasoline Engine Exhaust Emission Equipped withThree-Way Catalyst

Pollutant	Emissions, g/km
Total hydrocarbons	0.14
Oxides of nitrogen, total	0.38
NO	0.35
N_2O	0.031
NO_2	0.005
Carbon monoxide	1.8
Toxics, total	0.0094
Benzene	0.0069
1,3 butadiene	0.00052
Formaldehyde	0.0011
Acetaldehyde	0.00077

Figure 13.9 shows the tight control currently used on motor vehicles with three-way catalysts. Present practice is to burn the fuel at a precisely controlled, slightly rich mixture (ER = 0.98-0.99). This prevents excessive NO_x formation without

producing excessive amounts of CO and VOCs. The desired A/F ratio is provided by fuel injectors with computer-controlled fuel and air injection rates, which are monitored and adjusted by the computer to optimize the combustion mixture for different driving conditions.

Figure 13.9
Effect of the Air/Fuel Ratio on Conversion Efficiencies of a Three-Way Catalyst

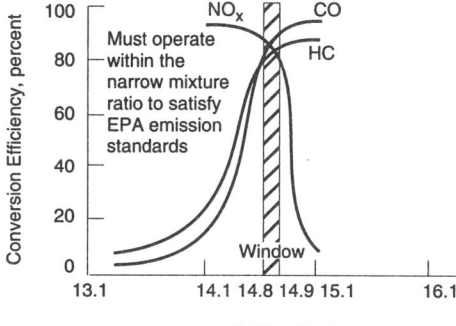

Source: Niepoth, Gumbleton, and Haefner, 1978. General Motors Corporation.

Typical design criteria for an auto exhaust catalytic converter are:

1) 90-99% overall destruction of CO, HC, and NO_x in a small, lightweight, package. The catalyst needs to be encased for easy installation underneath an automobile.

2) Destroy CO, HC, and NO_x at as low a temperature as possible. Typical catalysts do not begin to promote a reaction until they are heated by exhaust gas to their "light-off" temperature of about 350°C (662°F). Thus, they are inactive during the period of highest emissions; that is, during cold start, and this reduces the overall efficiency of the catalyst so that higher efficiencies are required during high temperature operation to offset the cold start emissions.

3) Control excessive surface temperatures to prevent excessive heat flow to the surrounding environment.

4) Perform satisfactorily for 100,000 miles (required by USEPA regulations).

Vehicle catalysts operate in the range of 250 to 600°C (480 to 1,100°F) with space velocities of 10,000 to 100,000/hr (volume flow of exhaust gas per hour divided by bulk volume of catalyst), depending on the operational mode of the vehicle (idle, cruise, or acceleration). Figure 13.10 illustrates the actual performance of a cata-

Figure 13.10
Performance of a Catalytic Converter with a Three-Way Catalyst

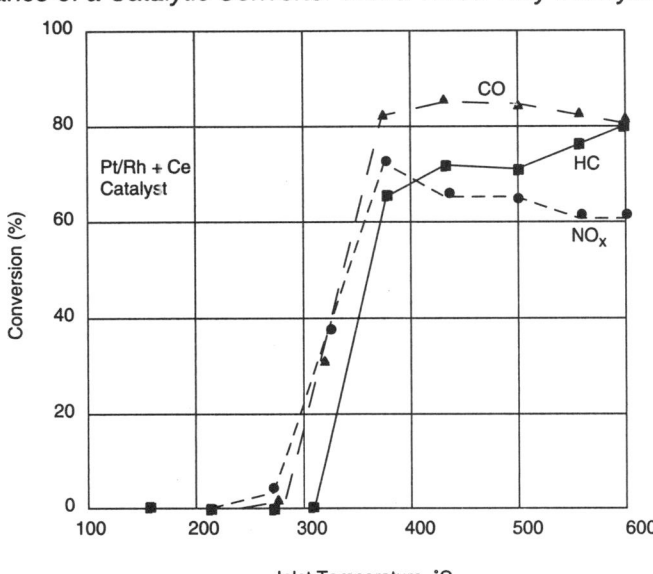

Source: From *Air Pollution, Its Origin and Control*, 3rd edition, by Wark, Warner, and Davis; Copyright 1998 by Addison Wesley Longman, Inc. Reprinted with permission.

lytic converter equipped engine that was tested on a dynamometer showing the conversion efficiencies for CO, HC, and NO_x (Wark, Warner, and Davis 1998). The catalyst and substrate volume was 1.68 liters (8 • 7 • 14 cm). The results show the typical response of a three-way catalyst. Below 300°C, the catalyst was inactive. The light-off temperature, defined as the temperature at which conversion efficiency reaches 50%, occurred at about 325°C above which the conversion efficiencies of CO, HC, and NO_x increased rapidly. Above 350°C, the efficiencies remained relatively stable.

Example 13.5 Catalytic Converter Conversion Efficiency

The flow rate of NO_x and HC measured at the exhaust manifold are 3.5 g/mile and 13 g/mile, respectively. Calculate the required efficiency of conversion that would be required of a catalytic converter to satisfy the emission standards in Table 13.8.

Solution:

From Table 13.8, the emission standard for NO_x is 0.4 g/mile.

$$\text{conversion efficiency} = \frac{3.5 - 0.4}{3.5} \cdot 100 = 88.6\%$$

The emission standard for HC is 0.25 g/mile.

$$\text{conversion efficiency} = \frac{13.0 - 0.25}{13.0} \cdot 100 = 98.1\%$$

13.8 Design of Catalytic Converters Based on Mass Transfer Considerations

The overall rate of catalytic reactions depends on two steps: transport of reactants to the catalytic surface and the reaction rate at the surface. The overall mass transfer coefficient, K_o, can be defined as a function of the gas phase mass transfer coefficient, K_m, and the reaction rate coefficient, K_r, as shown in Equation [13.10] (Retallick 1981):

$$\frac{1}{K_o} = \frac{1}{K_m} + \frac{1}{K_r} \qquad [13.10]$$

At steady state, the rate of mass transfer and the rate of reaction are the same. However the rate may be limited by either of the two steps. At temperatures above 525°K, mass transfer usually controls the rate of the process. Thus, the design of catalytic units reduces to specification of the proper length of bed to permit sufficient residence time, based on mass transfer rates, to achieve the desired degree of destruction.

For a channel in a catalytic reactor, as shown in Figure 13.11, the mass transfer rate of reactant A can be expressed as a function of x:

Figure 13.11
One Channel of a Catalyst Reactor

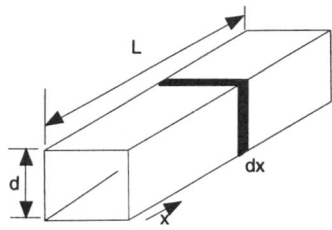

Note: This channel is part of a honeycomb catalyst used on present day motor vehicles.

When mass transfer is the limiting factor, the surface concentration of the reactants is nearly zero. For that case, a material balance of reactant, A, in a differential reactor volume is:

$$\frac{d[A]}{dl} = -\frac{aK_m[A]}{v} \qquad [13.11]$$

where:
l = the longitudinal distance along the channel
a = is the surface area per unit volume of the reactor
K_m = the mass transfer coefficient
v = the bulk velocity of the fluid

For a given reactor length, L, the concentration at the reactor outlet, C_L, is

$$C_L = C_o \exp\left[-\frac{aK_m L}{v}\right] \qquad [13.12]$$

When $L = v/K_m a$, $C_x = C_o e^{-1}$. By designating $v/K_m a$ as the length of a mass transfer unit:

$$L_m = \frac{v}{K_m a} \qquad [13.13]$$

Equation [13.13] provides the distance, L_m, the gas must travel in the catalyst to reduce the concentration by one mass transfer unit (63.21% reduction).

For a given reactor of length, L, the number of transfer units and C at the reactor outlet can be calculated:

$$N = \frac{L}{L_m}, \quad C = C_o e^{-N} = e^{-L/L_m} = \exp\left(-\frac{aK_m L}{v}\right) \qquad [13.14]$$

For catalysts with small tubes, the gas flow is laminar (i.e., Re < 2,100). In laminar flow, the gas film mass transfer coefficient, K_m, is related to the Sherwood Number, Sh, (which has the magnitude of 4.4 for cylindrical channels) by:

$$Sh = \frac{\text{Total mass transfer}}{\text{Molecular mass transfer}} = \frac{K_m D}{\mathcal{D}} \qquad [13.15]$$

where:
$\mathcal{D}$ = Diffusion coefficient
D = Effective diameter of the channel

$$K_m = \frac{4.4\mathcal{D}}{D} \quad [13.16]$$

The surface area, a (cm²/cm³), inside the tube of the honeycomb catalyst is given by:

$$a = \frac{L\pi D}{L\pi(D^2/4)} = \frac{4}{D} \quad [13.17]$$

Solving for the tube diameter D = 4/a

$$L_m = \frac{v}{K_m a} = \frac{v}{(4.4\mathcal{D}/D)\bullet(4/D)} = \frac{vD^2}{17.6\mathcal{D}} \quad [13.18]$$

The gas pressure drop across the catalyst can be calculated by:

$$\Delta P = \frac{2 f L v^2 \rho}{g D} \quad [13.19]$$

The Fanning friction factor, f, is 16/Re for circular tubes and 14/Re for square channels. The other variables are:

- L = total length of catalyst tube
- v = the gas velocity
- ρ = the gas density
- g = the gravitational constant
- D = effective diameter of the channel or tube.

Space velocity, SV, for catalytic converters is defined as the gas flow rate per unit volume of catalyst. This parameter is also often used for the design of catalytic combustion incinerators applied to industrial sources (see Chapter 11).

$$SV = \frac{v}{L} \quad [13.20]$$

SV = space velocity, hr⁻¹

and Equation [13.12] can be rewritten as

$$C/C_o = \exp\left(-\frac{aK_m L}{v}\right) = \exp\left(-\frac{aK_m}{SV}\right) = \exp-\frac{17.6\mathcal{D}}{D^2(SV)} \quad [13.21]$$

This equation allows the two design methods used for catalytic combustion [space velocity (see Chapter 11) and mass transfer] to be compared.

Catalytic combustors can be used upstream from gas turbines. Here, the advantages of catalytic combustion are threefold: less NO_x production because the combustion temperature is lower; less emission of unburned hydrocarbons and carbon monoxide; and greater heat release per unit volume. A design that improves heat release is the graded-cell support, in which hole size decreases in the direction of flow.

In practice, the calculated length of catalyst should be considered a minimum effective length for achieving the desired conversion. Doubling the calculated length will provide an adequate safety factor in most cases; but a greater safety factor may be needed if the exhaust gases contain dust or poisons that could blind the catalyst.

The design of catalytic oxidation units based on mass transfer considerations can be applied to the control of emissions from both industrial sources and motor vehicles, as shown by the following examples.

Example 13.6 Catalytic Converter Design

A 4.0 liter, V-6 engine that gets 20 miles/gallon of gasoline has uncontrolled hydrocarbon emissions of 13.0 grams hydrocarbons/mile traveled.

a. Design a catalytic converter to remove (oxidize) 99.99% of the VOCs (benzene, diffusivity = 5.29 • 10^{-4} ft^2/sec) in the exhaust gas stream. The catalyst (a monolith substrate with 400 holes/$in.^2$) will need to operate at 1,000°F. The open area of the catalyst perpendicular to the gas flow is 77%. Velocity in the catalyst will be 20 ft/sec at 1000°F, 1 atm pressure at 50 miles/hr. Determine the required catalyst length, cross-sectional area perpendicular to the gas flow, gas residence time, space velocity, and the VOC emissions from the catalytic converter in grams/mile.

b. Compare these emissions with the US EPA standards (98% required) and discuss the results.

Solution:

$e^{-N} = 0.0001$, $N = 9.21$

$D = 0.044$ in.$= 0.00367$ ft

Using Equation [13.16], $K_m = \dfrac{4.4 \mathcal{D}}{D} = \dfrac{4.4 \bullet 0.000529 \text{ ft}^2 \text{ / sec}}{0.00367 \text{ ft}} = 0.6342$ ft / sec

From Equation [13.17], $a = \dfrac{4}{D} = \dfrac{4}{0.00367 \text{ ft}} = 1089.9$ ft^2 / ft^3

Chapter 13: Control of Gaseous Emissions from Motor Vehicles

From Table 13.12, the percent open area for a 400 hole/in.^{2} catalyst is 77. Therefore,

$$v = (20 \text{ ft/sec}) / 0.77 = 26.0 \text{ ft/sec}$$

From Table 2.1: For air at 1,000°F

$\rho = 0.0275$ lb/ft^3

$\mu = 0.089$ lb/ft hr $= 2.47 \cdot 10^{-5}$ lb/ft sec

$$\text{Re} = \frac{\rho v D}{\mu} = \frac{0.0275 \text{ lb/ft}^3 \cdot 26.0 \text{ ft/sec} \cdot 0.00367 \text{ ft}}{2.47 \cdot 10^{-5} \text{ lb/ft sec}} = 106 \text{ (laminar)}$$

$$L_m = \frac{v}{K_m a} = \frac{26.0 \text{ ft/sec}}{0.6342 \text{ ft/sec} \cdot 1,089.9 \text{ ft}^2/\text{ft}^3} = 0.0376 \text{ ft} = 0.45 \text{ in.}$$

From Equation [13.18] $L = L_m \cdot N = 0.45 \text{ in.} \cdot 9.21 = 4.15 \text{ in.}$

For square channel $f = 14/\text{Re} = 14/106 = 0.132$

Using Equation [13.19]:

$$\Delta P = \frac{2 f L v^2 \rho}{gD} = \frac{2 \cdot 0.132 \cdot (4.15/12) \text{ft} \cdot (26.0 \text{ ft/sec})^2 \cdot (0.0275 \text{ lb}_m/\text{ft}^3)}{(32 \text{ lb}_m \text{ft}/\text{lb}_f \text{ sec}^2) \cdot 0.00367 \text{ ft}}$$

$\Delta P = 14.45 \text{ lb}_f/\text{ft}^2 = 2.9 \text{ in. H}_2\text{O}$

Using Table 13.5, and assuming air flow = 35 acfm:

$$\text{Cross-sectional area of catalyst} = \frac{35 \text{ ft}^3/\text{min}}{26.0 \text{ ft/sec}} \cdot \frac{1 \text{ min}}{60 \text{ sec}} = 0.0224 \text{ ft}^2$$

$$\text{Diameter of the catalyst} = \sqrt{\frac{0.0224 \text{ ft}^2}{\pi/4}} = 0.169 \text{ ft} = 2.03 \text{ in.}$$

The minimum catalyst needed = 4.15 inches long and 2.03 inches in diameter

$$\text{Gas residence time} = \frac{0.0224 \text{ ft}^2 \cdot 0.169 \text{ ft}}{35 \text{ acfm}} = 1.082 \cdot 10^{-4} \text{ min} = 0.0065 \text{ sec}$$

Using Equation 13.20, space velocity $= \dfrac{26.0 \text{ ft/sec}}{(4.15/12) \text{ft}} = 75 \text{ sec}^{-1}$ or $270{,}000 \text{ hr}^{-1}$

VOC emissions from catalytic converter =
$13.0 \cdot 0.0001 = 0.0013$ grams/mile

b. The USEPA standard requires an efficiency of 98% based on a preset driving cycle that includes time periods when the catalyst is inactive due to cold start (see Figure 13.10). The higher design efficiency used in this problem reflects the need to provide additional catalyst to account for catalyst inactivity during cold start and destruction of catalyst during the required 100,000 mile useful life of the vehicle.

Example 13.7 Catalytic Incinerator Unit Design

Exhaust air from a baking oven in an enamel-coating process contains 1,000 ppm methyl ethyl ketone (MEK). The exhaust velocity is 20 ft/sec, temperature is 1,000°F, and the diffusion coefficient of MEK in air is 0.00055 ft^2/sec at this temperature. The outlet concentration is required to be 10 ppm. The catalytic ignition temperature is 750°F so that mass transfer will control the reaction. Design a catalytic incinerator unit for control of MEK. Determine the space volume for the catalyst and compare the values to those provided in Table 11.6.

Solution:

Design of catalytic incinerator:

We may choose from three commercially available square-holed catalyst supports—200, 300, and 400 holes/in.^{2} (see Table 13.12). Using the manufacture's data on hole size and open area, we can calculate v and Re for flow through the holes in each type of support. Because all of the Re values are well below 2,000, we know that the flow is laminar. We can thus use the laminar flow equations to calculate L_m and the required reactor length ($L = NL_m$) for each type of support. The 10-ppm outlet concentration is 1% of the inlet concentration. Thus:

Example calculation for the 200 holes/in.^{2} of catalyst support:

$$e^{-N} = 0.01, N = 4.6$$

Using Equation [13.16], $K_m = \dfrac{4.4\mathcal{D}}{D} = \dfrac{4.4 \bullet 0.00055 \text{ ft}^2/\text{sec}}{(0.059/12)\text{ft}} = 0.492 \text{ ft/sec}$

From Equation [13.17], $a = \dfrac{4}{D} = \dfrac{4}{(0.059/12)\text{ft}} = 813.5 \text{ ft}^2/\text{ft}^3$

From Table 13.12, the percent open area for a 200 hole/in.^{2} catalyst support is 72. Therefore

$$v = (20 \text{ ft/sec})/0.72 = 27.8 \text{ ft/sec}$$

Table 2.1, for air at 1,000°F:

$\rho = 0.0275$ lb/ft^3,

$\mu = 0.089$ lb/ft hr $= 2.47 \cdot 10^{-5}$ lb/ft sec

$D = 0.059/12 = 0.00492$ ft

$$Re = \frac{\rho vd}{\mu} = \frac{0.0275 \text{ lb/ft}^3 \cdot 27.8 \text{ ft/sec} \cdot 0.00492 \text{ ft}}{2.47 \cdot 10^{-5} \text{ lb/ft sec}} = 152$$

$$L_m = \frac{v}{K_m a} = \frac{27.8 \text{ ft/sec}}{0.492 \text{ ft/sec} \cdot 813.5 \text{ ft}^2/\text{ft}^3} = 0.0694 \text{ ft} = 0.83 \text{ in.}$$

$L = L_m N = 0.83 \text{ in.} \cdot 4.6 = 3.8 \text{ in.}$

For square channel $f = 14/Re = 14/152 = 0.0921$

$$\Delta P = \frac{2 f L v^2 \rho}{gD} = \frac{2 \cdot 0.0921 \cdot (3.8/12)\text{ft} \cdot (27.8 \text{ ft/sec})^2 \cdot (0.0275 \text{ lb}_m/\text{ft}^3)}{(32 \text{ lb}_m \text{ft}/\text{lb}_f \text{ sec}^2) \cdot 0.00492 \text{ ft}}$$

$\Delta P = 7.9 \text{ lb}_f/\text{ft}^2 = 1.6 \text{ in. H}_2\text{O}$

Table 13.12 shows calculated reactor lengths and pressure drops for each type of support.

Table 13.12
Manufacturer's Data on Three Catalysts

	Holes/in.^{2} of catalyst support		
	200	300	400
Open area, %	72	65	77
Effective hole diamenter, in.	0.059	0.046	0.044
Gas velocity in holes, ft/sec	27.8	30.8	26.0
Reynolds Numbers	152	133	107
Length of 1 mass transfer unit, in.	0.83	0.54	0.42
Gas pressure drop, in. H$_2$O	1.6	2.0	1.5

Successful operation of current catalytic converters requires engines that are properly tuned and the use of nonleaded gasolines to prevent poisoning of the catalysts. The use of computerized controls and improvements in the catalysts have resulted in well-performing catalysts. Automobile manufacturers are testing Low Emission Vehicles (LEVs) that achieve a 50% reduction for NO$_x$ and 70% reduction of hydrocarbons over current standards. The vehicles will require lower sulfur fuel (<40 ppm) because studies show that sulfur can affect the LEVs' advanced catalytic converter and reduce its ability to control emissions.

13.9 Problem Set

13.1 Write out the chemical equation for the stoichiometric reaction of octane, C_8H_{18}, with air. Determine the theoretical air/fuel (A/F) ratio for this reaction on both a mole and mass basis.

13.2 An automobile burning C_8H_{18} fuel has an emission flow volume of 370 ft^3/min at 1 atm and 1,000°F. Consider that the exhaust emissions contain 2,000 ppm of CO, 1,000 ppm of HC, and 50 ppm of NO. Determine the amount in pounds per minute for CO, HC and NO emitted. (molecular weight of CO = 28, HC = 114 (C_8H_{18}), NO = 30)

13.3 A vehicle traveling along a highway at 55 mph emits CO and NO of 80 and 10 grams/liter of fuel burned, respectively. Assume that the vehicle travels 25 miles on the highway per gallon of gasoline burned. Calculate the CO and NO emission factors in grams/mile. Compare this to the controlled emission factors in Table 13.8.

13.4 The original hydrocarbon emission standard in 1970 was 180 ppm (2.2 grams/mile) of unburned hydrocarbons. Estimate the hydrocarbon emission reductions required for 1970 and for 1994 vehicles and compare to uncontrolled emissions.

Between 1970 and 1994, the number of cars has increased from 108 to 189 million, the yearly driving distance has increased by 11% per vehicle, and the gas mileage has increased from 20 to 28 miles/gallon. Have the emissions per car increased or decreased during this period? Have the emissions from all cars in the United States increased or decreased during this period?

13.5 Estimate the total emission of gasoline (gal/day) into the atmosphere from displacement losses when fueling at service stations in the United States. The United States consumption of gasoline is about $3 \cdot 10^8$ gal/day. Equation 9.3 may be used for calculation. Compare the displacement losses to the estimated HC exhaust emissions from automobiles (gal/day) when the average miles driven per vehicle is 12,000 miles/year and the emission factors are given in Table 13.8. (There were an estimated 189 million vehicles in the United States in 1994.)

13.6 A car with a piston engine is equipped with a catalytic converter to burn carbon monoxide and hydrocarbons in the exhaust gas. The catalyst temperature must be controlled at 600°F by adjusting the air intake to the converter. The heating value from burning the exhaust gas is 5.0 Btu/ft^3 at 600°F. During highway driving the exhaust rate is 70 cfm, the gas is at 800°F, and its molecular weight is 30. When the air temperature is 70°F, what catalytic converter air intake is required? Neglect heat loss from the converter to the surrounding air.

Chapter 13: Control of Gaseous Emissions from Motor Vehicles 505

13.7 A bulk terminal that transfers gasoline to service stations has submitted plans to install a carbon adsorption unit to control emissions from the filling of their tank trucks. You are given the following information:

- Average daily throughput is 120,000 gal/day of gasoline
- Maximum pumping rate is 2,000 gal/min
- Air flow to adsorber is 350 cfm
- Diameter of adsorber is 4 ft
- Height of carbon is 3 ft
- 2 beds, regeneration cycle time of 1/2 hour for each bed
- Emission factor is 5 lb/1,000 gal loaded
- Emission limit is 0.67 lb/1,000 gal loaded
- MW of gasoline is 68 lb/lb mol
- Ideal gas constant, $R = 0.732$ atm ft^3/lb mol °R
- T is 70°F
- Carbon density is 30 lb/ft^3
- Vacuum regeneration
- The following figure provides the adsorption capacity of activated carbon for gasoline vapors

Adsorption Isotherm for Gasoline Vapors

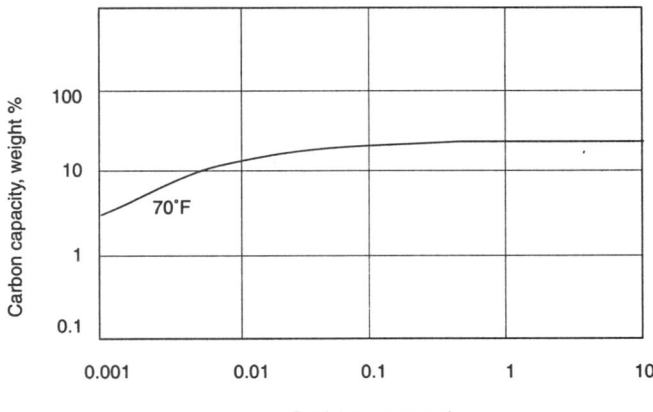

Note:

1. The velocity through adsorber should be between 20-100 ft/min.
2. The working charge of carbon may be estimated by doubling the amount of carbon at saturation capacity. (The saturation capacity is the ideal amount of solvent the carbon can hold and is read from the adsorption isotherm.) (See Chapter 10 for review of adsorption design).

Question:

1. What is the velocity through the adsorber?
2. How much carbon is required?
3. What is the bed depth?
4. Will the unit be in compliance?

13.8 The USEPA standard of performance for new power plants for natural gas-fired heaters and boilers is 0.2 lb NO_2/million Btu. What is their standard in terms of ppm (dry basis) in the flue gas at 3% oxygen measured in the stack? The heating value of the gas is 1,000 Btu/scf. (Treat NO_x as NO_2.)

13.9 The flow rate of a waste gas is 600 m^3/min at 300°K and 1 atm. The gas is preheated in a heat exchanger to 600°K and incinerated in a catalytic bed at that temperature to obtain 99.9% destruction of the toluene. At that temperature, and for this catalyst, mass transfer is the rate-limiting step.

The catalyst has circular channels with an effective diameter of 1.17 mm, and a fractional open area for flow of 65%. To avoid excessive pressure drop, the gas velocity (based on the total cross-sectional area of the incinerator), should not exceed 10 m/sec. Estimate the dimensions of the catalytic unit required. Assume that the transport properties of the combustion gases are similar to those of nitrogen.

13.10 Calculate the NO catalytic reduction rate constant, K_r. Given: 200 CPIS density, T = 350°C (620°F), Space velocity = 20,000 hr^{-1}, NH_3/NO = 0.95/1, Diffusivity of NO at 662°F = 0.8 cm^2/sec, Hole diameter = 0.059 in. = 0.00492 ft, Catalyst L = 4.5 ft, NO_x reduction = 95% for 42 ppm NO at catalyst inlet. Determine the magnitudes of K_m, K_r, and K_o.

13.11 A catalytic converter is being used to oxidize an inlet hydrocarbon concentration of 1,000 ppm to an outlet concentration of 10 ppm. The gas flow rate is 134.4 acfm at 1,000°F. The catalyst is 15 square inches in cross-section perpendicular to gas flow and is 0.5 ft in length in the direction of gas flow. 77% of the area perpendicular to the gas flow is open to the gas flow and the hole diameter in the catalyst is 0.1016 cm. The hydrocarbon diffusion coefficient is assumed to be 0.3839 cm^2/sec at 1000°F. Determine the overall mass transfer coefficient, K_m, and reaction rate constant, K_r. What is the oxidation rate controlling mechanism? What could be done to increase the rate of this process and further reduce the exhaust gas hydrocarbon concentration?

13.10 References

Cooper, C.D. and F.C. Alley. *Air Pollution Control: A Design Approach*. Waveland Press. Prospect Heights, IL. 1994.

DeNevers, N. *Air Pollution Control Engineering*. McGraw-Hill. New York. 1995.

Griffin, R.D. *Principles of Air Quality Management*. Lewis Publishers. Ann Arbor, MI. 1994.

Marchello, J.M. *Control of Air Pollution Sources*. Marcel Dekker. New York. 1976.

Retallick, W.B. *Design of Transfer-Limited Catalytic Incinerators*, Chemical Eng, Vol 88. 1981.

Wark K., C. Warner and W. T. Davis. *Air Pollution: Its Origin and Control*. 2nd. ed. Addison Wesley Longman. Menlo Park, CA. 1998.

Chapter 14

Air Quality Systems

14.1 Introduction

Air pollution control devices are one component of an overall air quality control system that captures the gas stream from a source, removes the pollutants, and emits the cleaned gas into the atmosphere. Figure 14.1 is a flow diagram of a generalized air quality control system. The exact type and size of each component in the system will depend on the specific application. The properties and characteristics of the particular gas stream and contaminants will dictate what type of control device will be used and what type of auxiliary equipment will be needed. This is illustrated in Figure 14.2. The figure shows the components of a scrubber-based air quality control system. Some components are common to all air pollution control systems and others are specific to a particular system. Depending on the specific situation, different types of hoods will be used to collect the emissions. Various methods can also be used to remove the collected contaminants. The hood and ductwork need to be large enough to capture the entire gas stream from the source. The fan and motor must be sized to provide the energy necessary to overcome the pressure drop in both the duct system and the air pollution control device. Hot processes may require pre-coolers before the control device. Removal of moisture from the gas stream may be required for proper operation. Selection of the

510   Fundamentals of Air Quality Systems

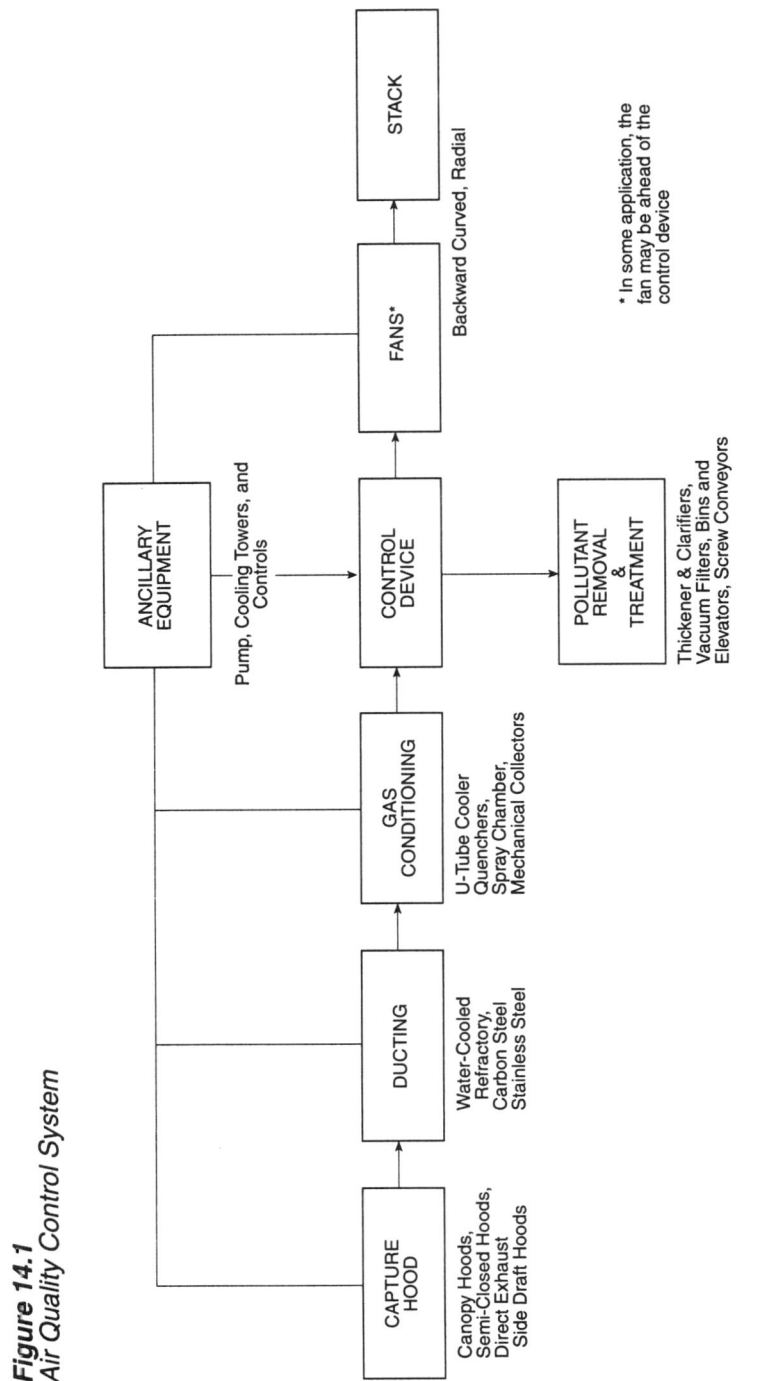

Figure 14.1
Air Quality Control System

Figure 14.2
Components of an Air Quality System used with a Wet Scrubber

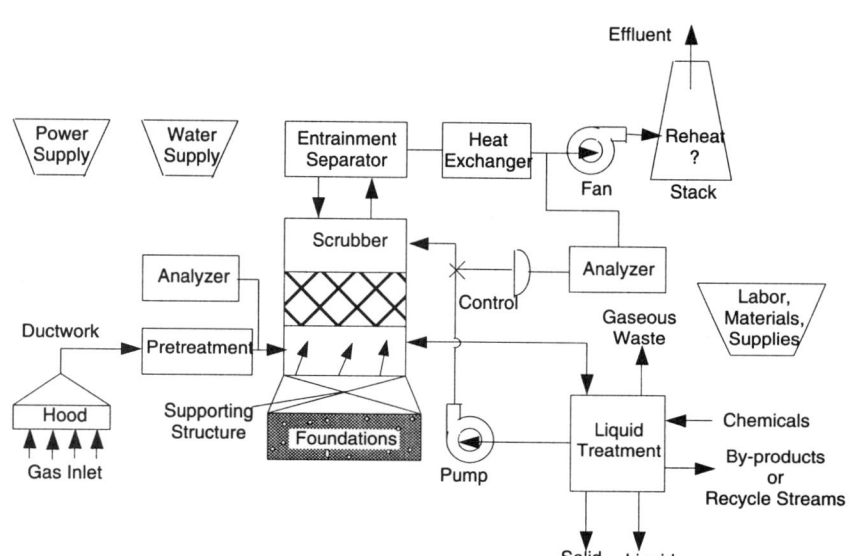

appropriate auxiliary equipment is directly related to the size and operating characteristics of the primary control device.

None of the air pollution control devices discussed in previous chapters can be applied universally. Adsorbers, absorbers, and refrigeration units are effective only with gaseous pollutants, while thermal and catalytic incinerators require combustible particulates or vapors for proper operation. Non-combustible, particulate-laden gas streams are controlled by cyclones, precipitators, scrubbers, or fabric filters. Precipitators and fabric filters are used solely for particulate collection, while scrubbers may be used for collection of both particles and gases (when used as a contactor/absorber).

In some cases, a control device may be compatible with the process, but not selected due to the particular plant location or cost of utilities. For instance, the potentially high cost of maintenance and repair for damage due to freezing with scrubber systems may preclude the use of these systems in colder climates. The high use of electrical power and utilities associated with venturi scrubbers would make these systems less attractive in areas where water is scarce or electrical power is costly. Combustion processes may present safety hazards in some areas.

14.2 Selection of Air Pollution Control Devices for Application to Industrial Sources

Many factors influence the choice of an air pollution control system (Figure 14.3). First, the nature of the contaminants themselves, and in particular their concentrations and characteristics, must be considered. In many instances, the behavior of the bulk (carrier) gas stream is more important than that of the contaminants during the initial selection phases of air pollution control systems because the corrosion potential, toxicity, and explosive nature of the carrier gas are critical to the proper selection and operation of air pollution control systems. Gas temperature is important because it affects the size of the collector. Temperature also affects solubility, adsorbability of gases, and resistivity of particles.

Figure 14.3
Factors to Consider in the Evaluation of Potential Air Pollution Control Systems Applied to Industrial Sources

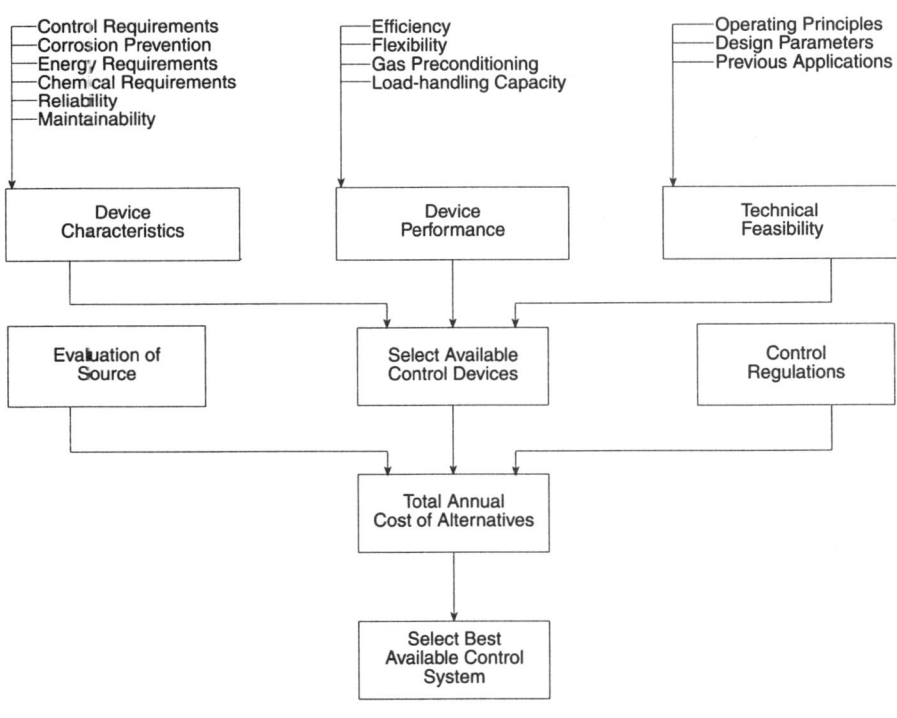

The choice between alternate control systems can be made by considering the technical feasibility and the performance characteristics of the devices themselves. Equipment reliability and maintainability are also important. The decisive factor to be considered in the selection of a control device is the need to meet the control regulations. However, the optimization of cost and effectiveness within the constraints of the control limitations are important. The energy required is a function of the characteristics of the various control approaches or apparatus and is of great importance. The pressure drop through a fabric filter may average from 4 to 8 in. H_2O gauge, while with a venturi scrubber for the same service it will be as high as 60 to 80 in. H_2O. Reliability and maintenance information should be available for evaluation if a system under consideration has been in use for an extended period of time. The effects associated with control equipment bypass or total or partial shutdown in order to accomplish maintenance also need to be considered. If the emissions are highly toxic, a backup control device may be required.

The selection of a control device for a specific application requires a detailed evaluation of the emission source. The objective of this evaluation is to obtain enough basic data about the processes involved and the potential or actual emissions, so that control system design criteria can be generated. The evaluation should trace emissions to their source. The various steps in the industrial process should be evaluated from a physical and operational point of view. Data are required on the exhaust rate, concentrations, and characteristics of all pollutants. For existing sources, emission testing is often required to determine the degree of control required to meet regulations. Specifications for the testing program should consider pressure and temperature of the gas stream, the total volume of the effluent, and concentration of pollutants and any fluctuations (batch or continuous) in these parameters.

The properties and characteristics of the particular gas stream will generally dictate which control option is appropriate. In some cases, several techniques may be suitable. Generally, the selection of one type in lieu of the others will be based on total costs (capital, maintenance, and operating). Table 14.1 provides an alphabetical list of industrial sources and applicable control devices that have been used to control emissions (Neveril 1978). In general, the control devices listed in this table have performed reliably and consistently under industrial conditions. If a suitable control device has been used, then it should receive serious consideration for application to a similar source. However, new technologies need to be evaluated for their effectiveness in meeting new air pollution control regulations.

14.3 Auxiliary Equipment

For each control device selected for a particular application, a certain amount of auxiliary equipment must be used with the control device for the efficient operation of the gas cleaning system. The types of auxiliary equipment required will depend on the application. Hot processes may require pre-coolers before the control device or the addition of moisture may be required for proper operation. In many instances, sources of pollutants are in open areas and the dusts and other

Table 14.1
Industrial Pollutant Sources and Typical Control Devices

Industry	Source	Control System	Typical Gas Temperature
Asphalt Roofing	1. Saturator and storage tank	1. Scrubber, precipitator, afterburner	1. 80-300°F
Basic Oxygen Furnace	1. Basic oxygen furnace 2. Charging hood	1. Precipitator, scrubber, bag house 2. Same as 1	1. 3,500-4,000°F 2. 150-400°F
Benzene Handling & Storage	1. Vents, storage tanks	1. Afterburner, adsorber, refrigeration	1. 70-100°F
Brick Manufacturing	1. Tunnel kiln 2. Crusher, mill 3. Dryer 4. Periodic kiln	1. Scrubber, baghouse 2. Baghouse, scrubber 3. Same as 1 4. Same as 1	1. 200-600°F 2. 70°F 3. 250°F 4. Same as 1
Castable Refractories	1. Electric arc 2. Crusher, mill 3. Dryer 4. Mold and shakeout	1. Baghouse, scrubber 2. Same as 1 3. Same as 1 4. Same as 1	1. 3,000-4,000°F 2. 70°F 3. 300°F 4. 150°F
Chemical Manufacturing Waste Disposal	1. Miscellaneous sources	1. Afterburners, flares	1. As required
Clay Refractories	1. Shuttle kiln 2. Calciner 3. Dryer 4. Crusher, mill	1. Baghouse, precipitator, scrubber 2. Same as 1 3. Same as 1 4. Baghouse, precipitator	1. 150-800°F 2. Same as 1 3. 250°F 4. 70°F
Coal-fired Boilers	1. Steam generator	1. Precipitator, scrubber, baghouse	1. 300-700°F
Conical Incinerators	1. Incinerators	1. Scrubber	1. 400-700°F
Cotton Ginning	1. Incinerators	1. Scrubber	1. 500-700°F
Degreasing	1. Degreasing tank	1. Adsorber, refrigeration	1. 70°F
Detergent Manufacturing	1. Spray dryer	1. Scrubber, baghouse	1. 180-250°F
Direct Firing of Meat	1. Smoke house	1. Afterburners, electrical precipitators	1. 120-150°F
Distilled Whiskey Processing	1. Distillation process	1. Adsorbers, afterburners	1. As required

Table 14.1 (cont.)
Industrial Pollutant Sources and Typical Control Devices

Industry	Source	Control System	Typical Gas Temperature
Dry Cleaning	1. Washer, extractor, tumbler	1. Adsorber	1. 70°F
Electric Arc Furnace	1. Arc furnace 2. Charging and tapping	1. Baghouse, scrubber, precipitators	1. 3,000 °F (direct exhaust) 2. 150 °F (canopy)
Feed Mills	1. Storage bins 2. Mills/grinders 3. Flash dryer 4. Tap Fume	1. Baghouse, scrubber 2. Same as 1 3. Same as 1 4. Same as 1	1. 70°F 2. 70°F 3. 170-250°F 4. 70°F
Ferroalloy Plants a. HC Fe Mn b. 50% Fe Si c. Hc Fe Cr	1. Submerged arc furnace (open) 2. Submerged arc furnace (closed) 3. Tap fume	1. Scrubber, baghouse precipitators 2. Scrubber 3. Same collector or baghouse	1. 400-500°F 2. 1,000-1,200°F closed arc 3. 150°F hood
Gasoline Bulk Terminal Storage	1. Vents, storage tanks	1. Afterburners, adsorbers, refrigeration	1. 70-100°F
Glass Manufacturing	1. Regenerative tank furnace 2. Weight hoppers and mixers	1. Baghouse, scrubber precipitator 2. Same as 1	1. 600-850°F furnace 2. 100°F mixers
Graphic Arts	1. Presses 2. Lithographics, metal decorating ovens	1. Adsorbers, afterburners 2. Afterburners	1. 100°F 2. 400-600°F
Gray Iron Foundries	1. Cupola 2. Electric arc furnace 3. Core oven 4. Shakeout	1. Afterburner-baghouse for closed cap, Afterburner-precipitator for closed cap, scrubber 2. Baghouse, scrubber precipitator 3. Afterburner 4. Baghouse	1. 1,200-2,200°F 2. -2,500°F direct exhaust 400°F hood 3. 150°F 4. 150°F
Industrial & Utility Boiler	1. Boiler	1. Precipitator, baghouse	1. 100-800°F
Insulation Wire Varnish	1. Spray booths 2. Flow coating machines 3. Dip tanks 4. Roller coating machines	1. Adsorbers, absorbers afterburners 2. Same as 1 3. Same as 1 4. Same as 1	1. 100°F

Table 14.1 (cont.)
Industrial Pollutant Sources and Typical Control Devices

Industry	Source	Control System	Typical Gas Temperature
Iron Ore Benefication	1. Crushing 2. Sinter machine	1. Baghouse, scrubber 2. Same as 1	1. As required 2. As required

Iron & Steel (Sintering)	1. Sinter machine
	  a. Sinter bed
	  b. Ignition fce.
	  c. Wind boxes
	2. a. Sinter crusher
	  b. Conveyors
	  c. Feeders
		1. Precipitator baghouse, scrubber
		2. Baghouse, scrubber
			1. 150-400°F sinter machine
			2. 70°F conveyor

Kraft Recovery Furnaces	1. Recovery furnace and direct contact evaporator	2. Precipitator, scrubber	1. 350°F
Lime Kilns	1. Vertical kilns 2. Rotary sludge kiln	1. Baghouse, scrubber, precipitator 2. Scrubber, precipitator	1. 200-1,200°F 2. 200-1,200°F
Maleic Anhydride	1. Benzene storage tanks, process vent & vac. refin. vent	1. Adsorbers, afterburners	1. 70-100°F
Miscellaneous Refinery Sources	1. Vents, storage tanks, etc.	1. Afterburner, flare adsorbers, absorbers refrigeration	1. As required
Municipal Incinerator	1. Incinerator	1. Scrubber, precipitator, baghouse afterburner	1. 500-700°F
Non-Metallic Minerals Industry	1. Miscellaneous sources	1. Scrubbers, baghouse	1. As required
Onshore Crude Oil Production	1. Vents, storage tanks	1. Adsorbers, afterburners, refrigeration	1. 70-100°F
Organic Chemicals	1. Miscellaneous sources	1. Scrubbers, adsorbers, absorbers, refrigeration, flares	1. As required
Paint Manufacturing	1. Vanish kettles	1. Afterburner	1. 500°F
Petroleum Catalytic Cracking	1. CO boiler from FCC	1. Afterburners, adsorbers, refrigeration	1. 500°F

Chapter 14: Air Quality Systems   517

Table 14.1 (cont.)
Industrial Pollutant Sources and Typical Control Devices

Industry	Source	Control System	Typical Gas Temperature
Petroleum Storage	1. Vents, storage tanks	1. Direct exhaust	1. 70-100°F
Pharmaceuticals	1. Reactors 2. Crystallizer 3. Centrifuge 4. Filter, dryer 5. Dist. column	1. Adsorbers, refrigeration, incineration 2. Same as 1 3. Same as 1 4. Same as 1 5. Same as 1	1. As required 2. As required 3. As required 4. As required 5. As required
Phosphate Fertilizer	1. Digester vent air 2. Filters 3. Sumps	1. Scrubber, baghouse 2. Same as 1 3. Same as 1	1. 150°F 2. Same as 1 3. Same as 1
Phosphate Rock Crushing	1. Crusher & screens 2. Conveyor 3. Elevators 4. Fluidized bed calciner, grinder & dryer	1. Baghouse, scrubber, precipitator 2. Same as 1 3. Same as 1 4. Same as 1	1. 70°F hoods 2. Same as 1 3. Same as 1 4. 600-1,500°F calciner
Polyvinyl Chloride Production	1. Process equipment vents	2. Adsorbers, afterburner	3. 15 to 130°F

Portland Cement	1. Rotary Kiln
	  a. Wet
	  b. Dry
	2. Crushers and conveyors
	3. Dryers
		1. Precipitators, baghouses
		2. Baghouses
		3. Precipitators, baghouses
			1. 150-850°F kilns
			2. 70°F crushers and conveyors
			3. 200°F

Primary Copper, Lead, Zinc Smelters	1. Roaster, converter	1. Precipitator, scrubber, adsorber	1. As required
Pulp and Paper	1. Fluidized bed reactor	1. Scrubber	1. 600-1,500°F
Refuse Waste Disposal	1. Furnace	1. Afterburner	1. 500-700°F
Rubber Products (Tires)	1. Rubber mill and mixers	1. Baghouse	1. 70°F
Secondary Aluminum	1. Reverbatory furnace 2. El. induction furnace 3. Crucible furnace 4. Chlorinating station 5. Dross processing 6. Sweating furnace	1. Scrubber (low energy) + baghouse, precipitator 2. Same as 1 3. Same as 1 4. Same as 1 5. Same as 1 6. Same as 1	1. 1,600 °F fluxing 600°F holding hearth 2. Base on type capture 3. Same as 2 4. Same as 2 5. Same as 2 6. Same as 2

Table 14.1 (cont.)
Industrial Pollutant Sources and Typical Control Devices

Industry	Source	Control System	Typical Gas Temperature
Secondary Copper Smelters	1. Reverbatory furnace 2. Crucible furnace 3. Cupola & blast furnace 4. Converter 5. El. induction furnaces	1. Baghouse scrubber precipitator 2. Same as 1 3. Same as 1 4. Same as 1 5. Same as 1	1. 2,500°F direct tap 2. Base on type capture 3. Same as 2 4. Same as 2 5. Same as 2
Service Stations	1. Loading rack	2. Adsorbers, refrigeration	3. 70-100°F
Sewage Sludge Incinerators	1. Multiple hearth incinerator 2. Fluidized bed reactor	1. Scrubber 2. Same as 1	1. 600 to 1,500 °F 2. Same as 1
Surface Coating-Spray Booths	1. Spray booth	2. Adsorbers, afterburner	1. 70°F
Vegetable Oil Processing	1. Solvent extraction process	1. Adsorbers, scrubbers	1. 100°F

Source: Neveril 1978

contaminants must be captured. Hoods and enclosures are used for this purpose. Often the gases from several sources are conveyed through a system of ducts to a single control device. Sufficient air must be provided by fans or blowers to capture and transport the contaminants. This section discusses the design of these components.

14.3.1 Hoods

Hoods are used to capture air contaminants. There are three general types of hoods: enclosures, receiving, and exterior. Enclosures surround the source of emissions, but may have one face partially or fully open for access. *Receiving hoods* are located to collect emissions that are ejected toward them, such as the dust particles from a grinder. *Exterior hoods* capture contaminants released outside of and at some distance from the hood.

Flow patterns into hoods and ducts (Figure 14.4) follow streamlines that depend strongly on the specific geometric shape and dimensions of each situation (Danielson 1973). Hoods must have adequate capacity to capture the emissions from the source. Generally, an attempt is made to establish a null point or null curve that locates or bounds the region that must be swept by the exhaust system to ensure capture of the contaminants.

Chapter 14: Air Quality Systems

Figure 14.4
a) Location of Null Points (b) Rectangular Hood Bounded by a Plane Surface

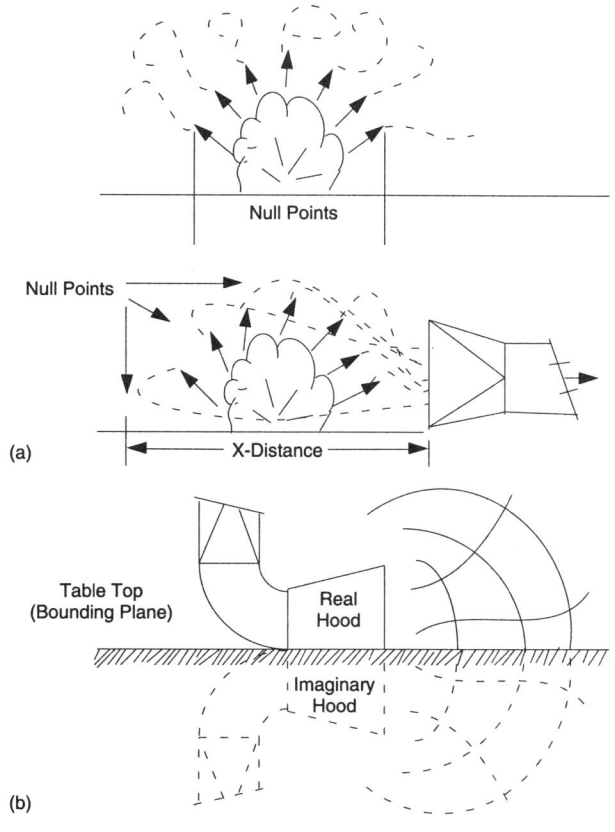

For any totally enclosed hood or for emissions generated inside a booth, the air flow into the hood can be computed by:

$$Q = vA \qquad [14.1]$$

where: Q = air flow (into hood face), cfm

v = air velocity required to capture the pollutant, ft/min

A = total open area of the hood, ft^2

The capture velocity is the velocity necessary (at any point) to overcome opposing air currents and to capture the air contaminants. Table 14.2 presents a range of capture velocities for various processes.

Table 14.2
Capture Velocities for Various Processes

Condition of Dispersion of Contaminant	Example	Capture veolcity (fpm)
Released with practically no velocity	Evaporation from tank; degreasing, etc.	50-100
Released at low velocity into moderately still air	Spray booths, intermittent container filling; low speed conveyor transfers; welding; plating; pickling	100-200
Active generation into zone of rapid air motion	Spray painting in shallow booths; barrel filling; conveyor loading; crushers	200-500
Released at high initial velocity in to zone of very rapid air motion	Grinding; abrasive blasting, tumbling	500-2,000

In each category above, a range of capture velocity is shown. The proper choice of values depends on several factors:

Lowers end of range
1. Room air currents minimal or favorable for capture
2. Contaminants of low toxicity or of nuisance value only
3. Intermittent, low production
4. Large hood-large air mass in motion

Upper end of range
1. Disturbing room air currents
2. Contaminants of high toxicity
3. High production, heavy used
4. Small hood-local control only

Source: Danielson 1973

For a free standing or unbounded hood, Equation [14.1] must be modified to account for the distance of a source from the hood face.

$$Q = v_x(10x^2 + A) \qquad [14.2]$$

where: Q = air flow, cfm

v_x = air velocity required for capture at some distance (x) from hood opening, ft/min

x = distance of contaminant source from the center of the hood face, ft

A = area of hood face, ft^2

For a square or rectangular hood that is bounded on one side by a flat surface (the floor or wall), Equation [14.2] becomes:

$$Q = v_x\left(\frac{10x^2 + 2A}{2}\right) \qquad [14.3]$$

Example 14.1 Paint Spray Booth Air Flow Requirement

Determine the total volumetric air flow rate of air entering a paint spray booth that is 10 ft wide and 7 ft high. Some objects are painted outside the booth but never a distance of more than 3 ft. Room air currents are kept low.

Solution:

The paint spray booth would be a plain opening bounded on one side by a flat surface (the floor). Therefore, Equation [14.3] applies.

The capture velocity can be estimated from Table 14.2. For paint spray emitted into a draftless area, the capture velocity should be 100 ft/min.

Substituting into Equation [14.3]:

$$Q = 100 \text{ ft / min} \left(\frac{10 \bullet (3 \text{ ft})^2 + 2 \bullet 7 \text{ ft} \bullet 10 \text{ ft}}{2} \right)$$
$$= 11,500 \text{ cfm}$$

Open surfaces are often controlled by slot hoods (Figure 14.5) (Danielson 1973). Slot hoods are commonly employed on tanks except when the contents are very hot and give rise to strong thermal updrafts. Generally, slots are provided along two parallel sides. The slots and exhaust system are usually designed to achieve an intake flow to the slots of 200 to 300 cfm per foot of length. The duct slot width is sized to give a linear velocity in the slot of about 4,000 ft/min. For a tank with two parallel lip hoods, the ventilation rate required and the lip width may be estimated from Figure 14.6.

Example 14.2 Degreasing Tank Vapor Removal

A degreasing tank 6 ft long and 4 ft wide is controlled by using parallel lip hoods along each of the 6 ft lengths. Determine the total exhaust rate required and the width of the lips.

Solution:

From Figure 14.6, the ventilation rate can be read as 860 cfm/ft of tank length.

The exhaust flow rate is :

$$Q = Q_v l_t$$

where: Q = exhaust flow rate, cfm

Figure 14.5
Open Top Tank Degreaser with Lip Exhaust (slots)

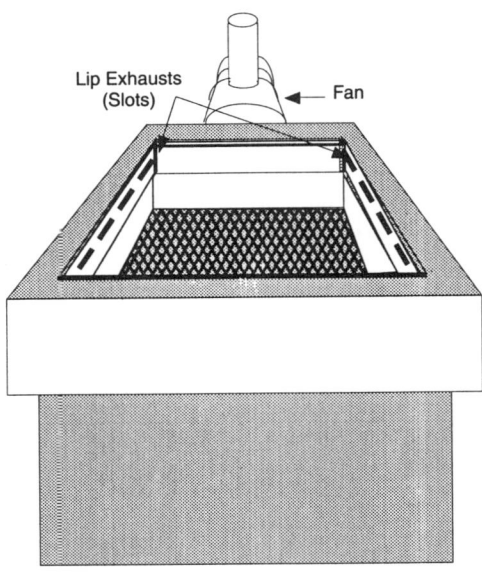

Source: Danielson 1973

Figure 14.6
Minimum Ventilation Rates for Lip Exhaust Hoods with Two Parallel Lip Hoods

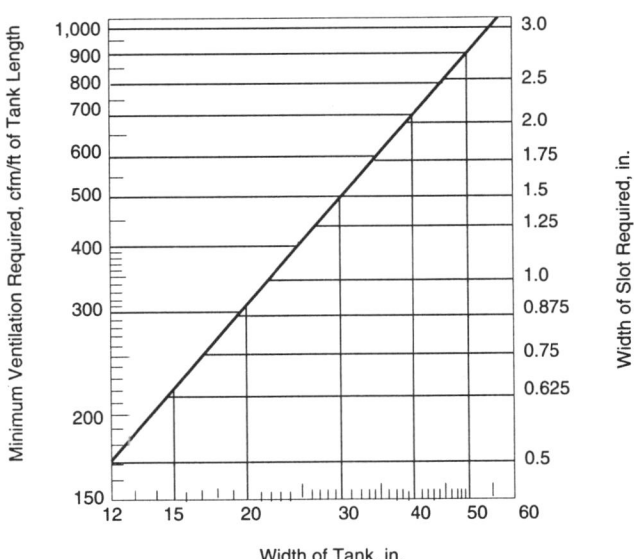

Source: US EPA 1973

Q_v = minimum ventilation rate per ft of tank length, cfm/ft

l_t = tank length, ft

then:

Q = 860 cfm/ft • 6 ft

= 5,160 cfm

Also, from Figure 14.6, the lip width is approximately 2.6 inches.

The design of hoods to capture gaseous emissions from a hot process is markedly different from that for a cold process. Expansion of the hot gases due to the thermal effect must be taken into account. As the hot gas rises, it mixes turbulently with the surrounding air. The higher the gas column rises, the larger it becomes, and the more it is diluted with ambient air. Danielson (1973) has discussed this problem in considerable detail.

When air flows into a hood or duct, pressure losses occur which result in a decreased flow rate. These losses are due to turbulence and are dependent on the shape of the hood or duct. The coefficient of entry, C_e, indicates the extent of these losses. For example, in a theoretically perfect hood with no turbulence loss, $C_e = 1.0$. Figure 14.7 lists the entry coefficients for some typical hoods. The coefficient of entry is used to determine the actual flow rate and pressure after the air enters the hood.

The head loss is related to C_e by

$$h_e = \left[\frac{(1-C_e^2)}{C_e^2} \right] VP \qquad [14.4]$$

where: h_e = entrance loss

C_e = coefficient of entry

VP = velocity pressure in duct

14.3.2 Calculation of Pressure Losses in Exhaust Systems

Pressure losses from air flowing in a pipe are due to friction and turbulence. Turbulent losses are due to rapid changes in direction or velocity. The sum of the friction and turbulent losses over a specific length of pipe is termed the *pressure drop*. The pressure drop of a system is determined by measuring the difference in total pressure at two points in the system (see Chapter 2).

The friction and dynamic losses can be divided into five categories. These losses are as follows:

Figure 14.7
Typical Hoods and their Entry Coefficients

Hood	Description	Entry Coefficient (C_e)
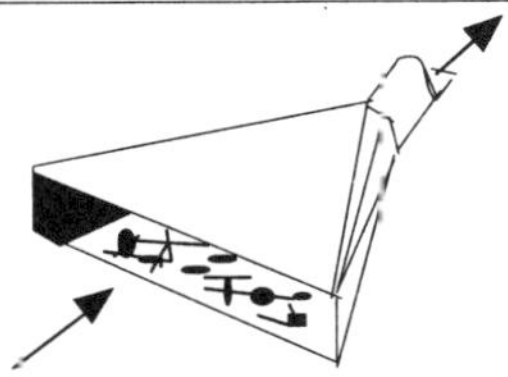	Slot	0.60
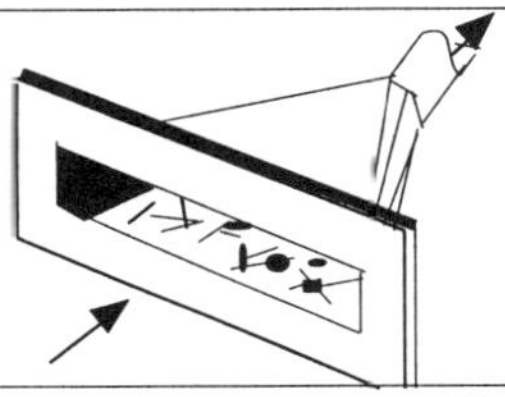	Flanged Slot	0.82
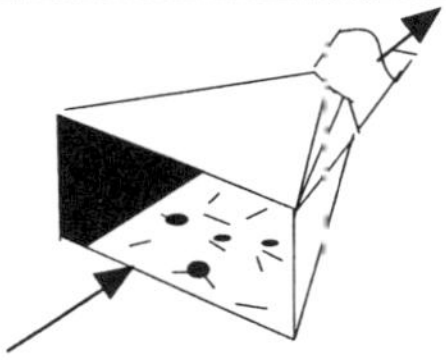	Plain Opening	0.72
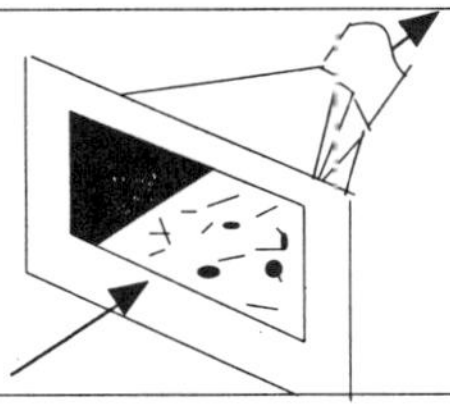	Flanged Opening	0.82
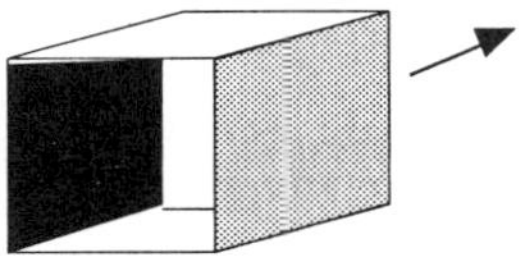	Booth	0.82
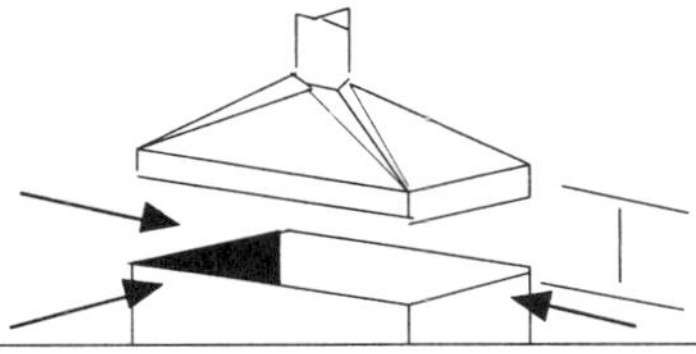	Canopy	0.82

Source: Danielson 1973

(1) *Inertial losses*: Energy required to accelerate a volume of air from rest (Table 14.3). Inertial losses are essentially the same as the velocity pressure.

(2) *Orifice losses*: Pressure losses at the hood or duct entrance due to turbulence. Orifice losses are dependent on the shape of the opening and are measured by the entry coefficient, C_e.

(3) *Straight duct friction losses:* Pressure losses due to air rubbing along the sides of the duct. Many charts and tables have been developed that give the friction losses in straight ducts. Figure 14.8 plots the friction losses in inches of water per 100 feet of duct versus air volume with duct diameter and velocity as parameters. If any two of the above quantities are known, the other two can be read from the chart.

(4) *Elbow and branch entry losses*: Pressure losses due to changed connection. These losses are computed as equivalent feet of straight duct that will have the same pressure loss. Equivalent lengths of elbows and entries are given in Table 14.4.

(5) *Contraction and expansion losses*: For gradual contractions in the cross sectional area of a duct, the pressure losses are small. Abrupt contractions are rare in well-designed exhaust systems except as outlets from chambers.

The Bernoulli equation (Chapter 2) may be applied to exhaust systems to determine operating conditions and design requirements. The losses in an exhaust system are due to inertial changes, flow changes, and friction. Inertial losses arise from energy required to accelerate the gas and from elbows and branches in the duct. Acceleration requirements may be obtained from

$$\frac{\Delta P}{\rho} = \frac{\left(v_1^2 - v_2^2\right)}{2g_c}$$ [14.5]

and are given in Table 14.3 (Danielson 1973).

Table 14.3
Acceleration Head for Air (70°F and 1 atm)

Velocity (ft/min)	Head (in. H_2O)	Velocity (ft/min)	Head (in. H_2O)
500	0.016	3,000	0.561
1,000	0.062	3,500	0.764
1,500	0.140	4,000	0.998
2,000	0.249	4,500	1.262
2,500	0.389	5,000	1.558

Source: Danielson 1973

Figure 14.8
Pressure Loss for Straight Circular Ducts

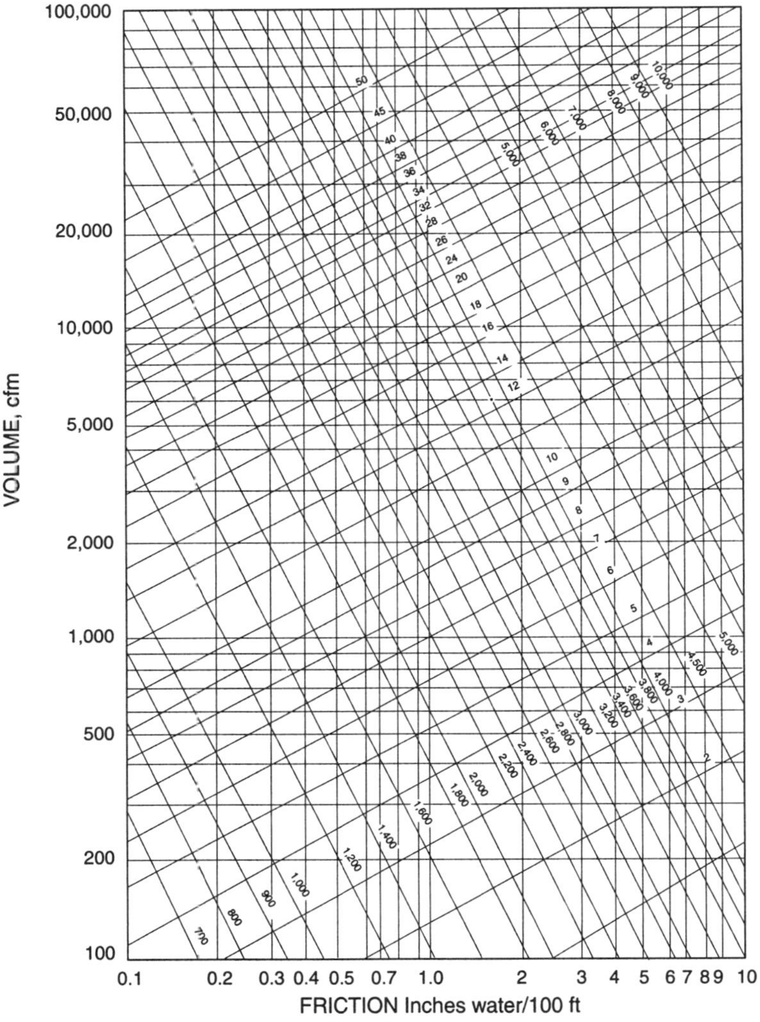

Source: Danielson 1973

Losses due to elbows and branches are generally expressed in equivalent length of straight duct (see Table 14.4) (Danielson 1973).

In straight ducts, the pressure loss due to friction per unit length of duct may be obtained from Figure 14.8 (Danielson 1973). This figure is for ambient air flow in circular ducts with moderate roughness. The required volumetric flow for exhaust systems may be determined from the design requirements of the hoods and

Table 14.4
Air Flow Losses Due to Elbows and Branch Energies

Duct Diameter (in.)	Losses in Equivalent Feet of Duct					
	Elbows[a]				Branch Entry	
	90° bend		45° bend		45° angle	30° angle
	$\frac{R}{D}=1$	$\frac{R}{D}=2$	$\frac{R}{D}=1$	$\frac{R}{D}=2$		
4	7	4	4	2	5	3
6	11	6	6	3	7	5
8	14	8	8	4	11	7
10	20	11	10	6	14	9
12	25	14	12	7	18	11
16	36	20	17	10	25	15
20	46	26	23	13	32	20
24	59	33	30	16	40	25
28	71	40	35	20	47	30
36	92	52	46	26	54	35
48	130	73	64	36	60	40

[a] R/D = turning radius/duct diameter
Source: Danielson 1973

enclosures. The conveying velocity in ducts depends on the nature of the contaminant, as shown in Table 14.5.

The design of a duct for an exhaust system uses a diameter which will handle the volume required by the hood at a velocity greater than the minimum transport velocity but smaller than that which would result in intolerable head loss or noise.

Table 14.5
Recommended Minimum Duct Velocities

Contaminant	Duct Velocity (ft/min)
Gas, vapors, smokes, fumes and very light dusts (zinc and aluminum oxide fumes, wood, flour and lint)	2,000
Dry, medium density dust (buffering lint; sawdust; grain, rubber and plastic dust)	3,000
Average industrial dust (sandblast, grinding, wood shaving and cement dust)	4,000
Heavy dusts (lead, and foundry shakeout dusts, metal turning)	5,000

Source: Danielson 1973

For multiple-branch duct systems, the head loss in two or more branches up to a junction of the branches will necessarily be equal; therefore, design of a branched system is concerned with obtaining the design flows in each branch, while balancing the head losses.

The balance in the head losses in a branched system may be obtained by either of two methods, namely, balanced duct design or blast gate design. In balanced duct design the dominant design parameter is the head loss in each of the branches. The diameters of the ducts are set so that the required flows are met with equal head losses in branches from a junction. In blast gate design, the ducts are sized to provide the minimum transport velocities at tolerable head losses and then the lower head loss branches are brought up to the higher ones through the use of blast gates or dampers in the ducts.

Example 14.3 Air Duct Design

Design a duct system for branches A and B of the layout shown with balanced head loss. The minimum transport velocity = 4,000 ft/min.

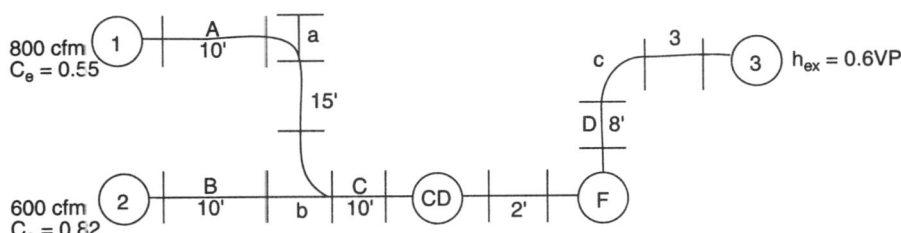

a, 45° entry; b, 45° entry; c, 90° elbow (2 D); CD, control device (head loss = 8 in. H₂O) F, fan; 3, exit bonnet; 1 and 2, entry hoods

Solution:

Branch A:

$$\text{Max. diameter:} \quad \frac{800}{\pi D^2 / 4} = 4{,}000 \text{ ft/min} \qquad D = 0.504 \text{ ft} = 6.05 \text{ in.} \approx 6 \text{ in.}$$

If a 6-in. duct is used: $V = 4{,}075$ ft/min, instead of 4,000

$$VP = \left(\frac{4{,}075}{4{,}002}\right)^2 = 1.04 \text{ in. } H_2O$$

$$h_e = \left[\frac{(1 - C_e^2)}{C_e^2}\right] VP = \left[\frac{(1 - 0.55^2)}{0.55^2}\right] \cdot 1.04 = 2.4 \text{ in. } H_2O$$

From Table 14.4, the equivalent length of a and b = 7.0 ft each entry

Entrance losses:

Total duct length = 10 + 15 + 14 = 39 ft duct

From Figure 14.8, 5 in. H_2O/100 ft friction loss

Loss in duct $\dfrac{5 \text{ in. } H_2O}{100 \text{ ft duct}} \bullet 39.0 \text{ ft duct} = 1.95 \text{ in. } H_2O$

Total loss in A = 1.04 + 2.4 + 1.95 = 5.39 in. H_2O

Branch B:

Max. diameter: $\dfrac{600}{(\pi D^2/4)} = 4,000 \text{ ft/min}$ $D = 0.437 \text{ ft} = 5.24 \text{ in.}$

For a 5.25-in. duct, V = 4,000, VP = $\left(\dfrac{4,000}{4,002}\right)^2 = 1.00$ in. H_2O

$h_e = \left[\dfrac{(1-C_e^2)}{C_e^2}\right] VP = \left[\dfrac{(1-0.82^2)}{0.82^2}\right] \bullet 1.0 = 0.49$ in. H_2O

Total duct length = 10 ft

From Figure 14.8: 5.8 in. H_2O/100 ft friction loss

Loss in duct $\dfrac{5.8 \text{ in. } H_2O}{100 \text{ ft duct}} \bullet 10.0 \text{ ft duct} = 0.58 \text{ in. } H_2O$

Total loss in B = 1.00 + 0.49 + 0.58 = 2.07 in. H_2O

Line B needs to have a blast gate installed to provide additional pressure drop in line B to equal the 5.39 in. H_2O in head loss line A.

14.3.3 Fan Systems

Fan Characteristics

One of the most critical parts of an air pollution control system is the air mover or fan. Its function is to move the desired amount of air through the system by overcoming resistances in the hoods, ducts, coolers, collection devices, stacks, and any other equipment present. The fan is also one of the most complex pieces of equipment in the ventilation system. Its performance depends on the type of fan employed, the parameters of its operation, the characteristics of the system it is used in, and the properties of the gas stream it operates on.

Fans designs can be classified as either axial or centrifugal. *Axial* fans are used to move large volumes of air against low resistances. They may be used for general ventilation or in low resistance industrial ventilation systems; however, they are not often used in air pollution control systems. Occasionally, an axial fan will be used in combination with the more common centrifugal fan to provide extra energy to overcome resistances.

The principal fan used in air pollution control systems is the *centrifugal* fan. The basic design of the centrifugal fan, as illustrated in Figure 14.9, employs a fan wheel or impeller mounted inside a scroll-type housing. Air is drawn into the inside of the impeller and then forced out through the housing. In general, centrifugal fans are distinguished by the design of the impeller. There are three basic impeller types (Figure 14.10): (1) forward curved, (2) radial, and (3) backward inclined (Crowder 1993).

Figure 14.9
Centrifugal Fan Components

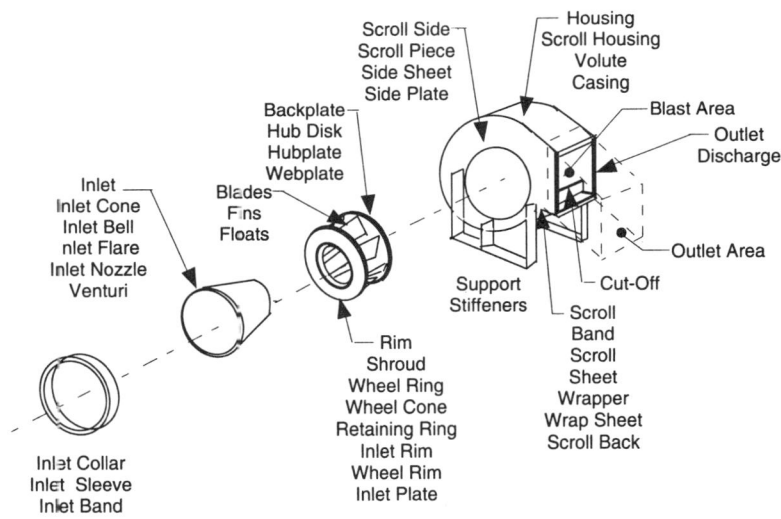

Forward curved impellers have blades that curve into the direction of rotation. These impellers are the smallest of all the centrifugal types and operate at the lowest speeds. As a result, they are quiet in operation but can only develop moderate static pressures. Because of this, they are not commonly used in air pollution control systems.

Radial impellers, shown in Figure 14.11, have blades that extend straight out from the hub in a radial direction. They are the simplest of all the centrifugal fans and the least efficient, but they are capable of developing high static pressures. For a

Figure 14.10
Vector Diagram of Forces with Forward, Radial, and Backward Curved Blades

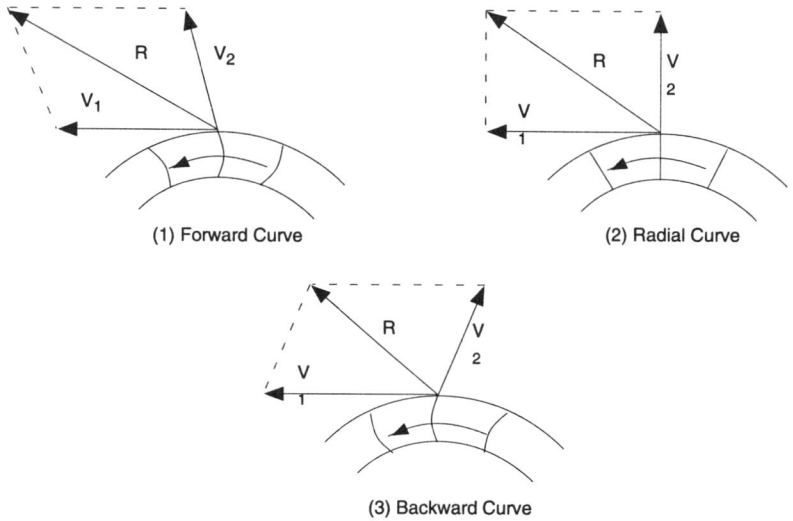

(1) Forward Curve   (2) Radial Curve

(3) Backward Curve

Note: The resultant force is largest with the forward curved fan and smallest with the backward curve fan. For the same pressure, the backward curved wheel mus toperate at a higher speed than the forward curve wheel.

Source: Crowder 1993

given duty, they operate at moderate speeds. The relationship shown in Figure 14.11 between the pressure developed and the air volume moved is called the *fan characteristic curve*. The highest mechanical efficiency is developed just to the left of peak static pressure, at about 30 to 40 percent of the wide-open volume. The horsepower requirement rises continually toward the free delivery volume. The radial blade shape is generally resistant to material buildup and may be used in systems handling either clean or dirty air.

The backward inclined impellers (see Figure 14.12) have airfoil-style blades that curve away from the direction of rotation. They have the highest efficiency of all the centrifugal fans and, for a given duty, will operate at the highest speed. Highest mechanical efficiency is developed to the right of peak static pressure, at about 50 to 60 percent of the wide-open volume. A unique characteristic of the backward inclined impeller is that the horsepower requirement reaches a maximum value near the point of peak efficiency and then declines toward the free delivery volume. For this reason, backward inclined fans are sometimes referred to as "non-overloading', since any variation from the optimum operating point due to a change in system resistance will result in a reduction in operating horsepower.

Figure 14.11
Radial Wheel Design and Performance

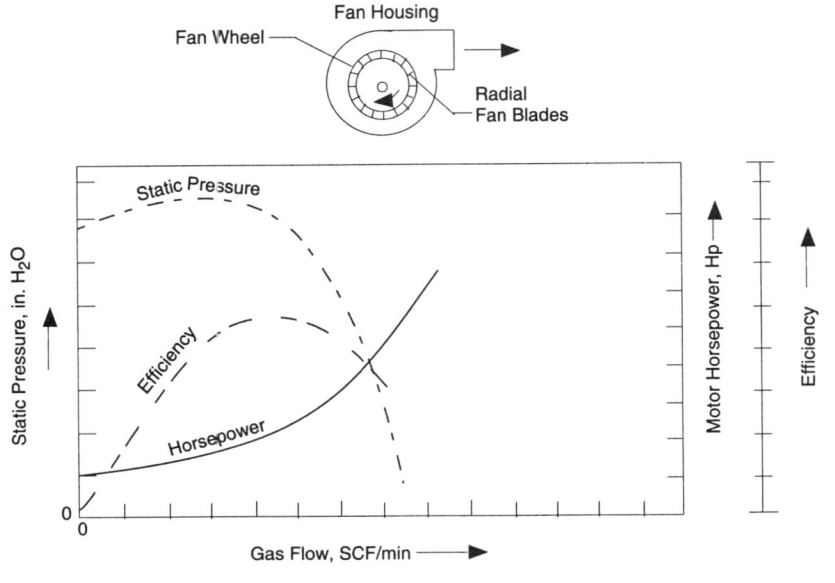

Source: Crowder 1993

Figure 14.12
Backward Inclined Standard Wheel Design and Performance

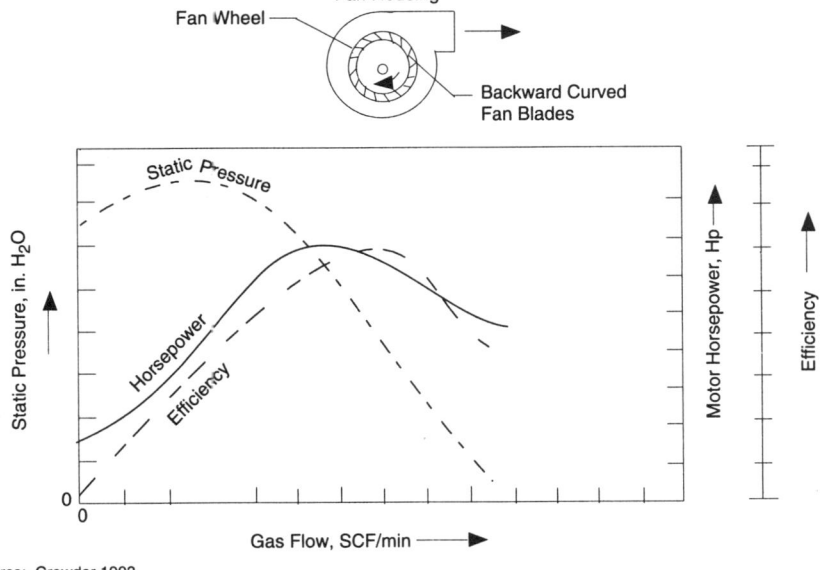

Source: Crowder 1993

Air Flow

When a fluid such as air is moved through a duct, a certain pressure or head is necessary to start and maintain flow. This total head or pressure has two components; static pressure and velocity pressure.

Static pressure is that pressure which tends to collapse or burst the duct, and is used in part to overcome the frictional resistance of the air against the duct surfaces as well as the resistance offered by such obstructions as coils, filters, dust collectors, elbows, etc.

Velocity pressure is that pressure necessary to maintain movement of the air, or, as the name implies, is that pressure necessary to give the air its velocity and is always considered positive in sign.

If a discharge or inlet duct on a fan is completely closed, only static pressure or potential energy is present on the discharge or suction side. The static pressure, velocity pressure, and total pressure are related as follows:

$$\text{Total pressure} = \text{Static Pressure (SP)} + \text{Velocity Pressure (VP)} \quad [14.6]$$

When pressures are measured on the discharge side of the fan, the static pressure is usually positive. When the pressure determinations are made on the inlet side of the fan, the static pressure is negative because it is less than atmospheric pressure. In determining total pressure, confusion can frequently be avoided by basing all calculations on absolute pressures. Absolute total pressure would then be the sum of the absolute static pressure and the velocity pressure. As previously defined, the absolute pressure is always obtained by adding atmospheric pressure to gauge pressure. The total energy developed by a fan may be calculated from the following expressions:

Total pressure developed by fan = Absolute Total Pressure at the fan outlet

— Absolute Total Pressure at the fan inlet

$$TP_f = [\text{Total Press. at fan disch.}] - [\text{Total Press. at fan suction}]$$

or $\quad TP_f = (SP_D + VP_D) - (- SP_S + VP_S)$

(* Static pressure readings are negative on the inlet side of the fan)

$$TP_f = SP_D + VP_D + SP_S - VP_S$$

$$\text{Static pressure of fan} = TP_f - VP_D \quad [14.7]$$

In the last two expressions above, it is not important whether the reading is positive or negative (i.e., above or below atmospheric pressure), since the sign in the equation properly takes this into consideration.

Example 14.4 Calculation of Fan Inlet and Discharge Pressures

Assume that an exhaust fan operates with a 3 in. H_2O static pressure at the fan inlet and a velocity pressure of 0.5 in. H_2O. The atmospheric pressure is 407 in. H_2O.

The absolute total pressure at the fan inlet will then be (407 - 3) + 0.5 = 404.5 in. H_2O.

Assume the static and velocity pressures on the discharge side are 2 in. H_2O and 1 in. H_2O, respectively.

The absolute total pressure at the fan discharge will then be 410 - 404.5 in. H_2O = 5.5 in. H_2O.

The static pressure of fan is then 5.5 - 1 (the velocity pressure at the fan outlet) = 4.5 in. H_2O.

Fan Laws

The fan laws that relate the performance variables for any homologous series is simply a range of fan diameters where all of the dimensional parameters are proportional. At the same relative point of operation on any two performance curves in a homologous series, the mechanical efficiencies will be equal. Under these conditions, the following relationships apply (size $\cong$ diameter):

$$Q_2 = Q_1 (size_2 / size_1)^3 (rpm_2 / rpm_1) \qquad [14.8]$$

$$P_2 = P_1 (size_2 / size_1)^2 (rpm_2 / rpm_1)^2 (\rho_2 / \rho_1) \qquad [14.9]$$

$$bhp_2 = bhp_1 (size_2 / size_1)^5 (rpm_2 / rpm_1)^3 (\rho_2 / \rho_1) \qquad [14.10]$$

The performance variables involved are flowrate, Q, fan size, size, rotational speed, rpm, pressure, P, gas density, ρ, and brake horsepower, bhp. In the above equations, the pressure may be represented by total pressure, static pressure, velocity pressure, fan total pressure, or fan static pressure.

In actual practice, the fan laws are typically used to determine the effect of changing only one variable at a time and are most often applied to a single fan size. The most common variable of interest is fan speed. For determining the effect of changing fan speed while operating on the same gas stream ($\rho_1 = \rho_2$), the fan laws become:

$$Q_2 = Q_1 (rpm_2 / rpm_1) \qquad [14.11]$$

$$P_2 = P_1 (rpm_2 / rpm_1)^2 \qquad [14.12]$$

$$bhp_2 = bhp_1 (rpm_2 / rpm_1)^3 \qquad [14.13]$$

Referring to the original equations, it is important to note that if only changes in gas density are involved, pressure capabilities and power requirements change proportionally, while flow rate is unaffected. This behavior is sometimes characterized by stating that "a fan is a constant volume machine", i.e., it moves volumes of air, not masses of air.

Example 14.5 Application of Fan Laws

A fan operating at a speed of 1,474 rpm delivers 10,200 cfm at 4 in. H_2O static pressure and requires 8.85 brake horsepower (bhp). What will be the new operating conditions if the fan speed is increased to 2,000 rpm.

Solution:

The new flow is:

$$\frac{Q_2}{Q_1} = \frac{rpm_2}{rpm_1}$$

$$Q_2 = 10,200 \text{ cfm} \cdot \frac{2,000 \text{ rpm}}{1,474 \text{ rpm}}$$

$$= 13,840 \text{ cfm}$$

The new static pressure of the system is:

$$\frac{P_2}{P_1} = \left(\frac{rpm_2}{rpm_1}\right)^2$$

$$P_2 = 4 \text{ in.} \cdot \left(\frac{2,000 \text{ rpm}}{1,474 \text{ rpm}}\right)^2$$

$$= 7.4 \text{ in.}$$

The horsepower required is:

$$\frac{bhp_2}{bhp_1} = \left(\frac{rpm_2}{rpm_1}\right)^3$$

$$bhp_2 = 8.85 \text{ bhp} \cdot \left(\frac{2,000 \text{ rpm}}{1,474 \text{ rpm}}\right)^3$$

$$= 22.1 \text{ bhp}$$

Note the drastic increase in horsepower necessary for increasing the fan speed.

Fan Performance

For a particular fan turning at a given rpm, there is one *and only one* fan curve. It represents all of the pressure/air volume combinations that the fan can produce when operating at that rpm. How the fan curve interacts with the ventilation system characteristics represented by the system curve determines the fan operating characteristics, i.e., volume and pressure.

An example showing normalized duct system curves for three systems is shown in Figure 14.13. These curves are the percentage of duct system resistance as a function of the percent of duct system flow rate. The system curves follow the general relationship:

$$P_2 = P_1(Q_2/Q_1)^2 \qquad [14.14]$$

Figure 14.13
Normalized Duct Curves

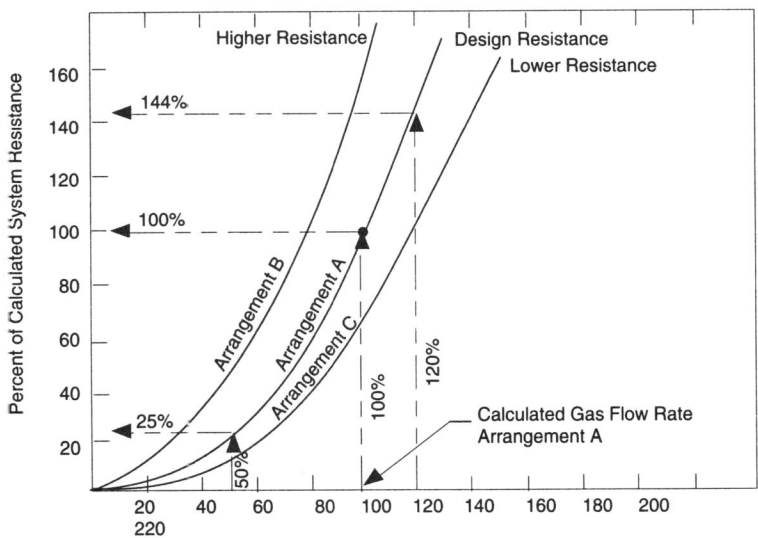

Source: Crowder 1993

The system curve is developed by first determining the resistance or static pressure for one flow rate through the system. Other points on the curve are then determined using Equation [14.14]. Thus, if the design point for system A were at 100 percent volume and 100 percent resistance, increasing the flow rate to 120 percent of the design flow would increase the resistance to 144 percent of the design resistance. Likewise, decreasing the flow rate to 50 percent of the design value would decrease the resistance to 25 percent of the design resistance.

The point of intersection of the system curve with the fan curve determines the actual fan performance. This is shown in Figure 14.14, where a normalized fan curve has been plotted with the system curves from the previous figure. The 100 percent design volume of System A has been arbitrarily selected to intersect at Point 1 with the 60 percent free delivery volume of the fan. Unless actions are taken to change either the fan curve or the system curve, the performance delivered will be that indicated by the intersection point.

Figure 14.14
Combination of System Curves and Fan Curves

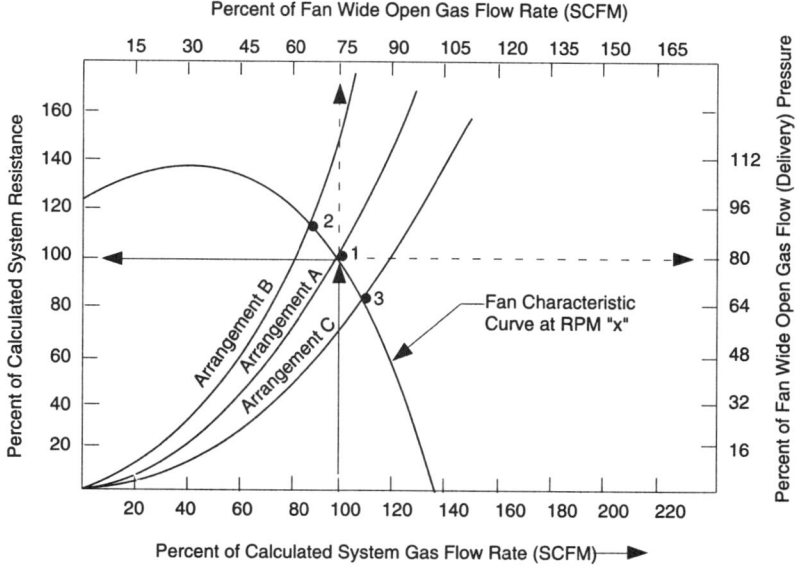

One way to change the flow rate is to change the system characteristics. This could be done by closing or opening a damper, producing a system with more or less resistance and changing the system curve. For example, referring again to Figure 14.14, the flow rate could be decreased to 80 percent of the design volume by closing a damper until the greater resistance represented by System B curve is obtained, shifting the intersection to Point 2.

Changes in flow rate could also be produced by changing the fan speed, thereby shifting the fan curve. This is illustrated in Figure 14.15, where a new fan curve representing a 10 percent increase in speed has been added. At this new speed, the point of operation shifts to point 2. Since flow rate is proportional to fan speed, this 10 percent increase in speed produces a corresponding 10 percent increase in volume. However, following the fan laws, this 10 percent increase in speed will require a 33 percent increase in operating horsepower.

Figure 14.15
Effect of Increased Fan Speed

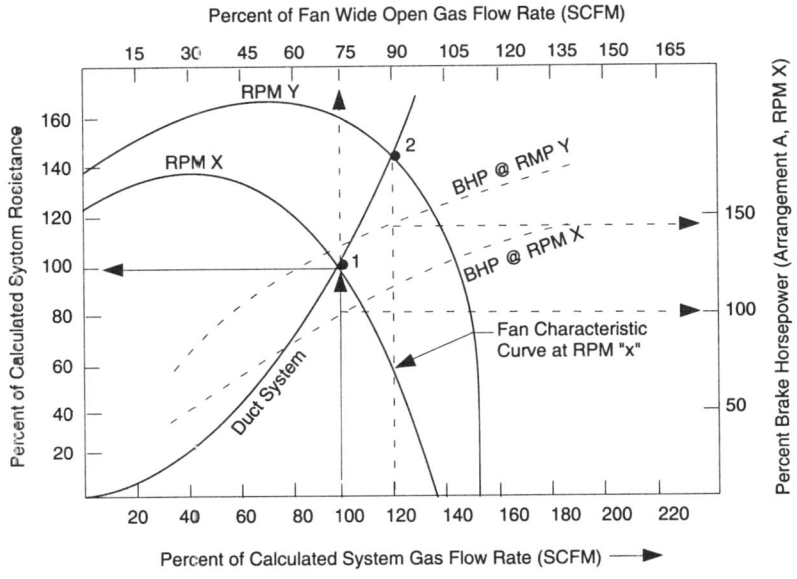

According to the fan laws, changing the gas density will also shift the fan curve. Additionally, since gas density affects the system resistance, the system curve will also be shifted. This is illustrated in Figure 14.16 for a density change from 0.0375 lb/ft³ to 0.075 lb/ft³. As previously indicated, the new operating point will deliver the same air volume but at double the resistance and double the horsepower requirement.

Fan Selection

Fans are typically selected using ratings tables published by manufacturers for their products. An example of one of these tables is shown in Table 14.6. In general, the rating table is entered along the row corresponding to the design volume and down the column corresponding to the design static pressure, including system effects. The point of intersection indicates the rpm that the fan would have to turn to deliver the required performance and the horsepower that would be needed to drive it. The point of maximum efficiency is usually underlined or printed in special type. These tables give the performance of a particular fan size for its range of capacities.

A ratings table indicates the performance of a fan when operating on air having a density of 0.075 lb/ft³. Since a given system may be handling air of a different density, some adjustments are involved before using the table. Remember, density

Figure 14.16
Influence of Gas Density

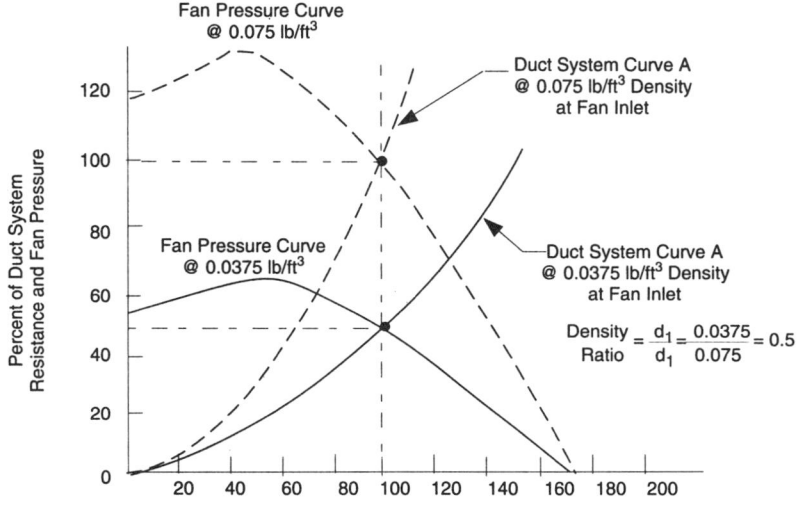

Source: Crowder 1993

does not affect fan volume, but it does influence static pressure conditions and horsepower requirements. The specific procedure involved in fan selection is as follows:

1. Determine the design air volume at actual conditions.

2. Calculate the fan static pressure at actual conditions, including system effects. Fan static pressure is defined as:

$$FSP = SP_{out} - SP_{in} - VP_{in} \quad (VP = \text{velocity pressure}) \quad [14.15]$$

In calculating fan static pressure, the sign of the static pressure is important and must be included. Some manufacturers rate their fans on fan total pressure. Fan total pressure is defined as:

$$FTP = TP_{out} - TP_{in} \quad [14.16]$$

The sign of the total pressure is again important and must be included.

3. Correct the fan static pressure to an equivalent value for standard air:

$$FSP_{equivalent} = FSP_{actual} \bullet (0.075 / \rho_{actual}) \quad [14.17]$$

Table 14.6
Typical Fan Rating Table

Wheel style: backward-inclined
Wheel diameter: 50 1/2 in.
Maximum fan speed: 1,134 rpm
Performances underlined are those at maximun efficiency

cfm	Ps OV	1/2 in. rpm	1/2 in. bhp	1 in. rpm	1 in. bhp	1 1/2 in. rpm	1 1/2 in. bhp	2 in. rpm	2 in. bhp	2 1/2 in. rpm	2 1/2 in. bhp	3 in. rpm	3 in. bhp
5,727	1,250	216	0.74	278	1.34	330	2.01	378	2.75	421	3.50	460	4.28
6,813	1,500	236	0.98	_291_	_1.67_	339	2.42	384	3.24	424	4.06	462	4.91
8,018	1,750	250	1.27	305	2.05	_352_	_2.87_	393	3.76	430	4.69	467	5.66
9,164	2,000	271	1.63	320	2.49	366	3.42	_405_	_4.35_	441	5.36	475	6.40
10,309	2,250	293	2.12	338	3.05	381	4.06	419	5.06	_453_	_6.14_	486	7.26
11,455	2,500	315	2.72	356	3.65	396	4.76	432	5.88	468	7.03	_499_	_8.19_
12,600	2,750	337	3.46	377	4.39	413	5.58	448	6.81	482	8.04	514	9.31
13,746	3,000	360	4.39	399	5.25	430	6.48	465	7.82	496	9.16	527	10.53
14,891	3,250	382	5.43	421	6.25	451	7.52	481	8.93	512	10.35	542	11.80
16,037	3,500	405	6.66	442	7.48	473	8.67	501	10.16	529	11.69	557	13.25
17,182	3,750	429	8.08	463	8.97	496	10.05	521	11.54	547	13.18	574	14.81
18,328	4,000	451	9.64	486	10.61	517	11.61	543	12.99	566	14.78	591	16.49
19,473	4,250	74	11.46	510	12.47	539	13.47	565	14.85	587	16.56	610	18.35
20,619	4,500	497	13.51	532	14.55	560	15.60	587	16.82	610	18.54	630	20.40
21,764	4,750	520	15.82	556	16.82	584	17.98	608	19.17	632	20.77	652	22.59

Note: performances based on standard air density of 0.0075 lb/ft^3

4. Enter the ratings table at the actual volume and the equivalent fan static pressure. Determine the rpm and horsepower requirements.

5. Correct the horsepower requirement to the conditions of actual operation:

$$\text{bhp}_{actual} = \text{bhp}_{equivalent} \cdot (\rho_{actual} / 0.075) \tag{14.18}$$

Since density varies inversely with temperature, corrections for operating conditions could also be made using a ratio of absolute temperatures. Because the exact input parameters may not be contained in the rating table, linear interpolation between the nearest values may be required.

In selecting the appropriate motor for a fan, the designer must give consideration to the possible air densities the fan may have to handle. For example, a system operating at an elevated temperature would require the horsepower as determined above. However, when starting the system, it may be necessary to handle colder, higher-density air, requiring more horsepower.

Chapter 14: Air Quality Systems 541

Should the system be located outdoors in an area that has extreme low temperatures, the horsepower required for startup could be considerable.

An alternate way to cope with cold air induced power requirements is to install the horsepower required for normal operation and then use an inlet or outlet damper, together with an amperage control system. When the fan is started on cold air, the amperage control system senses a high current flow and closes the damper to prevent or reduce air flow into the fan. The fan turning through the restricted air flow would heat the air, reducing its density and reducing the current draw. The amperage control system senses this change and opens the damper slightly to allow some air flow from the hot process. This process would automatically continue until the damper was fully open and the system was operating at design conditions.

Example 14.6 Design Review Problem

A 250 ton per day rotary kiln (lime stone calcining) operates as part of a lime manufacturing facility. A plant survey has determined the following:

- Exhaust gas from kiln: 30,000 scfm or 88,300 acfm at 1,100°F (the gas composition is N_2 60%, CO_2 24%, H_2O 15%, O_2 1% by volume)

- Pollutants are particulate matter with an MMD of 30 μm and 30 % less than 10 μm in size

- Control efficiency required is 99 %

- Control device must be located 200 ft from source

Provide a design and compare the cost for air pollution control devices that could be used to control the emissions from this source.

Solution:

Table 14.1 indicates that baghouses, scrubbers, and electrostatic precipitators have all been used on this type of source with exhaust temperature up to 1,200°F. Figures a), b), c) provide schematic flow diagrams of possible air pollution control systems that could be applied to this source.

A duct velocity of 4,000 ft/min. will be selected for the system.

Figure 14.17a) Venturi Scrubber System Design

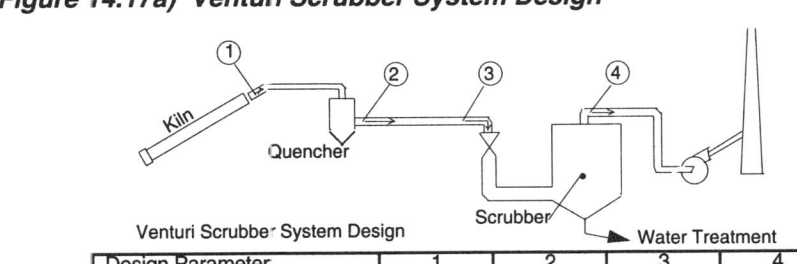

Venturi Scrubber System Design

Design Parameter	1	2	3	4
SCFM	30,000	40,200	40,200	40,200
Temperature	1100°F	220°F	190°F	690°F
CFM	88,300	52,000	49,700	48,200
Duct Diameter	64"	48"	48"	Neglect
Static Pressure (" WG)	Kiln Draft		-1"	-16"

Figure 14.17b) Electrostatic Precipitator System Design

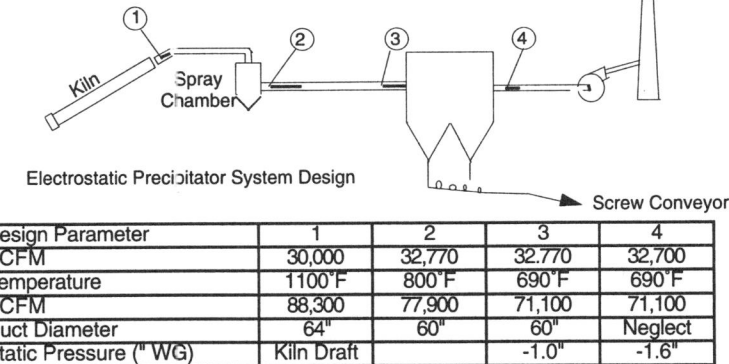

Electrostatic Precipitator System Design

Design Parameter	1	2	3	4
SCFM	30,000	32,770	32,770	32,700
Temperature	1100°F	800°F	690°F	690°F
ACFM	88,300	77,900	71,100	71,100
Duct Diameter	64"	60"	60"	Neglect
Static Pressure (" WG)	Kiln Draft		-1.0"	-1.6"

Figure 14.17c) Fabric Filter System Design

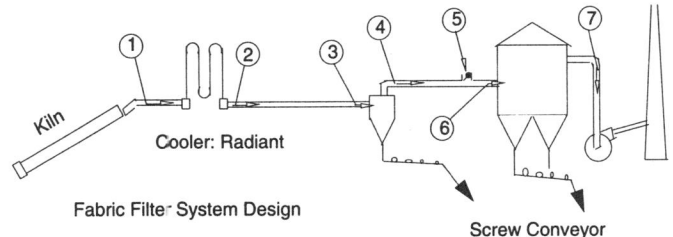

Fabric Filter System Design

Design Parameter	1	2	3	4	5	6	7
SCFM	30,000	30,000	30,000	30,000	38,600[b]	30,000[a] 68,600[b]	30,000[a] 68,600[b]
Temperature	1100°F	600°F	530°F	600°F	100°F[b]	500°F[a] 275°F[b]	500°F[a] 275°F[b]
ACFM	88,300	60,000	56,000	54,300	40,800[b]	54,300[a] 95,100[b]	54,300[a] 95,100[b]
Duct Diameter	64"	52"	52"	52"	52"	Neglect	Neglect
Static Pressure (" WG)	Kiln Draft	-2.1"	-2.1"	-8.7"	-	-	-14.7"

a = glass bag
b = polyester bag

Chapter 14: Air Quality Systems   543

A. Venturi Scrubber

1. Determine direct exhaust from kiln

$$\text{duct size} = \frac{88,300 \text{ acfm}}{4,000 \text{ fpm}} = 22.1 \text{ ft}^2/\text{ft}$$

64 in. duct (22.3 ft^2 / ft) with velocity of $\frac{88,300 \text{ acfm}}{22.3 \text{ ft}^2} = 3,960$ fpm

2. The quencher (see Figure 14.17a) is sized to cool 30,000 scfm from 1,100°F to 220°F.

   For exhaust of N_2 60 %, CO_2 24 %, H_2O 15 %, O_2 1% by volume

   molecular weight = (28 • 0.6) + (44 • 0.24) + (18 • 0.15) + (32 • 0.01)
   = 30.38 lb/lb mol

Assume the flue gas has a heat capacity the same as air. From Table 2.1 the heat capacity at 1,100°F was interpolated from the C_p at 1,200°F and 1,000°F as:

$C_{p(1100°F)}$ = (0.248+0.246)/2 = 0.247 Btu/lb °F

From Table 2.1: $C_{p(720°F)}$ ~ 0.24 Btu/lb °F

$C_{p(average)}$ = (.247+0.24)/2 = 0.2435 Btu/lb °F

From Table 2.2, molar volume for gas at standard condition (70°F, 1 atm) = 387 ft^3

The enthalpy difference is given by Equation [2.27] as $\Delta H = C_p(T_2 - T_1)$

From Equation [11.13]

heat transfer required in the quencher = (mass flow rate) • ΔH

$$\text{mass flow rate} = \frac{30,000 \text{ acfm} \bullet 30.38 \text{ lb}/\text{lb mol}}{387 \text{ ft}^3/\text{lb mol}}$$

= 2,355 lb / min

heat transfer required = 2,355 lb / min • 0.2435 Btu / lb°F • (1,100 – 220)°F
= 504,629 Btu / min

3. Determine the quantity of water required with inlet water at 60°F.
   From Table 2.3, the latent heat of evaporation for water at 60°F is 1,059.9 Btu/lb.

$$\text{Water needed for the quencher} = \frac{504{,}629 \text{ Btu/min}}{1{,}059.9 \text{ Btu/lb}}$$

$$= 476 \text{ lb/min}$$

$$= \frac{476 \text{ lb/min} \bullet 7.48 \text{ gal/ft}^3}{62.4 \text{ lb/ft}^3}$$

$$= 57 \text{ gal/min}$$

Since quenchers are normally supplied excess water, assume a water rate of 60 gal/min is supplied.

4. The gas volume after the quencher is:

$$\text{gas: } 88{,}300 \bullet \frac{680 \text{ R}}{1{,}560 \text{ R}} = 38{,}500 \text{ afm}$$

$$\text{water: } 60 \text{ gal/min} = \frac{60 \text{ gal/min} \bullet 62.4 \text{ lb/ft}^3}{7.48 \text{ gal/ft}^3} = 500 \text{ lb/min}$$

$$\text{water: } \frac{500 \text{ lb/min}}{18 \text{ lb/lb mole}} \bullet 387 \frac{\text{scf}}{\text{lb/mol}} \bullet \frac{680 \text{ R}}{530 \text{ R}} \approx 13{,}500 \text{ acfm}$$

Total volume flow rate = 38,500 + 13,500 = 52,000 acfm

5. Determine the duct size.

From Table 14.5, for average industrial dust (cement plant included), recommended duct velocity = 4000 ft/min.

$$\frac{52{,}500 \text{ acfm}}{4{,}000 \text{ ft/min}} = 13.0 \text{ ft}^2$$

use 48 in. (12.57 ft²/ft) duct

$$\frac{52{,}500}{12.57} = 4{,}140 \text{ ft/min}$$

6. Determine heat transfer for 170 ft of duct from quencher to scrubber.

(The spray chamber is assumed to be 30 ft long. Therefore, the length of duct from spray chamber to ESP = 200 - 30 = 170 ft)

Radiation cooling calculations: use Equation [2.38]

Assume ambient temperature = 100°F and total heat transfer across duct wall = 5.5 Btu/ft² hr°F

Radiated from duct = Heat transferred from gas stream

Chapter 14: Air Quality Systems

From Equation [2.38]:

$$U \cdot A \cdot \Delta T_m = m \cdot C_p \cdot \Delta T$$

duct area = 170 ft • 12.57 ft²/ft = 2,136.9 ft²

$$5.5 \frac{Btu}{ft°F\,hr} \cdot \frac{2{,}136.9\ ft^2}{60\ min/hr} \cdot \frac{(T_i - T_a) - (T_o - T_a)}{\ln \frac{(T_i - T_a)}{(T_o - T_a)}}$$

$$= (2{,}355 + 500) \frac{lb}{min} \cdot 0.24 \frac{Btu}{lb\ °F} \cdot (T_i - T_o)$$

$$195.88 \cdot \frac{(220 - T_o)}{\ln \frac{(220 - 100)}{(T_o - 100)}} = 685.2 \cdot (220 - T_o)$$

By trial and error: $T_o = 190°F$

7. Determine the new gas volume.

$$52{,}000\ acfm \cdot \frac{650\ R}{680\ R} = 49{,}700\ acfm$$

8. Determine sizing of the venturi scrubber for 99.3 % efficiency.
   30 % of the dust < 10 μm, MMD = 30 μm
   assume ρ_p = 2.0 g/cm³

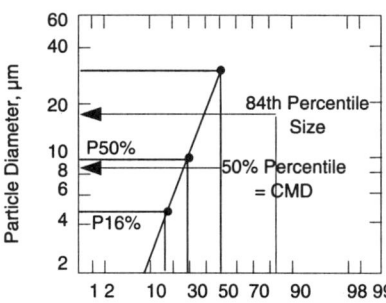

From the figure above, $d_{16\%} = 5$ μm

$$\sigma_g = \frac{d_{84\%}}{d_{50\%}} = \frac{d_{50\%}}{d_{16\%}} = \frac{30}{5} = 6$$

Equation [4.20], $\eta = 1 - \exp(-Ad_p^\beta)$

For the venturi scrubber, assume $\beta = 2$

From Figure 4.11, for $\eta = .993$ and $\sigma_g = 8.6$

$(d_{p50}/MMD) = 0.01$

$d_{p50} = 0.01 \cdot 30 = 0.3\ \mu m$

aerodynamic diameter $= 0.3\left(\dfrac{2}{1}\right)^{0.5} = 0.42 \approx 0.4\ \mu m$

From Figure 7.14, for $d_{p50} = 0.4\ \mu m$, a rectangular throat, and efficiency of 50%, the pressure drop is 16 in. H_2O. If Figure 7.18 is used and $f = 0.5$, the pressure drop is near 25 in. H_2O.

From Figure 7.17, Q_L/Q_G is selected as 10 gal/1,000 ft$^3 \cong 1.3$ litre/m^3

From Equation [7.22], $\Delta P = v_G^2 \rho_L \dfrac{Q_L}{Q_G}$

$16\ \text{in. } H_2O \cdot \dfrac{0.0361\ lb_f/in.^2}{1\ \text{in. } H_2O} \cdot \left(\dfrac{12\ in.}{ft}\right)^2$

$= v_G^2 \cdot 62.4\ lb_m/ft^3 \cdot \dfrac{10\ gal}{1,000\ ft^3} \cdot \dfrac{1\ ft^3}{7.48\ gal}$

$v_G^2 = 997.29 \dfrac{lb_f\ ft}{lb_m}$

$v_G^2 = 997.29 \dfrac{lb_f\ ft}{lb_m} \cdot g_c = 997.29 \dfrac{lb_f\ ft}{lb_m} \cdot 32.2 \dfrac{ft\ lb_m}{lb_f\ sec^2}$

$v_G = 179.2\ ft/sec$

A temperature drop of 20°F is estimated in the venturi scrubber.

9. Determine the fan size.

Capacity of fan $= 49,700\ acfm \cdot \dfrac{630\ R}{650\ R} = 48,200\ acfm$

B. Electrostatic Precipitator

1. Exhaust from kiln: same as venturi scrubber.

2. Add spray chamber to reduce temperature to 800°F to control resistivity with 60°F water (see Table 2.1)

    $C_{pair(average)} = (.247+0.244)/2 = 0.2455\ Btu/lb\ °F$

$\Delta H_{water} = (346.4 - 62.7) = 283.7$ Btu/lb

(Latent heat) • m_{water} + $m_{water}\Delta H_{water}$ = (mass flow rate)• ΔH_{air}

$1,059.9 \cdot m_{water} + m_{water} \cdot 283.7 = 2,355 \cdot 0.2455 \cdot (1,100 - 800)$

Water needed for the spray chamber = 129 lb / min

$$= \frac{129 \text{ lb / min} \cdot 7.48 \text{ gal / ft}^3}{62.4 \text{ lb / ft}^3}$$

$$= 15 \text{ gal / min}$$

3. Determine the gas volume flow rate.

   a) gas: $30,000 \cdot \dfrac{1,260 \text{ R}}{530 \text{ R}} = 71,320$ acfm

   b) water: $\dfrac{125 \text{ lb / min}}{18 \text{ lb / lb mole}} \cdot 387 \dfrac{\text{scf}}{\text{lb mole}} \cdot \dfrac{1,260 \text{ R}}{530 \text{ R}} = 6,389$ acfm

   Total volume flow rate = $71,320 + 6,389 \approx 77,900$ acfm

4. Determine the duct size.

   $$\frac{77,900 \text{ acfm}}{4,000 \text{ fpm}} = 19.5 \text{ ft}^2$$

   use 60 in. (19.6 ft²/ft) duct

5. Determine the heat transfer for 165 ft of duct from spray chamber to electrostatic precipitator. The spray chamber is assumed to be 35 ft long. Therefore, the length of the duct from the spray chamber to ESP = 200 - 35 = 165 ft).

   area = 165 ft • 19.6 ft2/ft = 3,234 ft²

From Equation [2.38]:

$$5.5 \frac{\text{Btu}}{\text{ft}^2 \, °\text{F hr}} \cdot \frac{3,234 \text{ ft}^2}{60 \text{ min / hr}} \cdot \frac{(T_i - T_a) - (T_o - T_a)}{\ln \dfrac{(T_i - T_a)}{(T_o - T_a)}}$$

$$= (2,355 + 125) \frac{\text{lb}}{\text{min}} \cdot 0.242 \frac{\text{Btu}}{\text{lb} \, °\text{F}} \cdot (T_i - T_o)$$

$$296.45 \cdot \frac{(800 - T_o)}{\ln \dfrac{(800 - 100)}{(T_o - 100)}} = 600.2 \cdot (800 - T_o)$$

By trial and error: $T_o = 690°$F

6. Determine the new gas volume.

$$77{,}900 \text{ acfm} \bullet \frac{1{,}150 \text{ R}}{1{,}260 \text{ R}} = 71{,}100 \text{ acfm}$$

7. Determine the precipitator size.

Select drift velocity from Table 8.2 for cement dust (dry process)

$w_e = 0.21$ ft/sec (average)

$$A = -71{,}100 \text{ acfm} \bullet \frac{\ln(1-.993)}{0.21 \text{ ft/sec} \bullet 60 \text{ sec/min}} = 28{,}000 \text{ ft}^2$$

8. Calculate the pressure drop.

Total pressure drop = 1.0 in. (chamber & duct) + 0.5 in. (precipitator)

= 1.5 in. H_2O

C. Fabric Filter

Requirements:

Use woven fabric: polyester (275°F), glass (550°F)

    A/C = 2 for glass: pressure air, insulated
    A/C = 3 for polyester: mechanical shaker

Design radiant cooler to low stack temperature without adding moisture to dust. The baghouse system needs a low relative humidity.

Use cyclone before baghouse to remove large particles.

Use dilution port to protect baghouse.

1. Exhaust from kiln: same as venturi scrubber.

    Design radiant cooler to lower temperature from 1,100°F to 600°F

    Try 5,000 ft/min through two 18 in. U tubes (1.767 ft²/ft) in series.

    = 88,300/1.767/5,000 = 10 pairs in parallel

    Assume L is the total length required for radiant cooler.

    Total area = L • 1.767 ft²

    From Table 2.1:

    $C_{pair(average)}$ = (0.247+0.242)/2 = 0.2455 Btu/lb°F

    From Equation [2.38]:

$$5.5 \cdot \frac{L \cdot 1.767}{60} \cdot \frac{(1,100-100)-(600-100)}{\ln\frac{(1,100-100)}{(600-100)}} = 2,355 \cdot 0.2455 \cdot (1,100 - 600)$$

$L \approx 400$ ft

For 10 parallel U tubes, the radiant cooler is 40 ft height.

2. Determine the new gas volume.

$$30,000 \text{ scfm} \cdot \frac{1,060 \text{ R}}{530 \text{ R}} = 60,000 \text{ acfm}$$

Duct diameter.

$$\frac{60,000 \text{ acfm at } 600°F}{4,000 \text{ fpm}} = 15 \text{ ft}^2 / \text{ft duct}$$

Use 52 in. duct (14.7 ft^2)

3. The cooler is assumed to be 30 ft long and the duct from the cooler to cyclone = 200 - 30 = 170 ft. Duct is 52 in.

$$\text{Surface area} = 170 \text{ ft} \cdot \pi \cdot \frac{52 \text{ in.}}{12 \text{ in./ft}}$$
$$= 2,315 \text{ ft}^2$$

Assume temperature drop by radiation = 0.4°F/ft

Temperature is found to be 530°F and the pressure drop is approximately 0.6 in. H$_2$O.

$$\text{New gas volume} = 60,000 \cdot \frac{990 \text{ R}}{1,060 \text{ R}} = 56,000 \text{ acfm}$$

4. Design a mechanical collector.

Stairmand high efficiency cyclone is applied.

Assume the pressure drop is limited to 6 in. H$_2$O

Table 5.1 for Stairmand cyclone: N_H = 6.4, a = 0.5D, b = 0.5D

From Equation [5.7]: 6 in. H$_2$O = $0.003 \cdot 0.0373 \text{ lb/ft}^3 \cdot v_g^2 \cdot 6.4$

v_g = 91 ft/sec

$a \cdot b \cdot v_g$ = 56,000 acfm/60

$0.1D^2 \cdot 91$ = 933

D = 10 ft

Use 10 ft diameter Stairmand cyclone for the mechanical collector

5. Determine the new gas volume. The cyclone can be assumed to have 30°F temperature loss.

$$56,000 \cdot \frac{960 \text{ R}}{990 \text{ R}} = 54,300 \text{ acfm}$$

6. Design the fabric filter.

   Select two cycle at 28,000 acfm with 6 in. pressure drop

   Dilution air for polyester bags is required (no dilution air is required for the glass bags)

   (30,000 scfm • 500°F) + (Dilution air) • 100°F = (30,000 scfm + Dilution air) • 275°F

   Dilution air required = 38,600 scfm

   Total gas volume is 30,000 + 38,600 = 68,600 sfcm or 95,100 acfm at 275°F

   Polyester 95,000 acfm at 275°F

   Glass 54,300 acfm at 500°F (no dilution air is needed)

   Cloth area:

   $$\text{Glass} = \frac{54,300}{2.0 \text{ A/C}} = 27,150 \text{ ft}^2 \text{cloth area}$$

   $$\text{Polyester} = \frac{95,100}{3.0 \text{ A/C}} = 31,700 \text{ ft}^2 \text{cloth area}$$

   ΔP for baghouse = 6 in.

   Total pressure

Radiation	2.1 in.
Duct	0.6 in.
Mechanical Collector	6.0 in.
Baghouse	<u>6.0 in.</u>
Total	14.7 in.

   Fan 54,300 or 95,100 acfm

Table 14.7 provides a comparison of costs for the four alternate air pollution control systems that have been designed to meet the emission requirements for the lime kiln. In this comparison, the costs of the duct work and stack have been kept constant. The water scrubber is considered to have a lifetime of 10 years while the other collection devices have a 20-year life. The venturi scrubber has a low capital cost, but a high operating cost. The dilution air required for the polyester baghouse makes this option more expensive than the filter with glass fabric.

Table 14.8 provides a summary of the cost estimate calculations. This problem presents only a preliminary design and cost estimate (± 25%) and is intended to aid in the selection of an air pollution control system for this source. This analysis suggests that a wet scrubber not be considered because of the additional cost for a wastewater treatment facility which has not been included in this cost estimate. A shaker baghouse with polyester bags would require insulation and dilution air to cool the system to 275°F to protect the polyester bags. The cost for this system is higher than a system using glass bags. The electrostatic precipitator has the lowest estimated cost. The cost estimate is 25% lower than for the fabric filter with glass bags. A more detailed analysis of cost for the electrostatic precipitator and fabric filter with glass bags should be undertaken.

14.4 Problem Set

14.1 What is the velocity head in a duct with an air velocity of 2,000 ft/min? What is the pressure drop if the duct contains 2 elbows (45° and 90°) and is 240 ft long? If the duct is handling 1,500 scfm, what size is the duct?

14.2 A 20 in. diameter duct with a velocity of 1,375 ft/min at 70°F has a pressure drop due to friction of 0.13 in. H_2O per 100 ft of duct. The process temperature is changed to 250°F. Determine the new pressure drop due to friction.

14.3 Show that

$$v, \text{ft/min} = (1,096.2)\sqrt{\frac{VP}{0.075}} = 4,005\sqrt{VP}$$

14.4 An acid cleaning tank is to be fitted with slot hoods to collect mist and vapors. The tank is 6 ft by 20 ft and must have slots along both of the 20-ft sides with a plenum and duct at one end leading to a scrubber. One-inch wide slots with a face velocity of 4,000 ft/min are to be used. The duct leading to the scrubber is 1 ft by 4 ft by 30 ft long. Pressure losses in the plenum and duct entry are equivalent to 50 ft of duct.

What is the required volumetric flow rate?

Estimate the total pressure drop of the slot and duct system.

14.5 Fan A is a 10-inch diameter fan that operates at 950 rpm and delivers 11,000 cfm. Fan B is from the same homologous series and operates at the same rpm except that its wheel diameter is 20 inches. Estimate the flow rate for Fan B.

14.6 A fan operating at a speed of 1,694 rpm delivers 12,200 acfm of flue gas at 5.0 in. H_2O static pressure and requires 9.25 bhp. What will be the new operating conditions if the fan speed is increased to 2,100 rpm?

14.7 Calculate the horsepower required to process a 6,500 acfm gas stream from an incinerator. The pressure drop across various pieces of equipment has been

Table 14.7
Cost Estimate Comparison — all amounts in $

Venturi Scrubber		Electrostatic Precipitator		Fabric Filters	Fiber glass	Polyester
Scrubber	30,000	E.P. Purchase	300,000	Baghouse	160,000	100,000
Quencher	30,000	Spray chamber	50,000	Insulation	0	60,000
Fan	10,000	Fan	10,000	Cooler	70,000	70,000
Motor	10,000	Motor	5,000	Cyclone	20,000	20,000
Duct	21,000	Duct	21,000	Fan	7,000	7,000
Stack	5,000	Stack	5,000	Motor	5,000	5,000
	106,000		401,000	Duct	21,000	21,000
				Stack	5,000	5,000
				Bags	15,000	10,000
					303,000	298,000
Instruments 10 %	10,600	Instruments 10 %	40,000	Instruments 10 %	30,000	30,000
Tax & Fee 8 %	8,500	Tax & Fee 8 %	32,000	Tax & Fee 8 %	24,000	24,000
	125,100		473,000		357,000	352,000
Installation (100 % eq)	125,100	Installation (100 % eq)	473,000	Installation (100 % eq)	357,000	352,000
Total Capital Cost	250,200	Total Capital Cost	946,000	Total Capital Cost	714,000	704,000
Annual Cost at 10 % and 10 years	40,000	Annual Cost at 10 % and 20 years	111,000	Annual Cost at 10 % and 20 years	84,000	83,000
Operation Cost		Operation Cost		Operation Cost		
Labor	20,000	Labor	4,000	Labor (25 shifts)	20,000	20,000
General Maintenance	35,000	General Maintenance	10,000	General Maintenance	35,000	35,000
Utilities		Electricity	30,000	Bag replacement & install	20,000	12,000
Fan 1.34 mKwh	60,000	Overhead	10,000			
Water pump	5,000			Electricity	31,000	60,000
Water usage	30,000			Overhead	30,000	30,000
Overhead	30,000					
	180,000		54,000		136,000	167,000
Total Annual Cost	220,000	Total Annual Cost	165,000	Total Annual Cost	220,000	250,000

Table 14.8
Cost Estimate Summary

	Scrubber	Electrostatic Precipitator	Filters	
			Glass	Polyester
Annual Cost	240,000	165,000	220,000	250,000

estimated to be 6.4 in. H_2O. The pressure loss for duct work, elbows, valves, etc., and expansion-contraction losses are estimated to total 4.4 in. H_2O. Assume an overall fan-motor efficiency of 63%.

14.8 Compare a 6 in. H_2O pressure drop multiclone separator for an incinerator discharging 1,000 ft^3/sec of 1,800°F gas for the various methods of cooling possible.

14.9 The rotary dryer emissions from an asphalt hot-mix asphalt concrete plant have been determined as follows:

Characteristics of rotary dryer emissions from an asphalt concrete plant

$$\begin{aligned}
\text{stack gas temperature} &= 93°C\ (199°F) \\
\text{pressure, atm} &= 1.0 \\
\text{gas flow rate, Nm}^3/\text{sec} &= 16.7 \\
\text{acfm} &= 43{,}455 \\
\text{gas density, lb/ft}^3 &= 0.0739 \text{ at } 25°C \\
\text{particle density, g/cm}^3 &= 2.0
\end{aligned}$$

Rotary Dryer Emissions, g/Nm^3

$$\begin{aligned}
\text{total uncontrolled} &= 55.5 \\
> 15.3\ \mu m &= 49.28 \\
15.3 - 12.9\ \mu m &= 1.66 \\
12.9 - 10.1\ \mu m &= 1.17 \\
10.1 - 7.28\ \mu m &= 1.22 \\
7.28 - 5.00\ \mu m &= 0.87 \\
5.00 - 2.5\ \mu m &= 0.90 \\
2.5 - 1.01\ \mu m &= 0.29 \\
< 1.01\ \mu m &= 0.11
\end{aligned}$$

The emission limit is 0.04 grains/scf.

Design a cyclone primary collector and a secondary collector selected from a preliminary design of fabric filter, venturi scrubber and electrostatic precipitator. The cyclone will have a maximum pressure drop of 5 in. H_2O and a collection efficiency of at least 90%. The secondary collector selected should be based on cost comparison of the preliminary designs.

14.5 References

Danielson, J. (Ed). *Air Pollution Engineering Manual*, AP40, EPA. Research Triangle Park, NC. 1973.

Neveril, R.B. *Capital and Operating Costs of Selected Air Pollution Control Systems*. EPA 450/5-80-002. Research Triangle Park, NC. 1978.

Crowder, J. *Emission Capture and Gas Handling System Inspection, Training Manual*. EPA 340/1-92-015a. Research Triangle Park, NC. 1993.

Marchello, J.M. *Control of Air Pollution Sources*. Marcel Dekker. New York. 1976.

Appendix A

Practice Problems for the Air Quality Portion of the P.E. Examination in Environmental Engineering (with Solutions)

The problems have been prepared as a study aid for the Air Quality Portion of the *Principles and Practice of Engineering Examination in Environmental Engineering*. The problems were prepared to cover the subject material contained in this book and are not actual problems that have appeared on previous examinations. Reference to specific equations and subjects in the book are provided along with the problem solution to aid in understanding the subjects that are addressed by the problems. The problems have been constructed to cover general topic areas. Each topic generally has four multiple-choice questions that address the problem area that is presented. The individual questions are not linked (the correct solution to a question is not required to solve other questions in the same problem).

1. Ideal Gas Law

Concentrations of atmospheric pollutants are expressed as parts per million by volume (ppm) or micrograms per cubic meter of air ($\mu g/m^3$). These two units can be compared using the Ideal Gas Law.

Equation [2.9] $$P \bullet V = \frac{M \bullet R \bullet T}{MW}$$

where: P = absolute pressure
V = volume of a gas
M = mass of gas
MW = molecular weight of a gas
T = absolute temperature
R = universal gas constant

The unit of R depends upon the units of measurements used in the equation, such as:

$$0.73 \frac{(atm)(ft^3)}{(lb\ mole)(°R)}, \quad 8.3 \bullet 10^{-5} \frac{(atm)(m^3)}{(g\ mole)(°R)}$$

The Gas Law applies to mixtures of gases as well as a pure gas.

1.1 Determine the concentration in μg/m³ of 1 ppm of SO_2 at 25°C and 1 atm.
   a) 32 μg/m³
   b) 800 μg/m³
   c) 1,000 μg/m³
   d) 2,600 μg/m³

1.2 The ozone concentration at an urban station is measured as 156 μg/m³. The temperature is 25°C at 1 atm pressure. Determine the concentration in ppm (MW of O_3 is 48 g/mole).
   a) 0.06 ppm
   b) 0.08 ppm
   c) 0.10 ppm
   d) 0.15 ppm

1.3 The SO_2 concentration in a stack is 600 ppm, the stack diameter is 10 ft, and the gas flow rate is 40 ft/sec. The gas temperature and pressure are 450°F and 1 atm. Determine the SO_2 mass flow rate in lb/hr.
   a) 650 lb/hr
   b) 1,350 lb/hr
   c) 3,250 lb/hr
   d) 6,800 lb/hr

1.4 In an industrial operation a ventilation air stream (1,000 cfm, 100°F) passes through a paint drying oven. Toluene (MW = 92) evaporates into the air stream at the rate of 0.25 lb/min from a painted surface. Determine the concentration of toluene in the air stream.
   a) 0.001 ppm
   b) 0.1 ppm
   c) 100 ppm
   d) 1,000 ppm

Appendix A   557

2. Cost Estimates

The cost of installing and operating air pollution control equipment is a function of many direct and indirect factors. The principal costs are associated with capital investment, installation, and operation. The usual basis for comparing the cost of control devices is the total annual cost.

2.1   Determine the annual cost of moving 1,000 cfm of air against a head of 1 in. H_2O if electricity costs are $ 0.015/KWh. The fan efficiency is 60%.

   a) 6.4 $/yr
   b) 15.4 $/yr
   c) 25.7 $/yr
   d) 51.4 $/yr

2.2   Determine the annual capital cost for an air pollution control device that costs $600,000 to purchase and $900,000 to install if the useful life is 20 years and the interest rate is 7%.

   a) $43,153/yr
   b) $95,130/yr
   c) $101,135/yr
   d) $109,910/yr

2.3   It has been estimated that it will cost an additional $0.10/gallon to provide gasoline with a lower sulfur content (40 ppm sulfur) that will prevent poisoning of new automobile catalysts in low emission vehicles (LEV). If a motor vehicle is driven an average of 15,000 miles per year and the gas mileage is 25 miles/gallon, determine the added cost per year for gasoline.

   a) $20/yr/vehicle
   b) $40/yr/vehicle
   c) $50/yr/vehicle
   d) $60/yr/vehicle

2.4   An electrostatic precipitator is to be upgraded from 95% efficiency to 99.99% by adding additional fields to the existing device. What will be the increase in cost for the control device? (Assume that the particle migration velocity does not change with efficiency for this industrial application and that there is no economy of scale).

   a) 2.5
   b) 4.6
   c) 5.0
   d) 6.2

3. Particle Motion and Collection Efficiency

Particles moving in the atmosphere experience a drag force caused by the resistance to particle motion due to contact with the air surrounding the particle. The magnitude of the drag force is related to the velocity of the particle and the flow pattern of the air around the particle. The particle Reynolds number is used as an indication of the flow pattern. From experiments, it has been observed that three fluid flow regimes exist that can be related to Reynolds number (laminar (Stokes), transitional, and turbulent). In the laminar flow range the form of drag, F_D, is

From Equation [3.8] $F_D = 3\pi\mu v d_p / C_c$

where: d_p = particle diameter
  v = particle velocity
  μ = fluid viscosity
  C_c = Cunningham Correction Factor

3.1 Determine the terminal settling velocity for a 1.0 μm diameter particle with a density of 6.2 g/cm³. The particle is considered to be in the Laminar flow range. The temperature and pressure are 20°C and 1 atm ($\mu = 1.8 \cdot 10^{-4}$ g/cm sec).

  a) 0.001 cm/sec
  b) 0.008 cm/sec
  c) 0.02 cm/sec
  d) 12.0 cm/sec

3.2 5 μm diameter particles with density of 2 g/cm³ are settling in a gravity collector that is 20 ft in both length and height. What is the maximum possible residence time for a particle in the gravity chamber if the flow field is laminar?

  a) 0.8 min
  b) 1.7 min
  c) 51 min
  d) 100 min

3.3 Which of the following parameters can be used to characterize the single particle collection efficiency by impaction and interception for 5 μm diameter dust particles collected by water drops in a gravity settling chamber?

  a) Diffusion Coefficient and Brownian Motion
  b) Gravity Settling and Cunningham Correction
  c) Stopping Distance and Relaxation Time
  d) Peclet Number and Reynolds Number

Appendix A

3.4 The particle migration velocity in a typical electrostatic precipitator collecting fly ash is between 5 and 10 cm/sec, and the distance from the charging electrode to the plate is 20 to 30 cm. The gas flow velocity is typically 1-2 m/sec (see Table 8.7). What is the ratio of particle collecting time to the average residence time for particles in a typical electrostatic precipitation?

   a) 0.01
   b) 0.1
   c) 1.0
   d) 10

4. Evaluation of Particulate Control Systems

An industry is currently operating an air pollution control system that consists of a cyclone followed by a fabric filter dust collector on a source that has an air flow rate of 10,000 acfm. The exit temperature is 130°F and the dust loading to the cyclone is 8 grains/scfm. The emission regulation is 0.08 grains/scfm. The cyclone has a collection efficiency of 80%.

4.1 Determine the overall required collection efficiency for the air pollution control system.

   a) 95%
   b) 99%
   c) 99.9%
   d) 99.99%

4.2 The overall pressure drop for the cyclone and fabric filter is 15 inches of water. Determine the horsepower requirements for the system.

   a) 10 HP
   b) 15 HP
   c) 25 HP
   d) 30 HP

4.3 The fabric filter has an air-to-cloth ratio of 2 ft/min and contains bags that are 8 inches in diameter and 12 feet long. Determine the number of bags in the baghouse.

   a) 18 bags
   b) 128 bags
   c) 200 bags
   d) 240 bags

4.4 The baghouse has a cleaning interval of 2 hours for removal of collected dust. Determine the amount of material collected on the filter in lb of dust/ft² of cloth area just before the bags are cleaned (the air-to-cloth ratio is 2 ft/min). The collection efficiency for the baghouse is 99% by weight.

    a)   0.054 lb/ft²
    b)   0.065 lb/ft²
    c)   0.22 lb/ft²
    d)   0.27 lb/ft²

5. Design of Air Pollution Control Devices

The emissions of particulate matter from industrial sources usually have a size distribution, and the pollution control devices used to collect the particles have a grade efficiency curve. The grade efficiency is the efficiency with which particles are collected as a function of particle size. In order to determine the removal efficiencies required to meet emission regulations, it is important to have the particle size distribution by mass, of the source to be controlled, and equations that allow the calculation of the grade efficiency curves for the control device.

5.1 The following table provides the reported source particulate size distribution and the grade efficiency for a proposed control device.

Size range	0-10	10-20	20-50
Emission (%)	20	30	50
Grade efficiency	40	60	90

Determine the collection efficiency of the control device.

    a)   50%
    b)   59%
    c)   71%
    d)   86%

5.2 An electrostatic precipitator is being considered as an air pollution control device. A pilot scale unit used on the source had an effective migration velocity of 5 cm/sec at 95% collection efficiency. The gas stream to be treated has an average flow rate of 100,000 acfm. Determine the total plate area required in the full scale electrostatic precipitator.

a) 1,100 ft²
b) 11,000 ft²
c) 30,500 ft²
d) 1,850,000 ft²

5.3 The required overall collection efficiency for a countercurrent, gravity wet scrubber to meet the local air pollution control regulations is determined to be 95%. A manufacturer has a standard scrubber that will be used to control emissions from the source. The liquid-to-gas flow rate in the scrubber is 7.5 gal/1,000 acfm, and the water drops are 500 μm in diameter. The particles to be collected have a 4 μm aerodynamic diameter. The single particle collection efficiency has been obtained for these particles from an efficiency curve (see Figure 7.3 as an example) to be 0.1. Determine the height of the scrubber required for this application. Consider that the water drops are only 50% effective due to wall losses and water evaporation. The ratio of the water drop terminal settling velocity to gas velocity will be maintained at 0.2.

a) 12 ft
b) 28 ft
c) 50 ft
d) 62 ft

5.4 A coal-burning power plant burns coal at a rate of 2,500 tons/day. The sulfur content of coal is 1.5% by weight. How many tons/day of $CaCO_3$ are needed to neutralize the SO_2 generated?

a) 37.5 tons/day
b) 58.5 tons/day
c) 75 tons/day
d) 117 tons/day

6. Energy Balance

Enthalpy (H) is a measure of the thermal energy of a substance. For air, a heat capacity ($C_{p(avg)}$) averaged over the range of temperature change is usually assumed and the relationship expressed as

Equation [2.27] $\quad \Delta H = \left(C_p\right)_{avg} \left(T_2 - T_1\right)$

A local industry has an air pollution emission of 32,000 cfm at 1,200°F that is essentially air containing a small amount of benzene (1,000 ppm).

6.1 Determine the enthalpy difference between 1,200°F and 250°F.

a) 200 Btu/lb
b) 230 Btu/lb
c) 250 Btu/lb
d) 300 Btu/lb

6.2 If the emissions are cooled from 1,200°F to 250°F by radiation and convective cooling in a duct system, determine the new gas volume.

a) 6,700 cfm
b) 13,700 cfm
c) 32,000 cfm
d) 74,800 cfm

6.3 If the emissions are cooled from 1,200°F to 250°F using dilution air at 70°F, determine the new gas volume.

a) 54,000 cfm
b) 65,000 cfm
c) 86,000 cfm
d) 168,000 cfm

6.4 If the emissions are cooled from 1,200°F to 250°F by spraying 70°F water into the air stream, determine the new gas volume.

a) 13,700 cfm
b) 18,100 cfm
c) 54,000 cfm
d) 86,100 cfm

7. Condensation

A 10,000 cfm emission stream (essentially air) at 250°F and atmospheric pressure contains 1,000 ppm benzene that will be cooled to 130°F in a shell and tube condenser. Water at a temperature of 70°F will be used for cooling. The heat transfer coefficient for the condenser is 150 Btu/hr ft^3 °F and the specific heat for benzene is 0.45 Btu/lb °F. The outlet water temperature will be maintained at 100°F. The required condenser heat load has been calculated as 968,000 Btu/hr ($Q = mC_p\Delta T_m$).

7.1 What is the required surface area of the condenser?

a) 43 ft^2
b) 77 ft^2

c) 159 ft²
d) 145 ft²

7.2 How much cooling water is required?
a) 65 gal/min
b) 100 gal/min
c) 125 gal/min
d) 286 gal/min

7.3 How much benzene will be condensed per hour?
a) 0 lb/hr
b) 1.5 lb/hr
c) 2 lb/hr
d) 2.5 lb/hr

8. Adsorption

Gasoline is loaded into trucks at a cargo terminal. An average of 25 trucks per day are loaded. The truck tanks have a maximum volume of 1,000 ft³ and the displacement gas has a flow rate of 200 ft³/min. Displacement losses from loading operations can be estimated as:

From Equation [9.3] $$D_L = 12.46 \left[\frac{S \bullet P \bullet MW}{T} \right]$$

where:
D_L = lb/10³ gallons transferred
P = true vapor pressure, psi (7 for gasoline)
MW = molecular wt, lb/lb mol (66 for gasoline)
T = Temperature of liquid, °R (50°F)
S = saturation factor, (0.6 for normal service)

8.1 A carbon adsorption unit is used to control 95% of the emissions. The carbon capacity is 0.3 g/g of carbon. Estimate the carbon required in the adsorber if it is regenerated once per day. The residual adsorbent after cleaning is 50% of the saturation capacity.

a) 5,700 lb
b) 8,500 lb
c) 12,800 lb
d) 19,400 lb

8.2 Determine the allowable cross-sectional area of the carbon bed if the maximum velocity through the adsorber is 30 ft/min.

a) 3.5 ft
b) 6.7 ft
c) 7.2 ft
d) 10.5 ft

8.3 A test was conducted on a pilot scale carbon adsorber with a bed depth of 2 ft to determine the length of the mass transfer zone (MTZ). The MTZ is that part of the carbon bed that displays a gradient in adsorbent concentration from zero to the equilibrium capacity. The saturation capacity was determined to be 28 lb solvent/ 100 lb of carbon. The breakthrough capacity was calculated to be 24.9% from the test data. The saturation capacity of the carbon in the MTZ can be assumed to be 50%. Determine the length of the mass transfer zone.

a) 2.0 inches
b) 3.2 inches
c) 5.3 inches
d) 6.4 inches

8.4 Estimate the concentration of gasoline in the vapor space of the truck tanks if the temperature is 95°F.

a) 52 ppm
b) 560 ppm
c) 1,250 ppm
d) 2,860 ppm

9. Combustion Kinetics

The overall combustion reaction in incinerators burning VOCs is usually considered first order because the concentration of VOCs to be burned are much less than the concentration of oxygen. The equation representing this reaction is

Equation [11.10] $C_{out} = C_{in} \exp(-kt)$

where k is the first order rate constant and t is the retention time in the combustion zone. The rate constant is a function of temperature and is usually given by the Arrhenius equation where

Equation [11.11] $k = Ae^{-E/RT}$

The activation energy, E, and the constant A are usually determined experimentally and provided in tables for various compounds.

9.1 For an incinerator temperature of 1,350°F and destruction efficiency of 95%, determine the retention time required for vinyl chloride (oxidation parameters for vinyl chloride are E = 63.3 Kcal/mol, and A = 3.57•10^{14} sec^{-1}).

   a) 0.01 sec
   b) 0.1 sec
   c) 0.5 sec
   d) 1 sec

9.2 If the US EPA-required destruction efficiency for vinyl chloride is 99.99% at 2200°F, determine the required residence time in the combustion zone.

   a) 0.00001 sec
   b) 0.00006 sec
   c) 0.0001 sec
   d) 0.0006 sec

9.3 Natural gas (1,050 Btu/scf) will be used to obtain a combustion reaction temperature of 1,250°F. Determine the required natural gas for a 10,000 scfm gas emission (essential air) that has an inlet temperature of 150°F. Heat capacity (C_p) of air will be 0.25 Btu/lb °F.

   a) 102 scfm
   b) 192 scfm
   c) 252 scfm
   d) 310 scfm

9.4 Determine the stoichiometric amount of air required to oxidize 100 lb of methane to CO_2 and H_2O.

   a) 64 lb
   b) 400 lb
   c) 1,717 lb
   d) 1,913 lb

10. Combustion Stoichiometry

The oxidation of combustible hydrocarbons will cause them to be converted to CO_2 and H_2O. The amount of oxygen required for complete combustion is known as the theoretical or stoichiometric oxygen. The equation for the complete combustion of propane is:

$$\text{Equation [11.1]} \quad C_3H_8 + 5O_2 \rightarrow 3CO_2 + 4H_2O + Q(\text{heat})$$

10.1 Determine the scf of air needed for complete combustion of 1 scf of propane.

a) 5 scf
b) 14.5 scf
c) 15 scf
d) 23.8 scf

10.2 The heat of combustion of propane is 2,590 Btu/ft^3 when the water formed is in the liquid state. Convert this to Btu/ft^3 of propane when the water is in the vapor phase. (The latent heat of vaporization for water at 60°F is 1,059 Btu/lb, Table 2.3)

a) 2,590 Btu/ft^3
b) 2,380 Btu/ft^3
c) 1,590 Btu/ft^3
d) 1,380 Btu/ft^3

10.3 Estimate the temperature required to destroy 99.9% of an organic compound using a residence time of 0.5 seconds. E (activation energy) and A (experimental constant) are 45.2 Kcal/g mol and $9 \cdot 10^{10}$ sec^{-1}.

a) 1,000°F
b) 2,000°F
c) 2,500°F
d) 3,000°F

10.4 Determine the scfm of natural gas (heat value = 1,050 Btu/scf) required to heat 8,500 scfm of a contaminated gas stream from 200°F to 1,400°F. Assume that there are no heat losses and the heat capacity of air is 7.5 Btu/lb mol °F.

a) 129 scfm
b) 192 scfm
c) 238 scfm
d) 328 scfm

11. Hazardous Waste Combustion

A hazardous waste containing chlorine is to be burned in an incinerator. The air pollution regulation requires 99.99% destruction efficiency of the hazardous material in an incinerator with operating condition of 2,200°F and 2 seconds retention time. The net heating value of the waste is 8,000 Btu/lb and the molecular weight is 120 lb/lb mole.

Appendix A

11.1 Determine the major combustion products that could be present in the emission stream when there is complete combustion of the waste.
   a) O_2, H_2O
   b) CO_2, H_2O, Cl_2, HCl
   c) H_2S, CH_4, Cl_3Cl
   d) CO, CO_2, H_2O

11.2 Determine the amount of excess air required to maintain the furnace temperature at 2,200°F.
   a) 1.1
   b) 1.5
   c) 2.0
   d) 2.1

11.3 Determine the size of the incinerator required to operate with an average energy release rate of 25,000 Btu/hr ft³ of incinerator volume. 500 lb/hr of waste with a heating value of 8,000 Btu/lb is to be combusted.
   a) 160 ft³
   b) 200 ft³
   c) 320 ft³
   d) 500 ft³

11.4 Determine the reduction in fuel (in percent) expected when a catalytic incinerator operating at 800°F is used to replace a thermal incinerator operating at 1,400°F.
   a) 25%
   b) 45%
   c) 55%
   d) 57%

12. Mass Transfer

The overall rate of a catalytic reaction depends on two steps: transport of reactants to the catalytic surface and the reaction rate at the surface. The overall mass transfer coefficient, K_o, can be defined as

$$\text{Equation [13.10]} \quad \frac{1}{K_o} = \frac{1}{K_m} + \frac{1}{K_r}$$

where: K_m = gas phase mass transfer rate
         K_r = reaction rate coefficient

At high temperature, the mass transfer rate usually controls the rate of the process. For this condition, the concentration at the outlet of the catalyst is given by:

Equation [13.14] $C = C_o e^{-N} = e^{-L/L_m}$

where: N = number of mass transfer units
L = length of reactor

$$L_m = \frac{v}{K_m \cdot a} \text{ for laminar flow conditions}$$

v = gas velocity
a = surface area per unit volume of reactor
($a = 4/D$ where D is the effective hole [channel] diameter)

12.1 Determine the Reynolds Number for flow through a catalyst with an effective hole size of 0.05 inches and gas velocity of 30 fps.

a) 110
b) 140
c) 280
d) 1,680

12.2 Determine the length of the catalyst (in inches) required to remove 99% of a pollutant from a gas stream flowing at 35 ft/sec through a catalyst with an effective diameter of 0.05 inches. (K_m was determined to be 0.5 ft/sec.)

a) 2 inches
b) 4 inches
c) 6 inches
d) 8 inches

12.3 A test on a pollutant source with a pilot scale catalytic reactor provides a removal efficiency of 95%. The catalytic reactor is 0.5 inches in length. Determine the required length of a full-scale catalyst (in inches) that is designed to remove 99.99% of the pollutant.

a) 1 inch
b) 1.5 inches
c) 2.5 inches
d) 3 inches

13. Motor Vehicle Emissions

Emissions from an uncontrolled motor vehicle are 13 g/mile of hydrocarbon and 4.0 g/mile of NO_x. The 1996 Light Duty Vehicle (LDV) exhaust emission standard for non-methane hydrocarbon is 0.25 g/mile and the standard for oxides of nitrogen is 0.4 g/mile. The emission standards are based on a driving cycle that includes various time periods of vehicle idle, acceleration, cruise, and deceleration.

13.1 Determine the efficiency of conversion of hydrocarbon that is required for a catalytic converter to meet the 1996 emission standards.
- a) 85%
- b) 90%
- c) 98%
- d) 99%

13.2 The LDV NO_x exhaust emission standard in 1971 was 4.0 g/mile. It has been calculated that the yearly driving distance for vehicles in the United States increased by 12% between 1971 and 1996 and the number of vehicles increased from 108 to 189 million. Determine the difference in NO_x emissions between 1971 and 1996 assuming all 108 million 1971 cars meet the 1971 standard and all 189 million 1996 cars meet the 1996 standard.
- a) increase by 10%
- b) increase by 60%
- c) decrease by 80%
- d) decrease by 90 %

13.3 Determine stoichiometric air required for the combustion of octane (C_8H_{18}) in pounds of air per pound of fuel.
- a) A/F = 12.5
- b) A/F = 13.5
- c) A/F = 14.5
- d) A/F = 15.0

13.4 Determine the difference in the evaporative fuel loss (in pounds) associated with refueling a vehicle with a 16-gallon tank capacity using Phase II displacement controls and uncontrolled refueling operations. Phase II controls refer to the capture of the vapors that are displaced from the vehicle fuel tank during refueling. Uncontrolled displacement losses are 9 lb/10^3 gal and controlled displacement losses are 0.9 lb/10^3 gal.
- a) 0.008 lb HC/veh.refueling
- b) 0.014 lb HC/veh.refueling
- c) 0.13 lb HC/veh.refueling
- d) 0.90 lb HC/veh.refueling

14. Absorption

Absorption is a mass transfer operation. The basic model for describing the absorption process is the two-film theory. The model proposes that a mass transfer zone exists to include a small portion (film) of the gas and liquid phases on either side of the interface. The rate of mass transfer is then equal to the amount of molecule A transferred times the resistance molecule A encounters in diffusing through the films:

Equation [12.3] and [12.4] $\quad N_A = k_g(p_{AG} - p_{AI})$ and $N_A = k_l(C_{AG} - C_{AI})$

where: $\quad N_A$ = rate of transfer of component A
k_g = mass transfer coefficient for gas film
k_l = mass transfer coefficient for liquid film

The mass transfer coefficients, k_g and k_l, represent the flow resistance the solute encounters in diffusing through each film respectively. At equilibrium, the overall mass transfer coefficients are related to the individual mass transfer coefficients by:

Equations [12.7] and [12.8] $\quad \dfrac{1}{K_{OG}} = \dfrac{1}{k_g} + \dfrac{H}{k_l}$ and $\dfrac{1}{K_{OL}} = \dfrac{1}{k_l} + \dfrac{1}{H k_g}$

H is Henry's Law constant. Henry's Law is used to predict solubility when the solute concentrations are vely low (dilute solution).

Equation [12.1] $\quad y = H \bullet x$

where: $\quad y$ = mole fraction in gas phase in equilibrium with liquid
H = Henry's Law constant, mole fraction in vapor/mole fraction in liquid
x = mole fraction in liquid phase in equilibrium with gas phase

14.1 The concentration of a solution of SO_2 in water in equilibrium with the gas phase is 0.25 mg/L. What is the concentration of SO_2 in the gas phase in ppm? (Henry's Law constant for SO_2 in water is 24 atm/mole fraction).

a) 0.08 ppm
b) 3 ppm
c) 215 ppm
d) 3,208 ppm

14.2 Most systems in the air pollution control field are gas phase controlled. For these conditions, which statement is not true?

a) H is very small
b) The gas is very soluble in the liquid

c) $K_{OG} \approx k_g$
d) The major resistance to mass transfer is in the liquid phase

An exhaust stream of 1,000 scfm (60°F) is known to contain 1% SO_2 by volume. It is determined that the SO_2 content needs to be reduced by 90% by scrubbing with water (the SO_2 initial concentration in the water is zero). Henry's Law constant = 24 mole fraction SO_2 in gas/mole fraction SO_2 in liquid. The flow diagram is shown below.

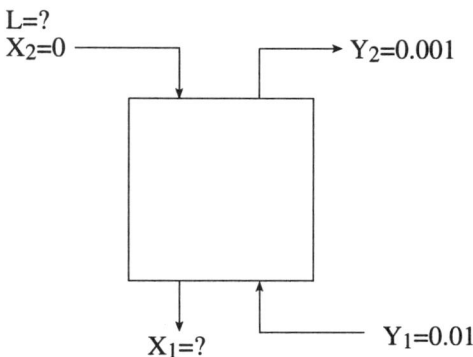

Exhaust gas flow rate = 1,000 scfm

14.3 What is the concentration of SO_2 in the effluent liquid?
   a) 21 mg/L
   b) 258 mg/L
   c) 398 mg/L
   d) 1,493 mg/L

14.4 Determine the fresh water flow rate, L, in gal/min required for a liquid-to-gas ratio (L/G) of 21.5 mol water/mol air.
   a) 125 gal/min
   b) 320 gal/min
   c) 485 gal/min
   d) 560 gal/min

1. Solutions — Gas Law

1.1 The correct answer is d)

From Gas Law Equation [2.9] $\rho = \dfrac{M}{V} = \dfrac{(P \cdot MW)}{R \cdot T}$

$P = 1$ ppm $= 1 \cdot 10^{-6}$ atm, $MW = 64$ g/mol, $T = 25°C$

$$\rho = \dfrac{1 \cdot 10^{-6} \text{ atm} \cdot 64 \text{ g/mol}}{8.3 \cdot 10^{-5} (\text{atm m}^3)/(\text{g mol}°K) \cdot (25+273)°K}$$

$\quad = 2.6 \cdot 10^{-3}$ g/m^3

$\quad = 2{,}600$ µg/m^3

1.2 The correct answer is b)

From Gas Law Equation [2.9] $\rho = \dfrac{M}{V} = \dfrac{(P \cdot MW)}{R \cdot T}$

$P = \dfrac{\rho \cdot T \cdot R}{MW}$

$\rho = 156$ µg/m^3 $= 1.56 \cdot 10^{-4}$ g/m^3, $MW = 48$ g/mol, $T = 25°C$

$$P = \dfrac{1.56 \cdot 10^{-4} \text{ g/m}^3 (25+273)°K \cdot 8.3 \cdot 10^{-5}(\text{atm m}^3)/(\text{g mol}°K)}{48 \text{ g/mol}}$$

$P = 8.0 \cdot 10^{-8}$ atm

$\quad = 0.08$ ppm

1.3 The correct answer is a)

Volume flow rate of $SO_2 = 0.006 \cdot 40$ ft/sec $\cdot \pi \cdot \dfrac{(10 \text{ ft})^2}{4} \cdot \dfrac{3{,}600 \text{ sec}}{\text{hr}}$

$\quad = 6{,}785.8$ ft^3/hr

From Gas Law: $M = \dfrac{P \cdot MW \cdot V}{R \cdot T}$

$P = 1$ atm, $MW = 64$, $V = 6{,}785.8$ ft^3/hr

$M = \dfrac{1 \text{ atm} \cdot 64 \cdot 6{,}785.8 \text{ ft}^3/\text{hr}}{0.73(\text{atm ft}^3)/(\text{lb mol}°R) \cdot (460+450)°R}$

$\quad = 653.7$ lb/hr

1.4 The correct answer is d)

$$P \cdot V = \frac{M \cdot R \cdot T}{MW}, \quad T = 100°F = 560°R$$

$$V = \frac{0.25 \text{ lb/min} \cdot 560°R}{1 \text{ atm} \cdot 92} \cdot 0.73 \frac{\text{atm ft}^3}{\text{lb mole}°R}$$

$$= 1.1 \text{ cfm (at } 100°F\text{)}$$

$$\text{mole fraction of toluene} = \frac{1.1}{1.1 + 1,000} = 0.001$$

$$= 1,000 \text{ ppm (V/V)}$$

2. Solutions — Cost Estimates

2.1 The correct answer is c)

From Equation [1.1] or [2.55] $\quad HP = \dfrac{Qh}{6,360E}, \quad E = 60\%$

$$= \frac{1,000 \cdot 1}{6,360 \cdot 0.6} = 0.262 \text{ HP}$$

$$\text{Annual Cost} = 0.262 \text{ HP} \cdot \frac{8,760 \text{ hr}}{\text{yr}} \cdot \frac{0.746 \text{ kWh}}{\text{HP hr}} \cdot \frac{\$0.015}{\text{kWh}}$$

$$= \$25.7/\text{yr}$$

2.2 The correct answer is c)

$P = \$900,000 + \$600,000 = \$1,500,000$

Equation [1.1] $\quad C_a = \left(\dfrac{\$1,500,000}{20}\right)\dfrac{(1+0.07)^{20}}{(1+0.07)^{20}-1} = \$101,135/\text{yr}$

2.3 The correct answer is d)

$$\text{added cost} = 15,000 \frac{\text{mile}}{\text{yr}} \cdot \frac{\$0.10/\text{gal}}{25 \text{ mile/gal}} = \$60/\text{yr}$$

2.4 The correct answer is d)

Equation [8.12] $\quad \eta = 1 - \exp\left[-\dfrac{wA}{Q}\right]$

$$\frac{1-\eta_1}{1-\eta_2} = \exp\left[-\frac{A_1}{A_2}\right]$$

$A_1/A_2 = \ln[(1-.9999)/(1-0.95)] = 6.2$

3. Solutions — Particle Motion and Collection Efficiency

3.1 The correct answer is c)

From Equation [3.13] $\quad v_T = \dfrac{\rho_p d_p^2 g}{18\mu}$

$$v_T = \frac{6.2 \text{ g/cm}^3 \cdot (1 \cdot 10^{-4} \text{cm})^2 \cdot 980 \text{ cm/sec}^2}{18 \cdot 1.8 \cdot 10^{-4} \text{ g/cm sec}}$$

$v_T = 0.02 \text{ cm/sec}$

3.2 The correct answer is c)

From Equation [4.8] $\quad t_c = \dfrac{H}{v_T}$

From Figure 3.5, $v_T = 0.2$ cm/sec

$t_c = \dfrac{20 \text{ ft}}{0.2 \text{ cm/sec}} \cdot \dfrac{30.48 \text{ cm}}{\text{ft}} = 3{,}048 \text{ sec}$

$t_c = 50.8$ min

3.3 The correct answer is c)

Refer to Chapter 3

3.4 The correct answer is c)

From Equation [4.8] $\quad t_c = \dfrac{H}{v} = \dfrac{20 \text{ to } 30 \text{ cm}}{5 \text{ to } 10 \text{ cm/sec}} = 2 \text{ to } 6 \text{ sec}$

From Equation [4.10] $\quad t_R = \dfrac{L}{v} = \dfrac{20 \text{ ft}}{1 \text{ to } 2 \text{ m/sec}} \cdot \dfrac{1 \text{ m}}{3.28 \text{ ft}} = 3.0 \text{ to } 6.0 \text{ sec}$

Therefore: $\quad \text{Ratio} = \dfrac{2 \text{ to } 6 \text{ sec}}{3 \text{ to } 6 \text{ sec}} \sim 1$

Appendix A

4. Solutions — Evaluation of Particulate Control Systems

4.1 The correct answer is b)

From Equation [4.5] $\eta = \dfrac{8 - 0.08}{8} \bullet 100 = 99\%$

4.2 The correct answer is c)

From Equation [2.55] $HP = \dfrac{Q \bullet TP}{6,360} = \dfrac{10,000 \bullet 15}{6,360} = 23.6 \text{ HP}$

4.3 The correct answer is c)

From Equation [6.1] $v_f = \dfrac{Q}{A}$

$$A = \dfrac{Q}{v_f} = \dfrac{10,000 \text{ acfm}}{2 \text{ ft/min}} = 5,000 \text{ ft}^2$$

Therefore: No. of bags $= \dfrac{5,000}{12 \bullet \pi \bullet (8/12)} = 198.9 \sim 200$ bags

4.4 The correct answer is a)

Loading to baghouse $= (1 - 0.8) \bullet 8 = 1.6$ grains/scf

From Equation [6.18] $w = C_i v_f t$

$w = 1.6 \dfrac{\text{grains}}{\text{ft}^3} \bullet 0.99 \bullet 2 \dfrac{\text{ft}}{\text{min}} \bullet (2 \bullet 60) \text{min}$

$= 380.2 \text{ grains/ft}^2$

$= 380.2 \dfrac{\text{grains}}{\text{ft}^2} \bullet \dfrac{1 \text{ lb}}{7,000 \text{ grains}} = 0.054 \text{ lb/ft}^2$

5. Solutions — Design of Air Pollution Control Devices

5.1 The correct answer is c)

(Refer to Chapter 4)

Size range	0-10	10-20	20-50	
(b) Emission (%)	20	30	50	
(c) Grade efficiency	40	60	90	
% collected (b) • (c)	8	18	45	Total = 71%

5.2 The correct answer is c)

From Equation [8.12] $\eta = 1 - \exp\left[-w\dfrac{A}{Q}\right]$

$$0.95 = 1 - \exp\left[-5 \text{ cm/sec} \bullet \dfrac{A}{100{,}000 \text{ ft}^3/\text{min}}\right]$$

$$2.996 = \dfrac{5 \text{ cm/sec}}{30.48 \text{ cm/ft}} \bullet \dfrac{A}{100{,}000 \text{ ft}^3/\text{min}}$$

$A = 30{,}439 \text{ ft}^2 \sim 30{,}500 \text{ ft}^2$

5.3 The correct answer is c)

From Equation [7.3] $\eta = 1 - e^{-f\eta_{drop}}$

$\eta = 0.95,\ \eta_{drop} = 0.1$

$0.95 = 1 - e^{-f \bullet 0.1}$

$f = 29.96$

From Equation [7.4] $f = \dfrac{6.12 \bullet 10^4 \bullet H}{D_D} \bullet \dfrac{Q_L}{Q_G}$

$29.96 = \dfrac{6.12 \bullet 10^4 \bullet H}{500} \bullet \dfrac{7.5}{1{,}000} \bullet (0.5 \text{ water use effective factor})$

$H = 62.27 \text{ ft}$

Equation [7.5] $H = z\left(\dfrac{v_t}{v_t - v_G}\right)$

$62.27 = z\left(\dfrac{1}{1 - 0.2}\right)$

$z = 50 \text{ ft}$

5.4 The correct answer is d)

$S + O_2 \rightarrow SO_2$

64

Equation [12.22] $SO_2 + CaCO_3 + \dfrac{1}{2}O_2 \rightarrow CaSO_4 + CO_2(g)$

64   100

$S \equiv CaCO_3$

32   100

$$CaCO_3 \text{ needed} = 2,500 \text{ tons / day (Coal)} \bullet \frac{1.5 \text{ ton (S)}}{100 \text{ ton (Coal)}} \bullet \frac{100 \text{ ton (CaCO}_3)}{32 \text{ ton (S)}}$$

$$= 117 \text{ ton / day}$$

6. Solutions — Energy Balance

6.1 The correct answer is b)

(Refer to Chapter 2)

$C_{p(\text{air at } 1,200°F)} = 0.248 \text{ Btu / lb°F}$, $C_{p(\text{air at } 250°F)} = 0.241 \text{ Btu / lb°F}$

$C_{p(\text{avg})} = 0.2445 \text{ Btu / lb°F}$

$\Delta H = 0.2445 \bullet (1,200 - 250) = 232 \text{ Btu / lb}$

6.2 The correct answer is b)

From Equation [2.7] $\dfrac{V_1}{T_1} = \dfrac{V_2}{T_2}$

$$V_2 = \frac{V_1 T_2}{T_1} = \frac{32,000 \bullet (460 + 250)}{(460 + 1,200)} = 13,687 \text{ cfm}$$

6.3 The correct answer is c)

From Equation [2.34] $Q_d = Q_e \left(\dfrac{T_e - T_f}{T_f - T_d} \right) \left| \dfrac{T_d}{T_e} \right|$

$$Q_d = 32,000 \bullet \left(\frac{1,200 - 250}{250 - 70} \right) \bullet \left(\frac{70 + 460}{1,200 + 460} \right) = 53,922 \text{ cfm}$$

Total gas volume at 250°F $= 32,000 \bullet \left(\dfrac{250 + 460}{1,200 + 460} \right) + 53,922 \bullet \left(\dfrac{250 + 460}{70 + 460} \right)$

$= 85,922 \text{ cfm}$

6.4 The correct answer is b)

From Equation [2.37] $M_a C_{p(\text{avg})}(T_a - T_f) = M_w [\Delta H_v + C_{pwv}(T_f - T_v)]$

$M_a = 32,000 \text{ cfm} \bullet \left(\dfrac{460 + 70}{1,200 + 460} \right) \bullet \dfrac{28.97 \text{ lb / lb mol}}{387 \text{ scfm / lb mol}} = 764.8 \text{ lb / min}$

$C_{p(\text{avg})} = 0.2445 \text{ Btu / lb°F}$, $\Delta H_v = 1,059.9 \text{ Btu / lb}$, $H_{250°F} = 85.5 \text{ Btu / lb}$

$764.8 \bullet 0.2445 \bullet (1,200 - 250) = M_w (1,059.9 + 85.5)$

$M_w = 155.1 \text{ lb / min}$

Total gas volume at 250°F = $32,000 \cdot \left(\dfrac{250+460}{1,200+460}\right) + \dfrac{155.1}{18} \cdot 387 \cdot \left(\dfrac{250+460}{70+460}\right)$

= 18,153 cfm

7. Solutions — Condensation

7.1 The correct answer is b)

From Equation [2.38] $\quad A = \dfrac{Q_{load}}{U \cdot \Delta t_m}$

$$\Delta t_m = \dfrac{\Delta T_1 - \Delta T_2}{\ln\left(\dfrac{\Delta T_1}{\Delta T_2}\right)}$$

$\Delta T_1 = 250 - 70 = 180$
$\Delta T_2 = 130 - 100 = 30$

$\Delta t_m = (180 - 30) / \ln(180/30)$
$\quad\quad = 83.7°F$

$A = \dfrac{968,000}{150 \cdot 83.7}$
$\quad = 77 \text{ ft}^2$

7.2 The correct answer is a)

From Equation [9.9] $\quad Q_{cw} = \dfrac{Q_{load}}{C_{pw}(T_{wo} - T_{wi})}$

$Q_{cw} = \dfrac{968,000}{1 \cdot (100 - 70)} = 32,267 \text{ lb/hr}$

$Q_{cw} = 32,267 \text{ lb/hr} \cdot 0.002 \dfrac{\text{gal/min}}{\text{lb/hr}}$

$\quad\quad = 64.5 \text{ gal/min}$

7.3 The correct answer is a)

Mass flow rate of benzene

From Equation [2.9]

$$M = 10,000 \bullet 1,000 \bullet 10^{-6} \text{ cfm} \bullet \frac{78 \text{ lb/mol}}{0.73 \left(\text{atm ft}^3/\text{mol}°R\right)(250+460)°R}$$

$$M = 1.5 \text{ lb/hr}$$

Boiling point of benzene = 175°F
From Figure 9.1, at 130°F, Vapor pressure of benzene = 6 psia
$$= 6/14.7$$
$$= 0.4 \text{ mole fraction}$$

0.4 mole fraction > $1,000 \bullet 10^{-6}$ mole fraction
Therefore: no benzene will condense
(see Figure 9.5, Efficiency of Condensors)

8. Solutions — Adsorption

8.1 The correct answer is b)

From Equation [9.3] $D_L = 12.46 \left[\dfrac{0.6 \bullet 7 \bullet 66}{(50+460)} \right] = 6.77 \text{ lb}/10^3 \text{ gallon}$

Gasoline loss per day = $25 \bullet 1,000 \text{ ft}^3 \bullet \dfrac{7.48 \text{ gal}}{\text{ft}^3} \bullet \dfrac{6.77 \text{ lb}}{1,000 \text{ gal}}$

$$= 1,266 \text{ lb/day}$$

Carbon needed = $\dfrac{1,266 \text{ lb/day}}{0.3 \text{ g gasoline/g carbon}} \bullet \dfrac{1}{0.5} = 8,440 \text{ lb} \sim 8,500 \text{ lb}$

8.2 The correct answer is b)

$$A = \frac{200 \text{ ft}^3/\text{min}}{30 \text{ ft/min}} = 6.7 \text{ ft}^2$$

8.3 The correct answer is c)

From Equation [10.12] $\text{MTZ} = \dfrac{1}{1-X_s} D \left\{ 1 - \dfrac{C_B}{C_S} \right\}$

$$\text{MTZ} = \frac{1}{1-0.5} \bullet 2 \left\{ 1 - \frac{0.249}{0.28} \right\}$$
$$= 0.443 \text{ ft} = 5.3 \text{ inches}$$

8.4 The correct answer is d)

From Equation [9.1] $\quad y_i = x_i \dfrac{P_i}{P}$

Liquid phase is pure benzene, $x_i = 1$

From Equation [9.2] $\quad \log p = A - \dfrac{B}{(T+C)}$

$$T = 95°F = 35°C$$

From Table 9.2, for benzene, $A = 6.90565$, $B = 1{,}211.033$, $C = 220.790$

$$\log p = 6.90565 - \dfrac{1{,}211.033}{(35 + 220.79)}$$

$\log p = 2.17$ mm Hg

Therefore: $\quad y_i = 1 \bullet \dfrac{2.17 \text{ mm Hg}}{760 \text{ mm Hg}} = 2.855 \bullet 10^{-3}$

$\qquad\qquad\qquad = 2{,}855$ ppm

9. Solutions — Combustion Kinetics

9.1 The correct answer is c)

$$T = 1{,}350°F = 732°C$$

From Equation [11.11] $\quad k = A e^{-\frac{E}{RT}}$

$$k = 3.57 \bullet 10^{14} e^{-\left(\frac{63.3}{1.987 \bullet 10^{-3}(732+273)}\right)}$$

$\qquad = 6.11 \text{ sec}^{-1}$

From Equation [11.10] $\quad \dfrac{C_A}{C_{Ao}} = \exp[-kt]$

$$(1 - 0.95) = \exp[-6.11 t]$$
$$t = 0.49 \text{ sec}$$

9.2 The correct answer is b)

$$T = 2{,}200°F = 1{,}204°C$$

From Equation [11.11] $\quad k = A e^{-\frac{E}{RT}}$

$$k = 3.57 \bullet 10^{14} e^{-\left(\frac{63.3}{1.987 \bullet 10^{-3}(1,204+273)}\right)}$$
$$= 153,275 \text{ sec}^{-1}$$

From Equation [11.10] $\quad \dfrac{C_A}{C_{A0}} = \exp[-kt]$

$$(1 - 0.9999) = \exp[-153,275t]$$
$$t = 0.00006 \text{ sec}$$

9.3 The correct answer is b)

$C_p(\text{air}) = 0.25$ Btu/lb °F

For air: molar volume = 379 scf , MW_{air} = 28.97 lb/lb mole

Mass flow rate $= \dfrac{10,000}{379} \bullet 28.97 = 764.4$ lb / min

From Equation [11.15] $\quad Q \bullet 1,050 = 764.4 \bullet 0.24 \bullet (1,250 - 150)$
$$Q = 192 \text{ scfm}$$

9.4 The correct answer is c)

From Equation [11.2] $\quad \begin{array}{cc} CH_4 + 2O_2 \rightarrow CO_2 + 2H_2O \\ 16 \quad\quad 64 \end{array}$

O_2 required for 100 lb CH_4 = (64/16) • 100 = 400 lb

in air: O_2 = 21% and N_2 = 79% V/V

O_2 in air(W / W) $= \dfrac{0.21 \bullet 32}{0.21 \bullet 32 + 0.79 \bullet 28} = 23.3\%$

air required $= 400 \bullet \dfrac{1}{0.233} = 1,717$ lb

10. Solutions — Combustion Stoichiometry

10.1 The correct answer is d)

Equation [11.1] $\quad C_3H_8 + 5O_2 \rightarrow 3CO_2 + 4H_2O$
5 scf of O_2 ≡ 1 scf of propane
O_2 in air = 21%
air = 5/0.21 = 23.8 scf

10.2 The correct answer is b)

Table 2.3: The latent heat of vaporization for water at 60°F = 1,059 Btu/lb

$C_3H_8 + 5O_2 \rightarrow 3CO_2 + 4H_2O$

379 scf of $C_3H_8 \equiv 72$ lb of H_2O

heat of vaporization of H_2O = 1,059 • 72/379
$= 210.18$ Btu/scf of C_3H_8

net heat of combustion = 2,590 - 210 = 2,380 Btu/ft³

10.3 The correct answer is d)

From Equation [11.11] $k = Ae^{-\frac{E}{RT}}$

$$k = 9 \bullet 10^{10} e^{-\frac{45.2}{1.987 \bullet 10^{-3} T}}$$

From Equation [11.10] $\dfrac{C_A}{C_{Ao}} = \exp[-kt]$

$(1 - 0.9999) = \exp[9 \bullet 10^{10} \exp\{-45,495.8/T\}]$

Therefore: T = 1,953.44°K = 3,000°F

10.4 The correct answer is b)

For gas at 60°F and 1 atm: molar volume = 379 scf/lb mol

From Equation [11.15] $Q \bullet 1,050 = \dfrac{8,500}{379} \bullet 7.5 \bullet (1,400 - 200)$

Therefore: Q = 192 scfm

11. Solutions — Hazardous Waste Combustion

11.1 The correct answer is b)

See Chapter 11 and Equations 11.20 and 11.21

11.2 The correct answer is a)

From Equation [11.37]

$$EA = \dfrac{[NHV]/\{0.3(T-60)\} - 1}{7.5 \bullet 10^{-4} \bullet NHV} - 1$$

$$EA = \dfrac{[8,000]/\{0.3(2,200-60)\} - 1}{7.5 \bullet 10^{-4} \bullet 8,000} - 1 = 1.1$$

Appendix A   583

11.3 The correct answer is a)
(Refer to Example 11.4)
$$V = \frac{8{,}000 \text{ Btu/lb} \bullet 500 \text{ lb/hr}}{25{,}000 \text{ Btu/hr ft}^3} = 160 \text{ ft}^3$$

11.4 The correct answer is b)
From Equation [11.14] $H_v \bullet Q_{800°F} = m \bullet C_p \bullet (800 - 60)$

$$H_v \bullet Q_{1{,}400°F} = m \bullet C_p \bullet (1{,}400 - 60)$$
$$Q_{800°F} / Q_{1{,}400°F} = 740 / 1{,}340$$
$$Q_{800°F} = 0.55 Q_{1{,}400°F}$$
$$\text{reduction} = \frac{Q_{800°F} - 0.55 Q_{800°F}}{Q_{800°F}} \bullet 100 = 45\%$$

12. Solutions — Mass Transfer

12.1 The correct answer is b)

From Equation [3.3] $Re = \dfrac{\rho v d}{\mu}$, $\rho_{air} = 0.0275 \text{ lb/ft}^3$,

$\mu_{air} = 2.47 \bullet 10^{-5} \text{ lb/ft sec}$

$$Re = \frac{0.275 \text{ lb/ft}^3 \bullet 30 \text{ ft/sec} \bullet (0.05/12) \text{ ft}}{2.47 \bullet 10^{-5} \text{ lb/ft sec}} = 139 \approx 140$$

12.2 The correct answer is b)

From Equation [13.16] $a = \dfrac{4}{D} = \dfrac{4}{(0.05/12) \text{ft}} = 960 \text{ ft}^2/\text{ft}^3$

From Equation [13.13] $L_m = \dfrac{v}{K_m a} = \dfrac{35}{0.5 \bullet 960} = 0.0729 \text{ ft} = 0.875 \text{ inch}$

$e^{-N} = 0.01$, $N = 4.6$, $L = 4.6 \bullet 0.875 = 4.0$ inches

12.3 The correct answer is b)

From Equation [13.14] $\dfrac{C}{C_o} = e^{-\frac{L}{L_m}}$, $(1 - 0.95) = e^{-\frac{0.5}{L_m}}$, $L_m = 0.167$ inch

$(1 - 0.9999) = e^{-\frac{L}{0.167}}$, $L = 1.54$ inches

13. Solutions — Motor Vehicle Emissions

13.1 The correct answer is c)

$$\eta = \frac{13 - 0.25}{13} \bullet 100 = 98\%$$

13.2 The correct answer is c)

Assume the yearly driving distance in 1971 is L mi/yr

NO_x emission in 1970 = (4.0 g/mi) • (L mi/yr) • 108 • 10^6 = 432L ton/yr

NO_x in 1996 = (0.4 g/mi) • (L • (120/100) mi/yr) • 189 • 10^6 = 90.7L ton/yr

$$\text{decrease} = \frac{432L - 90.7L}{432L} \bullet 100 = 79\% \sim 80\%$$

13.3 The correct answer is d)

From Equation [13.2] $\quad C_8H_{18} + 12.5O_2 \rightarrow 8CO_2 + 9H_2O$
$\qquad\qquad\qquad\qquad\qquad\quad$ 114 $\quad\quad$ 400

in air: O_2 = 21% and N_2 = 79% V/V

$$O_2 \text{ in air}(W/W) = \frac{0.21 \bullet 32}{0.21 \bullet 32 + 0.79 \bullet 28} = 23.3\%$$

$$\text{lb air / lb fuel} = \frac{400}{114} \bullet \frac{1}{0.233} = 15.0$$

13.4 The correct answer is c)

$$\text{Emission reduce} = \frac{(9 - 0.9)}{1,000} \bullet 16 = 0.13 \text{ lb HC / veh refueling}$$

14. Solutions — Absorption

14.1 The correct answer is b)

Calculate concentration of SO_2 in water in mole fraction.

$$\frac{0.45 \text{ mg}(SO_2)}{1 \text{ Liter}(H_2O)} \bullet \frac{(1 \text{ g} / 1,000 \text{ mg})}{(1,000 \text{ g} / \text{Liter})} \bullet \frac{(1 \text{ mol} / 64 \text{ g})}{(1 \text{ mol} / 18 \text{ g})} = 1.26 \bullet 10^{-7} \text{ mole fraction}$$

From Equation [12.2]

$$y = H \bullet x = (24 \text{ atm / mole fraction}) \bullet (1.26 \bullet 10^{-7} \text{ mole fraction})$$

$$= 3.0 \bullet 10^{-6} \text{ atm} = 3 \text{ ppm}$$

14.2 The correct answer is d)
Refer to Chapter 12

14.3 The correct answer is d)
From Equation [12.2]
$$Y_1 = X_1 H$$
$$X_1 = Y_1/H$$

$$X_1(0.01)/24 = 0.00042 \frac{\text{moles of } SO_2}{\text{mole } H_2O}$$

$$X_1 = \frac{0.00042 \text{ mol } SO_2}{1 \text{ mol } H_2O} \cdot \frac{(64 \text{ g/mol}) \cdot (1{,}000 \text{ mg/g})}{(18 \text{ g/mol}) \cdot (L/1{,}000 \text{ g})}$$
$$= 1{,}493 \text{ mg/L}$$

14.4 The correct answer is a)
(See Figures 12.6 and 12.7)

$$L/G = 21.42 \frac{\text{lb mole water}}{\text{lb mole air}}$$

mole flow rate of air = 1,000 scfm/(379 scf/lb mol) = 2.64 lb mol/min
mole flow rate of water = 2.64 • 21.42 = 56.5 lb mol H_2O/min

$$\text{gal/min} = 56.5 \frac{\text{lb mole}}{\text{min}} \left(\frac{18 \text{ lb}}{\text{lb mole}}\right)\left(\frac{\text{gal}}{8.34 \text{ lb}}\right)$$
$$= 122 \text{ gal/min}$$

Appendix B

Selected Symbols and Acronyms
(Descriptions are followed by the chapter number where the symbol/acronym is discussed initially or frequently.)

Ψ	(psi) Stokes' number, 3, 7
$\varnothing$	ratio of specific gravity of the scrubbing liquid to that of water, 4, 12
$\propto$	is proportional to
ϕ	(phi) absolute humidity, 2
α	(alpha) packing density of the fibers, 6
α and β	empirical constants which are determined from experiment, 7
β	(beta) a constant in the grade penetration equation, 4
β	mobility, 3
β	dimensionless affinity coefficient, 10
$\mathcal{D}$	diffusion coefficient, 3, 6, 13
θ	(theta) angle between baffle and flow direction, radians, 7
θ	average residence time, 11
θ_1	an apparent delay time (a lag prior to onset of the oxidation reaction), 11
ε	(epsilon) dielectric constant of the particle, 2
ε	porosity, 6
ε	pressure drop, 12

υ	(upsilon) kinematic viscosity, 2, 3
ρ	(rho) density, 2, 3
ρ_a	air density, 2
ρ_g, ρ_G	gas density, 2, 3, 7, 11
ρ_L	liquid density, 7, 10
ρ_p	particle density, 3, 6
ρ_w	water density, 2
$\mathcal{K}$	(kappa) a structural constant of the absorbent, 10
η	(eta) efficiency, 4, 11
η_{DC}	collection efficiency by diffusion and interception, 6
η_I	impaction efficiency, 3, 4
η_{ICD}	collection efficiency by impaction, interception and diffusion, 3
η_{drop}	single drop collection efficiency, 7
λ	(lambda) relative air-fuel ratio, 13
λ	latent heat of vaporization, 2, 3
η_T	total efficiency, 6
μ	(mu) viscosity of the fluid, 2, 3
μ	particle size, 1, 4
μ_L	liquid viscosity, 7
$\mu°$	viscosity at 0°C and prevailing conditions, 2
σ_g	(omicron) geometric standard deviation, 4
σ_L	liquid surface tension, 7
τ	(tau) unit shearing stress between adjacent layers of fluid, 2
τ	relaxation time, 3
τ_B	service time (adsorption time), 10
τ_c	chemical time, 11

Appendix B

τ_m	mixing time, 11
τ_r	residence time, 4, 11
ω	(omega) angular velocity, 5
ΔT	(delta) temperature change, 2
ΔT_1	inlet temperature differences between vapor and coolant, 9
ΔT_2	outlet temperature differences, 9
ΔT_{lm}	log mean temperature, 2, 9
$\Delta p, \Delta P$	change in pressure, 6
a, A	area, 2, 6
a	inlet height of a cyclone, 5
A/Q	specified collection area, 8
A_b	bag area, 6
A_d	cross sectional area of droplets, 7
b	baffle spacing, 7
b	width
bhp	brake horsepower, 1, 14
B	dust outlet diameter of a cyclone, 5
c, C	concentration, 2, 7
C_a	annual cost, 1
C_{A*}	equilibrium concentration, 11
C_B	breakthrough capacity, 10
C_c	Cunningham correction factor, 3
C_D	drag coefficient, 3, 7
C_p	average specific heat of a gas or liquid, 2, 11
d, D	diameter, 2, 3, 4
d_a	aerodynamic diameter, 3

d_g	geometric mean diameter, 4
d_p	particle diameter, 4
d_{gm}	mass median diameter, 4
d_s	Stokes diameter, 3
dv/dy	velocity gradient, 2
d_{p50}	the particle with 50% efficiency, 4
d_{50}	geometric mean diameter, 4
D_c	cyclone body diameter, 5
D_D	drop diameter, 7
D_f	fiber diameter, 6
E	emission factor, 9
E	mechanical efficiency of fan system, 11
E	equivalence ratio, 13
E_p, E_o	field strength, 8
F	packing factor, 12
F_{cent}	centrifugal force, 5
g	gravitational acceleration, 2
G	mass flow rates of gas, 2
G_E	flue gas from excess air, 11
G_f	total mass of flue gas, 11
G_m	molar flow rate of gas, 12
h	head loss, 1, 11
h	height
h	cylinder height of a cyclone, 5
h	bag height, 6
h	film coefficient, 2

Appendix B

h, H	specific enthalpies, 1, 2, 11
H	Henry's Law constant, 2, 12
H	overall height cyclone, 5
H	thickness of fibers, 6
HEEL	residual adsorbate, 10
H_{OG}, H_{OL}	height of transfer units based on overall gas or liquid film coefficient, 12
HP, hp	horsepower, 1, 2, 14
HTU	height of a transfer unit, 12
i	interest cost, 1
k	Boltzmann's constant, 6, 8
K	Kozeny constant, 6
K	configuration of a cyclone, 5
K	parameter of proper flow regime, 3
K_e	equilibrium constant, 13
K_r	reactive rate coefficient, 13
K_1	fabric resistance, 6
K_{OG}, K_{OL}	overall mass transfer coefficient, gas or liquid phase, 12
K_p	a modified impaction parameter, 4, 6
K_p	equilibrium constant, 11
l, L	length
L	mass flow rate of liquid coolant, 7
L/G	liquid/gas flow ratio, 7
LEL	lower explosive limit, 13
Li	L/G ratio, 7
m, M	mass, mass flow rate, 2, 10
m	slope, 12

MMD	mass median diameter, 4, 6
MTZ	mass transfer zone, 10
MW	molecular weight, 2, 9
n, N	number
n	an empirical exponent, 2, 5
n	refractive index, sodium D wavelength, 8
N	lifetime or amortization base for equipment, 1
N	Avogadro's number, 3
N	final concentration, 6
N_A	rate of transfer of component, 12
N_e	number of vortex turns, 5
NG	natural gas, 11
N_H	inlet velocity head of a cyclone, 5
N_i	ion concentration, 8
OP	optimization parameter, 5
p, P	pressure, 2
p^*, Pa	partial pressure of solute at equilibrium, 2, 12
ppm	parts per million by volume, 2
Pe	Peclet number, 3
P_G	power input from gas stream, 7
P_i	vapor pressure of pure component i, 9
P_t	penetration, 4
P_T	total contacting power, 7
q	charge on particle, 8
q_s	saturation charge, 8
Q	condenser heat load, 9

Appendix B

Q	volume flow rate of air or gas, 1, 2, 11
Q	heat transfer rate, 2
Q_{cw}	cooling water flow rate, 9
Q_d	dilution air flow rate needed, 2
Q_e	exhaust air flow rate to be cooled, 2
Q_f	net heating value of contaminant, 11
Q_G	gas flow rate, 7
Q_L	liquid feed rate, 7
r	rate
r_d	droplet radius, 7
rpm	rotational speed (revolutions per minute)
R	universal gas constant, 2, 3, 11
R	cyclone radius, 5
Re	Reynolds Number, 2, 3
Re_p	Reynolds number of particle, 3
RVP	Reid vapor pressure, 9
S	filter drag, 6
S	outlet length of a cyclone, 5
S	saturation factor, 9
$S_1, S_2, -$	individual component drag, 6
Sc	Schmidt number, 3, 6
SCA	specified collection area, 8
SP	static pressure, 2
surf	surface parameter of a cyclone, 5
t	time
T	temperature

TLV	threshold limit value, 9
TP	total pressure, 2
U	overall heat transfer coefficient, 2, 9
v	velocity
v_{TS}	terminal settling velocity, 3
V	volume
VP	velocity pressure, 2
w, W	width
w	migration velocity, 8
w_e	precipitation rate parameter, 8
W_e	adsorption capacity, 10
WK	work input rate, 2
x	distance
x, X	mole fraction, 12
x_{stop}	stop distance, 3
X_S	degree of saturation in the MTZ, 10
y^*, Y	mole fraction in gas phase in equilibrium with liquid, 12
z, Z	length
Z_i	mobility of the ions, 8

Appendix C

Conversion Factors

TO CONVERT	MULTIPLY BY	INTO
acre	4046.87	m^2
atmosphere (atm)	29.921	inch mercury (in. Hg)
atmosphere (atm)	406.793	inch water (in. H_2O)
atmosphere (atm)	14.6959	pound/square inch (psi)
barrel (Engl.)	0.1637	m^3
barrel (Petroleum barrel)	0.15876	m^3
Btu/sec	1.0548	kW
Btu/cubic foot	8.899	$kcal/m^3$
Btu/pound (Av.)	0.55554	kcak/kg
Btu/square inch	390.57	$kcal/m^2$
°C	°C 1.8+32	°F
centimeter (cm)	0.3938	inch (in)
cm^2	0.001076	square foot (ft^2)
cm^2	0.55	square inch (in^2)
cm^3	0.000035314	cubic foot (ft^3)
cm^3	0.061023	cubic inch (in^3)
cubic foot	28.3168	1
cubic foot	0.028317	m^3

TO CONVERT	MULTIPLY BY	INTO
cubic inch	16.3872	cm^3
cubic yard	0.764557	m^3
°F	(°F - 32) • 0.5555	°C
fluid ounce	29.573	cm^3
foot (ft)	0.3048	m
foot pound	1.3551	Joule (J)
foot pound (Av.)/sec	1.3551	Watt (W)
gram (g)	15.43236	grain
g	0.035274	ounce (oz)
g	0.0022046	pound (lb)
g/cm^3	62.42	pound/cubic foot (lb/ft^3)
grams/liter (g/L)	70.115	grain/gallon (Engl.)
g/L	58.416	grain/gallon (US)
g/L	0.010017	pound/gallon (Engl.)
g/L	0.008345	pound/gallon (US)
g/m^3	0.43701	grain/cubic foot
gallon (Engl.)	4.546	L
gallon (US)	3.785	L
gallon (Engl.)	0.16045	ft^3
gallon (US)	0.13368	ft^3
gallon (Engl.) min	0.273	m^3/h
gallon (US) min	0.227	m^3/h
gallon (Engl.)/min	0.0758	L/s
gallon (US)/min	0.063	L/s
grain (gr)	0.064798	g
grain/cubic foot	2.2883	g/m^3
horsepower (hp)	0.1782	kcal/s
horsepower (hp)	0.7453	kW
inch	25.4	mm
inch mercury	0.03342	atmosphere (atm)

TO CONVERT	MULTIPLY BY	INTO
inch water	0.0024583	atmosphere (atm)
Joule	0.7398	foot pound
kcal	3.9683	Btu
kcal/s	5.6142	horsepower (hp)
kcal/kg	1.8001	Btu/pound
kcal/m^3	0.11237	Btu/cubic foot
kg	0.0009842	long ton (Engl.)
kg	0.0011023	ton (US)
kg	35.274	ounce
kg	2.20462	pound (lb)
kg/m^2	0.20482	pound/square foot
kg/m^3	0.010017	pound/gallon (Engl.)
kg/m^3	0.0083445	pound/gallon (US)
km	0.53961	mile (nautical)
km	0.65137	mile (statute)
kW	0.3406	hp
kWh	860.38	kcal
kWh	0.3418	horsepower hour
liter (L)	0.035315	ft^3
L	61.0240	in.^{3}
L	0.2201	gallon (Engl.)
L	0.26418	gallon (US)
long ton (Engl.)	1016.047	kg
meter (m)	3.2808	ft
m	39.37	in.
m	1.0936	yd
m^2	0.00024711	acre
m^2	10.7639	ft^2
m^2	1,550	in.^{2}
m^2	1.19399	yd^2

TO CONVERT	MULTIPLY BY	INTO
m^3	6.2989	barrel (Petroleum barrel)
m^3	35.3165	cubic foot
m^3	61025	cubic inch
m^3	1.308	cubic yard
m^3	220.1	gallon (Engl.)
m^3	264.18	gallon (US)
mile (nautical)	1.8532	km
mile (statute)	1.60935	km
mkg	0.009	Btu
mkg	7.2330	foot pound
mm	0.03937	inch
mm Hg	0.0193368	pound/square inch
net ton = short ton (US)	907.185	kg
pound	0.4535924	kg
pound/cubic foot	0.016019	$g/cm^3 = kg/L$
pound/cubic inch	0.02768	kg/cm^3
pound/gallon (Engl.)	99.832	g/L
pound/gallon (US)	119.83	g/L
pound/inch	0.17858	kg/cm
pound/square foot	4.8824	kg/cm^2
pound/square inch	0.068046	Atmosphere
pound/square inch	51.7149	mm mercury
quart (US) dry measure	1.1012	L
short ton = net ton (US)	907.185	kg
square foot	0.092903	m^2
square inch	6.45163	cm^2
square yard	0.83613	m^2
ton	0.98421	long ton (Engl.)
ton	1.10231	short ton = net ton (US)
Watt	0.7375	foot pound/sec
yard	0.9144	m

Appendix C

Lengths (distances):				
Units	multipy by	Units	multiply by	Units
inches (in.)	25.4	millimeters (mm)	0.0394	inches (in.)
feet (ft)	0.305	meters (m)	3.281	feet (ft)
miles	1.609	kilometers (km)	0.621	miles

Volume (capacity)				
Units	multiply by	Units	multiply by	Units
cubic inches (in.^{3})	16.387	cubic centmeters (cm^3)	0.061	Cubic inches (in.^{3})
US quarts (US qt)	0.946	Litres (L)	1.057	US quarts (US qt)
Imperial gallons (Imp gal)	4.546	Litres (L)	0.22	Imperial gallons (Imp gal)
Imperial gallons (Imp gal)	1.201	US gallons (US gal)	0.833	Imperial gallons (Imp gal)
US gallons (US gal)	3.785	Litres (L)	0.264	US gallons (US gal)

Mass (weight)				
Units	multiply by	Units	multiply by	Units
ounces (oz)	28.35	grams (g)	0.035	ounces (oz)
pounds (lb$_f$; lb)	0.454	kilograms (kg)	2.205	pounds (lb)

Force:				
Units	multipy by	Units	multiply by	Units
pounds-force (lb$_f$; lb)	4.448	Newtons (N)	0.225	pounds-force (lb$_f$; lb)
Newtons (N)	0.1	kilograms (kg)	9.81	Newtons (N)

Pressure				
Units	multiply by	Units	multiply by	Units
pounds-force per square inch (psi) (lb/in.^{2})	0.070	kilograms per square centimetre (kg/cm^2)	14.223	pounds-force per square inche (psi) (lb/in.^{2})
pounds-force per square inch (psi) (lb/in.^{2})	0.068	atmospheres (atm)	14.696	pounds-force per square inche (psi) (lb/in.^{2})
pounds-force per square inches (psi) (lb/in.^{2})	0.069	bars	14.5	pounds-force per square inches (psi) (lb/in.^{2})
pounds-force per square inches (psi) (lb/in.^{2})	6.895	kilopascals (kPa)	0.145	pounds-force per square inches (psi) (lb/in.^{2})
kilopascals (kPa)	0.01	kilograms per square centimetre (kg/cm^2)	98.1	kilopascals (kPa)
millibar (mbar)	100	pascals (Pa)	0.01	millibar (mbar)
millibar (mbar)	0.0145	pounds-force per square inches (psi) (lb/in.^{2})	68.947	millibar (mbar)
millibar (mbar)	0.75	Millimeters of mercury (mm Hg)	1.333	millibar (mbar)
millibar (mbar)	0.401	Inches of water (in. H$_2$O)	2.491	millibar (mbar)
millimeters of mercury (mm Hg)	0.535	Inches of water (in. H$_2$O)	1.868	millimeters of mercury (mm Hg)
Inches of water (in. H$_2$O)	0.036	pounds-force per square inches (psi) (lb/in.^{2})	27.68	Inches of water (in. H$_2$O)

Appendix C

Torque (moment of force)				
Units	multiply by	Units	multiply by	Units
pounds-force inch (lb_f in.) (lb in.)	1.152	kilograms centimetre (kg cm)	0.686	pounds-force inche (lb_f in.) (lb in.)
pounds-force inch (lb_f in.) (lb in.)	0.113	Newton meters (N m)	8.85	pounds-force inche (lb_f in.) (lb in.)
pounds-force inch (lb_f in.) (lb in.)	0.083	pounds-force feet (lb_f ft) (lb ft)	12	pounds-force inche (lb_f in.) (lb in.)
pounds-force feet (lb_f ft) (lb ft)	0.138	kilograms meters (kg m)	0.868	pounds-force feet (lb_f ft) (lb ft)
pounds-force feet (lb_f ft) (lb ft)	1.356	Newton meters (N m)	0.738	pounds-force feet (lb_f ft) (lb ft)
Newton meters (N m)	0.102	kilograms centimetre (kg cm)	9.804	Newton meters (N m)

Power				
Units	multipy by	Units	multiply by	Units
Btu per hour (Btu/hr)	$3.93 \cdot 10^{-4}$	horsepower (hp)	2544.4	Btu per hour (Btu/hr)
Btu per hour (Btu/hr)	0.293	Watts (W)	3.411	Btu per hour (Btu/hr)
calories per second (Cal/sec)	14.286	Btu per hour (Btu/hr)	0.007	calories per second (Cal/sec)
calories per second (Cal/sec)	4.1884	Joules per second (J/s)	0.23875	calories per second (Cal/sec)
horsepower (hp)	745.7	Watts (W)	0.0013	horsepower (hp)
horsepower (hp)	2544.4	Btu per hour (Btu/hr)	0.293	horsepower (hp)
Watts (W)	3.411	Btu per hour (Btu/hr)	0.293	Watts (W)
Watts (W)	$1.34 \cdot 10^{-3}$	horsepower (hp)	745.7	Watts (W)

Speed

Units	multipy by	Units	multiply by	Units
feet per minute (ft/min) (fpm)	0.005008	meters per second (m/sec)	196.85	feet per minute (ft/min) (fpm)
feet per second (ft/sec) (fpsec)	30.48	centimeters per second (cm/sec)	0.0328	feet per second (ft/sec) (fpsec)
kilometer per hour (km/hr)	0.27778	miles per hour (mile/hr) (mph)	1.6093	kilometer per hour (km/hr)
kilometer per hour (km/hr)	0.54	knots	1.853	kilometer per hour (km/hr)
miles per hour (mile/hr) (mph)	0.868	knots	1.152	miles per hour (mile/hr) (mph)

Temperature

Units	multipy by	Units	multiply by	Units
degrees Fahrenheit (°F)	(°F-32) • (5/9)	degree Celsius (°C)	°C • (9/5) + 32	degrees Fahrenheit (°F)
degrees Fahrenheit (°F)	°F + 460	degree Rankin (°R)	°R - 460	degrees Fahrenheit (°F)
degree Celsius (°C)	°C + 273	degree Kelvin (K)	K - 273	degree Celsius (°C)

Viscosity

Units	multiply by	Units	multiply by	Units
centi poise (cp)	0.01	grams per centimeter second (g/cm sec)	100	centi poise (cp)
pounds-mass per foot hour (lb_m/ft hr) (lb/ft hr)	0.00413	grams per centimeter second (g/cm sec)	242	pounds-mass per foot hour (lb_m/ft hr) (lb/ft hr)
pounds-mass per foot hour (lb_m/ft hr) (lb/ft hr)	1.492	kilograms per meter hour (kg/m hr)	0.670	pounds-mass per foot hour (lb_m/ft hr) (lb/ft hr)

Appendix D

Physical Constants

Planck constant $6.6260755 \cdot 10^{-34}$ J sec

Boltzmann constant $1.380658 \cdot 10^{-23}$ J/K

Elementary charge $1.60217733 \cdot 10^{-19}$ C

Avogadro number $6.0221367 \cdot 10^{23}$ particles/mol

Speed of light $2.99792458 \cdot 10^8$ m/sec

Gas constant 1545.33 ft/lb_f lb mol°R

8.3143 J/g mol K   1.987 Btu/lb mol °R

Molar volume 22.41383 m^3/kmol

Faraday constant $9.64846 \cdot 10^4$ C/mol

Gravitational constant 32.2 ft lb_m/lb_f sec^2

Acceleration due to gravity 9.80665 m/sec^2

parts per million - (1 in 1,000,000)

parts per billion - (1in 1,000,000,000)

milligrams per cubic meter (1 thousandth of a gram per cubic meter of air)

micrograms per cubic meter (1 millionth of a gram per cubic meter of air)

conversion factor for CO is 1.165 at 20°C, ppm - $\mu g/m^3$

conversion factor for NO_2 is 1.913 at 20°C, ppb - $\mu g/m^3$

conversion factor for O_3 is 2.000 at 20°C, ppb - $\mu g/m^3$

conversion factor for SO_2 is 2.704 at 20°C, ppb - $\mu g/m^3$

Prefix

yotta	[Y]	1 000 000 000 000 000 000 000 000	$= 1 \cdot 10^{24}$
zetta	[Z]	1 000 000 000 000 000 000 000	$= 1 \cdot 10^{21}$
exa	[E]	1 000 000 000 000 000 000	$= 1 \cdot 10^{18}$
peta	[P]	1 000 000 000 000 000	$= 1 \cdot 10^{15}$
tera	[T]	1 000 000 000 000	$= 10 \cdot {}^{12}$
giga	[G]	1 000 000 000 (a thousand millions = a billion)	
mega	[M]	1 000 000 (a million)	
kilo	[k]	1 000 (a thousand)	
hecto	[h]	100	
deca	[da]	10	
		1	
deci	[d]	0.1	
centi	[c]	0.01	
milli	[m]	0.001 (a thousandth)	
micro	[µ]	0.000 001 (a millionth)	
nano	[n]	0.000 000 001 (a thousand millionth)	
pico	[p]	0.000 000 000 001	$= 1 \cdot 10^{-12}$
femto	[f]	0.000 000 000 000 001	$= 1 \cdot 10^{-15}$
atto	[a]	0.000 000 000 000 000 001	$= 1 \cdot 10^{-18}$
zepto	[z]	0.000 000 000 000 000 000 001	$= 1 \cdot 10^{-21}$
yocto	[y]	0.000 000 000 000 000 000 000 001	$= 1 \cdot 10^{-24}$

Index

A

absolute, humidity, 40, **42**
    pressure, 33
    temperature, 33
    viscosity, 25, 29
absorption, **433**, 457
activated carbon, 312, **348**
adsorbate, **347**
adsorbent capacity, 359
adsorption, 312, **347**
    equilibrium, 350, 352
    isotherm, **350**, 374, 505
    parameters, 356
aerodynamic diameter, 80, 104
air, ambient, 45, **46**, 311
    atmospheric, 25
    properties, 29, 69
air flow rate, 11, 327
air quality systems, 509
    worldwide, 477
air-to-fuel ratio, 410, 481
air-to-cloth ratio, 164, **169**
air/water vapor mixture, **41**
airborne particles, 67
aldehydes, 7

ammonia, 269
amortization, 12
annual operating cost, 12
Antoine equation, **313**
areal density, 173
aromatics, 311
Arrhenius equation, **390**
atmospheric air, **25**
atomizing impingement collector scrubber, 202
auxiliary equipment, 513
Avogadro's number, 87
available heat, **396**
avalanche multiplication, **267**
axial fan, 530

B

back corona, **268**, 284
baffle, **243**
bag sizing, 172
baghouse, **163**
bed, capacity, 349
    depth, **364**, 366, 377
Bernoulli's Equation, 58, 59, 525
blinding, **168**
blowby, 480, 492

Boltzmann's constant, 187, **192**, 274
Boyle's Law, **30**
breakthrough, **360**, 366
Brownian, diffusion, 207
    motion, **87**, 90

C

Calvert design method, **230**
capital investment, 10, 15
capture velocity, 520
carbon, absorber, 317, 322, 326
    black, 268
    dioxide, 309
    monoxide, 7, 309
carbonization, 348
Carman-Kozeny equation, **179**
cascade impactor, 104, 286
catalytic, converter, **494**, 502
    incinerator, 32, 327, 383, 401, 500
cement, 3, 279
centrifugal fan, 530
Charles' Law, **30**, 34, 35
chemisorption, **349**, 433
chlorinated solvents, **327**, 338, 409
chlorofluorocarbons, 16
Clapeyron equation, **340**
Clean Air Act, 16, 491
coal, 7, 269
coal-fired boiler, 3, 20, 283
cohesion, 268
Colburn diagram, **457**
collecting, force, 99
    surface, 100, 268, 288
collection, efficiency, 84, 87, 90, 93, 107, 112, 114, 137, 147, 186, 208, 232, 279, 285, 292, 295
    electrode, **266**
collector performance, 119, 120

combustion, **312**, 388
    kinetics, **386**
    of gasoline, 312
concentration unit, 35
condensation, **312**, 326
condenser, **312**, 322, 328
contact power theory, **237**
control devices, pollutant, 514
convection, 45, 51
conversion efficiency, 404
cooling, 45, 51, 55, 325, 332, 415
corona, discharge, 266
    power, 284
    voltage, 267
cost estimate, **10**, 15, 552
countercurrent spray, 235
counterflow scrubber, **216**, 465
Cunningham correction factor, **71**, 73, 84, 237
cupola, 49
cut diameter, **110**, 118, 134, 235
cyclone, collector, 108, 127
    design configuration, 130, 131
    efficiency, 133, 135, 138, 144

D

Dalton's Law, **36**
Damkoler number, **387**
Darcy's Law, **173**
degreasing, 347, 521
density, 29, **33**
destruction efficiency, 407
    kinetics, **391**
Deutsch-Anderson equation, **278**, 280, 288, 295
dew point, 45
dielectric constant, 273
diffusion, charging, 273, 274, 285
    coefficient, **87**

dilution, 45, 47
direct cost, 15
discharge electrode, **265**
    pressure, 534
distribution analysis, 105
drag, coefficient, 67, 79
    filter, **174**
    force, **67**, 78
drift velocity, **277**
drop diameter, 204
duct configuration, 53, 296, 298, 529
    velocity, 527
dust, filter resistance coefficient, **173**
    resistivity, 283

E

electric charging, 265
electric fields, **271**
electrostatic, force, **271**
    precipitator, 9, 10, 121, **265**, 277, 542, 546
emission factor, 6
emission standards, **488**
energy balance, 393
enthalpy, **39**, 41, 331
entry coefficient, 524
environmental impact, 18, 339
equilibrium, constant, 485
    diagram, 438
    vapor content, 312
equivalence ratio, 481
evaporation rate, **311**
excess air, **386**, 423
exothermic, 349

F

fabric, cleaning, 167
    filter, 9, **163**, 542, 548
face velocity, **169**

fan, characteristics, 529
    curve, 531, 536
    laws, 534
    performance, 536
    speed, 538
    systems, 540
Fanning friction factor, 499
field charging, 274
filter, drag/drop, **174**
    fibers, 91
filtration velocity, **169**
flame arrestor, 329
flooded-disk scrubber, **227**
flooding, 450
fly ash, 268, 283, 286, 290, 296
fouling, 401
fractional efficiency, 196, 286
Freundlich isotherm, **354**

G

gas, atomized drops, 203
    cleaning, 9
    density, 132, 539
    emission, 6
    film control, 442
    inlet velocity, 132
    laws, **30**
    velocity, 278, 296, 367
gas-liquid relationships, 37
gaseous effluents, 45
gasoline, 309, 317
    marketing, **315**
Gaussian function, 100
geometric, mean, 102
    standard deviation, 102
Gibbs-Dalton rule, 41
global warming, 337
Gore-Tex®, 167

grade efficiency, 99, 107, 108, 111, 133, 279
gravitational settling, 203
gravity, collector, 113
    settler, 115, 129

H

Hazardous Organic Industry, 16
hazardous waste, 412
head loss, 11
heat, balance, **330**
    capacity, **39**
    content, 395
    of combustion, 409
    of vaporization, **40**
    transfer coefficient, 330, **331**
    transfer curve, 51, 52
heel, **363**
Henry's Law, 38, 353, 437
high energy scrubber, 202
HON, 16
hoods, **518**
hopper, **266**
horsepower, 11, 13, 62
humid volume, **42**
humidity effects, 349, 369
hydrocarbons, 7, 490

I

ideal gas law, **30**
impaction, 86, 91
    efficiency, **85**
incinerator, 3, 312, 322, 326, **383**, 424
indirect cost, 15
inertial, forces, 27
    impaction, 137, 201, 214
    loss, 525
inlet, pressure, 534
    velocity, 133, 146

installed cost, 11, 13, 15
integrated penetration, 142
interception, 201, 214
interest, compound, 12
ion density, 273
ionization, 268
iron, 269
isotherm, 333, 352

J-K-L

Johnstone design method, **226**
kinematic viscosity, 27
kinetics, **403**, 411
Kleinschmidt equation, **205**, 228, 256
knitted wire configuration, 251
laminar flow, 69, 70, 77, 110, 133
Langmuir isotherm, 352
Lapple conventional cyclone, 133, 135
Leith/Licht, 139, 140
liquid film control, 442
liquid-to-gas ratio, 222, 444, 448
lithium, 269
loading loss, 318
log normal/probability distribution, 102, 106
log-mean temperature differential, 331
low energy scrubber, 202
low sulphur fuel, 8
lower explosive limit, **323,** 327

M

MACT standards, 16, 17
maintenance cost, 11
mass emission standard, 2
mass, median diameter, 104, **106**, 176
    transfer coefficient, 443
    transfer zone, 360
material balance, 46
Matts-Ohnfelt equation, **287**, 295

mean square displacement, 89
mechanical, efficiency, 11
    shaking, **167**
mesh entrainment separator, 252
migration velocity, **277**, 279
mist eliminator, 241
MMD, 117, 118, 176
mobile, air pollutants, 476
    packed beds, 436
mobility, 83
molar volume, 33
mole fraction, 37
molecular weight, **25**, 33
MTZ, **360**
multiclones, 127
multiple collecting fibers, **188**
multiple cyclones, 127, 144

N-O

National Emission Standard for Hazardous Air Pollutants, 1
natural fiber, 166
Navier-Stokes, 67, 115
Net Heating Value, 422
New Source Performance Standards, 1, 3
Newton's Law, 84
nitric acid, 3
nitrogen oxide emissions, 7, 484
Nukiyama-Tanaswa equation, **204**
null points, 519
OECD, 475
olefins, 311
operating line, 445
operation cost, 11
optimization parameter, 149
Organization for Economic Cooperation and Development, 475

orifice loss, 525
outlet concentration, 325, 328
overall collection efficiency, 107, 109, 116
ozone, 316, 337

P

packed towers, **434**, 449, 455
packing density, **189**
particle, charging, 273
    count, 100
    diameter, 82, 87, 89, 110
    diffusivity, 89
    re-entrainment, **143**
    size distribution, 100, 105, 229, 299
    size, 100, 135
particulate, collection, 99
    emission, 3, 99
    weight, 6
Peclet number, **89,** 387
penetration, 116, 287, 291
performance curve, 120
Peterson and Whitby cyclone, 131, 150
physical absorption, 349, 433
plate area, 292, 298
plug flow, 111
Polanyi potential theory, **355**
polar adsorbent, **348**
pollutant, sources, 514
    secondary, 478
pollution prevention, 13, 17
polydisperse dust, 286
polyethylene, 18
pore size, 348
potassium oxide, 269
potential flow, 87, 88
power cost, 11

precipitator rate parameter, 279, 281, **289**
pressure drop, **9**, 107, 132, 153, 172, 181, 232, 364, 452, 523
    head, 60
pressure, 58, **533**
process weight, 3, 5
psychometric chart, **43**, 44, 369
pulse jet, **167**

Q-R

quenching, 45, **47**, 49
radiant cooling, 53
radiation, 45, 51
    coefficient, 54
Raoult's Law, **37**
rapper, **266**, 296
Raschig ring, 435, 451
rate of heat transfer, 331
recycling, 14
Refractive Index, **358**
Reid Vapor Pressure, 316, 489
relative, humidity, **42**
    saturation, **41**
    settling velocity, 219
relaxation time, **82**
removal efficiency, 132, 326
residence time, 110
resistivity, **267**
reverse air flow, **167**
reverse-flow cyclone, 128
Reynolds Number, **27**, 52, 68, 86

S

saturation, capacity, 352, **360**
    charge, 277
    factor, 273, 316
    vapor pressure, **314**

Sauter surface mean diameter, **204**
Schmidt number, **89**
secondary pollutants, **478**
settling velocity, 76, 78, 79
shaker-type dust collector, 164
shearing stress, 27
shell-and-tube condenser, 329
Shepherd and Lapple cyclone, 131, 140, 150
Sherwood number, 498
sieving, **164**
single droplet collection efficiency, **206**, 214
single fiber collection 92, 93, **186**
size distribution, 100
slip correction factor, 72, 73, 84, 237
smog, 337, 478
SO_2, SO_3, 6, 269
soda ash, 269
sodium chloride, 269
solubility, **437**
space velocity, **402**, 499
sparkover, **267**
specific, cake resistance, 173
    collection area, **280**, 282
    heat, **39**
    humidity, 42
    volume, **33**
spray droplet wet scrubbers, **201**
spray tower, 203, 212
Stairmand cyclone, 131, 145, 150, 153
standard, air, **25**
    condition, 32
    deviation, 102
    gas conditions, 32
static pressure, 60, **533**
steam stripping, 362
stoichiometry, **385**, 408

Index

Stokes' diameter, 80
    law, 68, 70-75, 83, 115, 178, 228
    number, **84**
Stoneware correlation, 452
stopping distance, 83
stream lines, 91, 111
submerged fill, 317
sulfuric acid, 3, 269
sulphur, 269
Surf, 131, 149
surface, coating, **321**
    condenser, **335**
    polarity, **348**
Swift cyclone, 131, 150
synthetic fibers, 166

T

tangential, inlet, 225
    velocity, 129, 130
target collector, 100
tellerette, 435, 451
terminal settling velocity, 76, 212, 215
theoretical oxygen, **385**
thermal, conductivity, 29, 53
    incinerator, **312**, 322, 326, 383
    oxidation parameters, **392**, 413
threshold limit value, 323
total filtration time, **183**
total head, 60
transition flow, 70, 75
turbulent flow, 70, 75, 110, 112, 387, 523
two-film theory, **441**

U-V

unipolar corona, 273
United States EPA, 1, 6
universal gas constant, 30

vapor, emission, 320
    loss, 480
vapor pressure, **38**, 309, 318
vehicles in use, 476
velocity, gradient, 27
    head, 60
    pressure, 60, **533**
ventilation rates, 522
venturi scrubber, 9, 108, 203, **224,** 542
vertical flow adsorber, 365
vertical spray chamber, 219
viscosity, **26**
VOCs, **309**, 324, 416
volumetric flow rate, 32
vortex, 134

W-X-Y-Z

waste reduction, **14**
wave plate, 246
weir inlet, 225
wet bulb temperature, **42**
wet cyclones, 256
wet scrubber, 91, 201, 511
wire mesh eliminator, 251
working capacity, 375
working charge, **368**
zigzag baffle, 243
zone of influence, 110, 111